Immune Building Systems Technology

Wladyslaw Jan Kowalski, P.E., Ph.D.

Department of Architectural Engineering
The Pennsylvania State University

McGraw-Hill

New York   Chicago   San Francisco   Lisbon   London
Madrid   Mexico City   Milan   New Delhi   San Juan
Seoul   Singapore   Sydney   Toronto

The McGraw·Hill Companies

2003

Library of Congress Cataloging-in-Publication Data

Kowalski, Wladyslaw Jan.
   Immune building systems technology / Wladyslaw Jan Kowalski.
      p.   cm.
   Includes bibliographical references and index.
   ISBN 0-07-140246-2
   1. Buildings—Security measures.   2. Buildings—Protection.
3. Bioterrorism—Prevention.   4. Buildings—Environmental
engineering.   5. Decontamination (from gases, chemicals, etc.).
I. Title.
TH9705.K69   2002
696—dc21   2002029558

1 2 3 4 5 6 7 8 9 0   DOC/DOC   0 8 7 6 5 4 3 2

ISBN 0-07-140246-2

*The sponsoring editor of this book was Kenneth P. McCombs, the editing
supervisor was Stephen M. Smith, and the production supervisor was
Pamela A. Pelton. It was set in Century Schoolbook following the MHT
design by Deirdre Sheean of McGraw-Hill Professional's Hightstown, N.J.,
composition unit.*

Printed and bound by RR Donnelley.

McGraw-Hill books are available at special quantity discounts to use as
premiums and sales promotions, or for use in corporate training programs.
For more information, please write to the Director of Special Sales, McGraw-
Hill Professional, Two Penn Plaza, New York, NY 10121-2298. Or contact your
local bookstore.

 This book was printed on recycled, acid-free paper
containing a minimum of 50% recycled, de-inked fiber.

Contents

Preface   xi
Acknowledgments   xv
Symbols   xvii

Chapter 1. Introduction   1

    1.1   Introduction   1
    1.2   A Brief History of Chemical and Biological Warfare   3
    1.3   Bioterrorism Today   19

Chapter 2. Biological Weapon Agents   23

    2.1   Introduction   23
    2.2   Microorganisms   24
        2.2.1   Airborne Pathogens   27
        2.2.2   Vector-Borne Pathogens   37
        2.2.3   Food-Borne and Water-Borne Pathogens   40
    2.3   Toxins   41
    2.4   Bioregulators   47
    2.5   Weaponizing BW Agents   49
    2.6   Size Distribution of BW Agents   50
    2.7   Survival Curves of Microbes under UV Exposure 52
    2.8   Summary   59

Chapter 3. Chemical Weapon Agents   61

    3.1   Introduction   61
    3.2   Classification of CW Agents   61
    3.3   Design Basis CW Agents   66
    3.4   CW Agent Simulants   71
    3.5   Summary   71

Chapter 4. Dose and Epidemiology of CBW Agents   73

    4.1   Introduction   73
    4.2   BW Agent Dosimetry   75

4.3   CW Agent Exposure Dosimetry   81
4.4   CBW Agent Inhalation Dosimetry   85
4.5   CBW Agent Ingestion Dosimetry   87
4.6   Lethal Dose Curves for Toxins   89
4.7   Disease Progression Curves   89
4.8   Summary   96

Chapter 5. Dispersion and Delivery Systems   97

5.1   Introduction   97
5.2   Outdoor Dispersion Systems   97
5.3   Indoor Dispersion Systems   99
5.4   Dissemination of Food-Borne and Water-Borne Agents   110
5.5   Summary   112

Chapter 6. Buildings and Attack Scenarios   113

6.1   Introduction   113
6.2   Building Types and Relative Risks   115
      6.2.1   Commercial Buildings   119
      6.2.2   Government Buildings   123
      6.2.3   Food and Entertainment Facilities   127
      6.2.4   Healthcare Facilities   129
      6.2.5   Lodging Facilities   130
      6.2.6   Education Facilities   132
      6.2.7   Mercantile Facilities   134
      6.2.8   Assembly Facilities   136
      6.2.9   Special Facilities   138
6.3   Building Attack Scenarios   140
      6.3.1   Outdoor Releases   140
      6.3.2   Air Intake Release   141
      6.3.3   Indoor Explosive Release   144
      6.3.4   Indoor Passive Release   146
      6.3.5   Indoor Aerosolization Release   146
      6.3.6   Internal Release in Duct   147
6.4   Summary   150

Chapter 7. Ventilation Systems   151

7.1   Introduction   151
7.2   Types of Ventilation Systems   151
7.3   Ventilation Modeling   153
      7.3.1   Ventilation Modeling with Calculus Methods   156
      7.3.2   Ventilation Modeling with Computational Methods   157
      7.3.3   Ventilation Modeling with CONTAMW   162
7.4   Computational Fluid Dynamics Modeling   163
7.5   Summary   163

Chapter 8. Air-Cleaning and Disinfection Systems   165

8.1   Introduction   165

8.2   Filtration   167
      8.2.1   Mathematical Modeling of Filtration   172
      8.2.2   Filtration Applications   177
      8.2.3   Filtration Test Results for Microbes   184
      8.2.4   Filtration of Liquid Aerosols   185
8.3   Ultraviolet Germicidal Irradiation   188
      8.3.1   Mathematical Modeling of UVGI   189
      8.3.2   UVGI Applications   196
8.4   Combining UVGI and Filtration   200
8.5   Gas Phase Filtration   203
8.6   Summary   208

Chapter 9. Simulation of Building Attack Scenarios   211

9.1   Introduction   211
9.2   Baseline Building Attack Scenarios   213
9.3   Simulation of CBW Attack Scenarios   214
9.4   Simulation with CONTAMW   226
9.5   Simulation of Several Model Buildings   230
9.6   Multizone Simulation with CONTAMW   238
9.7   Overload Attack Scenarios   242
9.8   Sudden-Release Scenarios   244
9.9   CW Removal by Outside Air Purging   246
9.10 Summary   247

Chapter 10. Detection of CBW Agents   249

10.1 Introduction   249
10.2 Chemical Detection   250
      10.2.1 Chemical Detector Response Time   256
10.3 Biological Detection   257
      10.3.1 Air Sampling   258
      10.3.2 Biosensors   261
      10.3.3 Particle Detectors   263
      10.3.4 Mass Spectrometry and LIDAR   267
10.4 Summary   268

Chapter 11. Immune Building Control Systems   269

11.1 Introduction   269
11.2 Control Systems   270
11.3 Control System Architectures   271
      11.3.1 Detect-to-Alarm   272
      11.3.2 Detect-to-Isolate   273
      11.3.3 Detect-to-Treat   275
11.4 Emergency Systems   276
      11.4.1 Outside Air Purging   277
      11.4.2 Secondary System Operation   278
      11.4.3 Sheltering Zone Isolation   279
11.5 Building Automation for Immune Buildings   281
11.6 Summary   285

Chapter 12. Security and Emergency Procedures   287

12.1   Introduction   287
12.2   Physical Security Measures   288
12.3   Incident Recognition   293
12.4   Emergency Response   297
12.5   Disabling Devices   301
12.6   Emergency Evacuation   303
12.7   Sheltering in Place   304
12.8   Medical Response   306
12.9   Security Protocol   307
12.10 Personnel Training   309
12.11 Summary   310

Chapter 13. Decontamination and Remediation   311

13.1   Introduction   311
13.2   Decontamination by Physical Means   312
13.3   Decontamination by Chemical Means   314
13.4   Ozone   315
13.5   Chlorine Dioxide   322
13.6   SNL Foam   323
13.7   Summary   325

Chapter 14. Alternative Technologies   327

14.1   Introduction   327
14.2   Thermal Disinfection   329
14.3   Cryogenic Freezing   329
14.4   Desiccation   329
14.5   Passive Solar Exposure   330
14.6   Vegetation Air Cleaning   331
14.7   Antimicrobial Coatings   332
14.8   Electrostatic Filters   333
14.9   Negative Ionization   333
14.10 Ultrasonication   334
14.11 Photocatalytic Oxidation   336
14.12 Ozone Air Disinfection   337
14.13 Microwave Irradiation   340
14.14 Pulsed White Light   341
14.15 Pulsed Filtered Light   343
14.16 Pulsed Electric Fields   346
14.17 Gamma Irradiation   346
14.18 Electron Beams   347
14.19 Summary   348

Chapter 15. Economics and Optimization   349

15.1   Introduction   349
15.2   Selecting Performance Criteria   349

15.3   Economics of Filtration   350
15.4   Economics of UVGI   354
15.5   Economics of Carbon Adsorbers   363
15.6   Energy Analysis   365
15.7   Summary   366

Chapter 16. Mailrooms and CBW Agents   367

16.1   Introduction   367
16.2   Mailroom Contamination   367
16.3   Building Ventilation Systems   368
16.4   Building General Areas   369
16.5   Mail Handling by Employees   372
16.6   Contaminated Letters and Packages   373
16.7   Mail Processing Equipment   378
16.8   Delivery Vehicles   379
16.9   Mailroom Protocol   380
16.10   Summary   382

Chapter 17. Epilogue   383

17.1   Introduction   383
17.2   The Future of Bioterrorism   385
17.3   Collateral Indoor Air Quality Benefits   387
17.4   The Engineer and the Future of Disease Control   388

Appendix A   Database of Biological Weapon Agents   391
Appendix B   Database of Pathogen Disease and Lethal Dose Curves   439
Appendix C   Database of Toxins and Dose Curves   453
Appendix D   Database of Chemical Weapon Agents   469
Appendix E   UVGI System Sizes and Kill Rates   519
Appendix F   Source Code for Direct UVGI Field Average Intensity   525
Glossary   531
References   549
Index   571

Preface

This book is intended to meet the needs of engineers, architects, designers, building managers, and other professionals who are attempting to protect buildings against the growing threat of bioterrorism and the release of hazardous chemical agents. Due to the broad nature of the subject matter, this book also provides technical information of use to police, rescue teams, emergency medical response personnel, security personnel, firefighters, military personnel, epidemiologists, and microbiologists working on the problem of defending against chemical and biological weapons of mass destruction.

This book grew out of research that began in 1996 at the Pennsylvania State University into the field of aerobiological engineering, which is the science of controlling the aerobiology of indoor environments. This field is a hybrid of microbiology and architectural engineering, or building science, in relation to diseases that spread in indoor environments. These diseases are primarily, but not limited to, airborne or respiratory diseases.

Research into aerobiological engineering at Penn State underwent a transition in 1999 as a result of the Defense Advanced Research Projects Agency (DARPA) initiation of the Immune Building project. The DARPA project focuses on developing immune building technologies for governmental and military applications. Immune building research at Penn State has taken a parallel approach with a focus on providing cost-effective engineering solutions for the commercial and residential building industries. Though this research is currently directed toward the design of systems for defending against chemical and biological weapon (CBW) agents, it is merely a reapplication of existing technologies with more severe design basis scenarios. That is, immune buildings are aerobiologically engineered to protect against the intentional release of hazardous agents as well as the naturally occurring kind. The difference is not merely one of degree but, as will be seen herein, there are also many qualitative differences.

This book is designed to serve both as a reference handbook and as a text that can be read from beginning to end without significant dependence on outside sources. Relatively light treatment is given in this book to microbiological issues. Only information that is directly relevant to the CBW problem is addressed so as to spare the engineer from the more esoteric concepts that are often included in this sort of research. Matters of microbiology, physiology, genetics, chemistry, and physics that relate to this subject are treated in many of the references and these can be consulted by those who need more detailed information. The collateral nonengineering information in this book is introduced and structured in a manner that should require no specific background or outside sources, but it does require the reader to develop familiarity with the terminology of aerobiological engineering. To this end a glossary has been provided, and readers are encouraged not merely to know the definitions, but to understand the relevant or implied concepts.

The units of choice in this book represent a hybrid of common usage in both engineering and microbiology. They are predominantly metric (and mainly "SI") units, but English units are used where necessary and converted as appropriate. It is unfortunate that the United States has not made a full commitment to SI unit conventions, and as a result engineers today find themselves constantly obliged to make uncertain unit conversions. I encourage engineers to make the effort to become familiar with metrics as it is the preferred language of the sciences, even though many areas of industry still insist on using only English units.

I hope that microbiologists will not be too critical of this reduction of vast quantities of biological information to a form palatable to engineers. The division of such complex information into organized blocks and simple equations is, admittedly, an oversimplification. I understand as well as anyone that the field of microbiology does not lend itself easily to overly simplistic quantification and resists attempts to make it conform to engineering methodologies. Suffice it to say that these linear approximations of the nonlinear world of microbiology serve the purposes of engineering design and that the sum total of this quantitative approach results in considerable leverage on the general problem of controlling airborne disease. The fact is that this block micromanagement of microbiological data tends not to be sensitive to the small errors of approximation implicit in each of a large number of models. All in all, these errors center about a mean and have the effect of canceling each other out. In the end, the basic safety factor that every engineer uses instinctively tends to erase the numerous small errors and assumptions made in the process of getting to the final point of sizing a system for disinfecting air and protecting building occupants.

The handling and processing of such massive quantities of microbiological and chemical data invariably cause some errors to slip through.

These errors may be more apparent to those in the fields of microbiology or chemistry than they are to engineers, and I ask that anyone who identifies any problems with the data in this book contact me with corrections. These errata will be posted on the Penn State *Aerobiological Engineering Website* (Kowalski and Bahnfleth 2002b). Additional information on related subjects and available products is available online, and a number of these sources have been provided with website address URLs in the References section.

Finally, to preempt any concerns that this book gives away too much information, this matter has been given careful consideration and there is no detailed information here that would easily or directly facilitate people with harmful intentions. No information on how to develop biological or chemical weapons is provided here, although this information is, unfortunately, freely available elsewhere. Critical details of the hypothetical dispersion systems have been omitted, as have other details that would facilitate the construction or use of such devices. If some of the results of the modeling of attack simulations seem less specific than they should, it is because knowledge of such system vulnerabilities is best left to the engineers who evaluate them, and also because every building is unique and some results cannot be generalized. In any event, the information presented in this book should do more good in the hands of the right people than it would do harm if it were understood by the other kind.

I hope this book provides the necessary impetus to designers and managers to implement protective systems in buildings on a national scale so as to eliminate the catastrophic potential of the effects of biological and chemical weapons and simultaneously to immunize our buildings against the more mundane threat of everyday disease transmission, a result that may even drive some respiratory diseases to extinction.

Wladyslaw Jan Kowalski, P.E., Ph.D.

Acknowledgments

I gratefully acknowledge all those who provided assistance, reviews, images, and the technical information necessary to complete this work. They include all the following individuals and the companies and organizations they represent: William Bahnfleth, Amy Musser, Stanley Mumma, Chobi Debroy, Richard Mistrick, Brad Striebig, Jelena Srebric, Gretchen Kuldau, Daniel Merdes, John Brockman, Chuck Dunn, Mike Ivanovich, Henry St.Germain, Dave Witham, Stephen Stark, Bill Perkins, Philip Mohr, Art Anderson, Wilson Poon, John Buettner, Charles Dunn, Jack Matson, Bruce Tatarchuk, Jeffery Siegel, Jing Song, Thomas Whittam, Leon Spurrell, Clarence Marsden, Larry Kilham, Linda Bartraw, John Wellock, Dave Neufer, George Flannigan, Nancy Sabol, John Harris, Bill Jacoby, Jennifer Bowers, Bernice Black, Meryl Nass, Ann McDougall, Louis Geschwindner, Jack Kulp, Robert Gannon, Eric Peyer, George Walton, Michael Modest, Dick Huggins, John Garrett, J. Glenn Songer, Elisa Derby, Ed Westaway, Alex Wekhof, Roy D. Wallen, Leonard E. Munstermann, Brian Bush, Ken McCombs, Matt Holmquist, Steve Careaga, Jenny Moyna, Paul S. Cochrane, Camille Johnson, Neil Carson, Garry T. Cole, Basil Frasure, Amram Golombek, Ann McDougall, Hans Buur, Malcolm Jones, Chris Brown, Ariella Raz, John D. German, Cornelia Büchen-Osmond, Debra Short, Larry Schmutz, Cheryl Christiansen, Bill Sawka, Francois Elvinger, Bev Corron, Georgia Prince, Phil Sansonetti, James W. Kazura, Matthew Wolf, Scott Roberts, Teddy Pastras, Holly Krauel, Janice Bowie, Helen Lee, Sue Poremba, Eric Burnett, Franz Heinz, Christian Peters, Michael Sailor, Patrice Moore, Mik Pietrzak, Christine Bailey, Rob Crutcher, Chuck Murray, Richard Osborne, Jim Cordaro, Raimo Vartiainen, P. J. Richardson, Beth Huber, John Antoniw, Betty Wells, Chuck Haas, Jane Whitlock, Jerry Straily, Raj Jaisinghani, Harold James, Lucas Bacha, Ted Glum, Jeff D'Angelo,

Donald E. Woods, Eric Grafman, Steven J. Emmerich, Bill Thayer, Vincent Agnello, Ching-Hsing Liao, Cathi Siegel, Cynthia S. Goldsmith, Michael Bowen, Peter Charles, Dorothy Hsu, Rose Spencer, David Evans, Peter d'Errico, Carl Gibbard, and Kurt Behers.

And thanks to Amy Alving for developing the Immune Building concept.

Symbols

The following symbols represent the common or standard variables used in the type of engineering and microbiological calculations presented in this book. This list excludes the economic terms used in Chap. 16, which are not standard or used elsewhere. The metric or SI units shown are the favored units for these variables, although some of these variables have been used with English units in Chap. 16. Several duplicitous variables are used in this book, but they are used in a defined context. The exception is ID_{50}, which can be either infectious dose (BW agents) or incapacitating dose (CW agents) depending on how it is used. Also, the term "cfu" is used for all microbes, including viruses, even though "pfu" would be the correct unit.

a	arbitrary constant
a	filter media volume packing density (media/total volume), m^3/m^3
ACH	air change rate, 1/h
a_i	volume i packing density
b	arbitrary exponent
B_r	breathing rate, m^3/min or L/min
C, $C(t)$	concentration, cfu/m^3 for BW, $\mu g/m^3$ for CW
C_a	airborne concentration, cfu/m^3 for BW, $\mu g/m^3$ for CW
CD_{50}	mean casualty dose, $mg \cdot min/m^3$
Ch	Cunningham slip factor
D	duration, days or hours
D_{37}	37 percent of the duration
D_d	particle diffusion coefficient, m^2/s
d_f	fiber diameter, μm
d_{fi}	fiber i diameter, mm
D_L	logmean diameter, μm
D_{max}	maximum diameter, μm
D_{min}	minimum diameter, μm

Dose	dose, mg·min/m^3, mg, or cfu
d_p	particle diameter, μm
dP_F	filter pressure loss, in.w.g.
E	efficiency of a filter, fractional
e	symbol for exponentiation
E_D	diffusion efficiency, fractional
E_R	interception efficiency, fractional
E_s	efficiency of a single filter fiber, fractional
E_{si}	single fiber i efficiency
E_t	exposure time, s or min
E_{tot}	total UV power, μW
E_{uv}	UV power, μW
f	resistant population fraction, fractional
F_i	view factor of segment i
F_K	Kuwabara hydrodynamic factor
frac$_i$	fraction i fiber diameters
F_{tot}	total view factor
H	height, cm
hp	horsepower, hp
I_D, I_{Dn}	direct or incident intensity, μW/cm^2
ID$_{50}$	mean incapacitating dose, mg/kg or mg·min/m^3
ID$_{50}$	mean infectious dose, cfu/m^3
I_R, I_{Rn}	reflected intensity, μW/cm^2
I_S	UV intensity, μW/cm^2
k	Boltzman's constant, 1.3708×10^{-23} J/K
k	UVGI rate constant, cm^2/μW·s
KR	kill rate, fractional or %
KR$_i$	kill rate of component i, fractional
KR$_T$	total kill rate, fractional
l	length of lamp, cm
L	length, m or cm
L(Ct)$_{50}$	mean lethal dose (exposure), mg·min/m^3
LC$_{50}$	mean lethal concentration, mg/m^3
LD$_{50}$	mean lethal dose, mg/kg
l_g	arclength of lamp, cm
l_g, l_b	length of lamp segments, cm
m	resistant population UVGI rate constant, cm^2/μW·s
MDP	mean disease period, days or hours

MIP	mean infectious period, days or hours
N	number of room air changes per hour, 1/h
$N(t)$	number of microbes at time t
Nr	interception parameter
OA	outside air flowrate, m^3/min
P	power, W or μW
Pe	Peclet number
peak	peak of infection or contagiousness, days or hours
P_{lamp}	lamp total wattage, W
P_v	velocity pressure, in.w.g.
Q	airflow, m^3/min
r	radius of lamp, cm
RR	release rate, cfu/min or mg/min
S	Survival, fractional
$S, S(t)$	source release rate, $cfu \cdot min/m^3$ for BW, $\mu g \cdot min/m^3$ for CW
S, S_i	fiber projected area, fiber i projected area
SSC	steady-state concentration, cfu/m^3 or mg/kg
T	temperature, K
t	time, s, min, h, or days
U	media face velocity, m/s
V	volume, m^3
W	width, cm
x	distance from lamp axis, cm
x	dose, $mg \cdot min/m^3$, mg, or cfu, or arbitrary variable
X	height parameter
y	total number or percentage of cases, casualties, or fatalities
Y	width parameter
Δh	heat increase, W
ε	inhomogeneity correction factor
η	absorption efficiency
η	gas absolute viscosity, $N \cdot s/m^2$
η_{fan}	efficiency of fan, fractional
η_{motor}	efficiency of motor, fractional
λ	gas molecule mean free path, μm
μ	particle mobility, $n \cdot s/m$
π	pi, 3.14159...
ρ	reflectivity, fractional or %
σ	standard deviation

1

Introduction

1.1 Introduction

Buildings are enclosed environments with air distribution systems intended to make life habitable for the occupants, but they simultaneously disseminate any contaminants that might be internally released, whether from natural sources or otherwise (Fig. 1.1). The most severe challenge to the habitability of any building, besides fire and smoke, can come from the intentional release of either chemical or biological agents. Many such agents exist and the threats posed by the various ways in which they could be released are so complex that no single system can be relied on to provide total building protection.

Furthermore, the wide variety of buildings and facilities that may be subject to such releases precludes any "one size fits all" solution. Instead, the designer or engineer is faced with a set of options and decisions that are based on specifically identified threats. Defining the threats and establishing performance criteria are the first steps to take in immunizing a building against biological weapons (BWs), chemical weapons (CWs), or any form of unwanted indoor contaminants.

Ordinary buildings tend to assist the spread of airborne disease, and tend to disperse contaminants such that all occupants suffer similar risks. Immune buildings are those that have been designed or retrofitted to reduce the risks associated with airborne disease transmission. In the ideal immune building no disease would ever be transmitted by the airborne route. No common colds, no flus, not even childhood diseases would be exchanged between people in such a building, although transmission by other means, such as direct contact, might still occur.

Figure 1.1 Buildings are machines for living in, but they can also be engineered to protect occupants from airborne hazards. (*Photo courtesy of Art Anderson, Penn State.*)

The principles by which an immune building protects occupants against contagion from normal human sources apply equally well to any biological agents that are intentionally disseminated indoors. The only difference is the degree of protection that must designed into the facility since a terroristic act may involve considerably higher concentrations of bioaerosols. Furthermore, the lethality of the BW agents likely to be used in such acts is much higher than would be encountered naturally, and so extra levels of protection, or at least extra levels of analysis, may be required to assure the safety of any building occupants.

An immune building that protects against BW agents will also offer some degree of protection against CW agents. However, the removal of chemicals and gases generally requires separate technologies. Since BW agents are by far the more dangerous of the two, a two-stage design problem is posed. First, the system must be designed against the likely BW agents. Next, the system can be enhanced to deal with a likely array of CW agents.

The bottom line, given a defined set of performance goals, is always economics. The ideal air-cleaning system for any building is the most cost-effective system. It would be easy to spend a fortune installing protective systems in any building, but this is not necessarily a practical approach in today's cost-conscious engineering world.

Radiological releases are not addressed in this book, except as they relate incidentally to the subject matter. The problem of radiological releases is complex and requires extensive treatment of radiation exposure issues that are beyond the scope of this book. However, radiological contamination is controlled with essentially the same technologies as are chemical and biological weapon (CBW) agents, and any immune building system designed for CBW agents will likely have some capacity to protect against airborne radiological materials. In any event, a radiological release may be one of the least likely scenarios due to the difficulties of obtaining and handling such materials.

Before plunging into the details and design aspects of protecting buildings against chemical and biological agents, it can be most informative to review the history of poisons, toxins, chemical warfare, and biological weapons so as to understand how we arrived at the present situation, in which we are obliged to expend resources protecting ourselves against such manifest evils. This historical excursion will also provide perspective on the types of weapons used and the methods of delivery, as well as providing insight into the motivations of those who would use biological weapons.

1.2   A Brief History of Chemical and Biological Warfare

By the dawn of history a variety of poisons were well known to the ancients. Hellebore, aconite, hemlock, wild olive, and other naturally occurring plants had been identified by physicians like Hippocrates and Galen for their potential as medicines and their use as poisons. The Roman poet and scholar Lucretius referred to poisons as options for ending the pain of the elderly or the sick.

The use of poison predates the historical record. As shown in Table 1.1, poisoned arrows were used by ancient tribal peoples, and such techniques are still used for hunting by some South American tribes. Mythological references from as early as 1200 B.C.E. indicate that the use of poisoned arrows in war was known to the ancients. Besides poison-tipped arrows, poisons were also used on wells to deny fresh water to the enemy.

The first use of chemical weapons may date to 2000 B.C.E. in India, where rags soaked in flammable fluids were burned to create toxic fumes to harass the enemy (Hersh 1968). The Sung Dynasty in ancient China developed the technique of generating arsenical smoke. Then, as now, these techniques were probably not very effective, and were used only on rare occasions.

In the *Art of War,* written about 500 B.C.E. (Sawyer and Sawyer 1994), Sun Tzu cautioned against encamping troops downstream of the

TABLE 1.1   Historical Timeline of Chemical and Biological Warfare

Ancient world	
—	Prehistoric hunting use of poison-tipped arrows
2000 B.C.E.*	Use of toxic smokescreens in India
1250 B.C.E.	Use of poison arrows in war
600 B.C.E.	Greeks poison water supplies in war
500 B.C.E.	Use of poison in wars in ancient China
336 B.C.E.	First mass poisonings in Rome
Middle Ages	
1155	Barbarossa poisons wells with bodies
1346	Kaffa attacked with plague corpses
1422	Bodies catapulted into Carolstein during siege
1456	Toxic smoke used by Belgrade Christians
1500	Epidemics decimate Native Americans
Modern age	
1763	British Colonial Army defeats Native Americans by spreading smallpox
1764	Virginia Militia victorious after smallpox decimates Native Americans
1777	General Washington defends against British smallpox threat
1781–1783	Canadian traders spread smallpox among western Native Americans
1800s	U.S. Army spreads smallpox to Plains and West Coast Native Americans
1863	Wells poisoned with bodies in the U.S. Civil War
1890	British use chemical shells against Boers
1915–1916	Germans infect Allied horses with glanders
1915–1918	All-out chemical warfare in World War I
1930	Several countries begin research on biological weapons
1936	Italy drops mustard gas bombs on Abyssinians
1937–1945	Japan attacks China with chemical and biological agents
1942	Soviets release tularemia against Germans at Stalingrad
1943	German Nazis use poison gas against Jewish resistance fighters
1943	Czechoslovakian resistance uses anthrax against Germans
1944	Polish underground uses typhus against German troops
1948	Israel uses typhus to poison wells in Arab towns
1960–1980	Soviet Union weaponizes biological agents
1975–1991	Purported use of CBW agents in Afghanistan, Laos, Cambodia, and Cuba
1979–1980	Massive anthrax attacks on black tribal lands in white-ruled Rhodesia
1980	Apartheid South Africa uses cholera and anthrax against black insurgents
1980	Iraq uses chemical weapons against Iran
1984	Rajneeshee cult contaminants food with salmonella in Oregon
1988	Iraq uses mustard gas against Kurdish rebels
1991	Right-wing extremists develop ricin in Minnesota
1990–1993	Aum Shinrikyo releases anthrax and botulinum in Japan
1994	Aum Shinrikyo sarin gas attacks in Japan
1996	Shigella food contamination in Texas
1998	Yugoslav forces poison water supplies in Kosovo with chemicals
1998	Christian Identity/Aryan supremacists acquire plague and order anthrax
2001	Anthrax spore mailings in U.S.—unknown extremists suspected

*B.C.E. = Before the Common Era; an archaeological dating term that replaces B.C. (i.e., Jesus was born in 4 B.C.E.).

enemy, since they may "throw poison on the water to be carried down to us." He noted, however, that this idea came from earlier writers. This particular use of poison bears a resemblance to the way the Germans dispersed poison gases toward the French when they were downwind in World War I.

The ancient Greeks looked upon the use of poison arrows with contempt, considering it a dishonorable way to fight a war. As proof of the inferiority of the kinds of people who resorted to such methods, they pointed out that no society that used poisons in war had ever prevailed against the more civilized societies that did not. The Greeks did, however, occasionally poison wells they had to abandon to the enemy.

During sieges of ancient cities, it was typical of the attackers to attempt to burn down the walls or gates. Combinations of sulfur, pitch, quicklime, and naphtha were sometimes used as accelerants and the result often produced choking smoke and fumes, but this was a side effect. The use of these and other flammable materials in the form known as "Greek fire" could temporarily disorient or incapacitate the enemy (Fig. 1.2), but the intention was usually to set opponents' ships or equipment on fire, as described in Edward Gibbon's *Decline and Fall of the Roman Empire* (Gibbon 1995).

The ancient Greeks used poison to administer the death penalty to their more respected citizens, the most notable among them being Socrates, who was forced to drink hemlock at the age of 81 by democratic vote. Later, as the wealth of the expanding Roman Empire grew, poisons came into increasing use as a means of murder, usually with the aim of acquiring a person's wealth.

The first terrorist-style incident in history involving poisons occurred in 336 B.C.E. when the foremost men in Rome began dying off

Figure 1.2 Illustration of "Greek fire" being used to burn enemy ships at the siege of Constantinople in 687. (*From the Scylitzes manuscript, P. Oronoz, Madrid.*)

in large numbers from what was believed to be an epidemic, until authorities were informed by a servant that Roman matrons were systematically poisoning their husbands. The magistrates caught 20 women in the act of brewing large quantities of poison. These women claimed it was only medicine, and when challenged to take it themselves they obliged, and shortly died. Some 170 women were subsequently convicted of involvement in the murders, which were the first poisonings in Rome. The whole incident was ascribed to "mass madness" and was so incomprehensible to Roman males that future generations disbelieved it (Livy 24 B.C.E.).

The ancients were also aware of the possibility of poisonous gases, and numerous references were made to noxious fumes emanating from the earth. In Italy, there were well-known reports mentioned by Lucretius that in "the Avernian places" fumes rising from the earth could kill birds in the air (Latham 1983).

In 1155, Barbarossa used bodies of men and animals to poison wells at the battle of Tortona (Poupard and Miller 1992). The use of carcasses to poison wells has been known since ancient times, and this could be considered a crude form of biological warfare.

In the year 1346 in the Crimea, an invading Tartar army catapulted plague-infected cadavers into the city of Kaffa, now Feodosia, in Ukraine. The Tartars had suffered numerous casualties from the plague and were apparently trying to spread their misfortune to the enemy. The city suffered a great plague, and the defending army retreated. However, the catapulted cadavers and the subsequent plague were not necessarily related, since the fleas that carry the plague abandon cadavers, and so victory cannot be ascribed solely to this tactic (Christopher et al. 1997).

The first use of chemical weapons during the Middle Ages seems to have occurred when Christians defending Belgrade in 1456 used burning rags soaked in a toxic mixture to fend off Turkish attackers. It seems unlikely that this act would have killed any enemy soldiers or that it was the reason for the Christian victory, but it may have served to harass the enemy.

The first successful use of disease in the conquest of nations happened unintentionally and unexpectedly in the Americas. As the conquering armies of Spain marched across Central and South America, they were often preceded by mass epidemics. The spread of disease was exacerbated by natives who were fleeing the Spanish as well as the outbreaks. Epidemics of measles, smallpox (see Fig. 1.3), typhus, and other diseases to which the Native Americans had no natural immunity, swept across the Americas and within a few years wiped out tens of millions of people (Jackson 1994; Whitmore 1992). When the Europeans arrived to colonize North America, the native populations

Figure 1.3 Detail from one of the last Aztec tapestries. It shows the smallpox epidemic of 1538. In the center are bodies wrapped in blankets, and on the right two men are dying.

had been decimated to the point that many of their largest agricultural communities, especially in the Ohio valley, had been abandoned and most tribes had reverted to nomadic hunting lifestyles. Estimates of the pre-Columbian population in the Americas range from less than 20 million to over 100 million (Dobyns 1966). Figure 1.4 shows the combined results of some studies on the population decline in Mexico, which was estimated to have as many as 20 million people when the Spanish arrived.

The first premeditated use of biological weapons occurred in the Americas under British rule in 1763 (Hersch 1968; Hoffman and Coutu 2001). During the French and Indian War, which had been going badly for the British colonials in the Great Lakes area, General Jeffrey Amherst (Fig. 1.5) suggested causing a smallpox outbreak among the Native Americans. Under Colonel Henry Bouquet, blankets, clothes, and handkerchiefs were collected from a hospital that had been dealing with an outbreak of smallpox. These were given to Native American delegates who had been invited to Fort Pitt to discuss peace, presumably including a peaceful surrender of the fort. The effect was stunning—the confederation of tribes under Pontiac that had been fighting

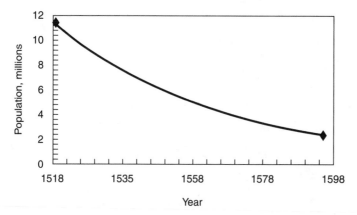

Figure 1.4 Estimated decline in Mexican population after Spanish conquest. Data points are shown for 1519 and 1595. [*Based on data from McCaa (1995).*]

Figure 1.5 General Jeffrey Amherst, the father of biological warfare, pictured most ironically as a knight in shining armor in this painting by Joshua Reynolds. (*From the National Archives of Canada.*)

British domination collapsed amid epidemics that killed large numbers of the natives, estimated by some at over 50 percent, in the local Delaware, Mingo, and Shawnee tribes in Ohio. Some of the Native American towns in Ohio disappeared completely, being either wiped out by smallpox or abandoned by the survivors (Tanner et al. 1986). The British, who had been entirely on the defensive throughout this war and had suffered a series of losses, were now able to launch a successful offensive and, by the spring of 1764, native resistance had all but ceased.

Hostilities had spilled over and continued in other frontier areas after 1763, particularly in Virginia where an army of 1000 men had been mobilized (Tanner et al. 1986). In September 1764, Colonel Andrew Lewis of Virginia wrote Colonel Bouquet that the local Indians they had been fighting were "dying very fast with the smallpox; they can make but little resistance..." (Tanner et al. 1986). The inference that these American colonials had learned the technique of spreading smallpox to Indians from Bouquet seems almost inescapable, but the epidemic may have also spread south from the Great Lakes as a result of the outbreaks caused by Amherst the year before.

Knowledge of the success of this form of biological warfare must surely have passed rapidly among the military elite in the British colonial forces and subsequently among the American Revolutionary Army. When American forces invaded Canada under General Thomas in 1776, they became overwhelmed by diseases, especially smallpox. It was even stated that "smallpox was ten times more terrible than Britons, Canadians, and Indians together," according to the Adams family letters (Freeman 1951). General Thomas died of smallpox on June 8, 1776. No suggestion was made at this time that the smallpox outbreaks were anything other than natural occurrences, but the British commander they were facing was surely aware of how to use it as a weapon. The Americans were equally aware that immunity to smallpox was desirable. A commissioned report at the time criticized the invasion of Canada, stating that "three-fourths of your army has not had the smallpox," implying that they had not been variolated, or inoculated, against smallpox (Freeman 1951).

The defensive nature of the biological aspects of the Revolutionary War appears repeatedly. During March 1776, General Israel Putnam assembled a force of 500 men who had experienced smallpox to cross British defenses and enter Boston (Freeman 1951). These men were known by their facial scars, and the fact of their immunity to smallpox was well understood by the commanders, including General George Washington (Fig. 1.6), who directed the operation.

In July 1776, Washington ordered that Continental soldiers only be sent to the banks of the Hudson after they were "free of smallpox." This expression was commonly used to refer to those who had been immunized, since it meant they were free of the smallpox-like illness caused by variolation, but could also mean that they had survived natural smallpox. In a 1777 document discussing the assignment of a spy to be sent into a British camp, it was noted that the agent had not been inoculated, and was immediately sent "to take the infection" (to get inoculated) and then returned "from the Small Pox" prepared for his espionage mission (Freeman 1951).

Figure 1.6 George Washington took the first recorded defensive measures known anywhere against biological war during the American Revolution. (*From an 1851 engraving by Geoffrey of Didier Publishing, Paris.*)

In 1777, during the siege of Boston, Washington received reports that the British were attempting to spread smallpox among his own troops. He initially disbelieved this until his troops began falling ill, and then ordered his army variolated against the disease (Gibson 1937). This was a drastic measure as variolation, a precursor to vaccination, tended to kill about 3 percent of those who were inoculated. Obviously it was well understood by 1776 that the British could induce such outbreaks, and the matter was taken with grave seriousness.

The dark legacy of General Amherst was transmitted directly to his officers and associates who, like him, desired the extermination of Native American tribes. Among those who carried out an apparently unofficial commission to wage a genocidal war against native tribes were Colonel Robert Rogers, Alexander Henry, and Peter Pond

(Hoffman and Coutu 2001). Rogers was an experienced Indian fighter renowned for his ruthless massacre of 140 men, women, and children at St. Francois during the French and Indian War. Henry was a supplier for General Amherst. Pond was a former officer in Amherst's army and an unscrupulous fur trader in the North West Trading Company. Both Rogers and Pond were present at Mackinac when Captain Ecuyer, under Bouquet's orders, gave smallpox-laced blankets to the Indian delegates.

Pond, Henry, and some others appear to have been using the fur trade as a front to disseminate smallpox among as many of the western tribes as they could before being discovered by civilian authorities (Hoffman and Coutu 2001). Pond and his partners embarked on a deadly mission in 1781 in which "expensive bed gowns, strouds, and blankets" were given away to indigenous tribes, including the Ojibway, the Cree, the Chipewyan, the Dakota, the Assiniboine, and possibly also the Ottawa, the Missauga, the Wyandot (Huron), and the Potawatomi. The trail they took and the stops they made during 1781–1783 form a perfect match with the pattern of smallpox outbreaks, as shown in Fig. 1.7.

As with the outbreaks during the French and Indian War, the fatality rate from these epidemics was so high that some entire villages disappeared. Considerable evidence exists to implicate Pond and his

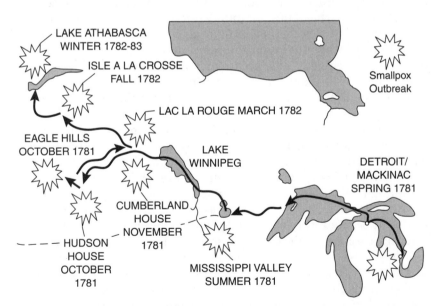

Figure 1.7 The trail of epidemics in the wake of Pond's mission to distribute smallpox-laced blankets to the fur-trading tribes of Native Americans in 1781–1783. [*Based on Hoffman and Coutu (2001)*.]

associates in these epidemics, but much of it is of a circumstantial nature. Critical passages and entire sections of log books and diaries were systematically destroyed in subsequent years. The reasons these men embarked on such a rampage could have been many, including profit and revenge, but whether they were acting on government orders or acting on behalf of the trading company remains unclear. It is likely these outbreaks spread well beyond the original foci, but little research has been done into the general patterns of smallpox spread among native populations during this period.

Before the epidemics in western Canada had finished running their course, the new United States had assumed control of the Colonies and was already eyeing westward expansion. The U.S. Congress discarded the treaties that the British had formerly honored with Native Americans and encouraged settlers to move west into Indian territories, and into harm's way. Each time the U.S. Army was sent into action against hostile tribes, smallpox seemed to follow. The end result was always the same—the decimated tribes were forced to cede their lands to the U.S. government, who then sold them off. In the early 1800s, smallpox was breaking out among numerous western tribes, particularly among the Plains Indians and any other tribes that resisted westward encroachment, and the pattern suggests deliberate depopulation.

Circumstantial and anecdotal evidence exists that implicates the U.S. Army in deliberate attempts to wipe out Indian resistance with smallpox during the 1800s. Certainly, the military had been privy to the knowledge of this method of fighting Native Americans since the Revolutionary War. According to Stearn and Stearn (1945), the U.S. Army distributed smallpox-laced blankets to Native American tribes during the 1800s, particularly targeting the Plains tribes and the tribes on the West Coast. It was government policy at this time to destroy the buffalo and wipe out the food supply of the Plains Indians, as well as to confiscate their lands. Captured Indians were sent to prison camps where medical attention was nil and the death rate high. The Indians Wars may have been the third war that was successfully fought using biological weapons, after the French and Indian War and the war against the Indians in Virginia in 1764.

By the Civil War, or the War Between the States as some historians refer to it, both military commands allowed disease take a toll among prisoners through neglect of sanitary conditions, denial of adequate medical service, starvation, and the exchange of diseased prisoners (Sartin 1993). In prison camps on both sides, soldiers died in alarming numbers from diseases like dysentery, smallpox, and influenza, just as captured Native Americans did in the prisons to which they were sent.

From as early as the Civil War, poison gases like chlorine or flammable forms of arsenic were under consideration as potential weapons of war

(Clarke 1968). A few decades later the British, exasperated in their bloody struggle against the first world's guerrilla fighters, the Boers, resorted to the unconscionable tactic of firing explosive shells that released poisonous clouds of lyddite. Ultimately this approach proved ineffective and the British turned to more successful methods like scorched earth, forced population relocation, marauding cavalry, and other such counter-guerrilla tactics to defeat the Boers.

The British had, however, set a dangerous precedent by firing chemical shells, and soon several nations were developing chemical weapons far more deadly than lyddite. In World War I, both sides were ready with stockpiles of deadly gases the likes of which had never been seen in the natural world. When conventional trench warfare produced nothing but a stalemate of bloody attrition, these weapons of mass destruction were loaded and fired, first by one side and then by the other in retaliation.

The Germans used chemical weapons first, firing shells that released clouds of chlorine gas on French troops in the low-lying areas at Ypres. Even though French intelligence knew the chemical attack was coming, they were completely inept at preparing their troops, who suffered 15,000 casualties and 5000 deaths. The French army retreated in panic, leaving a 4-mile long gap in the front lines. The Germans, not anticipating such success, or perhaps afraid of advancing into the killing fields, failed to follow up on this strategic advantage.

What followed next was a regular mutual exchange of chemical weapons, including participation by the late-arriving Americans. Troops on both sides adapted to gas warfare (see Fig. 1.8), and the stalemate remained but simply shifted to a new level of lethality. The soldiers even became adept at distinguishing between the varieties of poison gases by their smell and appearance.

The CW agents used in World War I consisted mainly of phosgene, chlorine, and mustard gas. Of the approximately 1,300,000 casualties caused by all the agents used, some 400,000 were caused by mustard gas alone. The rapid adoption of defensive measures, specifically gas masks, by both sides denied any advantage from the use of chemical weapons and they became useless—that is, until the innovative Americans came up with blister agents that could kill and incapacitate by being absorbed through the skin. The defensive response was, of course, full-body protection (Fig. 1.9), and so the cycle of technology development continued without influencing the outcome of the war.

Worldwide revulsion at the inhumanity of chemical weapons in World War I led to the Geneva protocol, which was ratified by all the former enemies and subsequently violated by several of them. Italy dropped mustard gas bombs against the defenseless Ethiopian nationalists in 1936, to the scorn of the world community (Clarke 1968).

Figure 1.8 U.S. troops training with gas masks under tear gas clouds sprayed from a plane. (*Photo by J. Delano from the U.S. Government Archives.*)

Chemical weapons research continued in secrecy, with newer, deadlier agents such as sarin, tabun, and soman being developed. Nerve agents were developed by the Germans, and Japan investigated hundreds of chemicals. By the 1930s, several countries, including the United States, Germany, Japan, and the Soviet Union, were pushing weapons research beyond chemicals into the arena of biological agents.

In World War II, the allies believed the Germans might use a toxin weapon, botulinum toxin, and carried along antidotes whenever they faced enemy lines. Japan used a variety of disease agents and chemicals on the helpless population of occupied China, apparently both for experimentation and in attempts to commit genocide. The Soviets, in desperation during the siege of Stalingrad, released tularemia into areas held by the Germans, but they caused more infections among their own people because of a change in the winds (Alibek and Handelman 1999). The Germans reported numerous infections among their troops in this time period, and it was enough to temporarily delay their progress, but they never suspected they had been the victims of a biological attack.

The German SS used poison gas to kill and drive Jewish resistance fighters out of their underground hideouts and from the sewers during the Warsaw Ghetto Uprising in 1943 (Fig. 1.10). Survivors were sent to concentration camps where many were put to death in the gas cham-

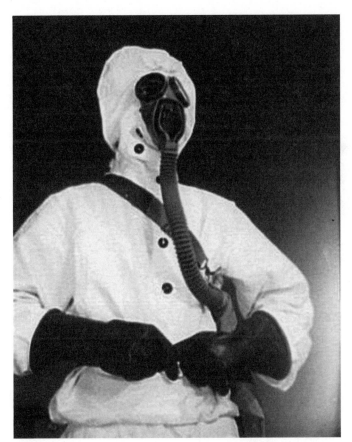

Figure 1.9 Blister agents developed during World War I led to the use of full-body protection. (*Photo by H. R. Hollem from U.S. Government Archives.*)

bers using various agents including the hydrogen cyanide (HCN) gas Zyklon B manufactured by the I. G. Farben Company.

During World War II, the Czechoslovakian resistance used anthrax as a weapon against German occupation forces by disseminating spores on envelopes. It is not known how effective this was. The Polish underground in Warsaw used a similar tactic with greater effect. They spread *Salmonella typhi* bacteria and caused typhoid fever epidemics among occupying Germans throughout the war, killing at least 256 German troops. The Poles were eventually caught with a variety of biological agents they had developed.

Shortly after World War II, Israel used typhus to contaminate wells in Arab towns they captured in 1948 in order to prevent the inhabitants from returning (Barnaby 2000). A similar attempt to poison wells in

Figure 1.10   German SS bodyguards protect Maj. Gen. Stroop during the Warsaw Ghetto Uprising in 1943. (*Photo from the Polish National Archives.*)

Egyptian Gaza with typhus and cholera failed when the soldiers sent on this mission were captured.

The BW agents used by the Japanese included bubonic plague, cholera, typhoid, typhus, and anthrax. According to some sources, the Japanese had set up a biological weapons factory in Manchuria and regularly used prisoners for testing, apparently making them stand in the open while crop-dusting planes flew over them, dispersing the agents.

In this period, researchers in various countries were developing dispersion systems, mainly shells, and selecting microbes capable of surviving the explosive release. Nonairborne agents such as cholera and typhus, which are water-borne and mosquito-borne, respectively, were tested for delivery by aerosol and other routes. The degree of success achieved is not known, but various sources note continued government research, production, or stockpiling of both these agents and their antidotes (Kortepeter and Parker 1999; NATO 1996; Siegrist 1999; Siegrist and Graham 1999; Wright 1990).

The most serious research in the field of biological weapons since World War II was conducted by the Soviet Union under the Biopreparat program. This heavily funded agency embarked on a Manhattan-style research project and developed everything from weaponized smallpox to genetically engineered Ebola virus (Alibek and Handelman 1999). If it had not been for an accidental release of anthrax spores at Sverdlosk in 1979 that caused over a hundred casualties, the United States would have been completely unaware of this program (Meselson et al. 1994).

The largest anthrax outbreak in history occurred during the struggle between the apartheid Rhodesian government and the black African people of what is now Zimbabwe (Barnaby 2000; Davies 1983, 1985; Nass 1992a,b). In October 1978, the government launched air attacks against guerrilla training camps in rural areas. Within weeks, major outbreaks of anthrax began occurring in all of the black-owned Tribal Trust Lands. A total of over 10,000 casualties resulted, with at least 182 fatalities (Davies 1985; Nass 1992b). Actual fatalities may have been higher due to the breakdown of reporting services in many areas of the country during the insurgency. So many cattle died that food shortages became a problem (Turner 1980). No significant outbreaks occurred in any white areas and no whites were affected. The Rhodesian military had apparently employed the crop-dusting approach, possibly with helicopters, to disseminate anthrax spores. The outbreaks began shortly after the Rhodesian military were sent on air raids into the Tribal Trust Lands, and they stopped as soon as the war ended. Although no detailed study of this event has yet been performed that directly implicates the Rhodesian military, the circumstantial evidence is overwhelming. Natural anthrax outbreaks occur sporadically and in endemic locations, with rarely more than a few people becoming infected at any given time. Figure 1.11 shows the Tribal Trust Lands and some of the locations of the anthrax outbreak. The claim by some former Rhodesian government ministers that this was a naturally occurring epidemic that "skipped" white lands leaves microbiologists and epidemiologists incredulous, to say the least.

At least one veteran of the Rhodesian conflict, Jeremy Brickhill, has come forward and stated that the Rhodesian Central Intelligence Organization (CIO) and the Selous Scouts used anthrax, cholera, contaminated foods, and organophosphorus (a chemical weapon) against the insurgents (Nass 1992). Rhodesian forces had also used cholera to contaminate wells in Mozambique (Barnaby 2000). In addition to these agents, American doctors examined large numbers of black prisoners from one training camp who were ill and dying with excessive blood loss, and determined that they were contaminated with the toxin warfarin, an anticoagulant. Although Rhodesia had not previously been suspected of having a bioweapons program, their close ally, apartheid South Africa, admitted to having a government-sponsored bioweapons program (Carus 1998; Leitenberg 2001).

South Africa developed anthrax, cholera, salmonella, and clostridium in a bioweapons program that officially began in 1980 and was administered by the Surgeon General's office of the South African

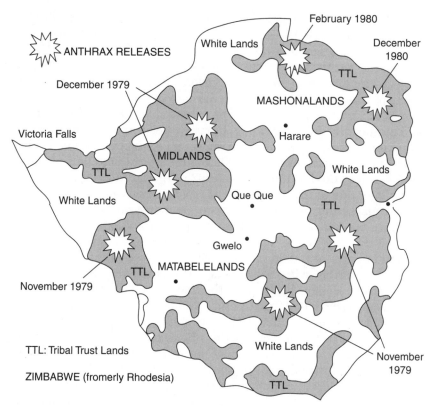

Figure 1.11 Map of Zimbabwe showing Tribal Trust Lands that were subject to major anthrax outbreaks during the 1978–1980 insurgency. Dates indicate epidemic peaks. Locations shown are approximate: Exact foci of most of the outbreaks are not known. [*Based on Nass (1992b) and S. Rhodesia (1957).*]

Defense Forces (SADF). They used anthrax for assassinations. Cholera agent was stored in 5-mL vials that were emptied into water wells by the SADF in areas of insurgency, including neighboring Namibia, but no details of the effects on the population are known (Leitenberg 2001).

Accusations by Cuba of bioterrorist attacks, including biological agro-terrorism, being practiced by the Cuban exiles in the Omega-7 terrorist group and U.S. agents, were refuted in a lengthy study funded by the U.S. Arms Control and Disarmament Agency (Zilinskas 1999a). In spite of these refutations, some questions remain regarding the contamination of Cuban agricultural products with biological agents. However, agro-terrorism is beyond the scope of this book and information on this matter can be obtained elsewhere (CNS 2002).

During their retreat in 1998, Yugoslavian armed forces and militia, the same people responsible for ethnic cleansing, dumped carcasses and chemicals like gasoline, paint thinner, and other hazardous materials into wells throughout Kosovo. Over 70 percent of the water supplies were poisoned and numerous people made ill, but no deaths were reported (Hickman 1999).

The ease with which both rogue governments and terrorist groups resort to the development and use of biological weapons of mass destruction is a portent of things to come, since almost every weapon ever built has been used. Furthermore, the level of scientific research and the degree of success achieved by programs like Biopreparat under the Soviets are frightening. A matter that may be of even greater immediate concern is the apparent disappearance from Soviet stockpiles of huge quantities of BW agents like tons of weapons-grade smallpox (Alibek and Handelman 1999). Perhaps only the paperwork was lost, but the doubt that remains makes smallpox one of the BW agents that poses a current threat.

It is hard to ignore the fact that throughout history a common denominator exists in virtually all the offensive uses of biological weapons. The perpetrators invariably hold their victims in extreme contempt and regard them as subhuman. The attitude of the British and Americans toward the Native Americans, of the Germans toward the Jews they sent to the gas chambers, of the Japanese toward the Chinese, of the white Rhodesians and apartheid South Africans toward black Africans, and of the Yugoslavian militia toward the Kosovo Albanians is the same—one of dehumanization. It is the same attitude that the Islamic extremists hold against what they call Western "infidels" and the "Great Satan." It is worth noting that this is the same attitude held by the numerous hate groups and extremists that proliferate in America today.

1.3 Bioterrorism Today

Worldwide disdain of weapons of mass destruction has not inhibited countries from continuing to research and develop improved CBW agents. For Third World countries threatened by war, these weapons offer an inexpensive alternative to high-technology weapons and nuclear arms (Bashor 1998; Cole 1996). For a fraction of the cost of building a nuclear bomb, any country with a modest microbiological research establishment could develop toxins or disease agents with a hundred times the killing potential of a nuclear weapon (Drell et al. 1999; Pavlin 1999; Pile et al. 1998; U.S. Congress 2000).

Iraq, for example, actively pursued the acquisition and development of both chemical and biological weapons (Zilinskas 1997). These may have even been used during the Gulf War. Other countries and rogue states have reportedly followed their example (Davis and Johnson-Winegar 2000; Smart 1997; U.S. Congress 2001).

Most of the incidents involving actual CBW agents in recent years have been the work of either right-wing extremist groups or fanatical religious cults (Stern 1999; Tucker 1999). The Rajneeshee cult in Oregon used salmonella to contaminate salad bars in a number of restaurants. In Minnesota, the right-wing Minnesota Patriots Council extracted the toxin ricin from castor oil beans and planned to use it on government agencies. The Aum Shinryko cult used several CBW agents in Japan, though their attacks were not all successful. Larry Wayne Harris, a white supremacist associated with the racist groups Christian Identity and Aryan Nations, obtained plague bacillus (*Yersinia pestis*) and several other bacteria before he was caught trying to obtain anthrax (Tucker 1999).

The events of September 11, 2001, and the subsequent dissemination of anthrax spores in the mail between October 2 and November 2, 2001, which infected 17 people and killed 5 (Atlas 2001), have brought home the sobering realization that even a superpower has vulnerabilities. With financing from America's enemies, dedicated terrorists could prepare for years and cause mass casualties by taking advantage of our own equipment and infrastructure. In this light, the specter of biological weapons looms heavier today than at any time in the past. The potential for the development of disease agents with specific properties that facilitate dissemination and cause high fatality rates exists implicitly in the Pandora's box of genetic engineering (Haar 1991; Henderson 1999a; Holloway et al. 1997). The possibility that lone individuals with a basic working knowledge of microbiology, and inexpensive or crude equipment, could develop weapons of mass destruction is very real.

As the trend indicated in Fig. 1.12 suggests, the likelihood of terrorist incidents occurring that involve chemical or biological weapons has been increasing over time. In the context of recent events, it may be naïve to assume such activities will blow over, and it would be prudent to anticipate the worst and design defensible buildings for such scenarios.

Figure 1.13 illustrates the total number of attacks on U.S. interests around the world in 2000. There were a total of 200 incidents that year (H.A.S.C. 2001), of which 178 were directed against American commercial interests, including businesses, buildings, and commercial infrastructure. Most of these attacks involved explosives or arson.

Engineers, designers, and microbiologists have been working for decades in various fields that have contributed to the development of healthcare technologies for the disinfection and cleaning of air and surfaces in buildings. These same technologies are needed today for defense against new biological threats, which are really the old biological threats, just ratcheted up one level by pernicious intent. The tech-

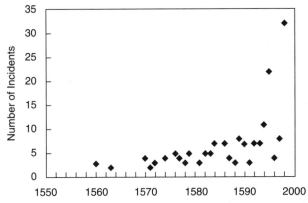

Figure 1.12 Number of terrorist incidents involving chemicals or biological agents in the United States. [*Based on data from Tucker (1999).*]

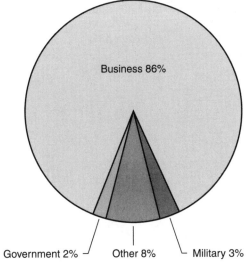

Figure 1.13 Total terrorist attacks against the United States in 2000. [*Based on data from H.A.S.C. (2000). Government total includes diplomatic assaults.*]

nologies that form the foundations of immune building design can be directly applied to the defense of buildings against BW agents. With additional design considerations, these technologies can be also applied against CW agents. It is to this end that the following chapters provide design information, methodologies, and guidance for today's engineers in the hope of protecting against tomorrow's dangers.

2

Biological Weapon Agents

2.1 Introduction

Several hundred microorganisms exist that may cause disease or adversely affect human health in various ways, but only a few dozen of these have characteristics that make them adaptable as biological weapon (BW) agents. In addition to the microbes themselves, there are dozens of toxins produced by them, and by other plants and animals that may also be used as biological weapons.

This chapter addresses those agents that have been previously identified as potential BW agents, and examines the properties that make them dangerous and the attributes by which they may be defended against during an intentional release. No extensive expertise in the area of microbiology is necessary, but some familiarity with basic terminology will be helpful, and definitions will be introduced as necessary. Consult the glossary to become familiar with the vernacular.

There are three basic types of microorganisms that can affect human health: viruses, bacteria, and fungi. A fourth category, protozoa, does not form a major airborne threat and so is not specifically addressed here. In addition to living microorganisms, there are also toxins, which are poisons produced by some microbes, plants, and animals. Both microbes and toxins are considered BW agents, but they are distinct from each other and are treated separately here.

Figure 2.1 illustrates the diversity of BW agents and how they may be subdivided for convenience. Additional subdivisions are possible, but the ones presented here are sufficient for the purposes of understanding the general nature of biological weapons.

A complete database of potential BW agents is presented in two parts in the Appendices. Appendix A summarizes microbes that have BW

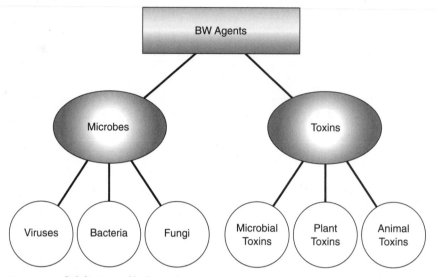

Figure 2.1 Subdivision of biological weapon agents.

agent potential. Appendix B summarizes most of the known deadliest toxins. A discussion of the individual agents, or groups of these agents, follows in this chapter in order to clarify their unique hazards and how they might be used in any intentional releases inside or outside of buildings.

2.2 Microorganisms

A considerable number of microorganisms have been identified in the literature that have been used or produced as bioweapons, or else have been studied for their potential as bioweapons. Not all of the pathogens identified in the literature necessarily make sense as BW agents, but for the sake of completeness they are all addressed here, either as specific species or as groups of closely related species.

Little distinction is made in the literature of biological warfare between those agents that could be disseminated by the airborne route, and those that must be ingested or enter the blood to produce symptoms. Food-borne and water-borne pathogens like salmonella are unlikely to cause illness when inhaled, although there has been little or no study on such matters. Similarly, pathogens that require a vector like mosquitoes to deliver them to the bloodstream of a host may not be able to cause an infection when inhaled, although there is some evidence to suggest that this may be possible. These distinctions may not be relevant to traditional discussions of the use of bioweapons, but in the context of terroristic acts against buildings they may make all the difference.

The distinction between these types of pathogens is shown in Fig. 2.2. Pathogens that are airborne often transmit by the direct contact route also. This has been well illustrated in the case of anthrax spores. Pathogens that are food-borne are also often water-borne and so this distinction is not critical either. In fact, food-borne pathogens will be considered to encompass water-borne pathogens throughout this text. The chart shown in Fig. 2.2 is a simplification that is relevant to the analysis of the use of bioweapons in buildings and may not perfectly define other disease transmission situations, such as hospital-acquired infections.

The use of Fig. 2.2 becomes apparent when it is realized that vector-borne pathogens are not a serious threat to enclosed buildings. Airborne microbes may pass through dust filters, but mosquitoes are unlikely to do so. Other vectors, like ticks, are equally unlikely to be found indoors. Food-borne and water-borne pathogens may not be an inhalation threat, and therefore these need special consideration in terms of protection against their use inside buildings.

A complicating factor is that the vector-borne pathogens might have been weaponized to enable airborne transmission. Such has been suggested in the case of Venezuelan equine encephalitis (VEE) and related viruses. If these have indeed been weaponized such that they are infective by aerosols, then it is conservative to include these as part of the BW agent database.

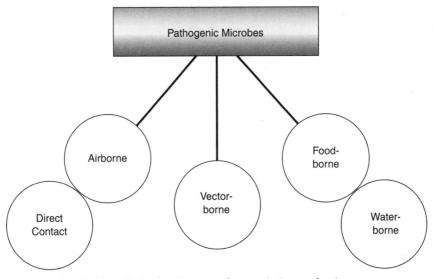

Figure 2.2 Subdivision of microbes in terms of transmission mechanism.

Another source has inferred that vectored pathogens have been weaponized on the basis that the U.S. Army has researched these pathogens and produces and stockpiles many of the vaccines for them (Wright 1990). However, it could equally well be argued that the Army is merely being prudent in anticipating that soldiers may be called to duty in areas where mosquitoes and ticks carry such diseases.

The food-borne pathogens are unlikely candidates as weaponized aerosols. However, their potential for contaminating both water and food supplies has been adequately demonstrated by history. Wherever soldiers have campaigned in tropical climates, the toll from dysentery and cholera, for example, has often been greater than the number of casualties suffered in battle. Again, the records of military research funding for the study of such pathogens can be ambiguously interpreted as either intended weaponization or defensive forethought.

In any event, such agents as salmonella have already been used in previous terrorist incidents, and they are included in the database as potential BW agents regardless of the means of their delivery or the manner of their transmission. It should merely be recognized that food-borne pathogens are only airborne hazards in that they can contaminate food and water, or might be ingested from fomites.

Table 2.1 summarizes all the microbes that have been cited in various studies as potential BW agents. These include airborne, food-borne, and vector-borne agents. The studies from which they have been cited are a representative cross section and do not necessarily include every major work on the subject. They merely provide a starting point for selecting the pathogens that form the final database in Appendix A.

A number of the pathogens identified in Table 2.1 have dubious value as bioweapons. *Histoplasma capsulatum,* for example, is practically innocuous to most people, and tends only to cause distress to those with impaired immune systems. Far more dangerous are the fungi *Paracoccidioides brasiliensis* or *Blastomyces dermatitidis,* which have received no mention in the literature at all.

Some pathogens have received notoriety due to their being the subject, actual or suspected, of genetic manipulation. Ebola, for example, which is not definitively an airborne pathogen, was reportedly weaponized by the Soviets. In fact, weaponization through genetic engineering exists as a potential method of enhancing the virulence of almost any pathogen. Therefore, pathogens that have been mentioned in this context have been included in the final database.

Because of genetic engineering, it is entirely possible that new pathogens may be developed which will render the database incomplete. It is also possible that old pathogens may develop new disease mechanisms or that previously unknown pathogens may adapt to new conditions, as *Legionella* has. Therefore, this database can only be considered representative of the known threats up to the present.

Airborne contamination is the most serious biological weapon threat. Airborne contamination would spread faster and infect or incapacitate more people than would a food-borne or vector-borne attack. Pathogens that spread by direct contact alone do not have as much potential for dissemination as do those that transport by both routes.

A discussion of each of the major pathogens in Table 2.1, or groups of pathogens where several bear strong similarities, is provided here to augment the database in Appendix A and to provide additional source information. These pathogens are broken down into the divisions of (1) airborne pathogens, (2) vector-borne pathogens, and (3) food-borne pathogens. It should be noted that since most of the food-borne pathogens may settle on foods by the airborne route, they could conceivably be intercepted by air-cleaning methods.

2.2.1 Airborne pathogens

The pathogenicity of most airborne microbes listed in Table 2.1 is fairly well understood, as is the means of their control. Many of these microbes are contagious, or communicable, and spread both by inhalation and by direct contact with infected individuals or with fomites. Fomites are infectious droplets or particles left on surfaces like towels and doorknobs. Such particles may remain infectious or viable from minutes to months, depending on the microorganism. The airborne pathogens are addressed in alphabetical order, except where they constitute part of a group with significant commonalities.

Bacillus anthracis is the bacterium that produces anthrax spores, one of the deadliest agents ever created by nature. It has often been cited as a potential BW agent, and has been in production for this purpose in several countries, including Japan, Iraq, and the former Soviet Union. As an example of the devastating effects a release of anthrax could have, an accident in the Soviet Union, at a place called Sverdlosk, in 1979 caused hundreds of casualties (Meselson et al. 1994). In this incident, a high-efficiency particulate air (HEPA) filter was removed for maintenance but not replaced. In the ensuing production shift, in which weaponized anthrax spores were being loaded into shell casings, a large quantity of spores was aerosolized and blew out the exhaust ventilation duct. Over the next several days, hundreds of innocent civilians downwind of the Biopreparat factory where the munitions were being made came down with anthrax infections and many of them died over the next few weeks.

One of the unique advantages that anthrax has over other bacterial and viral agents is that in the spore form it is particularly hardy and resistant to both heat and dehydration. A spore is essentially a seed for

TABLE 2.1 Checklist for Pathogenic Microorganisms with Biological Weapons Potential

Microbe	Type†	Source or reference for potential use as BW agent*														
		1	2	3	4	5	6	7	8	9	10	11	12	13	14	15
Bacillus anthrax spores	A	X	X	X	X	X	X	X	X	X	X	X	X	X	X	
Blastomyces dermatitidis	A															X
Bordetella pertussis	A												X	X		
Brucella	A	X	X	X	X	X	X	X	X						X	
Burkholderia mallei	A		X	X	X		X	X	X	X					X	
Burkholderia pseudomallei	A		X	X			X	X	X	X					X	
Chikungunya virus	V	X		X											X	
Chlamydia psittaci	A		X	X	X			X			X					
Clostridium botulinum	F		X													
Clostridium perfringens	F							X								
Coccidioides immitis	A	X	X	X											X	
Corynebacterium diphtheriae	A														X	
Coxiella burnetii	A	X	X	X	X	X	X	X	X	X	X	X	X	X	X	
Crimean-Congo hemorrhagic fever	A							X	X		X		X	X	X	
Dengue fever	V		X	X					X		X				X	
Ebola (GE)	A	X	X	X	X	X	X		X	X	X	X	X		X	
Francisella tularensis	A						X			X	X	X	X	X	X	
Hantaan virus	A										X				X	
Hepatitis A	Int								X							
Histoplasma capsulatum	A														X	
Influenza	V					X										
Japanese encephalitis	V							X			X				X	
Junin virus	A							X	X		X		X	X	X	
Lassa fever virus	A								X		X				X	
Legionella pneumophila	A												X	X		
Lymphocytic choriomeningitis	V										X		X	X	X	
Machupo virus	A													X	X	

28

Agent	Route	1	2	3	4	5	6	7	8	9	10	11	12	13	14	15
Marburg virus	A						X				X		X	X	X	X
Mycobacterium tuberculosis	A						X						X	X	X	
Mycoplasma pneumoniae (GE)	A													X		X
Nocardia asteroides	A													X		
Paracoccidioides brasiliensis	A													X		X
Rickettsia prowazeki	A	X				X			X	X	X	X	X	X	X	
Rift Valley fever	V			X	X		X		X	X	X	X	X	X	X	
Rocky Mountain spotted fever	V	X		X	X	X				X	X		X	X	X	
Russian spring-summer encephalitis	V			X										X	X	
Salmonella typhi (typhoid fever)	F	X		X	X	X	X			X	X		X	X	X	
Shigella	F	X		X	X	X	X		X	X	X		X	X	X	
Stachybotrys chartarum	A	X		X										X	X	X
Streptococcus pneumoniae	A						X		X	X				X	X	
Tick-borne encephalitis	V	X		X										X	X	
Variola (smallpox and camelpox)	A	X	X	X	X		X		X	X	X		X	X	X	
VEE (EEE, WEE)	V	X	X	X	X		X		X	X	X		X	X	X	
Vibrio cholerae (cholera)	F	X	X	X			X		X		X			X	X	
West Nile virus	V							X								
Yellow fever	V	X	X	X	X	X	X		X	X	X		X	X	X	
Yersinia pestis	A	X	X	X	X	X	X		X	X	X		X	X	X	

*Sources: 1: Wald 1970; 2: Clarke 1968; 3: McCarthy 1969; 4: Hersh 1968; 5: Cookson and Nottingham 1969; 6: SIPRI 1971; 7: Ellis 1999; 8: Canada 1991a; 9: Thomas 1970; 10: Wright 1990; 11: NATO 1996; 12: Kortepeter and Parker 1999; 13: Franz 1997; 14: Paddle 1996; 15: other sources.

†Notes: A = airborne; F = food-borne; V = vector-borne; Int = intravenous; GE = genetically engineered.

new bacterial growth. The spore is considered dormant, or not biologically active, and so is not vulnerable to factors that would destroy ordinary bacteria. It is a small, enclosed shell made of tough, compacted proteins that can remain dormant for indefinite periods of time. When conditions are right, the spore will germinate and return to bacterial form. The right conditions just happen to be the warm moist conditions of the human lungs, or sometimes human skin, where bacterial growth will proceed at a rapid pace.

Anthrax infections proceed so rapidly that unless treatment is undertaken within the first few days, the patient may be unable to recover. Symptoms and treatments, including methods of sterilization, are summarized in Appendix A. Additional information may be found in Chap. 8, which addresses methods of air cleaning specifically to remove anthrax spores, and in Chap. 16, which deals with chemical and biological weapon (CBW) agents in mailrooms.

Blastomyces dermatitidis is a fungus that can produce spores which may cause human infections. Very few fungal spores are capable of causing human infections, except in the immunocompromised, since they exist ubiquitously in the environment and human defenses have essentially adapted to them over past evolutionary epochs. In some cases, where a large dose has been inhaled, or where chronic exposure has produced an infection, *Blastomyces* may produce a serious lung infection called blastomycosis. This infection is potentially fatal if not treated.

Because of the limited danger posed by *Blastomyces,* it is not likely to be used as a BW agent, unless a variant is deliberately selected for virulence, or created by genetic engineering. Since most spores lend themselves to weaponization due to their hardiness, the choice of a natural spore for virulence selection or genetic manipulation implies that *Blastomyces* has the potential to be used as a weapon, but nothing more.

Bordetella pertussis is the bacterial cause of the common childhood illness called whooping cough. Infections are by direct contact or airborne transmission from asymptomatic carriers. Most of the infections occur in children, and a large number of adults carry lifelong immunity. Because of these factors, it is an unlikely choice as a BW agent.

Brucella (specifically including *Brucella melitensis, Brucella suis, Brucella abortus,* and *Brucella canis*) is a bacterium that may cause a serious illness called brucellosis by inhalation or ingestion. It can have a long incubation period but because of the severity of the symptoms and the small number of organisms required for infection (10 to 100) it is considered to be a viable BW agent by many sources. It could be used as either a bioaerosol or as a food contaminant.

Burkholderia mallei (formerly known as *Pseudomonas mallei* and sometimes *Actinobacillus mallei*) is the bacterial cause of glanders, a disease that can be contracted from contact with horses, but is not transmissible between humans. It can have an incubation period as short as 1 day, and can rapidly result in death. It is considered a viable candidate for use as a BW agent.

The organism *Burkholderia pseudomallei* (formerly known as *Pseudomonas pseudomallei*) causes a different infection known as melioidosis, which can sometimes become rapidly fatal. It has a short incubation period and for these reasons is considered by many to be just as likely to be weaponized as the related microbe *Burkholderia mallei.*

Chlamydia psittaci is a bacterium that can be found in bird droppings. The disease caused, called psittacosis, may take days or weeks to develop but can have severe symptoms. It is not transmitted between people and causes no secondary infections. Older studies cite it as a potential BW agent, but newer studies almost never do. It must be considered an unlikely candidate for such use.

Coccidioides immitis (Fig. 2.3) is a fungus that produces spores which, when airborne, can cause a flu-like infection. In some cases a

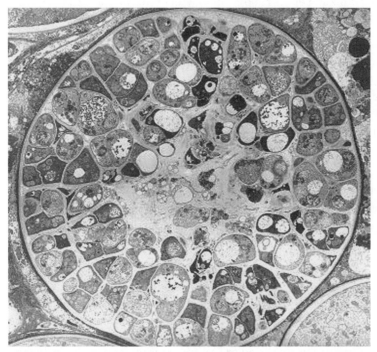

Figure 2.3 *Coccidioides immitis* in an early phase of sporulation. [*Reprinted from Bosche et al. (1993) with permission of G. T. Cole and Plenum Press.*]

severe lung infection is possible and, if meningitis develops, a 90 percent fatality rate can occur without treatment. This microbe has some potential for use as a biological weapon.

Corynebacterium diphtheriae is a bacterium that can cause diphtheria via the inhalation route and also produces a toxin. The toxin can result in a fatality rate as high as 10 percent. It has a relatively short incubation period, 2 to 5 days, but is also commonly the subject of immunization programs. Therefore, it is an unlikely candidate as a BW agent.

Coxiella burnetii is a member of a particular bacterial group known as *Rickettsiae* that also includes *Rickettsia prowazeki*. *Coxiella* causes Q fever, or Query fever, at low doses by the airborne route. It has a relatively long incubation period of up to a few weeks and causes fatalities in less than 1 percent of cases. It has moderate to low potential as a BW agent.

Ebola virus is one of the deadliest viruses known. It is closely related to Marburg virus. They are both transmitted by direct contact between humans, and between humans and primates, and cause severe hemorrhagic fevers that may have high fatality rates. Marburg is believed to be able to transmit by the airborne route while Ebola has only been shown to transmit as an aerosol in laboratory conditions (Jaax et al. 1995, 1996; Johnson et al. 1995). The Soviets manipulated Ebola genetically and demonstrated its capacity for airborne transmission in primates (Alibek and Handelman 1999). All things considered, these viruses must be considered strong candidates for use as BW agents.

Francisella tularensis is a bacterium that can transmit to humans by close contact with animals or by inhalation. It is infectious in very small doses, 5 to 10 organisms, and can incapacitate a person for weeks. It is not contagious between humans. The type A strain of the microbe can have a 35 percent fatality rate. It has long been studied as a bioweapon and is considered a prime choice as such.

Hantaan virus, also sometimes called hantavirus, exists in wild rodents around the world (Fig. 2.4). It is not normally communicable between humans but infections are caused by the inhalation of dried mouse droppings or urine. It can cause severe symptoms within days and has a high fatality rate. It is stable almost indefinitely in warm solution. Although not often cited as a BW agent, it has strong potential for such use.

Histoplasma capsulatum is a fungus that can produce airborne spores. It causes common asymptomatic infections in eastern and central North America. Since millions of Americans have been exposed to this naturally occurring environmental spore unawares, its use as a bioweapon is unlikely. On the other hand, if it were selected for virulence or otherwise manipulated to become a more dangerous pathogen, it could prove to have potential as a BW agent.

Figure 2.4 Hantaan virus is present in the urine of infected rodents. (*Image reprinted with permission of the Centers for Disease Control, Atlanta.*)

Influenza A virus regularly sweeps the world in 2-year pandemic cycles. Occasionally a virulent strain develops and causes millions of fatalities worldwide. The suggestion of its use as a bioweapon is apparently predicated on the deliberate cultivation of highly virulent strains, or even genetic manipulation. As such, it must be considered to have some potential for use as a BW agent.

Junin virus is one of the arenaviruses, along with Machupo and Lassa fever. These viruses have similar characteristics and all cause hemorrhagic fevers. They have fatality rates that range up to 30 to 50 percent. They are caused by inhalation of rodent feces or may be spread by person-to-person contact. Their incubation periods can be moderately long. They would presumably be aerosolized for use as bioweapons and therefore have a moderate potential for use as BW agents.

Legionella pneumophila is an environmental bacterium that causes the eponymous legionnaire's disease. It is normally only dangerous when it is aerosolized in large concentrations, and even then it is mainly a threat to elderly males. Even in large concentrations it is often not fatal to healthy individuals. It makes an improbable choice as a bioweapon, but can be found almost anywhere and might make a weapon that could target particular individuals or groups who may be vulnerable.

Lymphocytic choriomeningitis virus is one of the arenaviruses. It is similar to the arenaviruses that cause hemorrhagic fever, but instead it causes a mild flu-like illness with a fatality rate less than 1 percent. It can transmit both by the airborne route and by ingestion of contaminated food. It has a low to moderate likelihood of being used as a bioweapon.

Mycobacterium tuberculosis is the well-known tuberculosis (TB) bacillus, the cause of tuberculosis. Although it is infectious at very low doses, 1 to 10 bacilli, it can take months to develop an infection. It, too, is an improbable choice for a bioweapon, and yet it may have some potential as a weapon of terror due to the ease with which it may be obtained and the possibility that it may be deliberately selected for multidrug resistance.

Mycoplasma pneumoniae is a bacterium of unusually small size, approximately 0.2 to 2 microns, that can cause lung infections from inhalation. It has a very low fatality rate and a moderately long incubation period. It would seem an unlikely choice for a bioweapon except for the fact that anecdotal evidence suggests that it has been isolated in cases of Gulf War illness in an apparent genetically engineered form (Thomas 1998). Whether that is true or not, it may be a good candidate for genetic engineering because it is a bacterium of such small size, which allows for good aerosolization, and yet has a large genome that can carry considerable information. It could conceivably be modified to carry a virulence gene or even a gene for a toxin like botulinum.

Nocardia asteroides is a spore-forming bacterium that belongs to a group known as the actinomycetes. Being a bacterial spore gives this microbe inherent survivability as an aerosol. It is closely related to several other species that have similar characteristics and can all cause nocardiosis, a disease of the lungs. It is not known to cause fatalities, and the incubation period can last for months, so its potential for use as a bioweapon is very low.

Paracoccidioides brasiliensis is an airborne fungal spore that can cause an acute infection of the lungs called paracoccidioidomycosis. It is one of only four fungal species that cause true respiratory infections. It can be fatal if untreated but little is known about its infectiousness or incubation period. If selected for virulence, it could be a candidate for use as a biological weapon.

Rift Valley fever virus is a member of the bunyaviruses and is vector-borne by mosquitoes in Africa. It causes a febrile illness, but little else is known about it. It is similar to other vector-borne microbes like the flaviviruses and the alphaviruses, including Crimean-Congo hemorrhagic fever.

Stachybotrys chartarum is a fungus that produces spores that may contain dangerous mycotoxins. It is often found growing indoors under certain conditions and has been associated with "bleeding lung disease" in some infants (Reijula et al. 1996). Because of the fact that it is an airborne spore and has the potential to cause toxic illness or even death, it must be considered as having a moderate to low potential as a bioweapon.

Streptococcus pneumoniae is a bacterium that can cause sudden pneumonia and other infections. Case fatalities can be as high as 10 percent.

It transmits by direct contact or contact with fomites. Its ability to transmit by the airborne route is unproven but it may have such potential, and so has a low to moderate likelihood of being used as a bioweapon.

Variola virus is the cause of smallpox. Although it has been eradicated, it still exists in laboratories and was reportedly weaponized in large quantities in the former Soviet Union. It is similar in form and properties to vaccinia virus but is highly contagious by the airborne route and can cause fatalities in one-third of those infected. It must be considered a strong candidate for use as a biological weapon.

Venezuelan equine encephalitis (VEE) is an arbovirus or vector-borne virus found in Central and South America. It is carried primarily by mosquitoes and causes encephalitis. It is a togavirus in a group known as alphaviruses that includes Western equine encephalitis (WEE), Eastern equine encephalitis (EEE), and Chikungunya virus. VEE, WEE, and EEE can have fatality rates of 1, 3, and 60 percent, respectively. They are all considered to have a moderate potential for use as BW agents, except VEE, which has a high potential due to its reported weaponization.

Yersinia pestis is the cause of the well-known plague, or bubonic plague. Although it is essentially a vector-borne bacterium, it can also spread by the airborne route, in which case it is known as pneumonic plague. It has a short incubation period and a high fatality rate, 50 percent, making it a likely choice for use as a bioweapon.

Most of the airborne pathogens described above can be grown using commonly available equipment, such as the small incubator shown in Fig. 2.5. The materials needed to grow such BW agents are also commonly available and can be found on the shelves of college laboratories such as in Fig. 2.6. The skills needed to grow BW agents, however, are more critical as determinants of their possible use.

Considering each microbe individually, and the various characteristics of each, it is possible to compile a list of the most likely BW agents. The criteria on which to base such a selection have been summarized in at least one conference, and include availability, ease of production, lethality, stability in storage, infectivity, deliverability as an aerosol, effectiveness at low doses, high infection rate, and ability to cause severe disease (Zink 1999). The conference that produced these criteria selected anthrax spores and smallpox as the agents most likely to be used as bioweapons.

Ken Alibek, who headed the Soviet Biopreparat biological weapons program, was asked what were the 12 most effective pathogens by the House Armed Services Committee (H.A.S.C. 2000). He identified the BW agents the Soviets chose to produce as smallpox, plague, anthrax, glanders, tularemia, melioidosis, Ebola, Marburg, Machupo, Bolivian hemorrhagic fever, Q fever, and epidemic typhus.

Figure 2.5 Microorganisms can be cultured in small quantities in incubators like the one shown above.

Bolivian hemorrhagic fever is, in fact, Machupo virus. One agent seemingly missing from the Soviet list is Lassa fever virus, which the Soviets had attempted and apparently failed to obtain (Garrett 1994). Although epidemic typhus, from *Rickettsia prowazeki,* is normally vector-borne, it can also transmit by the airborne route, as can the variant of Ebola the Soviets developed. Therefore, there are no strictly vector-borne pathogens on this list.

Table 2.2 summarizes those pathogens from Appendix A that might prove most effective as terrorist bioweapons, based on the previous reviews and the Soviet list. Although there are, all told, some 18 microbial pathogens represented on this list, perhaps three or four could be selected as representative of the entire array in terms of physical size, toughness, and lethality. In addition, not all these pathogens have known properties, and the analyses that can be performed are somewhat limited. Therefore, some of the microbes that seem most suited for analysis, and most representative of the entire potential arsenal, would be anthrax spores, *Mycobacterium tuberculosis,* and variola (smallpox).

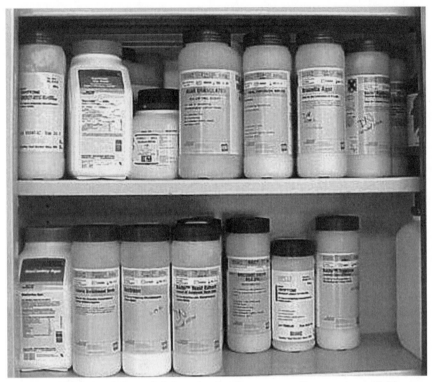

Figure 2.6 The materials needed for culturing dangerous pathogens are available commercially without restriction.

2.2.2 Vector-borne pathogens

Vector-borne pathogens are normally carried or delivered to their hosts by insects such as mosquitoes, lice, and ticks. They form a group of viruses known as arboviruses. The most important arboviruses in human disease are the togaviruses, the flaviviruses, and the bunyaviruses.

Chikungunya virus is closely related to Venezuelan equine encephalitis, Eastern equine encephalitis, and Western equine encephalitis. These are all alphaviruses, a subgroup of togaviruses, and have a similar pathology. They are all transmitted by mosquito bites and have short incubation periods. Chikungunya is rarely fatal but has the potential to rapidly incapacitate. It would presumably be aerosolized for use as a bioweapon (NATO 1996). It is considered to have moderate potential for use as a BW agent.

Crimean-Congo hemorrhagic fever and Rift Valley fever are vector-borne viruses with incubation periods as short as 3 days. They cause

TABLE 2.2 Airborne Microbes Most Likely to Be Used as BW Agents

Agent	Incubation period, days	ID$_{50}$*	LD$_{50}$*	Fatalities, %	Availability	Contagious
Bacillus anthrax spores	2–3	10,000	28,000	5–20	High	No
Brucella	5–60	Low	—	2	High	No
Burkholderia mallei	1–14	—	—	—	High	No
Burkholderia pseudomallei	2	—	—	—	High	No
Coxiella burnetii	14–21	10	Low	<1	High	No
Ebola	2–21	10	Low	50–90	Medium	Yes
Francisella tularensis	1–14	10	Low	5–15	High	No
Hantaan virus	14–30	—	—	5–15	High	No
Junin virus	2–6	—	—	3–15	Moderate	No
Lassa fever	2–10	—	—	10–50	Moderate	Yes
Machupo	7–16	—	—	5–30	Moderate	No
Marburg	3–7	—	—	25	Medium	Yes
Mycobacterium tuberculosis	—	10	—	—	High	Yes
Rickettsia prowazeki	7–14	10	Low	10–40	High	No
Variola (smallpox)	12	—	—	10–40	Low	Yes
VEE (EEE, WEE)	2–6	Low	—	60	High	No
Yersinia pestis	2–6	—	—	50	Medium	Yes

*ID$_{50}$ = mean incapacitating dose; LD$_{50}$ = mean lethal dose. Both terms are discussed in greater detail in Chap. 4.

severe symptoms and can cause fatalities in one-half of cases. As with other vector-borne diseases, they would presumably be aerosolized, although the feasibility of doing so has never been proven. Cases have occurred in which they have apparently spread by contact; therefore, they must be considered to have some real potential as BW agents.

Dengue fever virus is in a class of viruses called the flaviviruses, which also include St. Louis encephalitis, Yellow fever, West Nile virus, Murray Valley fever, Russian spring-summer encephalitis, Powassan encephalitis, and Japanese encephalitis. All of these are vector-borne viruses transmitted either by ticks or mosquitoes (Fig. 2.7). Fevers can be short-lived and have severe symptoms, and case fatality rates can reach 50 percent for some of these viruses. As vector-borne diseases, they would make ineffective BW agents for use against buildings unless they could be both aerosolized and made infectious by the inhalation or skin contact routes. Since such characteristics have been demonstrated in the case of VEE (Franz et al. 1997; Kortepeter and Parker 1999; NATO 1996; Wright 1990), it seems that the flaviviruses may have potential as bioweapons.

Rickettsia prowazeki, Rickettsia canadensis, and *Rickettsia rickettsii* are vector-borne bacteria that are related to *Coxiella burnetii* and have

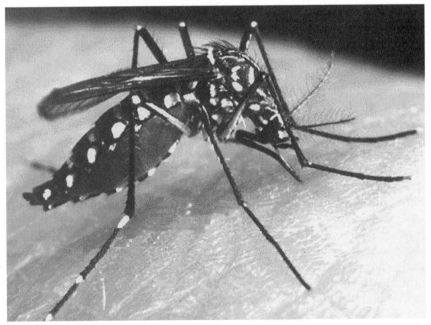

Figure 2.7 The mosquito *Aedes aegypti,* the Yellow fever mosquito, main vector for dengue/hemorrhagic fever. (*Image courtesy of Leonard E. Munstermann, Yale University.*)

some similar characteristics. These rickettsiae cause forms of typhus, including Rocky Mountain spotted fever, and they have moderate incubation periods. They are infectious at low doses. Use of these bacteria as BW agents would presumably involve aerosolization (NATO 1996), which may give them moderate value as bioweapons.

Tick-borne encephalitis is an arbovirus with a pathology similar to the other arboviruses previously addressed, including Chikungunya virus and VEE. Limited information is available on the pathogenicity or epidemiology of this virus, but it is considered to be encompassed by the other vector-borne viruses and so is excluded from the database in Appendix A.

2.2.3   Food-borne and water-borne pathogens

Food-borne or water-borne pathogens could be used as poisons but are not necessarily spread by the airborne route. Some food-borne pathogens can settle on foods and cause food poisoning; therefore, they could conservatively be included in any analysis of air-cleaning technologies. The computation of doses, however, cannot be performed since it depends entirely on the amount of food or water consumed. These microbes are included in Appendix A as a matter of completeness, but are excluded from Table 2.2, the select list of most likely airborne BW agents.

Clostridium botulinum is the bacterium that produces botulinum toxin, one of the deadliest substances known. Botulism is the acute toxic condition that results from ingesting contaminated food. Wound botulism is also possible. The microbe may be airborne when it settles on foodstuffs, and would presumably be aerosolized for use as a BW agent. However, it would be considerably more effective to aerosolize the toxin, botulinum, than to aerosolize the bacteria. The occasional citing of this microbe as a BW agent is probably due to confusion over the difference between the bacteria and the toxin produced, and so it must be discounted as a BW agent for use in aerosol form or as a food-borne pathogen. Because of its importance as a precursor agent, however, it is included in the database in Appendix A.

Clostridium perfringens is related to the previous organism, but has considerably different properties as a food-borne pathogen. The exceedingly short incubation time on food, some 6 h, and the small number of organisms needed to cause infection, 10 organisms per gram of food, give it potential as a BW agent for the poisoning of foodstuffs. Though rarely, if ever, fatal, it can incapacitate on the same day it is used. Therefore, it is considered to have moderate potential as a BW agent, one that might be disseminated as an aerosol to infect food supplies.

Hepatitis A virus, or HAV, can cause the enervating eponymous disease after a long incubation period. It is transmitted by the oral-fecal route, or by ingestion of contaminated food or water, making it essentially a food-borne pathogen. Fatalities are rare and many cases are asymptomatic. The choice of this virus for use as a biological weapon is dubious.

Salmonella typhi is a food-borne and water-borne bacterial pathogen that can communicate from person to person by contact, or through contact by an infected person with food. It could therefore be considered both a communicable disease and a food-borne disease, but the latter category is more appropriate because of its predominance as the mode of transmission. It can cause typhoid fever and severe symptoms, and has a case fatality rate as high as 16 percent. As a food-borne pathogen it has a moderate to high potential for use as a bioweapon.

Shigella represents several species of bacteria that may cause dysentery. It is a food-borne and water-borne pathogen that is transmitted by the fecal-oral route or by person-to-person contact. As such it can be considered both a communicable disease and a food-borne disease. Normal transmission is by contact with foodstuffs but, as with all food-borne pathogens, it may also settle on foods; hence it has the possibility of being aerosolized as a bioweapon. It can cause fatalities in 20 percent of cases. Because of the historical fact that it can cause more casualties in war than battle does, it could be considered a likely choice as a bioweapon in warfare. Its use as a terrorist weapon against buildings may be limited to water supplies and foods.

Vibrio cholerae is a food-borne and water-borne pathogen that causes cholera. It is not known to transmit by person-to-person contact but can have a fatality rate of 50 percent. Historically, in tropical regions, it can cause more fatalities than battle, and could be considered useful as a biological warfare agent. Its use as a terrorist weapon would be limited to poisoning water supplies; therefore, it has a low to moderate potential as a BW agent.

2.3 Toxins

Toxins are biological poisons and may include those produced by bacteria, called endotoxins, or those produced by fungi, called mycotoxins. Bacteria sometimes carry or create some toxins in their cell walls. Under certain environmental conditions, fungi will secrete toxins into their immediate surroundings. Usually this results from a shortage of nutrients, water, or space, and those microbes that generate toxins under these conditions are, in effect, waging biological warfare against their competitors.

Toxins can be extremely deadly to humans and other animals. The quantity necessary to kill a human is essentially invisible to the naked eye. Because of their concentrated toxicity, and the relative simplicity of manufacturing them, toxins are considered ideal candidates for use by terrorists. Only a simple laboratory and a basic working knowledge of microbiology would be necessary to create large quantities of such toxins. Since many of the microorganisms that produce toxins occur in the ambient environment worldwide, obtaining a sample for culturing would not be difficult.

Toxins affect human physiology in different ways and in different places. Table 2.3 shows some of the various sites in the body that are affected by certain toxins. The manner in which the toxin affects physiology is dependent to some degree on how it is received. Some snake venoms, for example, may have no effect on the stomach but can be fatal if they enter the bloodstream.

Table 2.4 identifies all the major toxins that have been cited in the literature on biological weapons. Many of these could be transmitted by the airborne route and would cause incapacitation or death if a sufficient quantity were inhaled. Some of these agents, particularly the venoms, may only be hazardous when they enter the blood. Therefore, such toxins may not be viable candidates for use as BW agents. Detailed knowledge on the effects of inhalation of venoms is not available, and so all of these toxins are conservatively included in the database in Appendix C.

Toxins come from a variety of sources like plants, bacteria, fungi, insects, shellfish, snakes, frogs, and even algae. These natural toxins are often used for defense against predators. They may be easy to obtain in small quantities, but not all are easy to produce in large quantities. A review of some of the major or representative toxins is provided here.

TABLE 2.3   Sites of Action in the Human Body of Some Toxins

Toxin	Site affected
Aflatoxin	Liver
β-Bungarotoxin	Nervous system
Botulinum	Nervous system
Ochratoxin	Kidneys
Palytoxin	Nervous system
Saxitoxin	Nervous system
SEB	Gastrointestinal tract
T-2 mycotoxin	Gastrointestinal tract
T-2 mycotoxin	Skin
Taipoxin	Nervous system

Aflatoxin is a mycotoxin that is exclusively produced by the fungi *Aspergillus flavus, Aspergillus fumigatus,* and *Aspergillus parasiticus.* There are at least five major variants of aflatoxin, called B1, B2, G1, G2, and M1, and they are often found in groups. Aflatoxin B1 is the strongest carcinogen, or cancer-causing compound, known. Aflatoxicosis is primarily a hepatic disease—it affects the liver. Aflatoxins are often found in animal feed, and consumption of excessive amounts can cause acute hepatic toxicosis, jaundice, and hemorrhaging. Acute aflatoxicosis in humans is characterized by abdominal pain, vomiting, pulmonary edema, and necrosis of the liver. The consumption of moldy corn with a concentration of 16 mg of aflatoxin per kg can cause almost 10 percent fatalities due to gastrointestinal hemorrhage (CAST 1989).

Anatoxin A, brevetoxin B, and microcystin are toxins produced in algal blooms. Brevetoxin B is associated with the "red tide" catastrophes that have killed massive numbers of fish and caused a number of human poisonings. Microcystin is associated with blue-green algae while anatoxin comes from the cyanobacteria *Anabaena flosquae.* Algae are generally single-celled species that grow mostly in water. They provide food for aquatic animals and play a part in natural water purification, but under favorable conditions algal blooms may spread widely and generate high levels of toxins. Fish and animals that normally feed on algae may be poisoned, and animals farther up the food chain, like seals, may also fall victim. Algal toxins can have varying effects, like central nervous system damage, neurotoxic effects, and paralysis.

Batrachatoxin is a neurotoxin that comes from the skin of the poison dart frog *Phyllobates,* which is native to the rain forests of South America. Neurotoxins are compounds that affect nerve tissue and they operate on the nervous system in a variety of ways. Batrachatoxin is the most potent neurotoxin among venoms and can kill a human within seconds.

Cobrotoxin, taipoxin, and β-bungarotoxin are neurotoxins that come from snake venoms (see Fig. 2.8). Cobrotoxin comes from the Chinese cobra *Naja naja atra,* taipoxin comes from species of the Taipin snake *Oxyuranus,* and β-bungarotoxin comes from the Chinese banded krait *Bungarus multicinctus.* There are literally hundreds of venomous snakes in the world, but these toxins are among the most potent venoms. Taipoxin and β-bungarotoxin are neurotoxins while cobrotoxin is a cytotoxin. A cytotoxin is a toxoid that affects general tissues.

Botulinum toxin is the most potent neurotoxin of all and comes from the bacteria *Clostridium botulinum,* a cause of food poisoning. It can cause flaccid paralysis and death. Fatalities from respiratory failure occur in 30 to 65 percent of cases of botulism. *Clostridium botulinum*

TABLE 2.4 Checklist for Toxins and Bioregulators with Bioweapons Potential

Toxins and bioregulators	Code†	Source or reference for potential use as BW agent*															
		1	2	3	4	5	6	7	8	9	10	11	12	13	14	15	16
Abrin	P								X				X				X
Aconitine	P												X			X	
Aflatoxin	M																X
Anatoxin A	N					X							X				
Angiotensin	B								X	X							
Apamin	V								X								
Atrial natriuretic peptide	B								X								
Batrachatoxin	N					X			X			X	X				X
Bombesin	B								X	X							
Botulinum	E	X	X	X	X	X			X	X	X	X	X	X	X	X	
Bradykinin	B								X	X							
Brevetoxin	N								X								
β-Bungarotoxin	N		X	X		X			X							X	
Cholecystokinin	B								X	X							
Ciguatoxin	N												X				
Citrinin	M																
Clostridium perfringens toxin	E												X				
Cobrotoxin	C															X	
Conotoxin	V															X	
Curare	P															X	
Diamphotoxin	V								X								
Diphtheria toxin	E					X			X				X	X		X	
Dynorphin	B								X	X							
Endothelin	B									X							
Enkephalin	B								X	X							
Gastrin	B								X	X							
Gonadoliberin	B								X	X							
α-Latrotoxin	N					X										X	

44

Toxin	Type	1	2	3	4	5	6	7	8	9	10	11	12	13	14	15	16
Maitotoxin	N											X					
Microcystin	P				X							X					
Neuropeptide	B																X
Neurotensin	B																X
Notexin	V				X							X					
Oxytocin	H				X							X					
Palytoxin	N								X								
Ricin	P				X				X			X			X	X	X
Sarafotoxin/endothelin	B											X					X
Saxitoxin	N				X				X			X			X	X	X
Sea wasp toxin	V								X								X
Shiga toxin	E											X					X
Somatostatin	B				X							X					
Staphylococcal enterotoxin A	E											X		X			
Staphylococcal enterotoxin B	E								X			X	X	X	X		
Substance P	B				X				X			X					
T-2 toxin	M				X							X			X		
Taipoxin	N				X							X					
Tetanus toxin	E								X			X			X	X	X
Tetrodoxin	N				X							X	X			X	X
Textilotoxin	N				X												
Thyroliberin	B				X							X					
Trichothecene toxins	M									X		X			X		X
Vasopressin	B				X							X					X

*Sources: 1: Clarke 1968; 2: McCarthy 1969; 3: Cookson and Nottingham 1969; 4: SIPRI 1971; 5: Ellis 1999; 6: Canada 1992; 7: Canada 1993; 8: Canada 1991b; 9: Canada 1991a; 10: Thomas 1970; 11: Wright 1990; 12: NATO 1996; 13: Kortepeter and Parker 1999; 14: Franz 1997; 15: Middlebrook 1986; 16: Paddle 1996.

†Notes: M = mycotoxin; V = venom; B = bioregulator; C = cytotoxin; E = enterotoxin; H = hormone; P = plant toxin.

Figure 2.8 Some toxins come from venomous snakes like the calon snake pictured here. (*Photo provided courtesy of Brian Bush, Snakes Harmful and Harmless, Australia.*)

can proliferate in inadequately prepared foods, including home-canned foods, fish, and meats (Freeman 1985). The symptoms of botulism can occur within 12 to 36 h of ingestion and can include nausea, abdominal pain, weakness, and blurred vision.

Clostridium perfringens toxin is also a bacterial toxin, one that can also occur in inadequately cooked meats and poultry, and especially in foods that are reheated. Although more common than botulism, food poisoning from *Clostridium perfringens* results in fewer fatalities, with death occurring in about 1 to 2 percent of cases. Symptoms include abdominal pain and diarrhea, and can occur within 8 to 12 h of ingestion.

Curare is a toxin that comes from certain species of South American trees. It is sometimes called "poison arrow alkaloid" since it is used by some tribes for hunting with either arrows or poison darts shot from blowguns. It consists of a number of toxic alkaloids and causes paralysis.

Diphtheria toxin comes from the bacteria *Corynebacteria diphtheriae*. It is an exotoxin that causes cell death. Diphtheria toxin causes damage to internal organs like the heart, the lungs, the liver, the kidneys, and the nervous system. Death usually results from cardiac arrest. This toxin is responsible for most of the effects of the disease diphtheria.

Ricin is a potent toxin that is extracted from the oil albumin of castor beans or castor oil. It is highly toxic by ingestion, or when it enters the eyes, the nose, or the bloodstream. Soviet secret agents used ricin for assassinations.

Saxitoxin is one of the causes of paralytic shellfish poisoning (PSP). It is an algal toxin and comes from certain dinoflagellates. It attacks the central nervous system and causes paralysis.

Sea wasp toxin is a venom produced by a number of jellyfish species, including the box jellyfish (Fig. 2.9). It affects the central nervous system and causes paralysis.

Staphylococcal enterotoxin B (SEB) is one of several toxins produced by the bacteria *Staphylococcus aureus*. Although normally a commensal microorganism, it can generate toxins when it contaminates food under the right conditions.

T-2 toxin is a trichothecene and a mycotoxin. It is produced by a species of fungi called *Fusarium sporotrichoides*. T-2 toxin causes cellular necrosis in various organs, including bone marrow, and can induce hemolysis and severe hemorrhaging (CAST 1989). Immunosuppression is also a side effect. There are many other trichothecene toxins, but T-2 is the most potent. For this reason both T-2 and the general group of trichothecenes have been given separate listings in Table 2.4 and in Appendix C.

Numerous other toxins exist in addition to those listed in Appendix C. These include toxins from spiders (Fig. 2.10) like the black widow spider, toxins from scorpions, toxins from bees and wasps, and toxins from a variety of other natural sources. Most of these toxins can be fatal in large doses, but the toxins that have been summarized here represent the most toxic agents.

2.4   Bioregulators

Bioregulators are a separate class of toxic agents that are normally used by organisms to control biological processes. They are biological

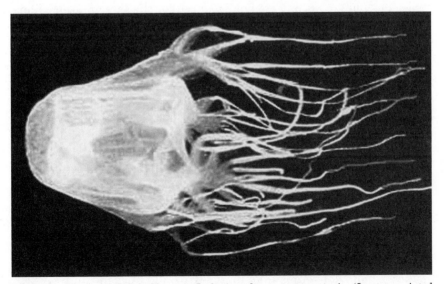

Figure 2.9   The box jellyfish *Chironex fleckeri* produces sea wasp toxin. (*Image reprinted courtesy of Rob Crutcher, BarrierReefAustralia.com.*)

Figure 2.10 Toxins are produced by some spiders, like this banana spider from Louisiana, which produces the toxin phoneutriatoxin.

molecules like toxins but are not produced for the same reasons. Some of these bioregulators have the potential to be used to incapacitate or even kill humans, and so are included in the database of toxins in Appendix C. Most of these compounds have not been studied to the point that the lethal doses are well understood, and so this information cannot be provided at present. A review of some representative bioregulators is presented here to familiarize the reader with the actions of these compounds and the ways in which they differ from toxins.

Angiotensin, also called "angiotonin" or "hypertensin," is a peptide that is found in the blood and is used to regulate blood pressure. Extremely small quantities can have lethal effects, but its actual mean lethal dose (LD_{50}) value is unknown.

Endothelin is a peptide hormone that is released by endothelial cells. It is a potent vasoconstrictor, but its LD_{50} value is unknown.

Substance P is a vasoactive intestinal peptide and is the most potent bioregulator found so far. It can also be found in spinal fluid. It induces vasodilation, and its LD_{50} value of 0.000014 mg/kg of bodyweight makes it more dangerous than most toxins.

Vasopressin is a peptide found in the pituitary glands that increases blood pressure and kidney water retention. It will constrict blood flow and clog the arteries. Its LD_{50} value is not known.

Bioregulators can be considerably more difficult to produce in large quantities than toxins, but the possibility exists that they can be produced in small quantities. They could, for example be extracted from organs like the pituitary glands of farm animals.

2.5   Weaponizing BW Agents

Weaponization of BW agents involves a variety of techniques for producing the deadliest forms of the agents that can be delivered and preparing them for delivery with various mechanisms. The weaponizing of pathogens usually involves selecting and culturing the most virulent strains and then selecting those variants that can withstand the delivery process. Explosive or aerosolization devices may be used for delivery, and the pathogens must survive the process in large enough numbers to make the weapon workable. Furthermore, the microbes must remain viable while aerosolized, and this may require additional selection and culturing. In liquid or powder form, biological agents may tend to clump, which reduces their ability to remain aerosolized. Compounds may be added to solutions or dry mixtures to prevent clumping.

The weaponizing of toxins primarily involves physical preparation. Once a batch of toxins has been processed, concentrated, and dried, it must be ground to a fine powder that will allow for easy aerosolization. Grinding machines for such applications are readily available and will produce powders with logmean diameters as low as 1 to 2 microns. The powder itself is suitable for dispersion, but it may also be mixed in a liquid to allow for easier aerosolization using spray-type aerosolizers. It may also be mixed with substances that prevent clumping, thereby facilitating aerosolization of individual particles.

Since most toxins would of necessity be ground to an aerosolizable size, or mixed in solution and aerosolized in droplets, all toxins could be considered to have the same size range, if we assume the most sophisticated grinding equipment is used. The only differentiating characteristic for toxins is the lethal dose. In fact, only two or three toxins need be selected to represent the entire database. The most toxic of these agents is undeniably botulinum, which is selected as a representative toxin. Two others can be used to represent the range of toxicity, and the ones selected are those that are best understood, ricin, and T-2 toxin. Together these three toxins span the entire range of toxicity shown in Appendix C.

2.6   Size Distribution of BW Agents

The BW agents identified so far have a wide range of characteristics. It is not difficult to design a general defensive system to deal with these threats, but it cannot be done cost-effectively with just a single component. Some of these pathogens are large and can easily be removed by filters. Others are in the most penetrating particle size range of filters but are susceptible to ultraviolet germicidal irradiation (UVGI).

Each microorganism occurs not in a single size but in a range of sizes that is practically unique for each species. The logmean diameters specified in the pathogen database in Appendix A are based on studies of the natural size distributions of microorganisms (Koch 1966; Kowalski et al. 1999). Microbes are distributed lognormally in size, instead of according to a normal bell curve. The reasons this occurs have been studied at length, and sufficient data exist to enable the prediction of the logmean diameters of most pathogens even though their actual size ranges have never been studied. As it turns out, the defining characteristic of lognormal curves, called the "coefficient of variation," remains in a narrow range for most microbes. Therefore, the logmean diameters can be estimated from the measured range of diameters with reasonable confidence.

Although toxins may be ground to a specific size, say 2 microns, it is inevitable that the resulting powder will consist of a range of particle sizes. The distribution of sizes will invariably be a lognormal distribution. Not only do all microbiological agents become distributed lognormally in size, but all micron-sized particles of inorganic origin end up distributed in this manner also. The lognormal size distribution of particles, whether biological or inorganic, is a natural phenomenon (Koch 1969, 1995).

The normal range of sizes necessary to result in aerosolizable particles is about 1 to 6 microns. The mean size of a particle in this range may be about 1.5 microns, since the distribution will be lognormal. Figure 2.11 shows a lognormal distribution for toxins in this size range, based on a standard deviation of approximately 0.5 and a coefficient of variation of 0.06.

Different equipment for producing powders can produce different size distributions. The more expensive units can produce powders in smaller size ranges, and this may enhance the ability to aerosolize the toxin. It is impossible to predict with certainty how an actual toxin size distribution may turn out, and so this size distribution should be considered representative only. Since this size distribution represents the capabilities of relatively expensive and high-quality grinding equipment, it could be considered conservative.

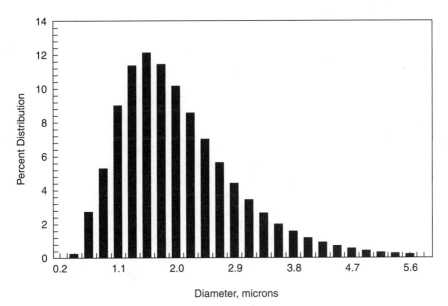

Figure 2.11 Lognormal size distribution of toxins.

Use of the complete size distribution, as shown in Fig. 2.11, will facilitate the most accurate prediction of the filterability of any toxin. However, an extremely good approximation can be obtained by taking the logmean value of the size range. The logmean diameter, D_{LM}, of any range of lognormally distributed particles can be computed as follows:

$$D_{LM} = \exp \{[\ln (D_{\min}) + \ln (D_{\max})] / 2\} \qquad (2.1)$$

In Eq. (2.1), $D_{\min}$ is the minimum diameter and $D_{\max}$ is the maximum diameter of the range of particle sizes.

Since microbes are distributed lognormally in size, their filtration rates can be estimated by using the logmean diameters instead of the average diameter. Use of the average diameter will produce errors in the predicted filtration efficiency. Shown in the pathogen database in Appendix A are the predicted filtration efficiencies for nominal filter sizes from a MERV 6 up to a MERV 15 filter.

Most microbes tend to be spherical or ovoid in shape and exist as populations that span a distinct size range. All submicron-sized collections of particles tend to distribute themselves lognormally, with smaller particles more numerous than larger ones. Because the size distribution does not form a normal bell curve, the average diameter is not representative and therefore the mean diameter of any species must be used for predicting filtration rates.

The actual size distributions have not been determined for most airborne microbes, and therefore the actual mean diameters cannot be known. Fortunately, the logmean diameter, computed by taking the average of the logarithms of the minimum and maximum diameters, provides an excellent approximation of the mean diameter (Kowalski et al. 1999).

Large differences may exist between the logmean diameter and the average diameter of microbes. For example, if the average diameter of *Chlamydia pneumoniae* (about 0.85 micron) were used in place of the logmean diameter (0.55 micron), the filter removal rate for a 90 percent dust spot efficiency (DSP) filter would have been predicted to be 95 percent instead of the actual 80 percent. Even greater discrepancies than this can result when considering nonspherical bacteria.

Nonspherical microbes must have their aspect ratio (the ratio of width to length) accounted for in the model. Spherical microbes like *Rhizopus* have an aspect ratio of 1.0, while *Stachybotris* has an aspect ratio that varies from about 0.3 to 0.5. Equivalent diameters can be established by various techniques to account for the filterability of nonspherical microbes. The mean diameters in Appendix A have been adjusted to include these factors (Kowalski et al. 1999).

Figure 2.12 illustrates the population size distribution for *Legionella pneumophila*, which clearly tends to be lumped at the lower end, and compares well with the predicted lognormal curve. This test was performed in water and only sampled sizes 0.6 micron and greater, but the lognormal pattern is well defined.

The BW agent database in Appendix A provides the computed logmean diameters for all BW agents. These can be used to determine the filter removal efficiency for any filter for which the performance curve is available. The estimated removal rates for several nominal-size filters are also provided.

2.7   Survival Curves of Microbes under UV Exposure

Microorganisms exposed to UVGI experience an exponential decrease in population similar to other methods of disinfection such as heating, ozonation, and the effects of ionizing radiation (Collins 1971; Riley and Nardell 1989; Sharp 1939). Some reports indicate that UVGI can break down toxins like aflatoxin and microcystin, but insufficient data are available to quantify these results (CAST 1989; Shepard et al. 1998). The population decrease due to exposure to UVGI is described by the characteristic logarithmic decay equation:

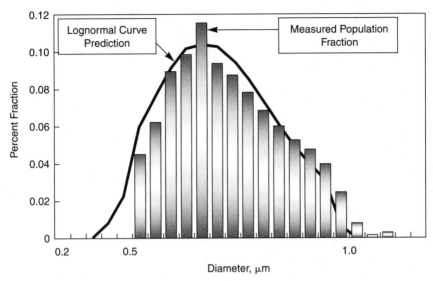

Figure 2.12 Measured size distribution of *Legionella pneumophila* compared with predicted lognormal curve. Data taken by the author with a Coulter counter, sizes limited to 0.6 micron (μm) and higher.

$$S(t) = e^{-k_o t} \tag{2.2}$$

In Eq. (2.2), $S(t)$ is the population fraction at time t in seconds, and k_o is the overall decay rate constant, in units of s^{-1}.

The exponential decay curve resulting from the above equation is known variously as the death curve, the survival curve, or the inactivation curve (Cerf 1977; Chick 1908). The overall decay rate constant k_o defines the sensitivity of a microorganism UVGI, and is unique for each species (Hollaender 1943; Jensen 1964; Rauth 1965).

The overall decay rate constant increases linearly as a function of UVGI dose (Antopol and Ellner 1979; Gates 1929; Rainbow and Mak 1973; Rentschler et al. 1941). Since the dose is a function of both time and intensity for any exposure time, it increases linearly with the intensity of the UVGI irradiation. The intensity, in units of μW/cm², can therefore be broken out as a separate parameter and the equation rewritten as

$$S(t) = e^{-kIt} \tag{2.3}$$

The intensity in Eq. (2.3) can be considered to represent either the irradiance on a flat surface or the fluence rate through the outer surface of a solid (i.e., a spherical microbe). In Eq. (2.3) the rate constant k is the decay rate per unit of intensity, in units of cm²/μW·s, and is therefore independent of the intensity. The rate constant k is the rate constant at an intensity of 1 μW/cm^2. Rate constants determined at some specific intensity other than 1 μW/cm^2 can be converted by dividing by the original intensity as follows:

$$k = k_a/I \qquad (2.4)$$

The preferred units for use in all of these calculations are microwatts/cm^2 (μW/cm^2) for intensity and microwatt·sec/cm^2 (μW·s/cm^2) for dose. The units of μW·s can also be given in SI terms as μJ. Table 2.5 has been provided to assist proper unit conversions. To find the equivalent units, read down to the gray box and then horizontally to the column of the desired units. For example, 1 W/m^2 is equivalent to 100 μW/cm^2; therefore, to convert 2 W/m^2 to μW/cm^2, multiply $2 \times 100 = 200$ μW/cm^2. A spreadsheet for this table is provided online (Kowalski and Bahnfleth 2002b).

Some of the rate constants in Appendix A are from water-based studies. A comparison of studies done in water and air or plates studies suggests that rate constants in air may be between 5 to 30 times higher than in water. Rate constants for plates are usually about 4 to 10 times lower than those in air, although the available data are not sufficient to establish a definite pattern (Kowalski et al. 2000). Theoretically, a microbe in air is exposed over its entire surface while a microbe on a flat surface receives only about one-fourth as much exposure, so a factor of 4 could be justified in lieu of further experimental evidence.

Figure 2.13 illustrates the comparative susceptibility to UVGI of the three microbial groups, viruses, bacteria, and spores. This chart is

TABLE 2.5   Units of Intensity and Dose

Conversion factors for intensity (read down from units to gray block and then horizontally for unit equivalence).

μW/mm^2	erg/mm^2·s	μW/cm^2	erg/cm^2·s	W/ft^2
W/m^2				
J/m^2·s				
1	10	100	1000	0.107639
0.1	1	10	100	0.010764
0.01	0.1	1	10	0.001076
0.001	0.01	0.1	1	0.000108
9.29	92.90	929.0	9290	1
J/m^2				
W·s/m^2				
μW·s/mm^2	erg/mm^2	μW·s/cm^2	erg/cm^2	W·s/ft^2

Conversion factors for dose; exposure time = 1 s (read up from units to gray block and then horizontally for unit equivalence).

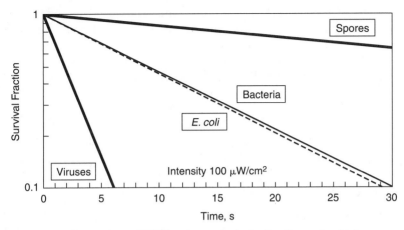

Figure 2.13 Comparison of UVGI rate constants for the three microbial groups. Averages for all available rate constants were used (see Appendix A for actual rate constants).

based on averaged rate constants, a procedure that would be incorrect for design purposes but that provides an illustrative comparison. The spores include both fungal spores and bacterial spores. The spores are seen to be resistant to UVGI while the viruses are highly vulnerable.

Also shown in Fig. 2.13 is the response of *Escherichia coli*, which is not an airborne pathogen but which is commonly used as a benchmark for UVGI design. The rate constant for *E. coli* is based on 90 percent kill at a dose of 3000 μW·s/cm². Equation (2.4), the product of intensity and time, It, is known as the dose and can be separately defined by

$$\text{Dose} = It \tag{2.5}$$

With the intensity specified in units of μW/cm², and the time in seconds, the dose will have units of μW·s/cm². Equation (2.5) applies when the intensity is constant, such as the exposure of bacteria on plates (petri dishes). In airstreams, the intensity distribution is a complex function of the distance and angle from the lamp, local reflectivity, enclosure shape, and lamp geometry.

If the average intensity can be determined or estimated, then the actual dose received by microbes in an airstream passing through an irradiation chamber will be given by

$$\text{Dose} = I_{\text{avg}} t \tag{2.6}$$

In Eq. (2.6), I_{avg} is the average intensity across the irradiation chamber, in units of μW/cm², and t = exposure time or travel time across the irradiation chamber in units of seconds.

Industry catalogs and design guides sometimes recommend using the lamp rating as the intensity across the UVGI chamber. This is a conservative approach as long as the dimensions of the chamber do not greatly exceed 1 m in any direction.

Incorporating the right half of Eq. (2.6) into Eq. (2.2), we have

$$S(t) = e^{-kI_{\mathrm{avg}}t} \qquad (2.7)$$

Equation (2.7) will be an accurate predictor of bulk average microbial response to UVGI because considerable mixing will always occur in any typical airstream, even over short distances. The kill rate, KR, will simply be the complement of S in Eq. (2.7), or

$$\mathrm{KR}\,(t) = 1 - e^{-kI_{\mathrm{avg}}t} \qquad (2.8)$$

In general, a small fraction of any microbial population is resistant to UVGI or other bactericidal factors (Cerf 1977; Fujikawa and Itoh 1996). Over 99 percent of the microbial population will normally succumb to initial exposure, but a remaining fraction will survive, sometimes for prolonged periods (Qualls and Johnson 1983; Smerage and Teixeira 1993). This effect may be due to factors like clumping (Davidovich and Kishchenko 1991; Moats et al. 1971) or dormancy (Koch 1995).

The two-stage survival curve can be represented mathematically as the summed response of two separate microbial populations that have respective rate constants k_1 and k_2. If we define f as the resistant fraction of the total initial population with rate constant k_2, then $(1 - f)$ is the fraction with rate constant k_1. The total survival curve is therefore the sum of the rapid decay curve (the vulnerable majority) and the slow decay curve (the resistant minority):

$$S(t) = (1 - f)\, e^{-k_1 It} + f e^{-k_2 It} \qquad (2.9)$$

In Eq. (2.9), k_1 is the rate constant for fast decay population, in units of cm²/μJ or cm²/μW·s, and k_2 is the rate constant for the resistant population. The factor f is the resistant fraction of the population.

Figure 2.14 shows data for *Streptococcus pyogenes* that display two-stage behavior. The resistant fraction of most microbial populations may be about 0.01 to 1 percent, but some studies suggest it can be a large fraction for certain species (Gates 1929; Riley and Kaufman 1972).

The initiation of exponential decay in response to UVGI exposure, or any other biocidal factor, is often delayed for a brief period of time (Cerf 1977; Munakata et al. 1991; Pruitt and Kamau 1993). Figure 2.15

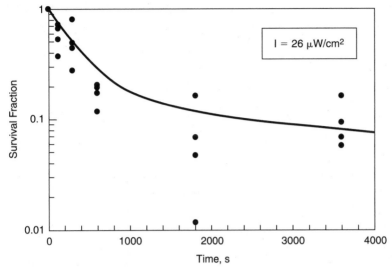

Figure 2.14 Survival curve of *Streptococcus pyogenes* showing two stages, based on data from Lidwell and Lowbury (1950).

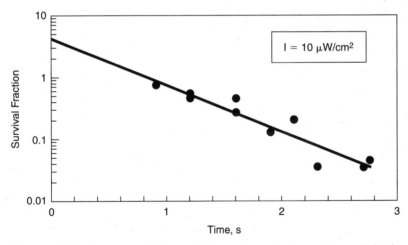

Figure 2.15 Survival curve of *Staphylococcus aureus* showing evidence of shoulder (Sharp 1939).

shows the survival curve for *Staphylococcus aureus,* where a shoulder is evident from the fact that the regression line intercepts the *y* axis above unity. Shoulder curves typically start out horizontally before developing a full exponential decay slope.

Various mathematical models have been proposed to account for the shoulder, including the multihit (or multitarget) model and others

(Casarett 1968; Harm 1980; Kowalski et al. 2000; Russell 1982). The multihit (multitarget) model tends to be the simplest to use and is described here. That model (Severin et al. 1983) can be written as follows:

$$S(t) = 1 - (1 - e^{-kIt})^n \qquad (2.10)$$

The parameter n represents the theoretical number of discrete critical sites that must be hit to inactivate the microorganism, and is unique for each species. In actuality, the number of discrete hits is probably in the hundreds of thousands, and so the parameter may be considered to be a representative fractional value and not necessarily the real physical number of target points.

In Eq. (2.10), the number of targets n is considered unique to each population fraction in a two-stage curve, since these behave as though they were independent. Therefore, by analogy to Eq. (2.10), the complete two-stage equation for the multihit model can be written as follows:

$$S(t) = (1 - f)\,[1 - (1 - e^{-kIt})^{n_1}] + f\,[1 - (1 - e^{-kIt})^{n_2}] \qquad (2.11)$$

In Eq. (2.11), n_1 represents the number of targets for the species in population 1, the fast decay population, while n_2 represents the number of targets in the resistant fraction. Figure 2.16 shows a shoulder

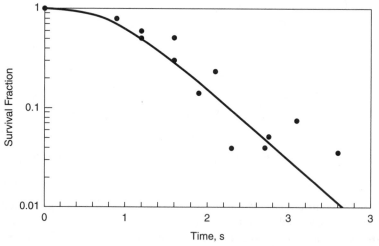

Figure 2.16 Classical shoulder and multihit model versus data for exposed plates of *Staphylococcus aureus*. Based on data from Sharp (1939). Test intensity estimated to be 1900 μW/cm^2.

TABLE 2.6 Shoulder Stage Parameters

Airborne pathogen	Reference	k original, cm^2/μW·s	Test intensity, μW/cm^2	Multihit model shoulder parameter n
Reovirus Type 1	Hill (1970)	0.001849	1160	1.29
Staphylococcus aureus	Sharp (1939)	0.000886	10	4.92
Mycobacterium tuberculosis	David (1973)	0.024530	810	2.34
	Riley (1961)	0.001137	85	1.83
Haemophilus influenzae	Mongold (1992)	0.000656	50	1.18
Pseudomonas aeruginosa	Abshire (1981)	0.000465	100	1.77
Legionella pneumophila	Antopol and Ellner (1979)	0.002503	50	1.67
Serratia marcescens	Rentschler et al. (1941)	0.001225	1	1.71
Bacillus anthracis (with spores)	Sharp (1939)	0.000509	1	2.63

curve generated by the multihit model using Eq. (2.10) compared against test data on *Staphylococcus aureus* irradiated on petri dishes.

Table 2.6 summarizes values of n, the number of targets, for the multihit model which have been derived from the original test data for several microbes (Kowalski et al. 2000). These can be used to generate single-stage shoulder curves. Insufficient data are available to determine the shoulder parameters for most BW pathogens and so several non-BW pathogens are provided for comparison.

2.8 Summary

This chapter has summarized the potential biological agents that may be used as weapons against buildings. Their properties as they relate to filtration and UVGI have been detailed. Additional details on these agents are provided in Appendix A. A suggested list of the most likely BW agents has been presented, and three agents, anthrax spores, smallpox virus, and TB bacilli, have been identified as ideal design basis microorganisms for use in the simulations of attack scenarios, which will be addressed in Chap. 9.

3

Chemical Weapon Agents

3.1 Introduction

Chemical weapons (CWs) have been used in both war (Fig. 3.1) and terrorist acts. They present a major potential threat in spite of the fact that biological weapons are more dangerous and less costly to use. Most CW agents are easier to make than BW agents and are therefore more available. The quantities needed, however, pose logistical difficulties to anyone intending to use them in large buildings.

The variety of chemical agents that could be used to harm humans seems almost unlimited. A bewildering array of dangerous and toxic chemicals has been identified by various organizations like the U.S. government's Occupational Safety and Health Administration (OSHA). What qualifies a chemical as a CW agent is primarily its potency, or how lethal it can be to people who are exposed for short periods to high concentrations.

Appendix D provides a summary of the various properties for these chemical agents, including information on chemical names, appearance, and smell. Many of the more common chemical agents are identified by letter codes, while others are recognized by their common names or by their basic chemical names. Some potential CW agents have no common names or letter codes, and these are identified by their full chemical names.

3.2 Classification of CW Agents

Several classifications exist for CW agents, but these are not clearly defined for every agent. Figure 3.2 shows the basic subdivisions for CW agents, based on criteria from various sources (CDC 2001; Ellis 1999;

Figure 3.1 World War II soldier training for gas warfare. (*Photo by Jack Delano, 1942, from the Library of Congress Prints and Photographs Division.*)

Irvine 1998; Miller 1999) and grouping all nonlethal agents into one category. Identified below the agent types are some examples of the more common agents in each category. Nonlethal agents may include incapacitating agents, vomiting agents, and tear gas. These agents are unlikely to be used in any terroristic attacks and so are excluded from consideration in attack scenarios. The incapacitating agents are included in Appendix D as a matter of completeness.

The four remaining categories of chemical agents have specific characteristics that may assist both in estimating dose and in terms of threat identification for emergency response. Some agents can be identified by their color or even by their smell.

The blood agents are a group of chemicals that interfere with cell respiration. They disrupt the exchange of oxygen and carbon dioxide between blood and tissues. Some examples are hydrogen chloride (HCL) and hydrogen cyanide (AC).

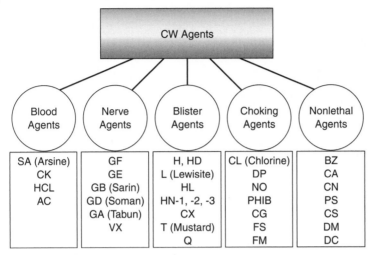

Figure 3.2 Breakdown of CW agents by type, with examples.

Nerve agents disrupt the functions of the central nervous system and can cause blurred vision, twitching, sweating, severe headaches, and other symptoms. Either inhalation or skin exposure can cause casualties. Some of the more common nerve agents include sarin, soman, and tabun.

Blister agents, also called "vesicants," are agents that can cause fatalities through skin exposure alone. They often blister the skin and sting the eyes. If inhaled, they will also burn the lungs. The more familiar blister agents are those that saw use in World War I, including mustard gas, nitrogen mustard, and lewisite.

Choking agents rely on inhalation alone to cause fatalities. Lungs may swell or fill with fluid, causing asphyxiation. Some choking agents that were used in the early days of World War I include phosgene and chlorine.

Nonlethal agents include chemicals like tear gas and vomiting agents that are used for riot control. These are not considered viable as chemical weapon agents and are not discussed further. They are included in the database in Appendix D for completeness and to dispel confusion about their effects.

Table 3.1 represents a survey of numerous sources that have identified CW agents. Additional sources for the information on these agents are provided in Appendix D. The more common or well-known agents, especially those that were used in World War I, are cited in most sources. Many other potential CW agents exist that can be highly lethal but have never actually been used. These latter agents are not

TABLE 3.1 Checklist for Chemicals with CW Agent Potential

Chemical agents	Code	Type†	Source or reference for potential use as BW agent*													
			1	2	3	4	5	6	7	8	9	10	11	12	13	14
Adamsite	DM	P		X	X							X				
Amiton							X	X		X						
Arsenic trichloride					X										X	
Arsine	SA	B														X
BZ	BZ	I		X				X								
Camite	CA	T			X											
Chlorine		C					X								X	X
Chloropicrin	PS	T									X			X	X	X
Chlorosarin		N						X								
Chlorosoman		N						X								
CN	CN	T		X	X						X			X		
CS	CS	T		X	X						X			X		
Cyanogen chloride	CK	B					X						X	X	X	X
Cyclohexyl sarin (GF or CMPF)	GF	N			X									X	X	X
Diethyl phosphite							X	X								
Dimethyl phosphite							X	X								
Diphenylcyanoarsine	DC	P			X											
Diphosgene		C											X			
GE	GE	N			X											
Hydrogen chloride		B						X			X			X	X	X
Hydrogen cyanide	AC	B				X	X	X		X			X	X	X	X
Lewisite	L	V			X	X	X	X		X					X	X
Methyldiethanolamine																
Methylphosphonyl dichloride	DF	V					X				X			X	X	
Mustard gas	H	V		X	X	X										
Mustard-T mixture	HT	V														X
Nerve gas	DFP	V	X													
Nitrogen mustard	HN-1	V				X				X	X		X		X	X

Agent		Type	1	2	3	4	5	6	7	8	9	10	11	12	13	14
Nitrogen mustard	HN-2	V											X		X	X
Nitrogen mustard	HN-3	V											X		X	X
Nitrogen oxide		C														X
O-mustard	T	V					X								X	X
Perfluoroisobutylene	PFIB	C				X	X			X		X	X		X	X
Phosgene	CG	C				X	X		X			X	X	X	X	X
Phosgene oxime	CX	V				X	X					X		X	X	X
Phosphorus oxychloride		V				X	X								X	X
Phosphorus pentachloride		V				X	X								X	X
Phosphorus trichloride		V				X	X								X	X
QL	QL	N				X								X		X
Sarin	GB	V	X	X	X	X	X	X	X		X	X	X	X	X	X
Sesqui mustard	Q	V				X										X
Soman	GD	N	X	X	X	X	X				X	X	X	X	X	X
Sulfur dioxide		C														X
Sulfur mustard (distilled)	HD	V				X	X			X		X		X	X	X
Sulphur monochloride		C														X
Sulphur trioxide		C														X
Tabun	GA	N	X	X	X	X	X				X	X	X	X	X	X
Thiodiglycol		C				X	X							X	X	X
Titanium tetrachloride		C					X									X
Triethanolamine/triethylamine	TEA														X	X
Triethyl phosphite						X								X	X	X
Trimethyl phosphite						X								X	X	X
VX	VX	N	X	X	X	X	X	X	X	X	X	X	X	X	X	X

*Sources: 1: Clarke 1968; 2: Hersh 1968; 3: Cookson and Nottingham 1969; 4: SIPRI 1971; 5: Canada 1992; 6: Canada 1993; 7: Canada 1991a; 8: Thomas 1970; 9: Wright 1990; 10: NATO 1996; 11: Paddle 1996; 12: Miller 1999; 13: Richardson and Gangolli 1992; 14: CDC 2001.
†See text for explanation of types.

often cited in the literature but may be more available or easier to produce and so are addressed in Appendix D.

The following codes are used in Appendix B and in Table 3.1 to distinguish the type of CW agent:

- N: Nerve agent
- B: Blood agent
- C: Choking agent
- V: Vesicant or blister agent
- I: Incapacitating agent (nonlethal)
- T: Tear gas (nonlethal)
- P: Vomiting agent (nonlethal)

3.3 Design Basis CW Agents

Table 3.2 shows chemical agents that have been selected from Appendix D as most likely to be used as CW agents based on a survey of the references and on their lethality. These agents form the design basis for future evaluations. They have properties that are mostly well known and are therefore suitable for use in simulating CW agent

TABLE 3.2 Design Basis CW Agents Ranked by Potency

CW agent	Code	Type	$L(Ct)_{50}$, mg·min/m^3	LD_{50}, mg/kg	Mol wt	Normal form
VX	VX	Nerve agent	10	6	267.37	Liquid
Cyclohexyl sarin	GF	Nerve agent	35	5	180.16	Liquid
Soman	GD	Nerve agent	70	0.01	182	Liquid
Sarin	GB	Nerve agent	100	0.108	140.11	Liquid
Tabun	GA	Nerve agent	135	0.01	162.13	Liquid
Sulfur mustard (distilled)	HD	Vesicant/blister agent	1,000	20	159.08	Liquid
Nitrogen mustard	HN-3	Vesicant/blister agent	1,000		204.54	Liquid
Lewisite	L	Vesicant/blister agent	1,200	38	207.32	Liquid
Phosgene oxime	CX	Vesicant/blister agent	1,500	25	113.93	Crystal powder
Nitrogen mustard	HN-2	Vesicant/blister agent	3,000	2	156.07	Liquid
Phosgene	CG	Choking agent	3,200	660	98.92	Gas
Cyanogen chloride	CK	Blood agent	3,800	20	61.47	Gas
Perfluoroisobutylene	PFIB	Choking agent	5,000	50	200	Gas
Hydrogen cyanide	AC	Blood agent	5,070	50	27.03	Liquid
Chlorine	CL	Choking agent	52,740		70.91	Gas

attacks on buildings. Some common CW agents have been omitted due to their similarity to agents on this list. They have been listed in order of potency based on the exposure dose. The normal form refers to the form in which it may be packaged prior to delivery as an aerosol.

The lethality of chemical agents is defined in terms of the dosage required to cause casualties or fatalities, and these dose values are relative to the route of exposure, the primary routes of interest being inhalation and skin exposure. The LD_{50} value is the ingested or absorbed dose required to cause 50 percent fatalities in a population. The $L(Ct)_{50}$ value is the dose due to exposure that causes 50 percent fatalities in a population.

The LD_{50} values are primarily established in terms of the effects on mammals like mice, rats, guinea pigs, and rabbits. The actual LD_{50} for humans is only known for a few agents that were used in World War I. Since the LD_{50} values for other mammals cannot be extrapolated exactly to LD_{50} values for humans, these values should be considered as estimates of potential lethality. For additional information on the animal sources for the LD values, see the indicated references in Appendix D, especially Richardson and Gangolli (1992).

The $L(Ct)_{50}$ is specified in units of mg·min/m³, or (mg/m³)·min. These units represent the number of minutes of exposure at an airborne concentration in mg/m³. The LD_{50} value is specified in units of mg/kg, and this dosage primarily relates to inhaled, ingested, or absorbed quantities. This latter dose limit requires some knowledge of the route of ingestion and such factors as breathing rate in order to estimate casualties. More will be said about dose calculations in Chap. 5.

The agents in Table 3.2 are individually reviewed here in their listed sequence to provide insight into their effects, modes of deployment, and detectability. The military designations for these agents are typically two- or three-letter codes. For the commercial chemical designations, the CAS number, or the HAZMAT ID number, refer to Appendix D. The descriptions of the odor or appearance in aerosol form come from the references and are assumed to represent lethal concentrations in air, although they may also have an odor when in nonlethal concentrations. These odors may not be apparent to everyone; some individuals, especially women, have a superior sense of smell and may recognize odors that others cannot detect.

VX gas is a nerve agent. When inhaled, it will cause breathing difficulty, sweating, drooling, weakness, nausea, vomiting, cramps, headache, involuntary defecation, twitching, convulsions, and loss of consciousness. It can be treated by injection of common nerve agent antidotes. It is less of a threat when only the skin is exposed. VX is heavier than air and is an odorless and colorless to straw-colored liquid that resembles motor oil.

Cyclohexyl sarin (GF) is a nerve agent. It is a strong irritant to the skin and eyes and may cause vomiting. In normal form, it exists as a crystal powder or, in solution, as dark brown liquid. It can be treated with the standard nerve agent antidotes. It is heavier than air in vapor form and is odorless.

Soman (GD) is a nerve agent. When inhaled in large concentrations, it will cause breathing difficulty, sweating, drooling, nausea, vomiting, cramps, involuntary defecation, twitching, convulsions, and loss of consciousness. It can be treated by injection of common nerve agent antidotes. In pure form, it is a colorless liquid with a fruity odor. It may also appear as an amber to dark brown liquid with a camphor odor. It is heavier than air.

Sarin (GB) is a nerve agent. When inhaled in large concentrations, it will cause breathing difficulty, severe congestion, sweating, drooling, nausea, vomiting, cramps, twitching, convulsions, and loss of consciousness. It can be treated by injection of common nerve agent remedies, including atropine. It can be neutralized with various forms of bleach, like NaOH solutions, or ethanol. Sarin is a colorless liquid that is odorless in pure form. It is heavier than air.

Tabun (GA) is a nerve agent. When inhaled, it can cause severe congestion, sweating, twitching, involuntary defecation, convulsions, and unconsciousness. It is less of a threat when only skin exposure is involved, but 1 to 2 h of skin exposure can be fatal. Tabun can be neutralized with various bleaches, like solutions of NaOH, and phenol or ethanol. It is heavier than air and is a colorless to brown liquid. Although odorless in pure form, it may have a faintly fruity smell.

Sulfur mustard (HD) is a blister agent and is one of the mustard gases, which includes H and HS. It can cause severe symptoms whether inhaled or by exposure through skin. When inhaled, it will cause severe irritation, coughing, sneezing, diarrhea, and fever. Exposure of skin and eyes will result in rapid blistering, erythema, stinging, and blindness. Effects may be delayed, depending on the dose. It can be neutralized with common bleaches, such as chloramine, and solutions of sodium hydroxide (NaOH). It is heavier than air and is colorless and odorless as a pure liquid. In weapons-grade form, it is yellow to dark brown or black in color and has a smell like faint burning garlic.

Nitrogen mustard (HN-3) is a blister agent and has effects similar to HN-1 and HN-2. It can burn the lungs, the skin, and the eyes, causing erythema. It is odorless in pure form but may have a bitter almond smell. It is more heat-stable than HN-2. It has a vapor density over seven times that of air.

Lewisite (L) is a blister agent, or vesicant, that need only be absorbed through the skin. If inhaled, it will cause immediate burning pain, nasal secretions, violent sneezing, coughing, and lung edema. It will sting the skin and eyes and cause blistering, burns, and blindness. Lewisite is a colorless, odorless oily liquid in pure form but may appear as an amber to dark brown liquid. It has a strong penetrating odor resembling geraniums. Several versions of lewisite exist, including L-1, L-2, and L-3. An antidote is available called British anti-lewisite (BAL), which can be applied as a topical coating to neutralize the agent.

Phosgene oxime (CX) is a blister agent that will cause pain to the skin and to the lungs when inhaled. It causes pulmonary edema and will destroy skin and eye tissue on contact. It is heavier than air when vaporized but is normally a yellow-brown liquid. In its pure form it is a crystalline powder.

Nitrogen mustard (HN-2) is a blister agent that can be absorbed through the skin or inhaled. It can burn the lungs, the skin, and the eyes, causing erythema. It is a colorless to yellow oily liquid that has a fruity or soapy odor. It has a vapor density over five times that of air. HN-2 is a military version of HN-1 and is similar to HN-3, and these are all referred to as "nitrogen mustards."

Phosgene (CG) is a choking agent that was used in World War I. It is corrosive to the lungs, and will cause a burning sensation, coughing, breathing difficulty, and a sore throat. Contact with the skin and eyes may cause redness, pain, and blurred vision. Phosgene is heavier than air and is a colorless gas with a distinctive odor. It is normally stored as a compressed liquid, in which form it has a colorless to yellow appearance.

Cyanogen chloride (CK) is a blood agent. When inhaled it may cause a sore throat, vomiting, nausea, drowsiness, confusion, and unconsciousness. The effects may be delayed depending on dose. It may cause redness and pain to the skin and eyes. The gas is heavier than air. It is normally stored as a compressed liquefied gas. It is a colorless gas with a pungent odor.

Perfluoroisobutylene (PFIB) is a choking agent that can cause redness and pain to the eyes and skin as well as damage to the lungs. When inhaled it will cause coughing, shortness of breath, and a sore throat; the symptoms may be delayed depending on dose and concentration. It is a colorless gas and is heavier than air.

Hydrogen cyanide (AC) is a blood agent and is also known as "prussic acid." When inhaled, it can cause shortness of breath, headache, nausea, confusion, drowsiness, and unconsciousness. It may be partially absorbed through skin exposure and will cause redness. It is slightly lighter than air and is normally stored in liquid form. It is a colorless gas or liquid with a characteristic odor.

Chlorine (Cl) is considered a choking agent. When inhaled, it is corrosive to the lungs and will cause coughing, headache, breathing difficulty, and nausea. The effects may be delayed, depending on dose. Chlorine may also cause eye pain and blurred vision. It may also burn the skin but is not absorbed well through the skin. Chlorine is often stored in liquid or solution, but as a gas it is heavier than air. It appears as a greenish-yellow gas with a pungent odor. It exists in a multitude of forms, including home cleaning agents and as an industrialized chemical. Weaponized chlorine is considerably more hazardous and should not be confused with the common or commercial varieties.

Many of the chemical weapon agents listed in Appendix D are formed from precursor chemicals that can be obtained from various commercial or other sources (Fig. 3.3). The formulas and methods used to create these chemical weapons are publicly available and anyone with a working knowledge of chemistry and the right equipment can synthesize them. Some of these chemical agents are used in certain industries, and although their use is controlled, it is conceivable they could be stolen or otherwise obtained illegally. It is difficult to rank chemical agents on the basis of their availability, but it could be reasonably assumed that their availability is roughly inversely proportional to their toxicity. That is, the more deadly an agent, the harder it might be to obtain or create it. In any event, chemical agents are easier to develop than biological agents, and so, although biological agents are considerably more dangerous, there is still a high degree of risk that chemical agents might be used by terrorists.

Information on dose, exposure, delivery mechanisms, and decontamination are provided in subsequent chapters and in Appendix D.

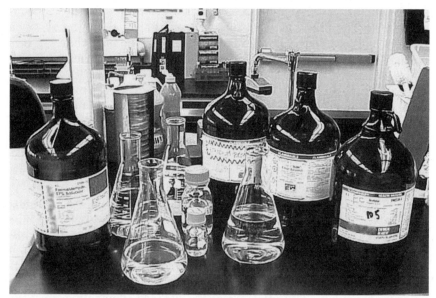

Figure 3.3 Commonly available chemicals and equipment can be used by those with the right skills to create chemical weapon agents.

TABLE 3.3 **Principal CW Agent Simulants**

CW agent simulants	CW agents
TMP	Sarin (GB), soman (GD)
DIMP	Sarin (GB), soman (GD)
DMMP	Sarin (GB), soman (GD)
Dihexylmethyl phosphonate	Sarin (GB)
Methyl salicylate	VX gas, general CW agents
2′,2″ Thiodiethanol	Mustard gas (HD)
Dihexylmethyl phosphonate	Sarin (GB), mustard gas (HD)
Chloroethyl ethylsulfide	Mustard gas (HD)
0-Ethyl-S-ethyl phenyl phosphonothioate	VX gas
Diphenyl chloro phosphite	G-agents (sarin, soman, tabun)

3.4 CW Agent Simulants

Several harmless chemicals exist that adequately simulate the properties of one or more CW agents for purposes of testing air-cleaning equipment or decontamination equipment. These innocuous chemicals are known as "simulants."

Table 3.3 summarizes some common simulants and the CW agents they can be used to simulate. Many additional simulants are available that are not completely harmless or nontoxic but can be used for testing purposes. The testing equipment for using such simulants includes aerosol generators and aerosol detection equipment (i.e., particle detectors or photometers).

3.5 Summary

In summary, the previous information on chemical weapons provides a basis for analyzing the effects of chemical releases inside buildings, in terms of the types and quantities that might be used and the casualties or fatalities that might be inflicted. Although it is the intention of this book to facilitate the design of protective systems for such attack scenarios, there is a considerable shortage of information on how well currently available technologies such as carbon adsorbers and filters can remove chemical agents from airstreams. Therefore, although a great deal is known about chemical agents and their effects, there is unfortunately insufficient information available to allow engineers to size equipment to protect against chemical weapons releases. On the other hand, biological weapons attacks are considerably more dangerous overall, and for these technologies exist that are sufficiently well understood to enable appropriate engineering design. These issues will be addressed in more detail in the upcoming chapters.

Dose and Epidemiology
of CBW Agents

4.1 Introduction

In order to estimate the infections, casualties, or fatalities that may occur in a building under various attack scenarios, it is necessary to determine the dose of an agent received by any building occupant. When a dose of an infectious agent is received by a building occupant, there is the additional consideration of whether or not the disease is contagious and how any secondary infections might spread. The study of the spread of a disease is known as "epidemiology."

The dose received of any CBW agent depends on the route of exposure. There are three possible routes that are relevant to exposure to CBW agents in a building—inhalation, skin exposure, and ingestion.

A variety of terms and factors are relevant to the study of dose and epidemiology, and these are defined in the following list. Additional information on dose is available from the references [see Cookson and Nottingham (1969), for example, or any epidemiology textbook].

ID_{50}: *Mean infectious dose* (BW agents). The dose or number of microorganisms that will cause infections in 50 percent of an exposed population. Applies only to pathogens. Units are always in terms of number of microorganisms, or, more correctly, the number of colony-forming units per cubic meter (cfu/m^3). Colony-forming units are the number of colonies of a microorganism that grow on plates or petri dishes after sampling a volume of air. Since there are few other methods to determine airborne concentrations of microbes, the cfu and number of microorganisms can be considered synonymous terms.

ID_{50}: *Mean incapacitating dose* (CW agents). The dose of a chemical agent that will incapacitate 50 percent of an exposed population. Applies to toxins or chemical weapons. The fact that this acronym is identical to the mean infectious dose is a convenience, and causes no confusion, since these are almost analogous processes and not primary parameters—in most cases, the mean lethal dose is the one of primary interest.

LD_{50}: *Mean lethal dose.*   The dose or number of microorganisms that will cause fatalities in 50 percent of an exposed population. Applies to microorganisms, toxins, or chemical weapons. For microorganisms, the units are number of microbes (or cfu/m^3). For toxins and chemicals, the units are mg/kg. It represents an absorbed, inhaled, injected, or ingested dose per bodyweight.

LC_{50}: *Mean lethal concentration.*   The concentration of chemical agents in the air that will cause 50 percent fatalities in the exposed population. Does not apply to microorganisms and is not used for toxins. The units are mg/m^3.

$L(Ct)_{50}$: *Mean lethal dose.*   The dose or product of concentration (C) times exposure time (t) that will cause fatalities in 50 percent of an exposed population. Units are $mg \cdot min/m^3$. Identical to LD_{50} expressed in $mg \cdot min/m^3$.

CD_{50}: *Mean casualty dose.*   The dose that will cause casualties in 50 percent of an exposed population. Applies primarily to incapacitating chemical agents such as tear gas. The units are normally $mg \cdot min/m^3$. This term will generally not be used since it is redundant to the ID_{50}.

The reason for the variation in the units for LD_{50} values is primarily because lethal doses can be determined for injected, ingested, or inhaled quantities on the basis of bodyweight of test animals, while lethal doses for blister agents have been determined only in terms of the concentrations that produce lethality, and not in terms of what was actually absorbed through the skin or other pathways. Because of the simplicity of determining LD_{50} values of airborne concentrations, these were often used even for nerve agents and blood agents.

In addition to the dose definitions above, some variations exist, such as CD, LD, LD_{25}, LD_{75}, and others. These are variations or nonstandard terms that have essentially the same meaning as the formally defined terms above. To avoid any confusion, such extraneous terminology will be avoided.

LD_{50} data are sometimes specified in terms of both a concentration and an exposure time. For example, mustard gas is stated to have an LD_{50} of 100 mg/m^3 for 10 min of exposure for inhalation by rats. This implies an LD_{50} of $100 \times 10 = 1000$ $mg \cdot min/m^3$. Such data would be useful if several points were specified, such as CD_{Lo}, LD_{Lo}, LD_{25}, LD_{50}, and LD_{75}, because then the slope of the dose-response curve could be elucidated.

Figure 4.1 summarizes the dose and units, and how they are associated with the types of CBW agents. In this chart, "cfu" represents "number of microbes" and "bodywt" is an abbreviation for bodyweight.

Because the potential for CBW agent use inside buildings is primarily limited to exposure and inhalation hazards, doses based on injected values, and toxins that only operate in the bloodstream (i.e., venoms), are not of concern. The doses based on ingested agents are of concern only for food-borne and water-borne agents, not airborne agents. Therefore, only three types of doses are used here: the dose from exposure, the dose from inhalation, and the dose from ingestion.

4.2 BW Agent Dosimetry

The dose from exposure to airborne pathogens applies to the inhaled dose. The dose from foodborne pathogens applies to the ingested dose. An infectious dose can cause infections in individuals but not necessarily any fatalities. The lethal dose is higher than the infectious dose, although not always by much. Many of the infectious doses and lethal doses for microorganisms are not known with certainty, or are based on animal studies. Sometimes the doses that will produce 50 percent infections or fatalities are defined by a range, which can be broad.

When a population is exposed to a range of doses, the result is typically a normal curve or bell curve such as the one shown in Fig. 4.2. Some members of that population will acquire infections at very low doses while others require a large dose to become infected. This is because some people may be susceptible, such as the elderly or the ill, while others may be in a healthy state or have strong immune systems.

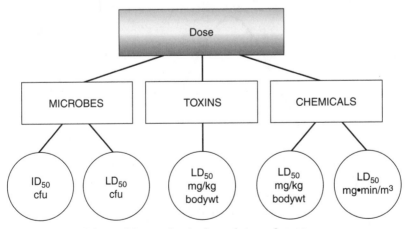

Figure 4.1 Breakdown of dose and units for each type of agent.

In Fig. 4.2 it can be observed that the ID_{50} is 20 and that it produces approximately 6.5 percent new infections. These 6.5 percent new infections, if added to the previous new infections for lower doses, will sum to 50 percent. Another way to view this is to consider integrating the curve—if it were integrated to the LD_{50} point of 20, it would produce a value for total infections of 50 percent. If the entire curve were integrated, it would produce 100 percent total infections.

If the concentration of an airborne microorganisms is approximately constant, as it might be in a building subject to a continuous release, the acquired dose is a linear function of time. Defining E_t = exposure time and C_a = airborne concentration, we can write this equation as follows:

$$Dose = E_t C_a \qquad (4.1)$$

If the airborne concentration is such that the LD_{50} is achieved at 4 h, then 50 percent of the population of the building will have been infected by that time. Plotting the infections against time as in Fig. 4.3 produces a bell curve that is identical to the one in Fig. 4.2. But in fact, the curves will not be exactly identical in an actual attack simulation because it takes some time for the airborne concentration to reach steady state.

The infections noted in Fig. 4.3 represent the infections caused at each point in time, or what are referred to as "new infections." These infections can be summed to produce the total infections at any point in time. Figure 4.4 shows a plot of both new infections and total infections over an 8-h period in the subject building. In this case, the total infections reach 100 percent after 8 h.

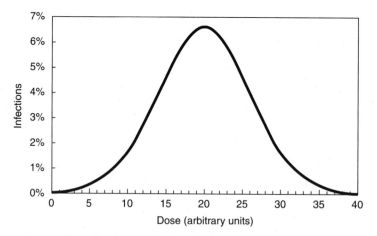

Figure 4.2 Normal distribution of new infections in a population receiving the same dose. The infections shown here correspond to those that occur additionally at each dose. The sum total of all infections will equal 100 percent.

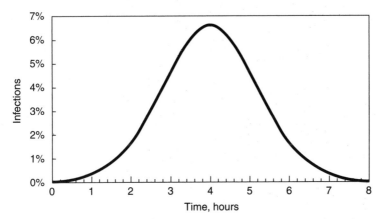

Figure 4.3 New infections over time in a building with constant concentration of an airborne pathogen.

The total infections noted in Fig. 4.4 represent the sum of the new infections. The mathematical relation for predicting the total infections in a building can be developed by beginning with the statistical definition of a normal bell curve. If y represents the number of new cases, the normal distribution curve can be defined as follows:

$$y = \frac{1}{\sigma \sqrt{2\pi}} e^{-0.5 \frac{(x - \mu)^2}{\sigma^2}} \tag{4.2}$$

In Eq. (4.2), σ is the standard deviation, μ is the mean, and x is the dose, with units depending on the units for the LD_{50} value.

In Eq. (4.2), the mean represents the LD_{50}, and the equation can be rewritten as follows:

$$y = \frac{1}{\sigma \sqrt{2\pi}} e^{-0.5 \left(\frac{x - LD_{50}}{\sigma} \right)^2} \tag{4.3}$$

It is now necessary to define the standard deviation in order to be able to use Eq. (4.3) to predict infections in a building subject to a BW agent release. The standard deviation is not known and may not be the same for all pathogens, yet it would be convenient to have some representative value that would produce reasonable results for all cases. The standard deviation could be any value between a small fraction of the mean and some multiple thereof. If it were too small, all the infections would occur in a narrow range of time, and this would be unnatural. If it were too large, the same number of infections would occur at time 0 as at the peak, and this also would be contrary to the observations of

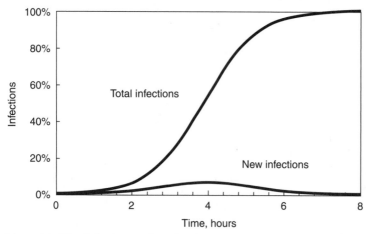

Figure 4.4 New infections and total infections in a building with constant concentration.

disease epidemiology. Trial and error indicate the standard deviation must be between about 0.25 and 0.5 of the mean to provide reasonable and elastic results.

Some data are available from Haas (1983), who tabulated dose-response data for several viruses and bacteria. The standard deviations ranged from 0.028 of the mean to as high as two times the mean. Some of the data sets were both limited and erratic, and if the extremes are discounted, there are some reasonable values, such as 0.527 for *Shigella dysenteriae* and 0.431 for echovirus. Based on this review, the above discussion, and some trial and error, the value of 0.5 is adopted. That is, the standard deviation is assumed to be 0.5 of the mean or LD_{50} value. Equation (4.3) can be rewritten as follows:

$$y = \frac{2}{LD_{50}\sqrt{2\pi}} e^{-2\left(\frac{x - LD_{50}}{LD_{50}}\right)^2} \tag{4.4}$$

Equation (4.4) can now be used to predict either new infections or new fatalities due to exposure to any dose of airborne pathogens. However, in order to compare equipment performance under simulated building attack scenarios, what is needed is an equation to predict the total fatalities or infections. In order to predict the total infections or casualties, it is necessary to sum the results of Eq. (4.4). Equation (4.4) could be integrated mathematically, but the resulting equation would include the error function, erf(x), which is not a simple function to resolve and requires the use of tables. Furthermore, the dose would not be a linear function as in the previous graphed examples, but would have a transient period and might even be subject to variations. As a result, Eq. (4.4) does not lend itself well to this application. What is needed is a closed form of the equation that will predict the total fatalities for any given dose.

The simplest form of equation that will satisfy this purpose is known as a "Gompertz curve" in engineering. It has the following general form, in which y is a function of x, and the parameters a and b are arbitrary constants:

$$y = a^{b^x} \tag{4.5}$$

Equation (4.5) is capable of producing a curve almost identical to that shown in Fig. 4.4 for total infections. It remains only to find the appropriate constants a and b to match the integrated form of Eq. (4.4). The constants a and b must be fractional to produce the Gompertz curve. Equation (4.5) must also be normalized such that it equals 0.5 (or 50 percent) at the normalized LD_{50} value. Therefore, the constant a must be 0.5, and the exponent x must be normalized around the LD_{50}. Inserting these parameters, the following equation is obtained:

$$y = 0.5^{b^{\left(\frac{x - LD_{50}}{LD_{50}}\right)}} \tag{4.6}$$

In order to establish a value for the constant b, it is necessary to consider the same criteria for an acceptable dose-response curve as discussed previously. That is, the fatalities or infections must be normally distributed over the range of doses such that low dose infections are nonzero and the slope in the midrange is not severe. This can be accomplished by fitting Eq. (4.6) to the graphs previously produced, since they were based on these criteria. Figure 4.5 shows a graph of Eq. (4.6) plotted against the dose-response curve from Fig. 4.4, and in which the constant b has been adjusted to a value of 0.1. Although the curve is not a perfect fit, the range of poorer fit is an area of uncertainty. The only area that is reasonably certain is the range near the LD_{50} value, and for this the curve is an excellent match.

Having found an appropriate constant, the final equation for prediction of total fatalities is as follows:

$$y = 0.5^{0.1^{\left(\frac{x - LD_{50}}{LD_{50}}\right)}} \tag{4.7}$$

Equation (4.7) is not intended to be used for epidemiological purposes, since the classic epidemiological model serves such purposes, but it is intended for the purpose of estimating fatalities to compare the

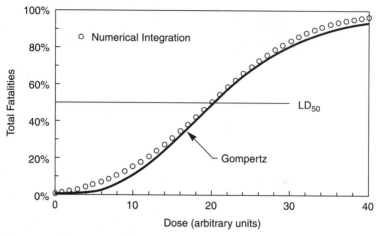

Figure 4.5 Comparison of fitted Gompertz curve with total fatalities computed by numerical integration.

performance of air-cleaning systems in building attack simulations. Since it is being used for comparative purposes, the inherent inaccuracy it may have as an epidemiological model may be a moot point. In any event, it compares reasonably well with dose-response data on inhalation anthrax from other sources (Druett et al. 1956; Hass 2002), and it should be able to predict fatalities accurately enough to determine how well equipment performs in reducing the fatalities. The actual application of this equation will become clear when the simulations are addressed in Chap. 9.

In summary, Eq. (4.7) provides a closed-form mathematical model that can be used to determine the fatalities or infections resulting from any simulated building attack, even if the concentrations of BW agents undergo transient variations. It can also be used, in this same form, to predict casualties or fatalities from CW agents, as will be shown in the following section.

Appendix B provides graphs of the disease progression and dose curves for all pathogens for which this information is available. Table 4.1 summarizes the data used for the dose curves in Appendix B.

The dose values in Table 4.1 represent the best estimates in the available literature. The references in Appendix A for each pathogen can be consulted for more detailed information or alternative values. The dose for smallpox is unknown but is assumed to be 10 to 100 virions (Zilinskas 1999b). Some doses, like those for anthrax, are a matter of debate, and it is thought that perhaps they should be much lower. In the event a new dose is published for this or any other microbe, the curves shown in Appendix B can be redrawn with Eq. (4.7).

TABLE 4.1   Doses for Biological Weapon Agents

Pathogen	Infectious dose ID_{50}, cfu	Lethal dose LD_{50}, cfu	Untreated fatality rate, %	Infection rate, %
Bacillus anthracis	10,000	28,000	5–20	None
Blastomyces dermatitidis	11,000	Unknown	—	None
Bordetella pertussis	(4)	(1,314)	—	High
Brucella	1,300	—	<2	—
Burkholderia mallei	3,200	—	—	None
Clostridium perfringens	10 per g of food	—	—	None
Coccidioides immitis	1,350	—	0.9	None
Coxiella burnetii	10	—	<1	None
Ebola	10	—	50–90	Unknown
Francisella tularensis	10	—	5–15	None
Hepatitis A	10–100	—	—	—
Histoplasma capsulatum	10	40,000	—	None
Influenza A virus	20	—	Low	0.2–0.83
Junin virus	Unknown	10–100,000	10–50	Low
Lassa fever virus	15	2–200,000	10–50	High
Legionella pneumophila	<129	140,000	39–50	<0.01
Lymphocytic choriomeningitis	Unknown	<1,000	<1	NA
Machupo	Unknown	<1,000	5–30	—
Mycobacterium tuberculosis	1–10	—	—	0.33
Mycoplasma pneumoniae	100	—	—	—
Paracoccidioides	8,000,000	—	—	None
Rickettsia prowazeki	10	—	10–40	—
Rickettsia rickettsii	<10	—	15–20	—
Salmonella typhi	100,000	—	10–40	—
Shigella	10–200	—	0.2	0.2–0.4
Variola (smallpox)	10–100 (?)	—	30	—
VEE	1	—	1–60	—
Vibrio cholerae	106–1,011	—	0.5	—
WEE, EEE	10–100	—	EEE 60, WEE 3	—
Yersinia pestis	100	—	0.5	Varies

4.3   CW Agent Exposure Dosimetry

The dose from exposure applies primarily to chemical agents that have been aerosolized and that are absorbed through the skin and eyes, but may also apply to inhalation of chemical agents if data are available. Table 4.2 summarizes all the CW agents from Appendix D for which exposure doses are known. Note that the casualty dose (CD) or incapacitating dose (ID) is provided mainly for incapacitating agents.

The interpretation of the LD_{50} exposure dose is fairly straightforward. For example, for nitrogen mustard HN-1, a blister agent, the LD_{50} dose due to exposure is stated as 1500 mg·min/m³. This means that at a concentration of 150 mg/m³ a lethal dose would be acquired in 10 min by 50 percent of the exposed population. This assumes a linear relationship between concentration and exposure time, which may not

TABLE 4.2   Exposure Doses for Chemical Weapon Agents

CW agent	Code	Type	CD, mg·min/m³	LD_{50}, mg·min/m³
Adamsite	DM	Vomiting agent		30,000
Camite	CA	Tear gas		3,500
Chlorine	CL	Choking agent		52,740
CN	CN	Tear gas	5	8,500
CS	CS	Tear gas	43,500	
Cyanogen chloride	CK	Blood agent		3,800
Cyclohexyl sarin	GF	Nerve agent	25	35
Diphenylchloroarsine	DA	Vomiting agent		15,000
Diphenylcyanoarsine	DC	Vomiting agent		30
Hydrogen cyanide	AC	Blood agent		5,070
Lewisite	L	Vesicant/blister agent		1,200
Mustard gas	H	Vesicant/blister agent		1,500
Nitrogen mustard	HN-1	Vesicant/blister agent		1,500
Nitrogen mustard	HN-2	Vesicant/blister agent		3,000
Nitrogen mustard	HN-3	Vesicant/blister agent		1,000
O-mustard	T	Vesicant/blister agent		400
Perfluoroisobutylene	PFIB	Choking agent		5,000
Phosgene	CG	Choking agent		3,200
Phosgene oxime	CX	Vesicant/blister agent		1,500
Sarin	GB	Nerve agent	35	100
Sesqui mustard	Q	Vesicant/blister agent		300
Soman	GD	Nerve agent	35	70
Sulfur mustard (distilled)	HD	Vesicant/blister agent	50	1,000
Tabun	GA	Nerve agent	300	135
VX	VX	Nerve agent	5	10

always be the case, but it is a reasonable approximation if the concentrations do not deviate significantly from the original test concentrations on which the LD_{50} is based.

This linear relationship can be quantified as follows:

$$LD_{50} = E_t C_a \qquad (4.8)$$

Equation (4.8) is known as the dose-response relationship and simply states that the LD_{50} equals the exposure time multiplied by the airborne concentration. This is, in fact, the standard definition of dose for any type of exposure, including toxic substances, heat, and nuclear radiation. Equation (4.8) can be rearranged to establish the limiting time of exposure before LD_{50} is reached for any concentration:

$$E_t = \frac{LD_{50}}{C_a} \qquad (4.9)$$

Equations (4.9) and (4.8) imply that the dose relationship is linear across the entire range, which may not always be true. At extremely low doses, the equation is known to be nonlinear (Heinsohn 1991). That is, there is some threshold concentration or dose above which these linear equations apply. It is possible that there is also some upper limit of dose beyond which the response is also nonlinear, with either more or fewer survivors than would be predicted. Phosgene, for example, operates on the lungs with a different mechanism in high concentrations than in low concentrations. The responses might behave in a linear fashion when the doses are near the LD_{50}, but be highly nonlinear far above or below the middle region.

The LD_{50} value for CW agents represents the midpoint of a normal distribution curve, just as it does for BW agents. It will be unique for each chemical agent listed in Table 4.1. Figure 4.6 shows the dose-response curve for phosgene, predicted with Eq. (4.7), based on an inhalation exposure LD_{50} of 3200 mg·min/m^3 for humans (Clarke 1968).

Figure 4.7 shows an example of a dose-response curve for sarin that combines casualties and fatalities. The LD_{50} is 1500 mg·min/m^3 [for monkeys, per Richardson and Gangolli (1992)] while the ID_{50} is 20 mg·min/m^3 (Clarke 1968). Other values have been reported for sarin that differ widely. These values are therefore representative only. The incapacitation represents casualties caused by the CW agent. It can be observed that sarin will incapacitate at low doses long before it causes fatalities.

A complete set of dose-response curves for all CW agents is provided in Appendix D. Due to variations in reported lethal doses, which can be extreme, these charts should be considered representative. When an

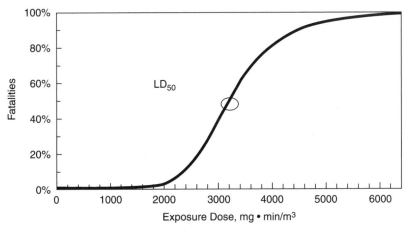

Figure 4.6 Dose-response curve for phosgene.

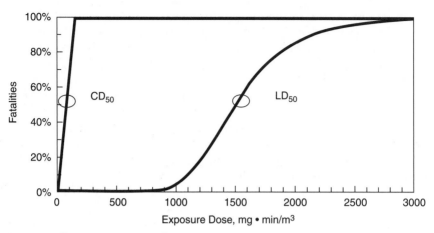

Figure 4.7   Dose-response curve for sarin.

LD_{50} value is determined that is more exact than those values shown in Appendix D, Eq. (4.7) can be used to generate a new curve.

As stated previously, most LD_{50} values are known only from tests with laboratory animals. The susceptibility of various mammals like rats, mice, rabbits, or guinea pigs can differ widely from that of humans. The reasons are many but primarily relate to physiological differences. Significant differences also occur between inhalation doses and doses absorbed dermally, or through the skin. Figure 4.8 illustrates the differences for dermal (skin) exposure to mustard gas, a blister agent, for three species including humans. Although the differences seem large, it should be noted they are well within an order of magnitude and that the LD_{50} for rabbits might well approximate the LD_{50} for humans. A caveat is that the dose for humans is stated as LD_{Lo}, mean a low lethality, rather than an LD_{50}, but this is, at least, a conservative value.

For many pathogens only the infectious dose, ID_{50}, is known, and the lethal dose, LD_{50}, is unknown. There is no way to accurately estimate the LD_{50} based on the ID_{50}, unless we know the fatality rate, and these are mostly not known with any certainty. Since Eq. (4.7) incorporates LD_{50}, x, and y, it should be mathematically possible to estimate the LD_{50}. Even though it would be mathematically correct, it is not necessarily going to produce the correct LD_{50}, but it could be considered a rough estimate in lieu of future data.

Modifying Eq. (4.7), we can write the predicted infections as

$$y = 0.5^{0.1^{\left(\frac{x - ID_{50}}{ID_{50}}\right)}} \tag{4.10}$$

In Eq. (4.10), x is the dose and ID_{50} is the mean infectious dose.

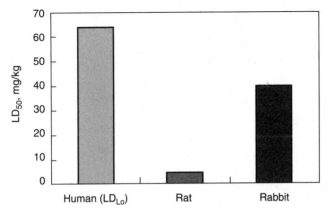

Figure 4.8 Mean lethal dermal exposure doses of mustard gas for three species.

4.4 CBW Agent Inhalation Dosimetry

Inhalation doses that are specified in terms of the dose per bodyweight are more complicated to compute since they depend on breathing rate and the airborne concentration. It is necessary to determine the airborne concentration first, and then to estimate the absorbed dose due to inhalation. Breathing rates can vary widely depending on level of activity and the physiological state of a building's occupants.

Table 4.3 shows the breathing rates associated with various forms of activity. Under normal conditions, the occupants of buildings might normally be engaged in light activity, and a breathing rate of about 0.032 m³/min should be representative. This may also depend on the type of building. In buildings such as factories or health clubs, a more representative breathing rate might be 0.05 m³/min or higher. Under emergency conditions, such as when a chemical release is apparent, breathing rate can increase considerably, which will exacerbate the doses received.

Given an airborne concentration, the inhaled dose is a function of the breathing rate. With E_t = exposure time, C_a = airborne concentration, and B_r = breathing rate, the dose will be

$$\text{Dose} = E_t B_r C_a \tag{4.11}$$

The actual fraction of inhaled vapors of any chemical agent that is absorbed is called the "absorption efficiency" of the agent. Determination of the absorption efficiency is a complex task that can involve many physiological characteristics as well as the characteristics of the agent itself. Much of this information is simply not known with any certainty. Equation (4.11) represents a conservative estimate

TABLE 4.3   Breathing Rates

Activity level	Breathing rate, min^{-1}	Volume	
		L/min	m^3/min
Rest	13.6	11.6	0.0116
Light activity	23.3	32.2	0.0322
Moderate activity	27.7	50	0.05
Heavy activity	41.1	80.4	0.0804

of the dose since it implies 100 percent absorption. In the event the absorption efficiency, η, of any chemical agent is known, the equation can be written more exactly as follows:

$$\text{Dose} = \eta E_t B_r C_a \qquad (4.12)$$

The LD_{50} values for CW agents that are toxic by the inhalation route are normally stated in terms of mg of agent per kg of bodyweight. Some blister agents are specified with these units when they also act through the inhalation route. These LD_{50} values can be converted to doses in terms of mg by assuming a bodyweight of 70 kg. Furthermore, by assuming a breathing rate and an arbitrary exposure time, the concentration representing an LD_{50} dose can be computed.

Table 4.4 summarizes all the agents for which inhalation LD_{50} values are known, excluding incapacitating agents and the agents addressed previously in Table 4.1, and shows the computed value for the LD_{50} in mg·min/m^3. This was arrived at by assuming the LD_{50} in mg was distributed over 8 h while being inhaled at a breathing rate of 0.032 m^3/min.

A breathing rate of 0.032 m^3/min was used in Table 4.4. Obviously the amount of any blister agent absorbed through the skin, or through the eyes, is not accounted for, and so the exposure doses for blister agents may not be representative. In fact, most of the computed doses appear unusually high compared to those for the major CW agents like GA and VX. Furthermore, this has not accounted for the actual absorption rate resulting from breathing the agent. That is, not all of the agent that is inhaled will be absorbed into the bloodstream. It has also been assumed that these agents will be lethal by the inhalation route, which is not necessarily the case since many of the LD_{50} values represent an injected dose. Appendix D summarizes the dose-response curves for the agents in Table 4.4.

As mentioned previously, many of the LD_{50} values for inhalation are based on tests with laboratory animals and these may not necessarily be accurate when applied to humans. It cannot be said with certainty that LD_{50} values for lab animals are even conservative, only that they

TABLE 4.4   Inhalation Doses for CW Agents

CW agent	LD_{50}, mg/kg	LD_{50} (for 70 kg bodywt.), mg	Concentration for 8 h, mg/m^3	$L(Ct)_{50}$, mg · min/m^3
Amiton	0.5	35	2	1,094
Arsenic trichloride	138	9,660	629	301,875
Chlorosarin	4.5	315	21	9,844
Diethyl phosphite	3900	273,000	17,773	8,531,250
Dimethyl phosphite	3050	213,500	13,900	6,671,875
Methyldiethanolamine	4780	334,600	21,784	10,456,250
Nerve gas DFP	1	70	5	2,188
Phosphorus oxychloride	380	26,600	1,723	831,250
Phosphorus trichloride	550	38,500	2,507	1,203,125
Thiodiglycol	4500	315,000	20,508	9,843,750
Triethanolamine	7200	504,000	32,813	15,750,000
Triethyl phosphite	3200	224,000	14,583	7,000,000
Trimethyl phosphite	1600	112,000	7,292	3,500,000

tend to be within a reasonable range. Figure 4.9 compares the inhalation doses for three species. The doses for humans and cats were not specified as LD_{50} values, merely as LD_{Lo}, which means some low lethality. Again, it can be seen that although there are large differences between the doses, they are essentially within an order of magnitude. Furthermore, the animal doses tend to provide conservative estimates.

Figure 4.10 shows an additional example of the comparative effects of the inhalation of a toxin, ricin, on four animal species. The duration of exposure was not specified, so only the concentrations are given. As in the previous comparisons, there are differences but they are not quite beyond one order of magnitude at the worst. The split columns for dog and rat exposure indicate the range between minimum and maximum concentrations that induced casualties.

4.5   CBW Agent Ingestion Dosimetry

Doses for ingestion apply mainly to toxins and pathogens. Although a considerable amount of information is available on ingestion of chemical agents, they are unlikely to be used in such a capacity. No equations or charts are necessary for computing dosages of ingested agents. The information provided in Appendix A for pathogens can be used directly for estimating casualties or fatalities using Eq. (4.7) or else the dose curves in Appendix B can be consulted directly.

Figure 4.11 provides a comparison of the effects of ingestion of a chemical agent, hydrogen cyanide, on three mammalian species. The doses for humans and pigs were stated as LD_{Lo}, meaning unspecified

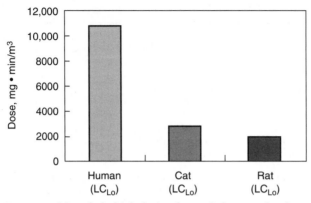

Figure 4.9 Mean lethal inhalation doses of phosgene for three species.

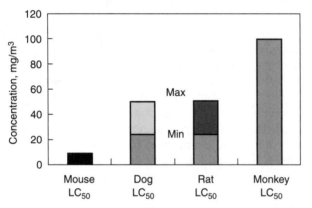

Figure 4.10 Comparison of LC_{50} concentrations of ricin for four mammalian species.

low fatalities. The doses for the three species are within one order of magnitude, similar to the spread of doses for inhalation and skin exposure.

Water-borne spread of chemical agents is less of a hazard than might be expected since much of the time chemical contamination may be obvious. A number of agents degrade in water and the half-lives for these are given in Appendix D. Others are only partially soluble. The problem of food-borne spread of chemical agents is perhaps equally unlikely, but the subject has not been studied in any depth.

Ingestion of biological agents is primarily a problem for food-borne or water-borne pathogens. The stomach has natural protection against many pathogens that are infectious by the inhalation route. Most respiratory viruses and bacteria are unlikely to cause any infections as a

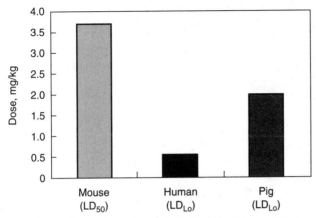

Figure 4.11 Comparison of mean lethal ingested doses of hydrogen cyanide for three species.

result of being swallowed, since stomach acids will destroy them, but to reach the stomach they may have to pass through the mouth, and there is a potential for any agent present in the mouth to reach the nasal mucosa or the lungs. The ingestion of toxins is a hazard, but this is treated in the following section.

4.6 Lethal Dose Curves for Toxins

Dose-response curves operate the same for toxins as for other agents. Toxins can be inhaled, ingested, or absorbed through the skin, but their toxicities are invariably specified as LD_{50} in units of mg/kg of bodyweight. Bioregulators are in the same class of toxins, at least in terms of their being biological weapons, and their dose-response curves are likely to be similar to toxins. Unfortunately, very little information is available on the LD_{50} values for bioregulators and so many of these are listed with the toxins in Appendix C, but without dose information.

Equation (4.7) can be used to determine the dose-response curve for any toxin or bioregulator. These curves have been generated for all toxins and bioregulators of interest for which dose information is available, and they have been summarized for convenience in Appendix C. Table 4.5 provides a summary of all toxins and bioregulators for which dose information is available.

4.7 Disease Progression Curves

The dose response of pathogens operates in the same fashion as other agents except for two factors. The first is that they may be contagious and cause additional infections outside of the initial area in which they

TABLE 4.5   Lethal Doses for Toxins and Bioregulators

Toxin	Type	Lethal dose, LD_{50}, mg/kg bodyweight
Abrin	Plant	0.00004
Aconitine	Plant	0.1
Aflatoxin	Mycotoxin	0.3
α-Latrotoxin	Neurotoxin	0.01
Anatoxin A	Neurotoxin	0.05
Apamin	Venom	3.8
Batrachatoxin	Neurotoxin	0.002
β-Bungarotoxin	Neurotoxin	0.014
Botulinum	Neurotoxin	0.000001
Ciguatoxin	Neurotoxin	0.0004
Citrinin	Mycotoxin	35
C. perfringens toxin	Enterotoxin	0.0003
Cobrotoxin	Cytotoxin	0.075
Curare	Plant	0.5
Diphtheria toxin	Exotoxin	0.0001
Maitotoxin	Neurotoxin	0.0001
Microcystin	Peptide	0.05
Palytoxin	Neurotoxin	0.00015
Ricin	Plant	0.003
Saxitoxin	Neurotoxin	0.002
Shiga toxin	Exotoxin	0.000002
Staphylococcal enterotoxin A	Enterotoxin	0.00005
Staphylococcal enterotoxin B	Enterotoxin	0.027
Substance P	Bioregulator	0.000014
T-2 toxin	Mycotoxin	1.21
Taipoxin	Neurotoxin	0.005
Tetanus toxin	Neurotoxin	0.000002
Tetrodoxin (TTX)	Neurotoxin	0.008
Textilotoxin	Neurotoxin	0.0006

were dispersed. The second difference is that they usually have a characteristic disease progression curve. Many pathogens have a complex pathogenicity and cannot always be generalized, but Fig. 4.12 illustrates the progress of a typical infection. Initially the symptoms may be subclinical but will usually become manifest within a few days, depending on the incubation period. Following the incubation period will be a period of potentially severe symptoms, during which the host is infectious if the disease is contagious. A long period of recovery may follow, during which infectiousness may diminish, or else infectiousness may persist even after the symptoms become subclinical. The curve shown is a general representation that applies to many, but not all, of the pathogens identified in Appendix A. Tuberculosis, for example, persists indefinitely without treatment.

The basic progression of the disease symptoms shown in Fig. 4.12 is measured as a percentage of the maximum symptoms that occur at the

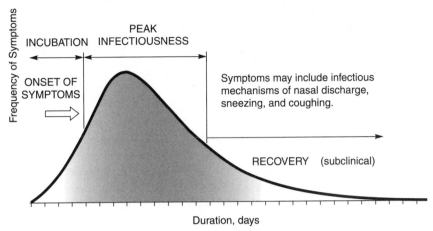

Figure 4.12 Generic representation of the progression of a disease.

peak. Diseases generally incubate rapidly in an exponential manner, and then decrease in severity over time as the immune system gets in gear. This progression could be easily modeled as two competing processes—one exponentially increasing function, and one overlapping exponentially decreasing function. However, the curve resembles a single lognormal curve, which would provide a simpler mathematical model overall. Modeling the progression of a disease as a lognormal curve will facilitate prediction of the average duration and the peak infection period. The peak infection period is of significance to epidemiological studies since it normally defines the period of contagiousness for communicable diseases.

The lognormal curve, used earlier in Eq. (4.2), can be adapted to the progression of any disease with some simple modifications. First, we delete the multipliers of the exponent to normalize the equation to a maximum value of 1 (or 100 percent), and this produces the following:

$$y = e^{-0.5\frac{(x - \mu)^2}{\sigma^2}} \tag{4.13}$$

In Eq. (4.13), x represents the time in days, μ is the mean, and σ is the standard deviation. The next step in lognormalizing this curve is to replace x with $\ln x$, the natural log of x, and replace the mean with the natural logarithm of the peak. The peak of the illness corresponds approximately to the peak of the lognormal curve, and so this conversion should be easy to grasp. This produces the following intermediate form:

$$y = e^{-0.5\left(\frac{\ln x - \ln \text{peak}}{\sigma}\right)^2} \tag{4.14}$$

Next, the standard deviation is defined as the natural logarithm of one-half of the duration D divided by the natural logarithm of the peak. This normalizes the duration in terms of the peak, and the definition, after simplifying, is

$$\sigma = \ln\left(\frac{D}{\text{peak}}\right) \tag{4.15}$$

However, the equation in this form does not decay fast enough, and it is necessary to multiply the exponent 0.5 by a factor of 10, which produces the correct type of rapid decay that characterizes most diseases. Of course, this is a simplistic generalized model and individual diseases may not conform precisely, but it serves the purposes of engineering design and helps overcome the limited knowledge of actual disease progression for many of the BW agents in the database. Simplifying the terms then produces the following equation for defining the progression of the disease in terms of y, the fractional percent of the symptoms:

$$y = e^{-5\left(\frac{\ln\,(x/\text{peak})}{\sigma}\right)^2} \tag{4.16}$$

Of some importance in the epidemiology of contagious diseases is the period of infectiousness. In general, the infectious period is coincident with apparent symptoms. This may not always be the case since asymptomatic diseases can be transmitted in some cases. But as a general rule of thumb, the infectious period can loosely be defined as any symptoms at about 37 percent or higher, based on the well-known D_{37} criteria used in radiation biology as a cutoff or dose limit (Casarett 1968; Harm 1980). The significance of the value of 37 is that it is the inverse of the natural exponent e, and that it has been found through experience in radiation biology to be a safe measure of the maximum dose to use in radiation therapy. Other than that it has no significance but serves as an approximation in lieu of more specific data on infectious diseases. Therefore, for purposes of estimating the infectious period, we can define it as the time when any value of y from Eq. (4.16) is greater than 37 percent, or

$$\text{Infectious period: whenever } y \geq 37\% \tag{4.17}$$

To determine the infectious period based on the model, we can invert Eq. (4.16) as follows:

$$x = \text{peak} \cdot e^{\left(\frac{\sigma}{\sqrt{5}}\right)} \tag{4.18}$$

Equation (4.17) will produce two possible values since the square root in the exponent could be positive or negative. The smaller number, when rounded off, will be the first day of infectiousness, while the larger will be the last day of infectiousness. These days, the first and last days of infectiousness, are defined as D_{37} and D_{37}', respectively. Furthermore, the period between these days can be defined as the mean disease period (MDP) for noncontagious disease. This term will be synonymous with mean infectious period (MIP) for contagious diseases. Equation (4.18) can be written as the following two equations by defining the first with a negative exponent:

$$D_{37} = \text{peak} \cdot e^{-\left(\frac{\sigma}{\sqrt{5}}\right)} \tag{4.19}$$

$$D_{37}' = \text{peak} \cdot e^{\left(\frac{\sigma}{\sqrt{5}}\right)} \tag{4.20}$$

The above equations may be more useful than merely computing the mean infectious period for contagious diseases. They may also be used to estimate the mean time until treatment measures can still be effective. That is, some treatments, like vaccines or antibiotics, may only be effective against a disease if they are administered in the early stages. Equation (4.17) provides an estimate of how advanced the disease is, and may be a useful indicator of the limits beyond which treatment may no longer be effective. This information has applications in predicting the limits of detection time for identifying pathogens in building air, a subject that will be addressed in more detail in Chap. 10. Because of the possible use of this parameter for all diseases, including noncontagious diseases, it is defined for all pathogens in the database. A summary of disease progression charts and parameters for all pathogens is provided in Appendix B.

Table 4.6 summarizes the parameters that define the progression of infections for pathogenic BW agents. The peak for each infection has been taken as the mean of the peak or as the mean of the incubation period from Appendix A. The duration has been estimated likewise.

Figure 4.13 graphically illustrates the progress of the various virus diseases from Table 4.6. These graphs are shown in terms of the percent of total symptoms or severity of symptoms based on the previous algorithms. Considerable diversity is evident in these curves, with some of short duration and others of extended duration. This chart illustrates the fact that an adequate amount of time may be available to detect pathogens and treat infections in most cases.

Figure 4.14 is an example from Appendix B. It shows the hypothetical disease progression for anthrax. The estimated MDP is identified as

4.6 Mean Value Estimates for Disease Progression

Airborne pathogens	Contagious?	Mean peak, days	Mean duration, days	Mean disease period, days
Bacillus anthracis	No	2.5	14	1.2
Blastomyces dermatitidis	No	23.5	37	19.2
Bordetella pertussis	Yes	10.5	35	6.1
Brucella	No	32.5	35	31.4
Burkholderia mallei	No	7.5	17.5	5.1
Burkholderia pseudomallei	No	8	14	6.2
Chikungunya virus	No	6.5	12	4.9
Chlamydia psittaci	No	10	21	7.2
Clostridium botulinum	No	1	1.5	0.8
Clostridium perfringens	No	0.625	1	0.5
Coccidioides immitis	No	17.5	28	14.2
Corynebacterium diphtheriae	Yes	3.5	10	2.2
Coxiella burnetii	No	17.5	21	16.1
Crimean-Congo hemorrhagic fever	Yes	2	12	0.9
Dengue fever virus	No	8.5	14	6.8
Ebola	Yes	11.5	61	5.5
Francisella tularensis	No	7.5	31.5	3.9
Hantaan virus	No	22	60	14.0
Hepatitis A	Yes	30	75	19.9
Histoplasma capsulatum	No	13	49	7.2
Influenza A virus	Yes	3.5	15.5	1.8
Japanese encephalitis	No	10	15	8.3
Junin virus	No	7	14	5.1
Lassa fever virus	Yes	10.5	17.5	8.4
Legionella pneumophila	No	6	14	4.1
Lymphocytic choriomeningitis	No	5	21	2.6
Machupo	No	11.5	28	7.7
Marburg virus	Yes	3	7	2.1
Mycobacterium tuberculosis	Yes	56	192	32.3
Mycoplasma pneumoniae	Yes	14.5	56.5	7.9
Nocardia asteroides	No	13.5	45	7.9
Paracoccidioides	No	7	9	6.3
Rickettsia prowazeki	No	10.5	14	9.2
Rickettsia rickettsii	No	8	14	6.2
Rift Valley fever	No	3	14	1.5
Salmonella typhi	No	2	35	0.6
Shigella	No	4	7	3.1
Stachybotrys chartarum	No	Unknown	Unknown	
Streptococcus pneumoniae	Yes	3	17.5	1.4
Variola (smallpox)	Yes	13	23	10.1
VEE	No	10	15	8.3
Vibrio cholerae	No	3	5	2.4
WEE, EEE	No	3	10.5	1.7
Yellow fever virus	No	4.5	6	4.0
Yersinia pestis	Yes	4	6	3.3

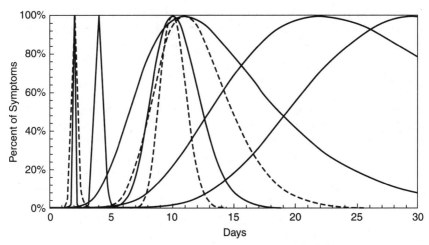

Figure 4.13 Combined disease progression curves for viruses in Table 4.6.

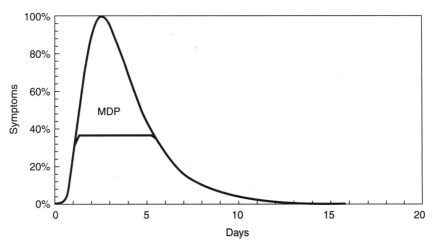

Figure 4.14 Mean disease progression curve for *Bacillus anthracis*.

the cutoff line above 37 percent of the disease symptoms. The disease may be asymptomatic for the first day or two, depending on the dose inhaled. After the D_{37} point, estimated to be 1.2 days on the average, the symptoms progress like those of a cold or flu, but will exceed them in severity within a day or two. In the case of anthrax, treatment must begin as soon as the diagnosis is made in order to assure recovery.

The individual curves and their parameters provided in Appendix B may be used to estimate the contagious period of any communicable disease. They may also be used to estimate the time available to detect a disease for the purposes of treating any infected occupants of a building that has been subject to a BW agent release.

4.8   Summary

In summary, this chapter has presented simple and functional mathematical methods for determining the fatalities due to inhaled or ingested doses for most chemical and biological weapon agents. The estimation of fatalities provides a basis for predicting the performance of air-cleaning systems, as will be shown in Chap. 9. The disease progression curves, also based on a relatively simple mathematical model, may provide a means of estimating the response times necessary for designing the control systems that will be addressed in Chap. 11, and may also assist in the identification of infections under the emergency protocols discussed in Chap. 12.

Dispersion and Delivery Systems

5.1 Introduction

Chemical and biological agents can be delivered by a variety of mechanisms. In World War I, chemicals were often fired into enemy lines in shells that exploded and dispersed clouds of gas. Sometimes they were aerosolized when the wind was blowing toward enemy lines. Chemical and biological weapon (CBW) attacks against buildings are more likely to take the form of internal releases, either by aerosolization or by spills.

Of the various ways in which a terrorist attack might be launched on a building, the explosion or release of CBW agents in the outdoor air is the least hazardous method, since only a fraction of the aerosol would enter the building. Both methods, outdoor dispersion and indoor releases, are discussed here for completeness.

5.2 Outdoor Dispersion Systems

Several methods might be used to release CBW agents outdoors, including explosive release, sprayers or aerosolizers, spilling or passive release, and the use of crop-dusting planes (Fig. 5.1). Most of these methods were used in World War I to disseminate chemical weapons, but sometimes with unpredictable results (Haber 1986). The use of the crop-dusting method to disseminate biological weapons (anthrax spores) was apparently used with success by the government of what was then Rhodesia (today known as Zimbabwe) against rebels during the massive outbreaks of anthrax in that country starting in 1978 (Nass 1992b). Crop-dusting with BW agents was also used by the

Figure 5.1 Crop dusting is one method that could be used to disperse CBW agents outdoors. (*Original photo provided courtesy of Basil Frasure.*)

Imperial Japanese against China in the periods before and during World War II (Chang 1997; Harris 1994).

These methods have the inherent problem that wind and air currents may disperse the agents so widely and rapidly that the resulting airborne concentrations may not be lethal. Such was the case when the Aum Shinryko cult attempted on several occasions to disperse anthrax spores with sprayers in Tokyo, Japan. Not a single reported infection was caused by these attacks, although on other occasions a number of deaths were caused using this method with other CBW agents.

Dispersion of CW agents during World War I had mixed effects that depended to a large degree on which way the wind was blowing and upon the terrain. Figure 5.2 shows the contours of airborne concentration for a hypothetical point-source release with winds blowing in a constant direction. The percentage of the lethal dose is seen to diminish rapidly downstream.

In World War I, releases were often the result of multiple shells, of lines of CW release cylinders, or of a plane spraying along a line parallel to the enemy troops. These approaches had the effect of widening the contours. Such attacks against buildings, however, would not be more effective than a single release upwind of a single building, but could cause contamination in a number of buildings simultaneously.

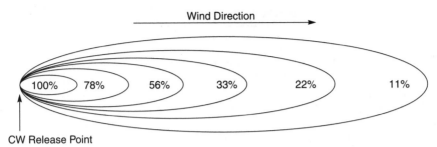

Figure 5.2 Contours for concentrations of CW agents from a wind-blown point source on the ground. Values represent percentages of lethal dose.

Table 5.1 summarizes the most lethal CW agents in terms of the quantities that might be required in open air, based on extrapolation of the results from a study by Calder (1957). The original study used toxins, and these results have been scaled to the lower toxicities of CW agents. Since the mean lethal dose (LD_{50}) values are mostly based on mammals, and since the agents were variously absorbed, inhaled, or ingested, these results must be considered illustrative only.

Figure 5.3 graphically illustrates the results from Table 5.1, showing the quantities that might be necessary to develop a lethal outdoor concentration of each CW agent sufficient to cover an area of 100 km². Obviously some very large quantities would be needed for an outdoor CW attack.

Table 5.2 summarizes the aerosol toxicity for some of the most well-known toxins, based on the study by Calder (1957). The open-air quantity is the estimated amount of toxin required to provide open-air exposure, under ideal meteorological conditions, across an area of 100 km². Figure 5.4 illustrates the range of quantities graphically.

Although considerable amounts of most of these toxins might be required for such outdoor dispersion, the most potent toxins (i.e., botulinum) can accomplish this coverage with a few dozen kilograms. Far lower quantities would be necessary for use against the volume of a typical large building.

The quantities of microbial pathogens that would be required for achieving such lethal outdoor coverage would be far less than needed for the most potent toxin, botulinum. In fact, an entirely new scale is needed to present the aerosol toxicity of microorganisms, as seen in Table 5.3 and Fig. 5.5.

5.3 Indoor Dispersion Systems

Three basic types of delivery mechanisms are possible for CBW agents—aerosolizers, explosive devices, and passive dissemination.

TABLE 5.1   CW Agents' Aerosol Toxicity in Open Air

Number	CW agent	LD_{50}, mg/kg bodyweight	LD_{50}, mg	Open-air quantity, kg
1	Tabun	0.01	0.7	19,691
2	VX	0.015	1	28,130
3	Soman	0.064	4.48	126,022
4	Sarin	0.1	7	196,910
5	Nitrogen mustard	2	140	3,938,200
6	Cyanogen chloride	20	1,400	39,382,000
7	Phosgene oxime	25	1,750	49,227,500
8	Lewisite	38	2,660	74,825,800
9	Hydrogen cyanide	50	3,500	98,455,000
10	Perfluoroisobutylene	50	3,500	98,455,000
11	Phosgene	660	46,200	1,299,606,000

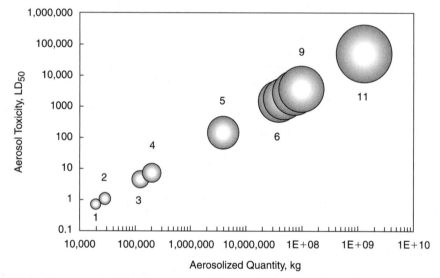

Figure 5.3 Open-air lethality and quantities for CW agents. Extrapolated based on results from Calder (1957).

Figure 5.6 illustrates these delivery methods, and Fig. 5.7 describes the delivery mechanisms. Each of these methods has unique advantages and disadvantages to the terrorist. Explosive release will usually alert building occupants and give them time to evacuate. Passive release may go undetected but may also take a considerable amount of time to have any effect. This may depend on the situation, however, since in cases like the Aum Shinryko subway attacks such passive spills caused over a hundred casualties and a dozen fatalities.

TABLE 5.2 Toxin Aerosol Toxicity in Open Air

Toxin or agent	Source	LD$_{50}$, µg/kg bodyweight	LD$_{50}$, µg	Open-air quantity, kg
Botulinum	Bacteria	0.001	0.07	2
Substance P	Bioregulator	0.014	1	28
Abrin	Rosary pea	0.04	2.8	79
Staphylococcus enterotoxin A	Bacteria	0.05	3.5	98
Diphtheria toxin	Bacteria	0.1	7	197
Palytoxin	Coral	0.15	10.5	295
Batrachotoxin A	Frog	2	140	3,938
Saxitoxin	Dinoflagellate	2	140	3,938
Ricin	Castor bean	3	210	5,907
Staphylococcal enterotoxin B (SEB)	Bacteria	27	1,890	53,166
Aflatoxin	Fungi	300	21,000	590,730
T-2 mycotoxin	Fungi	1210	84,700	2,382,611

Note: LD$_{50}$ is based on laboratory animals.

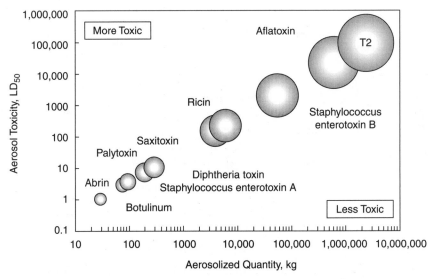

Figure 5.4 Open-air lethality and quantities for toxins. Extrapolated based on results from Calder (1957).

Bombs for disseminating chemicals or biological agents can be driven by two mechanisms—explosives or pressurized gas. The main problem with using explosives with biological agents is that the explosives can destroy them. The approach used by Iraq, the Soviets, and the Japanese when they constructed anthrax bombs was to use slow-detonating

TABLE 5.3   Microbial Aerosol Toxicity in Open Air

Pathogen	Group	ID_{50}	Open-air quantity, μg
Hepatitis A	Virus	55	3.73E-13
VEE (EEE, WEE)	Virus	105	4.40E-12
Ebola	Virus	10	1.10E-11
Influenza A virus	Virus	20	1.52E-11
Mycoplasma pneumoniae	Bacteria	100	8.98E-11
Francisella tularensis	Bacteria	10	1.29E-10
Coxiella burnetii	Rickettsiae	10	3.67E-10
Rickettsia prowazeki	Rickettsiae	10	3.67E-10
Brucella	Bacteria	1,300	2.93E-09
Mycobacterium tuberculosis	Bacteria	6	4.18E-09
Burkholderia mallei	Bacteria	3,200	4.96E-09
Yersinia pestis	Bacteria	3,000	5.72E-09
Bacillus anthracis	Bacterial spore	10,000	2.26E-08
Histoplasma capsulatum	Fungal spore	10	1.81E-07
Coccidioides immitis	Fungal spore	1,350	6.73E-07

Note: ID_{50} is based on laboratory animal testing.

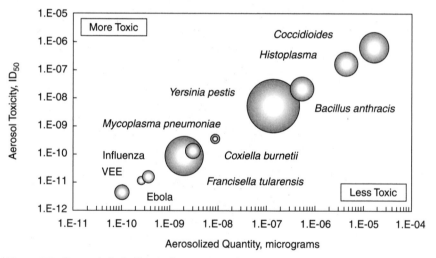

Figure 5.5   Open-air lethality and quantities for pathogens. Extrapolated based on results from Calder (1957).

explosives that do not explode with as much force. If a sufficient number of microorganisms survive such an explosive charge, the result is a successful dispersion of biological agents.

Explosive dispersion inside a building is one possible approach that terrorists may use, but this would cause immediate alarm and mitigate the effects. Aerosolization is probably the most dangerous method of

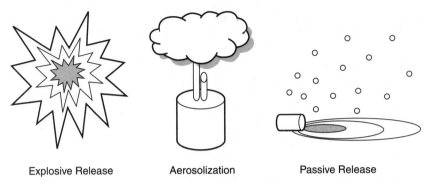

Explosive Release   Aerosolization   Passive Release

Figure 5.6 Illustration of the three basic types of indoor release mechanisms that are possible. A passive release can be a spill, or simply an opened container.

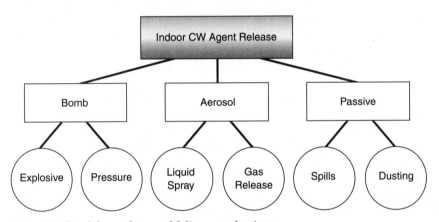

Figure 5.7 Breakdown of types of delivery mechanisms.

dissemination that can be anticipated, due to the potential for both rapid spread of an agent throughout a building and the lack of detectability.

All three types of CBW agents can be disseminated by the aerosol route—microbes, toxins, and chemicals. Chemicals in liquid form would be aerosolized by such mechanisms. Chemicals in gaseous form would be released directly from pressurized containers. Toxins in powdered form or in solution could be aerosolized by aerosol generators. Microorganisms can be released by all three mechanisms—explosive release, aerosolization, and passive release—especially if they have been weaponized.

The aerosolizer is a spraying device that comes in many forms and can be used to spray a liquid or a powder into the air. Aerosolizers are driven by pressurized gas that forces liquids through a nozzle and produces vapors. They can also use techniques such as ultrasonic atomizers or "spinning tops" to produce clouds of vapor. Aerosol nozzles, or venturi-style nozzles, can be driven by air pressure from building pneumatic sources, as they are in laboratories.

The various methods that are used to generate aerosols from liquid or powders are summarized in Table 5.4. The first method, condensation of supersaturated vapors, is strictly a laboratory technique for generating monodisperse aerosols. All the remaining methods could conceivably be used by terrorists to disperse one or more types of agents, chemicals, toxins, or microorganisms.

Combustion mechanisms for releasing aerosols were used in World War I but are an unlikely choice of terrorists because of the alarm that would be caused by the smoke, noise, and visible vapors.

The simplest aerosolizers are sprayers, like gas-powered sprayers, in which liquids or fine powders are hydraulically pumped through nozzles. This is essentially the principle used in hairspray cans, foggers for insecticides, some smoke generators (Fig. 5.8), and crop dusters. This is an imprecise method and may often produce condensation of droplets in addition to production of vapor. Such aerosolizers might be used to disseminate chemical agents and toxins, but because of the physical forces involved, they might destroy most bacteria or viruses in the process of creating aerosols.

Many aerosolizers use venturi nozzles, in which compressed air is ejected at high speed into a liquid stream emerging from a nozzle. Paint spray guns are an example of this, as well as some asthma inhaling equipment. Such equipment might be used to disseminate toxins or microorganisms. The forces involved, however, may destroy some bacteria and viruses. Of course, spores may survive such a process in large numbers. In all cases, it is necessary that only some minimum amount survive the aerosolization process.

Atomizers are high-precision devices that are used in laboratory experiments or for the efficient generation and control of relative humidity in indoor or other environments. Typical of these is the spinning disk or spinning top generator, in which a carefully controlled stream of liquid impinges on a spinning surface. The centrifugal action flings out droplets of a highly controllable size, depending on the system operating characteristics. Sizes in the micron range are achievable with extreme consistency. Such equipment could conceivably be used for the dissemination of microorganisms, especially since some of these methods provide a relatively gentle means of aerosolization that ensures a high degree of survivability of such microorganisms.

The dispersion of fine powders such as toxins can be accomplished by breaking up beds of material with aerodynamic or other forces.

TABLE 5.4   Techniques for Generating Particulate Clouds

- Condensation from supersaturated vapor
- Chemical reaction in combustion
- Disruption of liquids by physical means
- Hydraulic atomization in nozzles (sprayers)
  - Air blast or aerodynamic atomizer (venturi nozzles)
  - Centrifugal action of spinning devices
  - Electrostatic atomizer
  - Acoustic or ultrasonic atomizer
- Dispersion of fine powders

Figure 5.8 Smoke generator (at lower left) releasing test aerosol inside a tunnel. (*Photo provided courtesy of Concept Smoke Systems, U.K.*)

Presumably the powder will have already been ground to some size range (i.e., weaponized) that will permit aerosolization. Furthermore, such particles tend to agglomerate or become attached to each other, and so compounds may be added that will limit the cohesive and adhesive forces that operate between micron-sized particles. Given a powder of this sort, almost any of the means for dispersing liquids will be effective.

Figure 5.9 depicts a nebulizer driven by compressed air. Such nebulizers normally use a venturi nozzle and, depending on the nozzle, the liquid, and the compressed air source, will produce an aerosol that varies from large droplets to pure vapor.

The compressed air source might be a compressed air or pneumatic air source such as is typically available in many laboratory or industrial facilities. It could also come from a compressed air tank or directly from a compressor. Some portable aerosol generators, sprayers, and foggers include compressors and compressed air tanks. Figure 5.10 depicts the minimum equipment necessary to operate a portable aerosolizer or sprayer. The system could run with a compressed air tank alone, and would not need a compressor or a power source. The system could also run on a compressor alone, but would need a power source.

Portable aerosolization units that include compressors or compressed air tanks or both are commercially available in suitcase-like

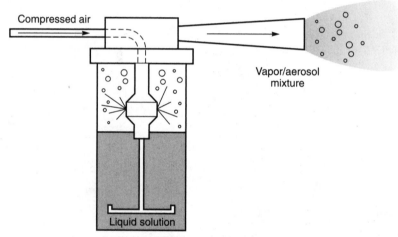

Figure 5.9 Diagram of typical nebulizer for aerosolizing liquids.

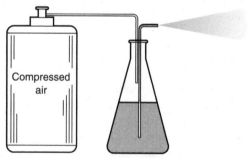

Figure 5.10 Hypothetical device for aerosolizing liquid or powder.

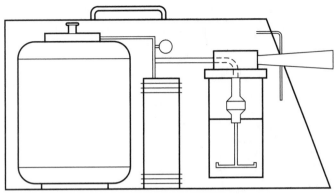

Figure 5.11 Portable compact suitcase-style aerosol generator with a compressor and a compressed air tank.

arrangements (Fig. 5.11). Some of these are compact and open or expand to larger size for use.

Dissemination of volatile chemicals can be accomplished passively, since they will evaporate. Dissemination of extremely small quantities of toxins and pathogenic microorganisms can be accomplished with very little energy input. In fact, the negative pressure inside a return air duct or air-handling unit may be sufficient to drive the aerosolization process.

Aerosolizers can be driven by a moving airstream, such as in the high-velocity ductwork at the entrance or exit of an air-handling unit or fan. Figure 5.12 shows how this might be attempted. In such locations, rubber tubing from a hole drilled in the ductwork could be connected to a nozzle on a container and this would allow a slow release of the agent into the building ventilation system. Such passive release scenarios pose unique threats and highlight the vulnerability of air-handling equipment to tampering. Clearly, protection of equipment rooms is paramount.

Although atomization might reduce water to molecules, microbes would be reduced to elementary bacteria, viruses, or clumps of the same. Bacteria and viruses can survive such aerosolization processes, to a degree that can vary from species to species, and even from variant to variant.

Aerosolizers can also be used to disseminate chemicals. Most chemicals evaporate under normal room temperature and pressure at a rate that depends on their vapor pressure and the local airborne concentrations. Both approaches, passive release and aerosol spraying, were used by the Aum Shinryko cult in Japan. One problem with these methods is that they may also contaminate the terrorists themselves, as happened in one of the subway incidents in Japan.

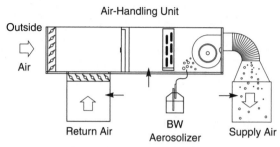

Figure 5.12 BW aerosolizer with a venturi nozzle attached to an air-handling unit. Arrows show alternative locations.

More rapid dissemination of chemicals can be obtained with bombs. Using explosives to blow up a tanker truck full of chemicals might be one approach that could be used to contaminate large outdoor areas.

Passive dissemination of toxins and pathogens is also a possibility. An agent like anthrax might be dusted on interior surfaces like carpets, and normal pedestrian traffic would be sufficient to cause aerial dispersion. The result would be that a number of people passing through would become infected.

Passive dissemination of a toxin or pathogen might be used in locations where people pass in large numbers, such as malls or enclosed sports stadiums. The agents could also be widely dispersed in such locations to maximize casualties. In large malls or sports stadiums, terrorists may release powdered agents continuously as they walked through crowded areas and corridors, but they would certainly contaminate themselves as well.

Certain locations come to mind as being particularly vulnerable to such passive dispersion. The interconnected infrastructure of Minneapolis or Montreal, for example, may allow teams of terrorists to contaminate not only indoor pedestrian thoroughfares, but also the interconnected buildings. Although the passageways and bridges in these cities are air locked, it is, of course, simple to enter and contaminate the main thoroughfares of each individual building (Fig. 5.13).

A group of individuals could disperse large quantities of an agent throughout an interconnected infrastructure in a day. The number of fatalities possible could exceed all previous terrorist attacks put together. Similar approaches might be used in any large mall or other indoor public assembly facility.

A more likely approach would be to use an aerosolizer to release the agents either into a ventilation duct or into the general areas of a building. Release of a BW agent on a single floor would heavily contaminate that floor, as illustrated in Fig. 5.14, but the remaining floors would see much lower concentrations since the agent would arrive from the supply duct in lower concentrations. Both the mixture of normal outside air and the passage through the air-handling unit would tend to reduce supply air concentrations.

Dispersion of the agent directly into the return or supply air duct by a slow-release aerosolizer would likely produce a far greater effect. Such an attack would infect a larger number of people before it was realized that an attack had occurred.

The effect of any building ventilation system is to recirculate airborne contaminants and purge them over time. Most buildings bring in about 15 to 25 percent outside air and mix it with recirculated air. Most buildings do not use any kind of filters, other than dust filters, or any air treatment systems, leaving them vulnerable to a ventilation system release.

Reduction of airborne spores normally occurs in air-handling units, especially through cooling coils. In a typical mechanically ventilated building, reductions of up to 40 percent of airborne spores may be possible (Fisk 1994; Seigel and Walker 2001). Viruses and bacteria, however, may not be removed at the same rates, if at all.

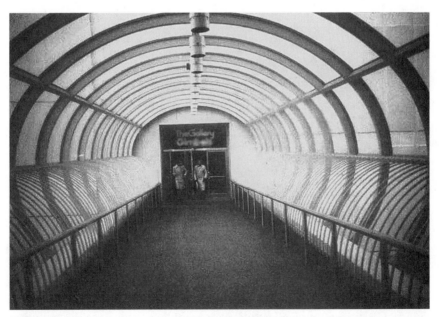

Figure 5.13 Passageways and tunnels between buildings may be vulnerable to passive dissemination of CBW agents.

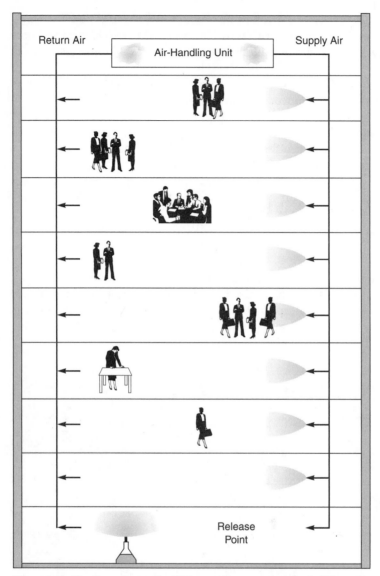

Figure 5.14 Passive release of a BW agent in one zone of a building may result in heavy local contamination but have a lesser effect on other floors.

5.4   Dissemination of Food-Borne and Water-Borne Agents

A separate hazard is posed by the potential use of food-borne or water-borne CBW agents. Building water supplies may be susceptible to large-scale contamination if water is stored, as in many older tall buildings (Hickman 1999). Contamination with CBW agents could occur at the reservoir, intake structure, or treatment plants or within the water distribution system (Lancaster-Brooks 2000). The contamination of food supplies is likely to be a threat only in buildings that are used for food storage or production, or in restaurants and cafeterias.

Chemical agents are unlikely to be used in a food- or water-borne attack. More likely is the use of toxins or food pathogens such as *Shigella,* cholera, and *Salmonella* (Hickman 1999). Each of these threats is implemented somewhat differently. Toxins may simply be dusted on food supplies or dumped into drinking water. Food-borne pathogens require appropriate conditions for survival and growth (Cliver 1990). Typically this means food at room temperature or warmer. Under such conditions the food becomes an amplifier of the pathogen, since it will multiply continuously and reach dangerous concentrations. Water-borne pathogens require appropriate conditions also, but generally do not multiply in water.

Several previous bioterrorist attacks have occurred in which food-borne pathogens were used. In 1985 the Rajneesh cult sprayed *Salmonella* bacteria on salad bars throughout The Dalles, Oregon, causing 750 people to fall ill. In South Africa, the apartheid government

Figure 5.15 Some buildings process their own potable water with filtration or distillation systems like the one shown here. Such systems need to be tamperproof.

used cholera in 5-mL vials to contaminate water wells in insurgent areas (Leitenberg 2001). In an incident in Israel in 1978, citrus fruits were injected with mercury, causing 12 illnesses and economic losses (Zilinskas 1999b). Recently in Chile, cyanide was used by terrorists to poison grapes (Zilinskas 1999b).

A variety of technologies exist for the sterilization of food in food production facilities, including thermal inactivation, pulsed electric fields, and other approaches. There is little in the way of protective engineering that can be done for food supplies in nonproduction facilities, other than security measures and monitoring.

Stored water supplies can be protected through security measures and controlled access to storage areas. Technologies for disinfecting water exist, including ultraviolet (UV) systems, charcoal adsorption systems, filters, etc. The practicality of using such technologies for what may be a low-risk threat needs to be evaluated on a case-by-case basis. Water filtration and distillation systems must also be protected (Fig. 5.15).

5.5 Summary

The preceding chapter has summarized a variety of methods and hypothetical devices that might be used to disperse biological or chemical contaminants in or around a building. This information will be used to define and analyze attack scenarios and evaluate building vulnerabilities in subsequent chapters. It will also provide a basis for developing security protocols in Chap. 12.

6

Buildings and Attack Scenarios

6.1 Introduction

Buildings and structures that house people are potential targets for any CBW attack. The intent may be to cause fatalities or to render the buildings unusable. The method of attack may depend on the type of building. In order to engineer an appropriate defensive system, it is necessary to consider the building type, size, occupancy, and vulnerabilities.

An almost endless variety of buildings exist in the modern world, and categorization is an indefinite science. Buildings can be classified by the function they perform, their physical size or floorspace, their total occupancy or occupancy per unit of floorspace, the relative importance of the building, or even the importance of the particular people employed or housed in it. Buildings can also be classified by the type of ventilation system and relative susceptibility or immunity to bioterrorist events. Even the location of a building is a factor since major cities are considered the most likely targets. A large building in a major city like Boston (see Fig. 6.1), for example, presents different risks than a government facility in a suburban or rural area. All of the factors mentioned need to be considered when designing a system for protection of buildings.

This chapter is devoted to classifying building types and CBW attack scenarios, and to identifying the relative risks of buildings to CBW attacks. The classifications used are representative only, and not every building will necessarily fit exactly into every category. Some buildings will fit no specific category while others may fit more than one. Most building sizes have been approximately defined, and yet there may be buildings larger or smaller than those identified, or that have occupancies

Figure 6.1 The John Hancock building in Boston. (*Photo courtesy of Art Anderson, Penn State.*)

outside the limits stated. The categories presented here should suffice to classify most buildings for the purpose of assessing the threat or risk of CBW attack, and should enable the designer to establish a basis for selecting an appropriate design approach to reduce the level of vulnerability of any particular building and render it immune to most of the defined attack scenarios.

6.2 Building Types and Relative Risks

Table 6.1 provides a breakdown of buildings classified by type and defined by function. Representative ranges for building occupancies and median floorspace have been assigned based on data from surveys of modern buildings (Bell 2000; EIA 1989).

Some attempt has been made to subdivide these buildings according to their relative risk to a bioterrorist attack, but the risk level or target value may not necessarily apply to every building of the specified type. Healthcare clinics, for example, might be considered a low risk for a bioterrorist incident, but numerous women's clinics have been attacked in the past by domestic terrorists, making these, but not all, clinics likely targets.

Certain buildings, like nuclear power plants, are omitted from Table 6.1 because of their special nature. These are addressed separately at the end of this section, but are not treated as specific building types for reasons that will be explained.

TABLE 6.1   Building Types, Sizes, Occupancies, and Risk Levels

Category	Building type	Occupancy, persons/1000 ft^2		Median size, ft^2		Relative risk level
		Min	Max	Min	Max	
Commercial	Commercial office	13	7	2,500	500,000	High
	Banks	20	7	2,500	25,000	Low
	Manufacturing	10	3	2,500	500,000	Medium
	Airports	20	10	10,000	500,000	High
Government	Courthouses	20	7	2,500	100,000	High
	Municipal	20	7	2,500	200,000	Medium
	Police station	10	2	2,500	100,000	Medium
	Post office	10	2	2,500	200,000	High
	Prisons	20	3	5,000	250,000	Low
	Military facilities	50	2	5,000	250,000	High
Food and entertainment	Food processing	10	3	2,500	200,000	Medium
	Restaurants	67	20	2,500	25,000	Medium
	Nightclubs	67	20	2,500	25,000	High
Healthcare	Hospitals	20	7	5,000	250,000	Low
	Clinics	20	7	2,500	10,000	Medium
Lodging	Residential	5	2	1,000	10,000	Low
	Apartments	10	3	5,000	100,000	Medium
	Hotels	10	5	5,000	200,000	High
Education	Schools	50	33	5,000	250,000	Low
	Libraries	33	10	5,000	250,000	Low
	Museums	33	10	5,000	250,000	Low
Mercantile	Department stores	67	13	5,000	250,000	Medium
	Supermarkets	20	10	5,000	100,000	Medium
	Malls	20	10	10,000	500,000	High
Assembly	Auditoriums	200	50	10,000	100,000	High
	Stadiums	200	50	10,000	200,000	High
	Churches	200	50	2,500	10,000	Medium

Buildings in each category are discussed in the following sections and their relative risks and vulnerabilities are addressed in general terms. Specific details of ventilation systems will be addressed in Chap. 7.

The risk levels given in Table 6.1 should be considered rough general estimates and may not apply to every individual building in the stated category. The problem of risk assessment, and the problem of assessing the risk of biological terrorism in general, is an uncertain art and one that has only recently been given any serious attention (Clarke 1999; Siegrist 1999; Stern 1999; Tucker 1999). The following method is proposed as a means of quantifying the relative risk of a building or building occupants to the hazard of a CBW attack.

The risk of a bioterrorist attack is considered to be a function of the hazard, or degree of the danger posed, multiplied by the exposure of a population (Zilinskas 1999c). For buildings, the exposed population is the occupancy, and the general function can be written as follows:

$$\text{Risk} = f(\text{hazard} \times \text{occupancy}) \tag{6.1}$$

The risk of an attack on a building is also a function of the building profile and building vulnerability. The general form of the relative risk of a building can be written as follows:

$$\text{Risk} = f(\text{hazard} \times \text{occupancy} \times \text{profile} \times \text{vulnerability}) \tag{6.2}$$

The hazard, needless to say, is high, since all CBW agents are potentially fatal. It would be fruitless to subdivide the CBW agents into their degree of hazard, although some are more deadly than others, and so the general class of CBW agents can be considered to have a hazard of 1 (or 100 percent). The occupancy of any building can be classed as low, medium, or high, and these terms can be correlated with approximate numerical values of, say, 0.25 (25 percent), 0.50 (50 percent), and 0.75 (75 percent), respectively. These choices are somewhat arbitrary but allow for further refinement in the case of particular buildings. For example, the highest-occupancy facilities (i.e., the Superdome during games) might be given a value of 0.99, or 99 percent, while single-occupant residential homes might be given, say, a 1 percent value.

Similarly, the profile of a building could be subjectively defined as low, medium, or high, with the same numerical values of 0.25, 0.50, and 0.75 to represent low, medium, and high, respectively. Again, there is room for refining these values for particular buildings. Buildings like the Capitol in Washington, D.C., might be considered to have a 99 percent risk value profile, for example.

The vulnerability of a building or facility can be defined in the same terms as the other factors. Assessing the vulnerability might be the most difficult, yet it can be approximated for most structures.

The vulnerability of most military installations is low, although the profile might be high, because they are generally defensible. Any high-security facility can be considered to have a low vulnerability, while any publicly accessible building can be considered to have a high vulnerability.

The use of numerical values here will allow the computation of an overall risk, and then this number can be reinterpreted as low, medium, or high. First, for mathematical convenience the following risk variables can be defined:

- R = relative risk
- H = hazard
- O = occupancy
- P = profile
- V = vulnerability

Equation (6.2) can then be written as follows:

$$R = f(H \times O \times P \times V) \tag{6.2}$$

The factors H, O, P, and V are multiplied as part of the function defining the risk. The risk is, of course, synonymous with probability, and the overall probability of these four mutually independent probabilities can be normalized by taking the fourth root of the product, which will produce the overall risk of a building to the dangers of a CBW attack as follows:

$$R = \sqrt[4]{H \times O \times P \times V} \tag{6.3}$$

Equation (6.3) can be generalized to account for as many additional factors as might be used to define the risk of any facility, including transient occupancy, or even the stated threats of terrorists. The general function could be written as:

$$R = \sqrt[n]{F_1 \times F_2 \times \cdots \times F_n} \tag{6.4}$$

In Eq. (6.4), n represents the number of risk factors, and F represents the value of each risk factor. It should be noted here that alternative methods of computing risks are possible, such as summing the risks and dividing by the number of risk factors. These will produce a risk that is relative to the method used.

Consider the following example of the risk of a high-profile commercial building with high occupancy and public access. The hazard = 1, the occupancy = 0.75, the profile = 0.75, and the vulnerability = 0.75. The risk computed from Eq. (6.3) will be the following:

$$R = \sqrt[4]{1 \times 0.75 \times 0.75 \times 0.75} = 0.81 \qquad (6.5)$$

It can be seen that the overall risk of such a building, 81 percent, would be considered high (as a verbal reinterpretation), as would be expected. Note that the risk has been increased above 75 percent by the hazard. That is to say, the danger due to CBW agents is so high that it increases the risk that such a building could be attacked.

Now consider the same building with the addition of immune building technologies to give it a low value for vulnerability. The relative risk would be computed as follows:

$$R = \sqrt[4]{1 \times 0.75 \times 0.75 \times 0.25} = 0.61 \qquad (6.6)$$

The relative risk has been reduced from 81 percent to 61 percent. Of course, the presence of immune building technologies will be unknown to the would-be terrorist, or at least it should be, and so the risk of attack might be the same, but the relative risk to the occupants has been reduced. Furthermore, security measures could be implemented to reduce the possibility of an attack, but ideally both security and immune building technologies would be implemented to reduce the overall threat to the building occupants. It is possible to further subdivide the vulnerability into a "security" factor, which is the perceived vulnerability, and an "immunity" factor, which is the immune building technology or actual vulnerability, but the present model should be sufficient for most risk analysis applications. The flexibility of this proposed method for assessing building relative risk should be apparent, since it goes beyond defining the threat of an attack to defining the threat to the occupants. The definition of a building's relative risk, or R in Eq. (6.3), can therefore be stated as the risk to the occupants of a building from a CBW attack in relation to other buildings rated by the same method.

For convenience, Table 6.2 has been prepared to allow a quick assessment of the relative risk of a building based on subjective or actual definitions of the profile, occupancy, and vulnerability. The values of low, high, and medium are 0.25, 0.5, and 0.75, respectively, as defined previously, and the relative risk computed from Eq. (6.3) is given as a percentage value, rather than as one of the subjective terms, because of the fine gradations that result. This approach may not be perfect, but it is, at least, a starting point for building managers to assess which buildings to immunize first and, perhaps, the degree of immunization needed.

There are a number of other factors that might contribute to increasing or decreasing the relative risk of a building to a CBW attack, such as the importance of persons occupying the building, but most of these are not generally applicable to most buildings. These

TABLE 6.2 **Estimates of Building Relative Risk**

Occupancy	Profile	Vulnerability	Relative risk, %
Low	Low	Low	35
Low	Low	Medium	42
Low	Low	High	47
Low	Medium	Low	42
Low	Medium	Medium	50
Low	Medium	High	55
Low	High	Low	47
Low	High	Medium	55
Low	High	High	61
Medium	Low	Low	42
Medium	Low	Medium	50
Medium	Low	High	55
Medium	Medium	Low	50
Medium	Medium	Medium	59
Medium	Medium	High	66
Medium	High	Low	55
Medium	High	Medium	66
Medium	High	High	73
High	Low	Low	47
High	Low	Medium	55
High	Low	High	61
High	Medium	Low	55
High	Medium	Medium	66
High	Medium	High	73
High	High	Low	61
High	High	Medium	73
High	High	High	81

additional factors are discussed in the following section in relation to particular building types.

6.2.1 Commercial buildings

Commercial buildings encompass the largest and most diverse type of buildings in the country today. They consist of office buildings, manufacturing facilities, and businesses of every variety. Airport facilities have been included in this category because of their physical similarity to large commercial buildings and their comparable vulnerability to terrorist attacks, even though some airports might more correctly be identified as municipal or government facilities.

Commercial office buildings come in every shape and size and use every type of ventilation system. Large, high-profile office buildings are prime candidates as targets for bioterrorists (Fig. 6.2). The World Trade Center towers were among the highest-profile buildings in the world, in one of the largest cities in the world. Similar well-known buildings may be at risk for bioweapon attacks. Not all commercial office buildings are potential targets, but because large office buildings

Figure 6.2   Commercial office buildings are the single largest risk group for bioterrorism.

are such an integral part of modern cityscapes, their risk is considered high. Small office buildings have limited risk, but the type of work performed and the importance of the personnel in such smaller buildings are also factors in estimating the risk level (Fig. 6.3).

The type of ventilation system used in an office building is a factor in its vulnerability. Most office building ventilation systems are of standard design and incorporate either constant-volume airflow designs or variable-air-volume (VAV) designs that adjust the percentage of out-

Figure 6.3 Federal buildings can be high-profile targets. (*Photo courtesy of Art Anderson, Penn State.*)

side air depending on load and outside air conditions. Modern office buildings tend to have no operable windows and are fairly airtight. Ventilation systems in office buildings tend to have similar sizes in proportion to the building size, and such buildings have similar occupancy levels. These buildings normally lack filters or other air quality control technologies. More will be said about ventilation systems in Chap. 7.

Almost all office buildings have one or more internal or external equipment rooms. These are security concerns and should be provided with appropriate physical protection and/or security. Access needs to be restricted and doors locked.

One of the most critical access points of any office building ventilation system is the outside air intake ducts. These are often located in or near the equipment room, and may be on the roof or at ground level. Ground level air inlets are particularly vulnerable to CBW agent releases if they are accessible from the outside.

Although the tallest office buildings may be prime targets for CBW attacks, such structures often have inherent defensibility due to the fact that they have multiple ventilation systems served by separate equipment rooms. The Sears Tower in Chicago, for example, has eight separate equipment rooms. Although the areas served might not be completely independent, there is some high degree of protection for the majority of the building occupants if the attack is confined to one area or one air intake. In cases where a building has distinct and isolated sections, each section could be treated as a separate building in any analysis.

Banks are essentially small office buildings, and many of the same considerations apply, except that banks are unlikely targets for bioterrorists. There are few high-profile banks, although there are some large bank buildings, so the target value is not high. Ventilation systems for banks are basically the same in form and operation as those of typical small office buildings, and the same protective measures apply. One additional consideration for banks is the possibility that terrorists or criminals may use incapacitating agents in the commission of a robbery, and therefore defenses against such chemical agents could be considered.

Manufacturing facilities exist in a diverse range of sizes and types, but manufacturing facilities are, in general, not likely to be prime targets of any bioterrorism attacks. For one thing, the occupancy levels tend to be lower as such facilities are devoted to production equipment of various sorts. Manufacturing facilities, such as factories for cars, airplanes, electronic equipment, and other sorts of hardware, are rarely high profile and are often located outside of urban areas. The security levels at such facilities also tend to be higher, and access is often restricted.

Manufacturing facilities often have large volumes with high ceilings to accommodate production equipment and usually employ high-volume ventilation systems designed for equipment load more than for personnel (Fig. 6.4). As a result, any chemical or biological release would be subject to considerable dispersion and reduction of concentration.

Certain manufacturing facilities may have target potential of a different sort. Food processing facilities are potential targets of any attempt to use food-borne pathogens and are therefore considered to be under the Food and Entertainment category of Table 6.1. Manufacturers

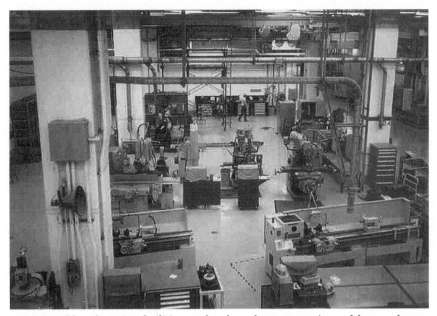

Figure 6.4 Manufacturing facilities tend to have low occupancies and large volumes, making them unlikely targets for CBW attacks.

of military equipment might be potential targets, and these could be considered as military facilities under the Government category in Table 6.1.

Airports, whether privately or municipally owned, are categorized as commercial buildings due their physical similarity to large office buildings, although they could just as well be included as malls, to which they bear the most similarity (Fig. 6.5). They have occupancy levels that can be high; however, they often have larger volumes than malls, rendering their occupancy density similar to that of malls on a per volume basis even when crowded. Large airports are high-profile targets. On the other hand, their volume as the result of typically high ceilings makes them somewhat less vulnerable than office buildings to CBW attacks. Airports are unique structures that do not fit easily into a category and must be considered on the basis of each individual facility in terms of both risk and the defensive measures that need to be taken against CBW attacks.

6.2.2 Government buildings

Most buildings that fall into this category are likely targets because of what they represent and because of the potential importance of their occupants. The Pentagon and the White House are, understandably,

Figure 6.5 Airports and malls are at risk for releases of CBW agents in general open areas.

very high profile targets. Courthouses, state capitols, and police stations have, in the past, been the targets of various sorts of attacks, and it is entirely likely that they will continue to be so in the future. Not all government buildings, however, are likely targets for CBW attacks, and so the type of building and its vulnerabilities must be considered for each and every facility.

Courthouses like the one shown in Fig. 6.6 have a risk as potential targets depending on their rank and prominence. State Supreme Court buildings and federal court buildings are often high profile. Certainly they could be viewed this way by domestic terrorists and also by criminals who perceive themselves as victims of injustice. As a matter of necessity, courthouses tend to have high levels of security, and this alone diminishes the likelihood of an attack. Security considerations in some facilities may need to refocus on the defense of not merely the occupants but also the air-handling equipment.

In form and function, most courthouses resemble office buildings and have comparable ventilation systems. The engineering of courthouse ventilation systems to defend against bioterrorism or the use of chemical weapons will, in general, be handled the same as the design of such systems for office buildings. Certain areas of courthouses may require special defensive consideration, such as the offices of the chief justices.

Figure 6.6 A typical county courthouse. (*Photo courtesy of Art Anderson, Penn State.*)

In such cases, these areas can be isolated and put on air supply systems separate from that of the main building.

Municipal facilities include city and local town government buildings like the one shown in Fig. 6.7. In general, these are not likely to be major targets of any CBW terrorists, but there are some exceptions. High-profile city government buildings in large cities might be potential targets. Domestic extremists or criminals with gripes against particular city governments might resort to the use of CBW agents. As with other buildings, each facility needs to be evaluated individually to determine its relative risk.

Police stations are traditional targets of both extremists and disaffected citizens or vengeful criminals, both for what they represent and because the police are often the front line of security against all forms of urban strife. On the other hand, police stations have inherently high security, making them difficult targets in spite of their high profile.

Police stations and fire stations have ventilation systems and occupancy levels that resemble those of small- or medium-size office buildings. Defensive recourses against CBW attacks can be approached in a similar manner but perhaps with new focus on protecting the ventilation systems.

Post offices would seem to be unlikely targets of CBW attacks if it were not for the fact of the potential for the dissemination of contaminated mail, as occurred in the 2001 anthrax mailings. Although the post offices may not have been the targets of these attacks, the effect is the same,

Figure 6.7   Boston City Hall. (*Photo courtesy of Art Anderson, Penn State.*)

since the facilities became contaminated as the result of the mail passing through. In addition, the seemingly endless series of disgruntled postal workers returning to wreak vengeance on their former employers and coworkers make these facilities higher-risk targets overall.

The U.S. Post Office has recently implemented some high-tech mail disinfection technologies in some of the larger facilities, giving these facilities some degree of protection from contaminated mail, but it is unlikely such expensive systems will be implemented in all post offices. As a result, it is important to review the ventilation system capabilities and associated vulnerabilities of these buildings for the protection of postal workers. One advantage that is found in most post office buildings is the large volume due to the high ceilings. This offers a certain degree of inherent protection against the dissemination of airborne CBW agents.

Prisons are, understandably, at low risk for any type of terrorist attack with CBW weapons. Furthermore, many prisons already have air disinfection systems in place to protect prison populations that historically have higher levels of diseases like tuberculosis. Security in prisons is, furthermore, among the highest of any facilities outside the military, and so CBW attacks are doubly unlikely.

Military facilities are high-profile targets for CBW attacks, although this depends to some degree on the exact type of military facility.

National Guard depots and storage facilities, for example, are extremely unlikely to be subject to CBW releases. The Pentagon, military command and control centers, and buildings on large military bases are high-risk targets, but they are also high-security areas that are defensible and often have systems in place to protect building occupants from outdoor releases (i.e., in times of war). Typical barracks and training centers, however, do not always have such protective systems, and most facilities that are protected were not necessarily designed for indoor releases, especially of the surreptitious kind. Therefore, all such military buildings should be reviewed for their ability to protect occupants against indoor CBW releases.

6.2.3 Food and entertainment facilities

Buildings in the food and entertainment industry have unique concerns related to the use of CBW agents. First, there is the fundamental threat of the use of such agents against crowded places like nightclubs. Second, there is the potential for the use of food-borne pathogens and toxins which, although they do not necessarily constitute an airborne hazard, could be used in any attempt to cause large-scale casualties. Both threats are considered here even though there is little that can be done in terms of engineering defenses against food-borne or water-borne pathogens.

Food processing facilities range in size from small family-owned bakeries to expansive automated production facilities. In terms of the potential for CBW attacks against occupants, such facilities must rank as low on the list as any general manufacturing facilities. Because of their inherently low occupant density, the often large volume of major food processing facilities, and the high-volume ventilation systems normally provided for cooling equipment, these buildings already have a certain degree of protection against airborne releases of CBW agents. Furthermore, many such facilities employ air filters and sometimes ultraviolet germicidal irradiation (UVGI) for the purpose of controlling food contamination and mold. As a result, some of these facilities already enjoy relative immunity from lower-level internal biological releases.

The real concern with food processing facilities is not protecting building occupants from airborne releases of CBW agents, but protecting the processed food from potential contamination with toxins or food-borne pathogens. Chemicals are unlikely to be used for this purpose because of their obviousness in terms of quantities needed, taste, and smell, but toxins and pathogens would be virtually undetectable to unsuspecting consumers. Admittedly, food processing facilities already have systems in place for guaranteeing the quality and safety of their foods, but terrorists intent on causing casualties, panic among consumers, or economic disruption may find a way to bypass existing safety measures. Terrorists might also target brand names of food products that are icons of the American consumer (Stern 1999). Therefore, buildings in the food processing industry are considered to be at medium risk for CBW attacks.

Restaurants and cafeterias are relatively unlikely targets for release of CBW agents, except that they have previously been the target of at least two incidents involving food-borne pathogens (Fig. 6.8). In 1984 the Rajneesh cult used cultures of *Salmonella typhimurium* to contaminate food in 10 salad bars in restaurants in Oregon, causing 751 casualties (Torok et al. 1997; Tucker 1999). In 1996, *Shigella dysenteriae* was used to contaminate food and cause an outbreak of disease in Texas (Kolavic et al. 1997). Such attacks launched with deadlier pathogens on a large scale have the potential for causing large numbers of casualties. Increased security around and monitoring of open foodstuffs in restaurants is one of the only ways of defending against such a threat. There is, as stated previously, little that can be done in terms of building engineering to prevent food-borne contamination, and so this problem is mainly addressed for the sake of highlighting the potential CBW threat.

Nightclubs have historically been targets of terrorists using bombs due to their accessibility and large crowds. For the same reasons,

Figure 6.8 Restaurants make unlikely targets for CBW agent releases but may be candidates for food tampering. (*Photo courtesy of Art Anderson, Penn State.*)

nightclubs may be prime targets for terrorists using CBW agents. The larger the nightclub, the higher the target value may be. Nightclubs in high-profile districts like Beverly Hills or Hollywood may be subject to such risks. Nightclubs that cater to a specific clientele may become targets of terrorists with specific causes. Military clubs, gay clubs, and others may be targeted with CBW agents by terrorists or religious extremists, as they have been in the past with more conventional weapons.

Ventilation systems in nightclubs typically deliver higher volumes of air than those in office buildings. This characteristic poses two competing considerations in an attack scenario. First, any disseminated agent may spread faster. Second, any disseminated agent may be removed faster. The actual degree of spread and dispersion depends on the details of the ventilation system and the building volume, and these must be analyzed on a case-by-case basis.

6.2.4 Healthcare facilities

Hospitals and other healthcare facilities like the one shown in Fig. 6.9 have perhaps the lowest target value of any facilities besides prisons. Even terrorists need to draw a line somewhere so that in their own eyes, at the very least, they are not completely despicable. This and the fact that there has never been an intentional attack on a hospital give these facilities a low rating in terms of risk of a CBW attack. There is still some possibility, however, that an individual with a personal vendetta against a person or a facility would use such agents, but hospital ventilation systems have a relatively high degree of engineered protection against airborne disease. Hospitals often use high volumes of outside air for continuous purging, multizone systems, and filters. Many hospitals use UVGI to irradiate the air in isolation wards and operating rooms as well. Perhaps the worst CBW attack scenario for hospitals is an outside air chemical release. Another scenario is the situation in which a large number of victims enter a hospital with an unknown infection. If this were a contagious disease, like smallpox, the hospital might become contaminated and have to be quarantined. In any event, any review of a hospital for defensibility should consider the ventilation and building details specific to the facility in order to establish the need for additional defensive measures.

Clinics are essentially small healthcare facilities that often cater to specific healthcare needs. In general, these are unlikely to be targets of a CBW attack, except that numerous clinics that provide women's services (for example, abortion clinics) have been attacked by religious extremist groups in the past, a trend which has shown no signs of abating. The homegrown terrorist Eric Rudolph, for example, bombed a

Figure 6.9 Healthcare facility in Stanton, VA. (*Photo courtesy of Art Anderson, Penn State.*)

women's clinic in Birmingham, a gay lounge in Atlanta, and the crowded Centennial Park at the Atlanta Olympics in 1996, killing two and wounding hundreds. Since these types of extremists have shown themselves to be willing to commit murder in the past, and considering the hundreds of anthrax hoaxes that have been directed against these clinics since the September 11 events, women's clinics should be considered to be at high risk for CBW attacks.

Clinics have no specific common ventilation system design. Some facilities have been built to hospital standards, while others are sometimes office buildings or even residences that have been converted. As a result, no general recommendations can be given and each facility should be retrofited in accordance with its own unique ventilation system configuration.

6.2.5 Lodging facilities

Apartment buildings may be at risk for CBW attacks depending on a number of factors, including their size and occupancy, the presence of

Figure 6.10   Large apartment buildings may be at risk for CBW attacks if they are served by central air-handling units. (*Photo courtesy of Art Anderson, Penn State.*)

any central ventilation systems, and whether or not they are considered to be high-profile buildings (Fig. 6.10). Because most apartment buildings, at least the older ones, often have individual air conditioning units and operable windows, they may be at low risk for an interior CBW agent release.

Residential buildings are among the least likely buildings to be subject to a CBW attack since occupancies are typically low. In spite of this fact, random acts of CBW terrorism, such as the outdoor sarin release in Matsumoto, Japan, in 1994, which caused hundreds of casualties, are a possibility.

There are some individuals who, by virtue of their importance, may be at risk. Certain residences may be at some risk due to their locations. Houses located near prime targets could conceivably see exposure to CBW agents delivered against those targets. In such cases, the attack scenario might be equivalent to an outdoor release. Design alternatives would then be limited to cleaning or disinfecting the outside air, or possibly isolating the interior.

Hotels are similar to apartment buildings except that they are generally higher-profile buildings and more open to public access. Many hotels have individual air conditioners and no central ventilation, giving them a certain high degree of immunity to CBW attacks. Newer hotels with central air-handling systems are at the same risk as office buildings. In such hotels, the defensibility can be enhanced by air-cleaning technologies and physical security measures, but the use of security protocols suffers from the limits to which such measures could be effected in any publicly accessible building (Fig. 6.11).

6.2.6 Education facilities

School buildings, whether K–12 facilities, high schools, or universities, are most improbable sites for CBW agent releases. Besides the fact that attacks against youth are generally frowned upon even by terrorists, schools are not perceived as being part of the military-industrial complex, nor specifically as part of the government, nor as part of any corporate structure.

Even so, there have been some terrorist incidents at schools, including a shooting by a domestic neo-Nazi at a Jewish grade school, and a bombing campaign targeting certain universities by the now-imprisoned Unabomber Ted Kaczynski. In addition, the interminable string of school shootings and bombings by crisis-stricken students against their own schools could be extrapolated to the situation in which such individuals might be clever enough to develop poisons, toxic chemicals, or even deadly pathogens for use against such facilities.

College and university buildings resemble small office buildings or apartment buildings in structure and form, as in Fig. 6.12, except that they often have higher occupancies. The approach to improving the immunity of such buildings against CBW agents is much the same as for office buildings, but consideration must be given to the potentially higher casualty rates. That is, a casualty rate of, say, 8 percent might be acceptable in an office building with 200 people, but might be unacceptable in a similar-sized college building with 1000 students. Again, the parameters unique to each building must be considered when developing a protective design.

Figure 6.11   Large hotels may be vulnerable because of their accessibility. (*Photo courtesy of Art Anderson, Penn State.*)

Libraries are considerably less likely to be targets of any CBW attack than college or office buildings. Although they have potentially high periodic occupancies, they have extremely low profiles. In the event such buildings are considered for immune building technology, they could be treated much like office buildings.

Museums are not much more likely to be targets of CBW attacks than libraries. Some museums, however, have cyclical occupancies that can be high, and these particular museums often contain priceless art and artifacts of value to national pride that give them high profiles. Although the contents of such museums are not necessarily susceptible to CBW agents, the number of visitors in popular museums combined

Figure 6.12 Typical college dormitories. (*Photo courtesy of Art Anderson, Penn State.*)

with a high profile may put them at high risk. The implementation of immune building technologies in museums may be subject to some special considerations. Art and artifacts in such facilities often place demanding limits on indoor temperature control and relative humidity. The addition of filters, a UVGI system, charcoal adsorbers, or other types of equipment could impact the ability of the ventilation system to control interior conditions. These factors should, of course, be reviewed in the process of designing any retrofit of air-cleaning systems, but they are of particular importance in the case of museums.

6.2.7 Mercantile facilities

Mercantile facilities range in size from very small to very large. The comparative risk of each for CBW attack is, to some degree, a function of its size and occupancy. The special case of food contamination is a concern for any facilities that serve or distribute foods, especially open foods such as fresh vegetables at grocery stores and salad bars.

Department stores tend to be midsized buildings with occasional high occupancies (Fig. 6.13). Because of their open public access, they may be at some risk of CBW releases. Such buildings often have high ceilings and large volumes, which gives them a certain degree of protection. Otherwise, their ventilation systems are similar to office buildings, and standard measures can be taken to improve their immunity to CBW attack, including the retrofiting of air-cleaning systems and

Figure 6.13 Crowded department stores and malls are potential targets for open area CBW releases. (*Photo courtesy of Art Anderson, Penn State.*)

the installation of protective devices for equipment rooms and air-handling equipment. Security protocols are less likely to be effective because of the number of shoppers and the packages customers carry.

Supermarkets are small to midsized buildings with relatively moderate or low occupancies. They would seem to be at low risk for any type of CBW release involving the airborne dispersion of CBW agents designed to cause casualties. However, they are end-of-line distributors of food products and this makes them potential targets for food contamination by means of poisons, toxins, or biological pathogens. As with the *Salmonella* incidents in Oregon (Tucker 1999), the contamination of open vegetables and salad bars at numerous stores could cause large numbers of casualties. The food-borne pathogen problem is, however, not controllable by means of building ventilation systems, and so other measures such as monitoring and security are the only feasible recourse.

Malls are large structures that house numerous shops and have large open areas that often have high occupancies. Because of their public access, these are potential targets for CBW terrorist acts. Malls are similar to airports in form and ventilation system design, and engineering approaches to improving their defensibility will normally include similar measures if undertaken. They often have huge volumes, which gives them a certain degree of protection due to dispersion effects, but at the same time the ventilation systems may recirculate air and cause widespread contamination. Security measures should be implemented to protect air-handling equipment and air intakes, but security protocols may be difficult to implement because of their public access and the fact that many occupants carry packages.

6.2.8   Assembly facilities

All areas of public assembly are subject to a higher risk of CBW agent releases by terrorists due their high occupancy levels. Even small facilities such as theaters can be the targets of terrorist attacks, as has been the case in some countries in the Middle East, where packed theaters have been firebombed by religious extremists. Although no such attacks have ever occurred in theaters in the United States, there are plenty of domestic fanatics who may be capable of doing so.

Auditoriums include any buildings or indoor facilities used for public assembly or entertainment. Theaters, public halls, concert halls, school auditoriums, etc., are all areas of occasional high occupancy and general access that are at risk for terrorist releases of CBW agents. Ventilation systems for auditoriums are generally of a simple, high-volume design that can be modified to accept air-cleaning and air disinfection technologies without much difficulty. The problem, however, is that with their large, open volumes and concentrated occupancies, an interior release of a CBW agent in an auditorium may not be affected by the ventilation system until it is too late to avoid casualties. Unlike office buildings, in which an internal release through the ventilation system is the most serious scenario that engineering design can protect against, auditoriums present different possibilities for terrorists, and the options for protecting against those possibilities are somewhat limited. Innovative engineering solutions might be required to address the hazards of an internal release on the main floor, such as the microdistribution of supply air, as opposed to the high-volume overhead and perimeter supply air system most often used. Under-the-floor supply air is another option, as is massive purging.

Stadiums (see Fig. 6.14) are often structures built for outdoor sports or other events. Some stadiums are enclosed, which makes them essentially large auditoriums. Open-air stadiums often have interior spaces for pedestrian traffic and shops, which makes them similar to malls or airports. Many stadiums like old baseball parks have little or no actual ventilation in interior spaces. Because of their high occupancy levels, stadiums are potentially high-risk targets for CBW attacks.

The same standard engineering design approaches used for malls and large office buildings can be taken to immunize stadiums against CBW agent releases, including steps related to air-cleaning systems and the physical protection of air-handling equipment. Security protocols for stadiums are fraught with the same difficulties as those for any area with general public access since it may be difficult to spot suspicious individuals or packages. Perhaps the best security in such facilities is to educate the audiences to look out for suspicious packages and persons. Of course, CBW agents might fit into someone's pocket, and so even watching for packages is not a guaranteed defense.

Churches, synagogues, mosques, and temples, like other places of public assembly, are potential targets for CBW attacks depending on both their size and also possibly their denomination (Fig. 6.15). Numerous churches across the country have been bombed or burned

Figure 6.14 The Silver Dome in Pontiac, Michigan. (*Photo courtesy of Art Anderson, Penn State.*)

Figure 6.15 Churches have been the targets of terrorist attacks in the United States. (*Photo courtesy of Art Anderson, Penn State.*)

down in the past few decades by racist groups such as the Ku Klux Klan (KKK). Some members of these groups have been implicated in attempts to use CBW agents such as ricin, plague, and anthrax (Tucker 1999). It seems likely that, if not contained, such groups will eventually take to using CBW agents against these same churches.

6.2.9   Special facilities

Some facilities fall outside the defined categories of Table 6.1 or are not included for specific reasons. Most buildings, whether identified on the list or not, can be treated in the same manner as any buildings on the list to which they are similar. Certain special facilities are addressed below.

Nuclear power plants (Fig. 6.16) normally consist of a collection of buildings including maintenance and equipment facilities, main power production buildings, and office buildings. Such facilities are unlikely targets for CBW agents and, in fact, often have inherent protection against such agents since they were designed to protect against radiological releases. The technologies used to control radiological hazards are similar to those used to control CBW agents (i.e., filtration and carbon adsorption). Nuclear facilities are also often designed to protect against various potential chemical releases from surrounding facilities, railroads, or highways. Not every building on nuclear power plant sites has such protection, but the level of security is extremely high at these

Figure 6.16 Nuclear power plants are unlikely locations for a CBW release, although the cooling towers in any such location may be vulnerable as outdoor dispersion mechanisms.

plants, making any attempt by outsiders to infiltrate most improbable. The possibility of insiders perpetrating such an act is not entirely beyond comprehension, and the possibility of outside air releases of CBW agents is also a minor potential threat, so consideration should be given to protecting plant office buildings with immune building technology.

Cooling towers, regardless of whether they are at nuclear power plants or on top of apartment buildings, are vulnerable to being used as dispersion systems for CBW agents. The dispersion of *Legionella pneumophila* from cooling towers is a well-known phenomenon, but other agents, including radiological agents, might be dumped in the cooling water and allowed to aerosolize in populated areas (Matson 2001).

News media facilities could conceivably be high-profile targets, but they are essentially small, medium, or large office buildings and so can be treated in a similar manner to office buildings if it is desired to install or retrofit immune building systems.

National monuments are often icons of national pride, and may be subject to attacks of destructive potential, but not necessarily CBW agents. Obviously, CBW agents are unlikely to do damage to the monuments themselves, but they may cause high casualties among visitors in those with high traffic. These facilities are often unique and must be evaluated on an individual basis.

6.3 Building Attack Scenarios

The type of attack scenario in which CBW agents would be directed against buildings for the purpose of causing casualties directly affects the engineering approach selected for defensive measures. Several possible scenarios exist in which CBW agents could be used against buildings. Military attack scenarios generally involve outdoor releases like those of World War I. Attack scenarios for buildings are more likely to involve releases at the air intakes or internal releases. All of these types are shown in the breakdown in Fig. 6.17 and addressed in the following sections.

6.3.1 Outdoor releases

Outdoor releases are a credible scenario on the battlefield but are not the most likely approach to be used against buildings. The reason is that outdoor air dispersion tends to diminish the concentration, and most buildings only take in a fraction of outdoor air, resulting in further dilution. Such releases are subject to variations in wind direction and airflow patterns around buildings, making any attempt to deliver an agent to a building air intake from a distance an unpredictable venture.

In Japan, the Aum Shinrikyo cult attempted on several occasions from 1990 to 1993 to release botulinum toxin and anthrax spores

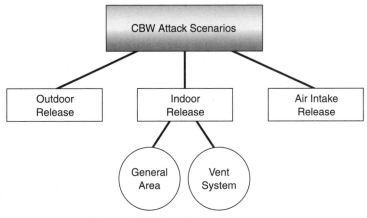

Figure 6.17   Breakdown of CBW attack scenarios.

outside of the Diet and other government buildings in Tokyo by driving around a truck with a spraying device, or by releasing the agents directly from a midrise office building window (Christopher et al. 1997; Olson 1999). Although some odors and pet deaths were reported, no actual human casualties occurred. On June 27, 1994, the same cult drove a modified truck around a residential neighborhood in Matsumoto and released sarin gas. This attack killed seven people and caused 500 casualties. Obviously, this latter sort of outdoor release entails a high risk to residential buildings, but it can still be dealt with through the use of immune building technologies.

Outdoor releases in crowded areas (Fig. 6.18) clearly have deadly potential, although they are perhaps the least dangerous of the various possible attack scenarios for any particular buildings. The reason, as illustrated in Fig. 6.19a, is that only a portion of the outdoor containing the agent will be drawn into the outside air intakes. Far worse would be the situation shown in Fig. 6.19b, in which the release occurs directly inside the outside air intake, in which case most of the agent will be drawn into the supply airstream. The general engineering approach to protection against outdoor air releases is to process any outside air that is brought in by the ventilation system or to isolate the interior spaces.

6.3.2   Air intake release

Releases in the air intakes of buildings have been hypothesized as one possible means of contaminating a building. Air intakes at ground level are considered vulnerable points where agents could be released either by spilling, explosive release, or placement of an aerosol generator (Fig. 6.20). Building air intakes are often located near ground level, making them accessible.

Figure 6.18 Outdoor attack scenarios involve either releases in crowded open areas or releases in urban areas where aerosols would be drawn into air intakes.

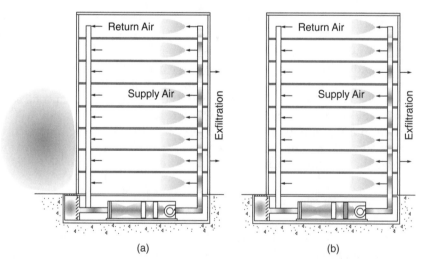

Figure 6.19 (a) Outside air release drawn into the air intakes, and (b) a release inside the outside air intakes.

Figure 6.20   Outdoor intakes at ground level, like the one shown above, are points of vulnerability for some large buildings.

Outside air intakes for typical buildings draw in about 15 to 25 percent of outside air for mixing with building return air. Assuming that all of the agent released is drawn into the ventilation system, the effect is the same as if the agent were released in the air-handling unit. Analysis of such a release event is therefore identical to that of an internal air-handling unit release, if there are no filters.

The spilling of a liquid into the air intake of a building or the placement of an aerosol-generating device at the air intakes has similar effects. The main difference is that the rate of release may be much higher from a powered or pressurized device. The release of a gas from a pressurized tank has the same effect as the use of an aerosol generator and can be considered an identical release.

The use of an explosive device at an air intake may drive some of the agent outward and away from the air intake. The worst-case scenario is, therefore, the spilling of an agent into an air intake or the use of a device or aerosolizer at an air intake. The likelihood of such an attack is even higher in a scenario in which multiple buildings are attacked simultaneously. If terrorists had a large enough quantity of a CBW agent, they could maximize casualties by moving from building to building and contaminating many air intakes instead of focusing an attack on only one large building.

6.3.3 Indoor explosive release

The detonation of a munitions-driven chemical or biological weapon is a concept left over from the days when military attacks were the only design-basis scenarios. Although such attacks are possible against buildings, they have certain shortcomings. First, the explosion would alert the occupants, and some of the vapors produced might be visible (Fig. 6.21), possibly giving most people sufficient forewarning to escape unharmed. Second, such a release in an open area would primarily contaminate that area and the immediate surrounding areas only. The explosive might also trigger the sprinkler system, which would tend to suppress particulate aerosols, although it might have little effect on gases.

An explosive release of a CBW agent would contaminate the immediate area to a much higher degree than might be necessary. That is, the concentrations produced might be much higher than the LD_{50} for the agent. After the initial release, the normal ventilation would both spread the agent and purge it at the outside air flowrate. Figure 6.22 illustrates the recirculation effect of a release in a general area. This approach on the part of a terrorist would be a comparatively inefficient way of trying to maximize casualties in a typical office building. In an

Figure 6.21 Explosive release or aerosolization of chemical agents in a general area might produce visible vapors. Most CBW agent releases would be invisible. (*Photo courtesy of Concept Smoke Systems, UK.*)

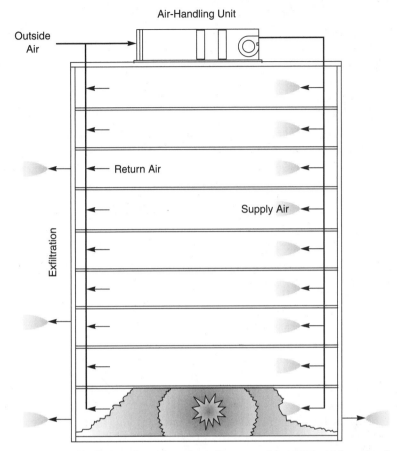

Figure 6.22 Explosive release in a general area would result in high contamination and exfiltration from that area. Other areas would receive diluted concentrations by recirculation.

auditorium, however, with a single large volume and dense occupancy, this approach might be the worst-case scenario.

One situation in which the explosive release might be effective in causing casualties is if the agent were fast-acting, like some of the chemical agents, and if it were detonated inside a large, crowded stadium. In this case, the dispersive effect of the explosion would have the effect of covering large numbers of occupants. At the same time, it would also alert everyone and allow many to seek immediate safety and treatment.

All things considered, the explosive release is an unlikely scenario and not the worst-case scenario for a typical large building. The surreptitious modes of release are far more likely to cause large numbers of casualties since, by the time people realized what had happened or the problem was diagnosed, it might be too late for effective treatment. This could be the case for certain pathogens like Ebola or other infectious agents like anthrax.

6.3.4 Indoor passive release

Passive releases are those that involve no equipment for driving the aerosolization process. They can involve simply spilling a liquid CW agent and allowing it to evaporate, or dusting toxins and dried spores on surfaces where air currents would aerosolize them or where physical contact might be sufficient to cause casualties. The use of anthrax spores in mail can be considered a form of passive release since either direct contact or the act of opening the envelope might be sufficient to aerosolize a lethal dose of spores.

The pouring of a liquid or powdered agent into ductwork would be one form of passive release. The airstream would gradually aerosolize the agent and distribute it throughout a building. The dusting of a carpet or floor in a high-traffic area of a building would be another form of passive release. The pedestrian traffic might be sufficient to cause aerosolization. A liquid could be disseminated in this fashion, but it might draw attention and might even have an odor.

The pouring of a liquid or powder directly into the high-velocity airstream at the outlet of a fan is another possible approach, but in this case the distinction between passive and forced aerosolization becomes somewhat blurred.

To the perpetrator, the advantage of passive release is its simplicity. The disadvantage is that there is no guarantee of widespread dispersal, and the actual dispersion pattern might be highly unpredictable. The person committing such an act would also be at great risk of contaminating himself or herself, although in light of the types of suicide attacks that occurred on September 11, this probably is a moot point.

6.3.5 Indoor aerosolization release

The most likely approach, and the one that could produce the most casualties in any given attack on a building, is the use of a sprayer or aerosolization device to contaminate a building internally. The locations inside a building where such a device could be placed include the general areas of the building (e.g., the main floor), the return air ducts or inside the air intakes, the supply air ducts, or inside the air-handling unit. Placement inside the outside air intakes, the return air ducts, or the air-handling unit would produce essentially the same result.

The placement of an aerosolizer in the supply duct will have different effects depending on where in the supply duct the aerosolizer is located.

Placement in the supply duct close to the air-handling unit is essentially the same as placement directly in the air-handling unit. Placement in the furthest supply air duct outlet is essentially the same as placement in a room. Placement anywhere in between will have effects between these two extremes. In any event, placement of an aerosolizer in the return air duct or air-handling unit will normally represent the worst case. An exception is when the forced return air is exhausted, in which case the air-handling unit placement is still the worst-case scenario.

Aerosolization of chemical agents sufficient to contaminate a large building would require some fairly bulky quantities, on the order of several gallons or more for most of the chemical agents listed in Appendix D. Most toxins require smaller quantities, possibly in amounts that could be concealed inside clothing or small packages. Pathogens require quantities that would fit in a pocket, although an airtight container would be necessary.

The equipment necessary for aerosolization, however, may not be quite small enough to escape notice. An Erlenmeyer flask loaded with a toxin or pathogen could be driven by a small pressurized tank. The tank could be the size of a canister such as is used for mixing drinks. Indeed, a soda mixer might even work in such an application if it could hold enough pressure. The pressurized tank could just as well be a small compressor except that it would need a power supply.

If the same device were placed in an open room, the room itself would experience high concentrations of contaminants, as illustrated in Fig. 6.23, but the remainder of the building would see lower concentrations than through placement in a duct. This is partly because room air mixing is never perfect and much of the agent may remain locally in or around the room. In addition, the ventilation system will be drawing in outside air and diluting the supply air concentration. The outside air that is being brought in is displacing exfiltrated air, which is either lost from the rooms to the outside or forcibly exhausted. The nature of the ventilation system determines where the outside air is coming from but the effect is the same—some of the contaminant will be exhausted on each pass through the building and ventilation system.

The resulting effect on the rest of the building due to a release in a general area or room would be similar to that depicted in Fig. 6.22 for an explosive release. The main difference is that the aerosolizer release might go unnoticed, and might therefore cause considerably more casualties.

6.3.6 Internal release in duct

The worst-case scenario, in terms of casualties, would occur with the placement of an aerosolization device in the air-handling unit itself. Placement in the return air duct would have the identical effect if there were no forced exhaust of return air to the air-handling unit, which is usually the case.

If a device like that depicted in Fig. 6.23 were placed inside the air-handling unit or return ductwork, the effect would be to contaminate the supply airstream, which would lead to progressive contamination of the entire building. Because this would distribute the contaminant more evenly throughout the building, it would be likely to cause maximum casualties.

The various placement locations can now be summarized in terms of the risk they pose to occupants, or their rank in terms of potential casualties caused. Figure 6.24 shows a schematic of a building with a ventilation system consisting of supply and return air duct, and an outside air intake near the roof.

Table 6.3 summarizes the points identified in Fig. 6.24 in terms of the relative risk of casualties that might be produced by placement of a CBW release device at those locations. This table assumes that the dust filter normally present in the air-handling unit inlet would cause some minor reduction in airborne levels of toxins or bioaerosols. This would not, of course, be true for most chemical agents since these would be vapors or gases and would pass through the filter mostly unattenuated. Not all building ventilation systems have ducted return air ducts, and for such buildings these points do not apply.

Table 6.3 only applies to buildings of the commercial or office building type with constant-volume minimum outside air systems. Buildings and facilities with large volumes, such as malls, airports,

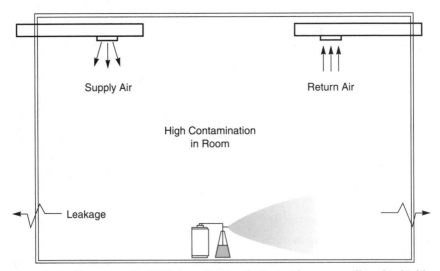

Figure 6.23 Placement of a CBW aerosolization device inside a room will tend to highly contaminate that room only.

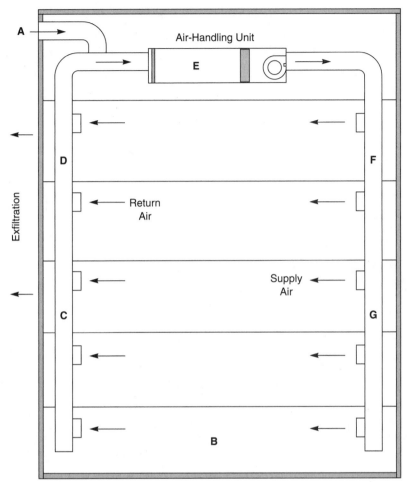

Figure 6.24 Potential locations of CBW aerosol disseminators in a building with supply and return ducts.

stadiums, and auditoriums, may have very different air dispersion characteristics that depend both on their sizes and the nature of their ventilation systems. For these facilities, the previous generalizations may not apply and only a specific analysis may be able to establish the scenarios of greatest risk.

Table 6.3 furthermore may not apply to buildings with other types of ventilation systems. Buildings with VAV systems have outside air rates that vary with climatic conditions. For VAV systems, the risk value of points C, D, F, and G may actually vary over time.

Buildings with 100 percent outside air systems will not have return ducts, and so points D and C will not exist. In addition, points F, G, and

TABLE 6.3   Relative Risk for BW and CW Agent Release Points

Point	Location	BW agents		CW agents	
		Rank	Relative risk	Rank	Relative risk
E	Air-handling unit	1	Highest casualties	1	Highest casualties
A	Outside air intake	2	High casualties	1	Highest casualties
C	Return air duct	2	High casualties	1	Highest casualties
D	Return air duct	2	High casualties	1	Highest casualties
F	Supply air duct	3	Low casualties	2	Low casualties
G	Supply air duct	4	Lower casualties	3	Lower casualties
B	Indoor general area	5	Lowest casualties	4	Lowest casualties

B may have an even lower risk relative to those for buildings with constant-volume systems. In all cases, however, point E, inside the air-handling unit downstream of the filter, will pose the greatest risk.

6.4   Summary

In this chapter, the variety of buildings has been described and associated vulnerabilities and relative risks in terms of CBW attacks identified. A mathematical method has been presented for assessing the relative risk of any building, and managers should be able to use this approach to decide which buildings may need immunization against the threat of CBW attacks, and perhaps how much to invest in protecting such buildings. Although the relative risk may not be of use in determining the budget needed to retrofit a building with immune building technologies, it will be shown in later chapters that this may be a moot point since the size of air disinfection systems is limited, or even defined, by the characteristics of the ventilation systems and buildings involved.

Ventilation Systems

7.1 Introduction

Ventilation systems are often as unique as the buildings they service. A wide variety of ventilation systems exist and their operating parameters can vary considerably. When a chemical or biological agent is released in an indoor environment, the ventilation system becomes, in effect, a delivery system for the agent. The type of ventilation system installed in a building can impact the dispersion of contaminants that are released in indoor environments, and it is important to review the ventilation system type and understand its operation in order to effectively protect the building against possible CBW attacks. In this chapter the differences between ventilation systems and their common characteristics are summarized, and a modeling method is presented. Alternative modeling methods are also discussed.

7.2 Types of Ventilation Systems

Although buildings may have a wide variety of ventilation systems, most of the differences involve the type of cooling or heating used. In terms of the differences between ventilation systems that may impact the release of CBW agents, there are but a handful of system types, and these are identified in Table 7.1.

Natural ventilation is common among many residential buildings and some other types of older buildings, especially in mild climates. There may be little that can be done to protect these buildings short of adding recirculation units.

TABLE 7.1 Types of Ventilation Systems

Type	Description
Natural ventilation	No forced air, no ductwork or fans
Constant volume	Recirculation systems with 15–25% outside air
Variable air volume	Recirculation with variable percentage outside air
100% outside air	Full-time 100% outside air, no recirculation
Dedicated outside air	Minimum 15–25% outside air supplied, no recirculation

Constant-volume systems comprise the majority of ventilation system types and are the primary type of system addressed here for purposes of simulating buildings and attack scenarios.

Variable-air-volume (VAV) systems alter the amount of outside air in response to either outside air temperature or outside air enthalpy. When conditions are right, the amount of outside air drawn in is maximized. The purpose of these systems is to save energy by taking advantage of mild air conditions. Normally, these buildings have some fixed amount of minimum outside air that may be between 15 and 25 percent. Under such conditions, this kind of system will operate almost identically to constant-volume systems. For this reason, and the fact that operating conditions may be unpredictable, VAV systems are not specifically analyzed here in terms of the effects of various attack scenarios. Instead, such buildings should be considered to lie between the extremes of constant-volume systems and 100 percent outside air systems.

Systems that use 100 percent outside air are found in some hospitals, healthcare facilities, clinics, and laboratories. They are often used in situations where the risk of internal contamination outweighs the costs associated with using large volumes of outside air in winter or summer.

Dedicated outside air systems (DOASs) are distinct from 100 percent outside air systems (referred to as "all-air" systems) in that they only deliver the minimum outside air needed for ventilation, or about 15 to 25 percent as much airflow as that of an all-air system (Mumma 2001a). Typically, the thermal loads are accommodated with hydronic systems rather than air systems. DOASs are relatively new in the United States but have a variety of features that give them some inherent protection against dispersion of contaminants (Mumma 2001b). Since they use 100 percent outside air, there is no recirculation of contaminants, and since they operate at lower total airflows, the total air that is filtered is only about 20 percent of conventional all-air systems, resulting in significant savings in operating costs (Mumma 2002). The air-cleaning rate may also be lower.

7.3   Ventilation Modeling

Most buildings normally take in a certain amount of outdoor air for cooling or ventilation purposes. Buildings with forced-air systems, like many modern apartment or office buildings (Fig. 7.1), will typically take in a volume of outside air of at least 15 percent of the total supply air. Naturally ventilated buildings take in an amount of outside air that depends on various factors unique to each building. The amount of outside air taken in displaces an equal amount of internal air that is exhausted or exfiltrated. The effect of this volume of outside air is to purge the indoor air of contaminants, but because it is typically mixed, or diluted with return air, the process is usually referred to as "dilution ventilation."

Outdoor air is naturally sterilized by sunlight, variations in temperature, variations in relative humidity, and the effects of other factors including pollution. At the same time, the outside air carries a variety of environmental microorganisms and contaminants, including pollen, fungal spores, bacteria, and bacterial spores. Most of these airborne microbes are relatively harmless to healthy humans in naturally occurring concentrations but can be a threat to those who are immunocompromised or who have allergies.

Figure 7.1   Many modern apartment buildings take in at least 15 percent outside air, often from air-handling units located on the roof.

The outdoor air rarely, if ever, contains pathogenic microorganisms of the sort that can cause serious infections. As a result, the outdoor air can be considered effectively sterile for the purposes of dilution ventilation. Pathogenic microorganisms like viruses normally hail from human or animal sources. They can also be deliberately introduced at dangerous concentrations. In either case, purging the indoor air with outdoor air is a viable approach to controlling indoor concentrations of any pathogenic microbes or chemical contaminants. The possibility of the outside air intake being contaminated is one that must be protected against, and this is addressed further in Chap. 9.

In typical buildings, dilution with outdoor air occurs at a rate of between 15 and 25 percent. This level of dilution ventilation offers a certain amount of protection against the release of CBW agents. However, the actual rate of dilution can vary considerably from room to room or from floor to floor. The effect that dilution will have on removing contaminants can only be estimated based on simplistic models. More sophisticated methods such as computational fluid dynamics (CFD) can be used to study the intricate details of airflow in specific situations, and these will be addressed at the end of this chapter.

Some buildings, like certain areas of hospitals or laboratories, use 100 percent outside air, which may offer a high degree of protection against CBW agent releases due to the continuous purging effect. Systems with 100 percent outside air may be inherently defensible against CBW agent releases even without the addition of filtration, UVGI, or other technologies, provided the air intake is protected. Retrofitting a system with 100 percent outside air, however, may not be economical due to the high costs of cooling or heating in climates that require a lot of either.

Many modern buildings use VAV systems that adjust the amount of outside air they bring in depending on the temperature, enthalpy, or internal cooling demand. In such systems, the outside air flowrate can sometimes reach 100 percent of the total supply air volume. Retrofitting modern air-handling units, like the one shown in Fig. 7.2, is often facilitated by the presence of walk-in access around the cooling coils.

Dilution ventilation removes all three CBW agents at essentially the same rate. That is, dilution ventilation does not discriminate based on the type of agent. The agents may be removed by other mechanisms, such as plate out, or adsorption, on buildings' surfaces, but these are factors that can be conservatively ignored when estimating the purging effect of outside air.

The rate at which indoor air contaminants are purged depends on the volume of outdoor air, the total volume of supply air, and the degree of mixing in internal areas. Mixing is very dependent on local conditions

Figure 7.2 Large commercial buildings often have trains of equipment with walk-in access to filters and cooling coils.

and difficult to estimate without actual test data. It can, however, be approximated by assuming complete mixing. The assumption of complete mixing is not necessarily conservative, but it may be representative for most buildings in general.

Another assumption necessary for developing a mathematical model of dilution ventilation is that the outside air is sterile. This is assumed in spite of the fact that there will always be some ambient environmental bacteria or pollutants. The concentration of fungal spores in outdoor air, for example, is often between 100 and 1000 cfu/m^3. Such contaminants are not considered a threat during a CBW incident, and so they are ignored in the mathematical model. Even so, they will likely be removed by any air-cleaning system designed for CBW defense.

The rate of removal of airborne pathogens, toxins, or chemical agents depends on two factors, the air change rate (ACH) and the ventilation effectiveness (or degree of air mixing). If plug flow (piston flow) were assumed, then one air change would completely remove all pathogens that were initially present in a room. However, this is rarely the case, except when it is by design (ASHRAE 1991).

Complete air mixing causes the indoor concentration of airborne pathogens to decrease in an exponential manner. This represents the limiting case for normal buildings, and is a simple model to use for evaluating the removal rate of airborne pathogens. Given the assumption of complete air mixing, the primary factor determining the removal of airborne pathogens is the air change rate.

7.3.1   Ventilation modeling with calculus methods

Calculus methods are one approach to estimating the effects of mixing, and they are addressed here prior to detailing computational methods. The classic model for mixing is a linear first-order differential equation (Boyce and DiPrima 1997; Heinsohn 1991). Figure 7.3 shows a single-building volume into which outside air flows at a rate of Q and which has a contaminant source discharging into the room at a rate of S.

The change in concentration over time can be found by first writing the mass flow balance for the concentration. The change in concentration over time $C(t)$, will equal the amount of contaminants entering from the outside air, C_a, minus the amount of contaminants exiting through the exhaust air, $C(t)$, plus the release rate of the sources:

$$\frac{d(VC)}{dt} = QC_a - QC + S \tag{7.1}$$

In Eq. (7.1), V is the volume, C is the concentration, C_a is the concentration of contaminants in the outside air, Q is the flowrate, and S is the source or release rate of the contaminant. Taking the outside air concentration to be zero, or $C_a = 0$, produces the following:

$$V\frac{dC}{dt} = -QC + S \tag{7.2}$$

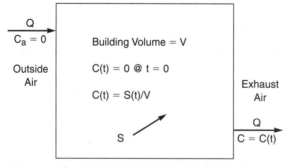

Figure 7.3   Single-volume model for a building with a constant source of contaminants and an initial concentration of $C(t) = 0$.

Rearranging Eq. (7.2) for integration over time t results in the following:

$$\int_0^{C(t)} \frac{dC}{S - QC} = \int_0^t \frac{1}{V} \, dt \tag{7.3}$$

The solution obtained for the concentration as a function of time, $C(t)$, is as follows:

$$C(t) = \frac{S}{Q} \left[1 - \exp\left(-\frac{Q}{V} t \right) \right] \tag{7.4}$$

In Eq. (7.4), S/Q represents the steady-state concentration. It can be rewritten in a slightly more convenient form by defining the number of room air changes per hour, N, as the airflow, Q, divided by the room volume, V, as follows:

$$N = \frac{Q}{V} \tag{7.5}$$

Rewriting the equation in its simplified form produces the following:

$$C(t) = \frac{S}{Q} \left[1 - \exp\left(-Nt \right) \right] \tag{7.6}$$

Figure 7.4 illustrates the generic response of the room concentration over time for a constant source and for various air change rates.

If the initial concentration is some value C_0, as in the case of a sudden release of an agent, the decay of the building concentration can be derived in a similar fashion and written as the following:

$$C(t) = C_0 \left[\exp\left(-Nt \right) \right] \tag{7.7}$$

Figure 7.5 shows the results of Eq. (7.7) for the same generic conditions as Fig. 7.4.

7.3.2 Ventilation modeling with computational methods

For more complex buildings with multiple spaces, it is generally easier to determine the building concentrations computationally, using either spreadsheets or programs. Computation of the concentration of airborne contaminants is accomplished using a sequence of time steps, in which a finite volume of fresh air replaces an equal volume of room air.

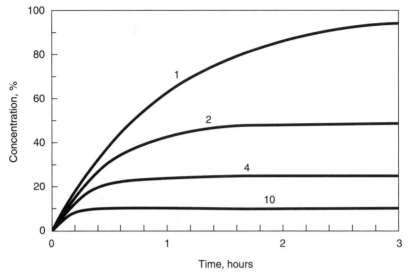

Figure 7.4   Effect of various air change rates on building concentrations from a contaminant released at a constant rate.

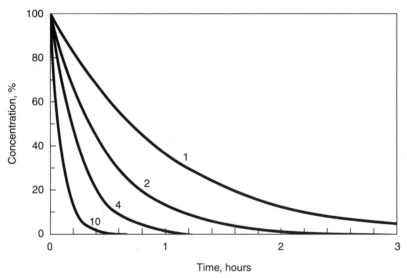

Figure 7.5   Decay of building concentrations after a sudden release of an agent for different air change rates.

This approach is sometimes known as "numerical integration." The choice of the time step can influence the concentration, but a value of 1 min will provide a close approximation of continuous flow over a period of several hours.

The removal rate of microorganisms is given by the indoor concentration, $C_i(t)$, at any given time t, multiplied by the outside air flowrate

OA, which is the volume of air displaced into and out of the building volume. For chemicals or toxins, the units might be micrograms or grams, and the concentration would have units of $\mu g/m^3$ or g/m^3, or else $\mu g/L$ or g/L, whatever units are most convenient. For microbes, we can use the single colony-forming unit (cfu), which may be spores, bacteria, or virions (virus particles), which gives us a concentration unit of cfu/m^3. The latter unit will be used in the description of the general computational method that follows.

For each minute of the analysis, the number of microorganisms exhausted, N_{out}, is determined from the indoor concentration, $C_i(t)$, at time t, multiplied by the outside air flowrate OA:

$$N_{out} = C_i(t) \cdot OA \qquad (7.8)$$

The number of microorganisms added from the outside air is assumed to be zero. Note that, for an outside release scenario, the outside air concentration would have some nonzero value and the indoor source would be zero. This situation is not specifically addressed here since this scenario, as explained previously in Chap. 6, is the less likely case, but it is simple to adapt the model to such a scenario. An agent released into the outside air intakes, however, could be modeled with the following computational method just as if it were released into an internal duct.

The rate of microbes released internally is specified per minute, and is designated P_{in}. The total population of microbes, $N(t)$, that will exist in the building for any given minute t will then be the previous minute's population plus the current minute's additions, minus the number exhausted to the outside air:

$$N(t) = N(t-1) - N_{out} + P_{in} \qquad (7.9)$$

In the model, complete mixing is assumed, and therefore the building microbial concentration will be defined as the building microbial population divided by the building volume V, for any given minute t, as follows:

$$C(t) = \frac{N(t)}{V} \qquad (7.10)$$

In Eq. (7.10), the term V refers to the total building volume in cubic meters, or whatever other units might be used. The value of the building concentration calculated per Eq. (7.10) is then used to compute the values for the next minute. This incremental calculation process, carried out for several hours, provides a close approximation of continuous flow, with the result being the same exponential decrease in the building microbial concentration over time, for an initial concentration, as seen with Eq. (7.7).

For example, assume a sudden release at time $t = 0$ for a building with a volume of $V = 5000$ m^3. Assume the initial concentration of microbial contamination is 0 cfu/m³ and that an airborne population of 10,000 cfu occurs at the end of the first minute. Assume the exhaust air flowrate, with 15 percent outside air, is 0.15(5000) = 750 m³/min. Therefore, at the end of the second minute, the following values can be computed:

$$\text{Concentration} = (10{,}000 \text{ cfu}) / (5{,}000 \text{ m}^3) = 2 \text{ cfu/m}^3 \qquad (7.11)$$

$$\text{Microbes exhausted} = (2 \text{ cfu/m}^3)(750 \text{ m}^3/\text{min}) = 1500 \text{ cfu} \qquad (7.12)$$

The total number of microbes in the building volume can now be computed for minute 2 by adding the previous minute's input to the next minute's input and subtracting the previous minute's exhaust amount as follows:

$$\text{Total microbes} = 10{,}000 + 10{,}000 - 1{,}500 = 18{,}500 \qquad (7.13)$$

The calculations for the first 20 min are summarized in Table 7.2. The results are virtually identical to those that would be obtained from using Eq. (7.7). The results in Table 7.2 are graphed in Fig. 7.6. Note that the concentration achieves near steady-state conditions in a comparatively short time. This is because the volume of this model

TABLE 7.2   Numerical Computation of Concentration

Time, min	Microbes released, cfu	Total microbes, cfu	Concentration in volume, cfu/m³	Microbes exhausted, cfu
0	10,000	0	0	0
1	10,000	10,000	2	1500
2	10,000	18,500	4	2775
3	10,000	25,725	5	3859
4	10,000	31,866	6	4780
5	10,000	37,086	7	5563
6	10,000	41,523	8	6229
7	10,000	45,295	9	6794
8	10,000	48,501	10	7275
9	10,000	51,226	10	7684
10	10,000	53,542	11	8031
11	10,000	55,510	11	8327
12	10,000	57,184	11	8578
13	10,000	58,606	12	8791
14	10,000	59,815	12	8972
15	10,000	60,843	12	9126
16	10,000	61,717	12	9257
17	10,000	62,459	12	9369
18	10,000	63,090	13	9464
19	10,000	63,627	13	9544
20	10,000	64,083	13	9612

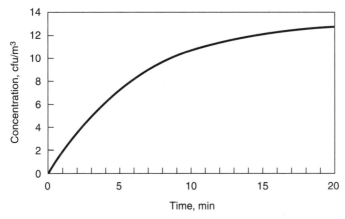

Figure 7.6 Increase in concentration over time from a constant source, computed by numerical integration.

building is comparatively small, and was assumed so to demonstrate the system response.

The method previously described can be adapted to account for any possible scenario, including a sudden release, a time-varying release, and buildings with multiple zones with varying zone airflows, stairwells, and elevator shafts. To deal with multiple zones, it is necessary to calculate each of their concentrations individually, and then sum their exhausts to determine the total exhaust rate for the entire building. Often, this level of detail is not necessary if it is merely intended to examine the effect of different technologies on the overall building. Multizone modeling is best handled by software like CONTAMW (NIST 2002).

Elevator shafts and stairwells in tall buildings can cause stack effects that may redistribute contaminants that are localized. Figure 7.7 illustrates a case in which a CBW agent is released on one floor of a building with a leaky elevator shaft. Even without the ventilation system operating, the contaminants may distribute throughout the building via exfiltration from the shaft or stairwell. Whether the gas or vapor will travel upward, downward, or both, and how much it will exfiltrate with or without ventilation, will depend on factors that include air temperature, building leakage, and wind pressure. For complex buildings with multiple zones, stairwells, elevator shafts, and leakage between zones, more sophisticated tools are available, such as the CONTAMW program (NIST 2002), which will be discussed in Sec. 7.3.3.

Filters and UVGI systems can be incorporated into the basic numerical integration model presented above, but it is first necessary to know

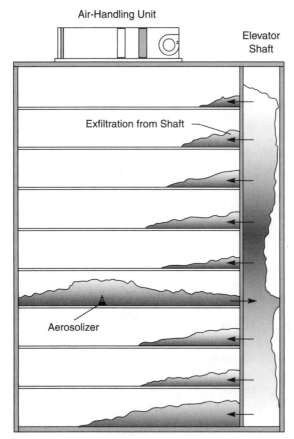

Figure 7.7 Illustration of how an elevator shaft might impact the release of an aerosol in a local area of a high-rise building.

what the kill rates or removal rates are against the specific microbes under study. The methods for determining these kill rates and removal rates are presented in Chap. 8.

7.3.3   Ventilation modeling with CONTAMW

The previous methods for modeling buildings are single-zone models that treat the entire building as one large area. The single-zone approach should be adequate for modeling most buildings for basic attack scenarios like an air intake release, or a release in the air supply or return duct. For improved modeling realism, or where multiple internal zones may have different airflows and volumes, the best approach is to use a multizone model such as the CONTAMW (NIST

2002) program already mentioned. Other programs are also available for multizone modeling, but the CONTAMW program is in the public domain and has been used in a number of published studies on contaminant dispersion inside buildings.

The CONTAMW program simulates the release and distribution of contaminants in a building with a ventilation system in much the same way as the previously described numerical integration or spreadsheet model. It has a variety of advantages, including the ability to model multiple zones, varying infiltration and exfiltration rates, and wind pressure or buoyancy effects (NIST 2002). It has the ability to model complex building systems, and allows for the easy manipulation or relocation of the contaminant sources under consideration.

Sufficient literature is available on multizone modeling methods from a number of sources (Musser 2000, 2001; Musser et al. 2001; Walton 1989; Yaghoubi et al. 1995), and especially through the NIST website (NIST 2002), so this information will not be repeated here. Instead, some examples will be provided in the simulations in Chap. 9.

7.4 Computational Fluid Dynamics Modeling

Computational fluid dynamics studies are the ultimate in airflow modeling realism and can predict the movement of air currents through passageways and around objects in a room. They can assist in the location of supply and exhaust ducts to maximize air mixing in a zone and to facilitate the flow of contaminants out of the zone. CFD is a computationally intensive approach that can be time-consuming but will provide more accurate predictions of air movement than are possible with the previously discussed methods. The question of whether such modeling power will be useful for any given situation depends on the problem to be solved and how much detail is required.

If, for example, it is necessary to design a room air supply and exhaust system to rapidly remove contaminants, CFD provides what is perhaps the only alternative besides tracer gas testing to design supply and exhaust registers and air velocities. Figure 7.8 shows an example of a pair of rooms complete with furniture and other typical appurtenances. A passive release of contaminants occurs in the hallway between the two rooms and the air currents from the displacement ventilation system carry it through the rooms, as shown in Fig. 7.9.

7.5 Summary

In summary, this chapter has described the various ventilation systems that are found in modern buildings, and has provided details on

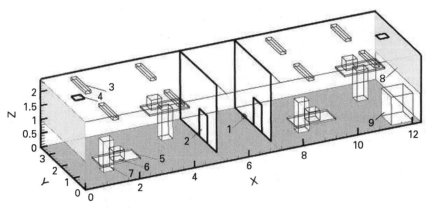

Figure 7.8 CFD model for two offices with a corridor in between: 1—passive contaminant source; 2—displacement air supply diffuser; 3—fluorescent lamp; 4—exhaust diffuser; 5—table; 6—computer; 7—occupant; 8—window; 9—copy machine. [*Image created with the FLUENT program (FDI 1988), courtesy of Jelena Srebric, Penn State.*]

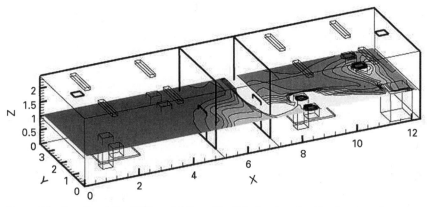

Figure 7.9 Results of the CFD model from Fig. 7.8 showing the distribution of contaminants resulting from ventilation air movement. [*Image created with the FLUENT program (FDI 1988), courtesy of Jelena Srebric, Penn State.*]

two methods for modeling building ventilation systems to effectively guard against CBW attacks—a calculus method and a numerical integration method or spreadsheet model. These modeling methods will enable engineers to model single-zone systems. For multizone modeling, the CONTAMW program is available, and for detailed modeling of room airflows, CFD programs like FLUENT or FIDAP (FDI 1998) are available. The spreadsheet model and the CONTAMW program will be revisited in Chap. 9 and used for simulating various attack scenarios.

8

Air-Cleaning and Disinfection Systems

8.1 Introduction

Air-cleaning and air disinfection technologies are fundamental components of any integrated immune building system, and operate in conjunction with the building envelope, the building ventilation system, and any CBW detection systems. The ventilation system performs the function of an air-cleaning component when it brings in outside air and exhausts or exfiltrates contaminated air. When operating with 100 percent outside air and sufficient volume, the ventilation system may even perform this function unassisted, except for the case of any outside air releases or releases in the air intakes (Skistad 1994). For most buildings, though, it is necessary to have additional components to process the air to remove chemical, biological, or toxin agents.

In addition to ventilation purging or dilution, several technologies are currently available that have an established ability to remove airborne contaminants, including filtration, ultraviolet germicidal irradiation (UVGI), and carbon adsorption. Each of these technologies has its own strengths, weaknesses, and costs. Table 8.1 summarizes the technologies, their primary applications, and their advantages and disadvantages.

All of the technologies in Table 8.1 are sufficiently well understood that predictions can be made as to their effectiveness in actual operation against specific contaminants, except for gas phase filtration. There is insufficient information on the removal rates of chemical agents to make any accurate performance predictions for carbon adsorbers or other forms of gas phase filters.

Figure 8.1 provides a graphic breakdown of four air-cleaning technologies and the bioweapon agents against which they are generally effective.

TABLE 8.1   Common Air-Cleaning Technologies and Applications

Technology	Application	Advantages	Disadvantages
Dilution ventilation	Purging of interior contaminants	Effective against all CBW agents	Requires high air exchange rates to be effective; can be costly
Filtration	Removal of airborne particulates	Effective against large airborne particles	Effectiveness depends on efficiency; high efficiency can be costly
UVGI	Disinfection of airborne pathogens	Effective against viruses and many bacteria	High power required for spores; can be costly
Carbon adsorption	Removal of gases and vapors	Effective against airborne chemical agents	Has little effect on microbes

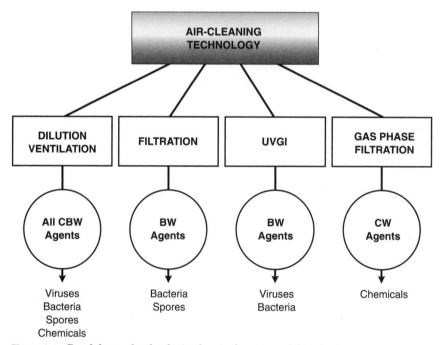

Figure 8.1   Breakdown of technologies for air cleaning and disinfection.

Several other technologies exist in a developmental stage that can also be applied to the cleaning or disinfection of air, including photocatalytic oxidation (PCO), ionization, pulsed light, pulsed electric fields, ozone, microwaves, plasma fields, gamma irradiation, impregnated filters, and others, but not all of these are well understood or feasible to use in commercial or residential applications. These developmental technologies are discussed in Chap. 14. Dilution ventilation was addressed in Chap. 7. This chapter focuses strictly on the three basic, affordable technologies that are sufficiently well understood to be applied in the defense of buildings against biological and chemical releases—filtration, UVGI, and carbon adsorption.

8.2   Filtration

Filtration is one of the primary means by which engineers can control indoor air contaminants (ASHRAE 1992). High-efficiency filters, such as the one shown in Fig. 8.2, protect against a wide variety of airborne pathogens in indoor environments. Toxins can also be removed by filtration, but filters have no effect on gaseous chemical agents. Chemical agents in liquid aerosol form may be partly filterable. The filtration of microorganisms is a topic that requires special focus because of the unique characteristics of both filters and airborne pathogens in the submicron size range.

This section addresses the critical aspects of filter sizing and provides a methodology for predicting performance against airborne

Figure 8.2 Example of a high-efficiency pleated filter. (*Image courtesy of Donaldson Company, Inc., Minneapolis, MN.*)

microorganisms and toxins. Results of studies on the filtration of viruses and bacteria are compared with theoretical filter performance. Some test results are reviewed to demonstrate how filtration might remove some chemical agents that are in liquid aerosol form.

Filters have been classified by various testing standards and methods to provide a basis for defining their performance, including dust spot (DSP) efficiency, total arrestance, and the European "EU" standard. High-efficiency particulate air (HEPA) filters are tested according to a separate method—penetration of dioctyl phthalate at a size of 0.3 micron—as a result of their use in the nuclear industry. ASHRAE Standard 52.2 was issued in 1999 to provide a uniform method of testing filters and establishing performance in more detailed terms (ASHRAE 1999a). This standard reclassified filters according to their minimum-efficiency reporting value (MERV). However, as of this printing, insufficient published data on MERV are available to define the general performance for all MERV-rated filters of interest. Therefore, the filter models provided here are based on a mixture of the more commonly used DSP efficiencies and MERV ratings. Estimated equivalent MERV ratings are provided for DSP-rated filters. The MERV filters used here are based on limited data sets and their performance is not necessarily the same for all filters of the same MERV rating, since performance will vary between manufacturers' models. HEPA filters do not require MERV ratings since they are tested by a separate standard.

Table 8.2 lists several basic types of filters, including dust filters, high-efficiency filters, HEPA filters, and ultra-low-penetration air (ULPA) filters (ASHRAE 1999a). This table also compares filters with their approximate or estimated MERV ratings. Six filters have been selected for model development in this chapter: MERV 6 (approximately 20 percent DSP), MERV 8 (approximately 40 percent DSP), MERV 11 (60 to 65 percent DSP), MERV 13 (approximately 80 to 85 percent DSP), and MERV 14 (approximately 90 to 95 percent DSP). These six filters provide a basis for sizing air-cleaning equipment. Models cannot, at the present, be developed for all filter types because of a lack of available data, but the ones presented here should be more than adequate for the intended purposes.

Fungal and bacterial spores range from about 1 to 20 microns in size, and can easily penetrate dust filters. High-efficiency filters, MERV 7–15 (25 to 95 percent DSP), remove spores at relatively high rates.

Bacteria range in size from about 0.2 to 2 microns and are removed at high rates by MERV 13–15 filters (80 to 90 percent DSP). Viruses, the smallest microbes, span a size range from 0.01 to 0.3 micron. A HEPA filter can remove viruses at high rates, and even a MERV 11 (60 percent DSP) filter can remove up to half of some viruses. However, it is not necessary to depend on filters for the interception of viruses since

TABLE 8.2   Filter Types and Typical Ratings*

Filter type	Applicable size range, μm	Dust spot efficiency, %	Total arrestance, %	MERV rating (estimated)
Dust filters	>10	<20	<65	1
		<20	65–70	2
		<20	70–75	3
		<20	75–80	4
High efficiency	3–10	<20	80–85	5
		<20	85–90	6
		25–35	>90	7–8
	1–3	40–45	>90	9
		50–55	>95	10
		60–65	>95	11
		70–75	>98	12
	0.3–1	80–90	>98	13
		90–95	na	14
		>95	na	15

*Note: na = not applicable.

UVGI destroys viruses more efficiently and will be seen to complement filter operation.

Appendix A shows the mean diameters for over 100 airborne pathogens and allergens. Also provided in Appendix A are the estimated removal efficiencies for five filter types in the MERV 6–15 range.

The effectiveness of a filter against airborne microbes depends primarily on the filter characteristics, the air velocity, the particle size, and the type of microbe being intercepted. Filter characteristics tend to vary from one manufacturer or model to another, and so the results developed here must be considered general. The manufacturer's performance curve, or the MERV test results, should be used for more detailed analysis if they are available.

Filters do not act like sieves for micron-size particles since the gaps between fibers may average 20 microns or more (Fig. 8.3). Probabilities determine filtration efficiency—the massive number of microscopic fibers in any filter ensures that the probability of a particle being intercepted by a fiber increases exponentially with the thickness of the filter.

A performance curve defines the removal rate versus particle size for any particular manufacturer's filter model. Performance curves are available from filter manufacturer catalogs but these curves stop short at a lower limit of 0.1 to 0.3 micron, and therefore cannot be used to predict the filtration of viruses and small bacteria. It is necessary to develop a mathematical model for filters so that performance curves can

Figure 8.3 Most air filters are composed of fine glass fibers with diameters between about 1 and 20 μm and with gaps of 20 μm or more. (*Image courtesy of Filterite, Timonium, MD.*)

be extended into the virus size range. Also, since filters are often operated above or below their specified air velocity, the mathematical model is useful for predicting microbial filtration rates at nondesign velocities.

Filter performance is defined in terms of efficiency versus particle size, as shown in Fig. 8.4, which is based on catalog data from one major vendor. This filter curve can be used to predict the removal rate of microbes down to 0.3 micron in size at the design velocity of the filter. Beyond these conditions, the curve is of no use, and modeling the filters is the only recourse.

Various mathematical models have been developed over the past century to predict the performance of filters. All of these models incorporate classical concepts that usually describe three basic filtration mechanisms: interception, diffusion, and impaction. Figure 8.5 graphically illustrates the range in which these processes predominate. It can be observed that the dip in the curve will produce a "most penetrating particle size range" (Matteson et al. 1987).

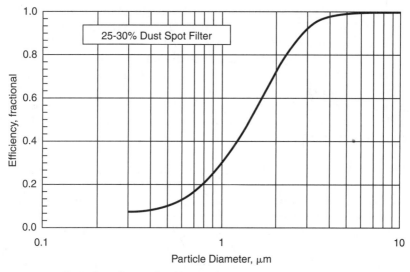

Figure 8.4 Typical vendor catalog filter performance curve.

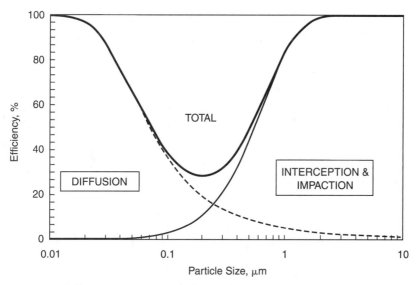

Figure 8.5 Diffusion, interception, and impaction form the major components of the filter model and define the performance curve.

Three primary mechanisms operate in filtration—impaction, interception, and diffusion. Impaction occurs when the particle inertia is so high that it crosses the air streamlines around a fiber, impacts, and remains attached. This process is only significant for large particles, often 5 microns and larger, and is often neglected in many filtration models since the interception parameter satisfactorily accounts for it (Brown and Wake 1991; Matteson et al. 1987; Stafford and Ettinger 1972; Suneja and Lee 1974; VanOsdell 1994; Wake et al. 1995).

Interception occurs when the particle diameter is large enough that the normal airflow streamline carries it within contact range of a fiber, as illustrated in Fig. 8.6. For particles larger than about 1 micron, this process is nearly inevitable in any high-efficiency filter. Basically, any particle that is large enough to be captured by impaction is large enough to eventually be captured by interception, at least in any medium- to high-efficiency filter.

Diffusion is a removal process that dominates for particles smaller than about 1 micron (Brown 1993; Davies 1973). Since these particles are subject to the effects of Brownian motion, they tend to contact, and stick to, fiber surfaces whether the airflow streamlines bring them within a particle diameter distance or not. This process is illustrated in Fig. 8.7. In the case of diffusion, a higher velocity will reduce the effect, since there is less time for a passing particle to diffuse into the fiber surface.

In the cases of impaction and interception, increased air velocity will increase particle removal efficiency, if it is above the most penetrating particle size range, since it increases inertial effects. Since diffusional efficiency peaks at small particle sizes and interception peaks at large particle sizes, the minimum efficiency, or most penetrating particle size, will occur between these extremes.

8.2.1 Mathematical modeling of filtration

The mathematical model of filtration is too complex to provide in a single equation and requires a careful definition of the fundamental relation-

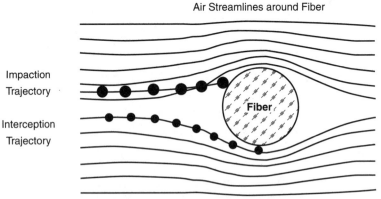

Figure 8.6 Illustration of the process of impaction and interception on a filter fiber.

Air Streamlines around Fiber

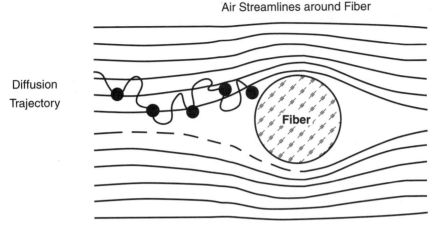

Diffusion
Trajectory

Figure 8.7   Illustration of the process of diffusion on a filter fiber.

ships. The efficiency of a fibrous filter is defined in terms of the efficiency of a single fiber (Brown 1993; Davies 1973) as follows:

$$E = 1 - e^{-E_s S} \qquad (8.1)$$

where E_s = single-fiber efficiency, fractional
S = fiber projected area, dimensionless

The fiber projected area is a dimensionless constant combining the three factors that determine filter efficiency—filter thickness (length along the airflow direction), filter packing density, and fiber diameter (Brown 1993; Davies 1973). It can be represented in forms that are more physically intuitive, but the mathematical definition simplifies to the following:

$$S = \frac{4 \times 10^6 \, La}{\pi d_f} \qquad (8.2)$$

where L = length of filter media in direction of airflow, m
d_f = fiber diameter, μm
a = filter media volume packing density, m³/m³

These three parameters can be used to define filter performance over almost any range of velocity and particle size. The single-fiber efficiencies for diffusion and interception are summed to obtain the total single-fiber efficiency as follows:

$$E_s = E_R + E_D \qquad (8.3)$$

where E_R = interception efficiency, fractional
$\quad\quad E_D$ = diffusion efficiency, fractional

The single-fiber diffusion efficiency is defined by Lee and Liu (Matteson et al. 1987) as follows:

$$E_D = 1.6125 \left(\frac{1 - a}{F_K} \right)^{1/3} \text{Pe}^{-2/3} \qquad (8.4)$$

where F_K = Kuwabara hydrodynamic factor
$\quad\quad$ Pe = Peclet number, dimensionless

The Peclet number characterizes the intensity of diffusional deposition, and an increase in the Peclet number will decrease the single-fiber diffusion efficiency. The Peclet number is defined as follows:

$$\text{Pe} = \frac{1 \times 10^{-6} \, U d_f}{D_d} \qquad (8.5)$$

where $\quad U$ = media face velocity, m/s
$\quad\quad D_d$ = particle diffusion coefficient, m²/s

The Kuwabara hydrodynamic factor (Matteson et al. 1987) is defined as follows:

$$F_K = a - \frac{a^2 + 2 \ln a + 3}{4} \qquad (8.6)$$

The particle diffusion coefficient is a measure of the degree of diffusional motion of a particle, and is similar to the diffusional motion of a molecule. It is defined as a function of the particle mobility by the Einstein equation:

$$D_d = \mu k T \qquad (8.7)$$

where μ = particle mobility, N·s/m
$\quad\quad k$ = Boltzman's constant, 1.3708×10^{-23} J/K
$\quad\quad T$ = temperature, K

The particle mobility is defined as

$$\mu = \frac{C_h}{3 \times 10^{-6} \, \pi \eta d_p} \qquad (8.8)$$

where $\quad \eta$ = gas absolute viscosity, N·s/m²
$\quad\quad C_h$ = Cunningham slip factor, dimensionless
$\quad\quad d_p$ = particle diameter, μm

The Cunningham slip factor accounts for the aerodynamic slip that occurs at the particle surface, and is defined as

$$C_h = 1 + \left(\frac{\lambda}{d_p}\right)(2.492 + 0.84e^{(-0.435d_p/\lambda)}) \qquad (8.9)$$

where λ = gas molecule mean free path, μm

The gas molecule mean free path in air at standard conditions is 0.067 μm (Heinsohn 1991).

The single-fiber interception efficiency E_R represents the fractional efficiency with which a single fiber removes particles from a passing airstream. Lee and Liu (Brown 1993; Davies 1973) have defined it in their model as

$$E_R = \frac{1}{\varepsilon}\left(\frac{1-a}{F_K}\right)\left(\frac{N_r^2}{1+N_r}\right) \qquad (8.10)$$

where N_r = interception parameter, dimensionless
F_K = Kuwabara hydrodynamic factor, dimensionless
ε = correction factor for inhomogeneity, dimensionless

The interception parameter is

$$N_r = \frac{d_p}{d_f} \qquad (8.11)$$

In Eq. (8.11), the correction factor $1/\varepsilon$ accounts for filter media inhomogeneity, or the deviation of filter fibers from the geometry of the mathematical models. Yeh and Liu (Raber 1986) have determined experimentally that the value of ε is approximately 1.6 for polyester filters. The inhomogeneous factor for glass fiber filters is not known, but the same value, 1.6, provides acceptable agreement with empirical data.

Manufacturers of filters and filter media today incorporate multiple fiber diameters, varying the proportion of each fiber size in order to obtain the required efficiency. The fiber diameters range from 0.65 to 6.5 microns. For multiple fibers, Eq. (8.2) becomes

$$S_i = \frac{4 \times 10^6 \, La_i}{\pi d_{f_i}} \qquad (8.12)$$

where d_{f_i} = fiber diameter i
a_i = volume packing density i
S_i = filter projected area i

The volume packing density, or packing fraction, for each fiber of a multifiber model is

$$a = \Sigma a_i \text{frac}_i \tag{8.13}$$

where frac refers to the fraction (or pecentage/100) of each of the fiber diameters.

The final equation defining the efficiency of a filter is then

$$E = 1 - e^{-\Sigma E_{S_i} S_i} \tag{8.14}$$

where E_{S_i} = single-fiber efficiency for fiber i

The single-fiber efficiency E_{S_i} for each of the fibers is calculated via Eqs. (8.2) through (8.12) as if each were a separate filter, and the results are combined in Eq. (8.14). The fractions in Eq. (8.13) can be adjusted to obtain almost any filter efficiency desired, while holding all other parameters constant, including the total packing fraction a and the filter length L. For HEPA and ULPA filters, however, an increase in length L is required to achieve the order-of-magnitude increase in efficiency.

Some manufacturers define three diameters and an associated percentage for each. This provides a reasonable approximation of continuous fiber sizes and allows the filtration model to be separated into three distinct sections based on the packing density at each fiber diameter. This approach has been used in this model to match the Ensor et al. (1991) data shown in Fig. 8.8.

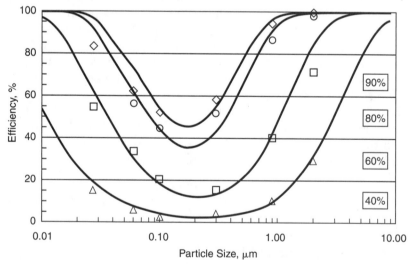

Figure 8.8 Modeled performance curves for five filters extended into the virus-size range, and compared with data from Ensor et al. (1991).

The filter performance was modeled on spreadsheets using the parameters shown in Table 8.3. Although the filter models do not match the data exactly, they are well within the accuracy that could be expected for these filters.

Figure 8.9 shows a generic HEPA filter model compared against test data for viruses from several studies (Harstad and Filler 1969; Jensen 1967; Roelants et al. 1968; Thorne and Burrows 1960; Washam 1966). Note the excellent agreement between the HEPA filter model and the test results in the low-size range. No test data exist for the larger viruses, for the other filters, or for velocities outside the design range.

Proprietary information on fiber diameters and their associated percentages was used to adapt the filter models and to match filter performance data from Ensor et al. (1991). A summary of the values for the five filter models is given in Table 8.3, along with other design parameters that act as constraints, such as filter face velocity and media area/filter face area ratio.

8.2.2 Filtration applications

Retrofitting a filter in an existing ventilation system may affect the fan performance and system airflow. These factors must be addressed and engineered as necessary to ensure overall system performance is not adversely impacted.

All filters have the following general characteristics:

- Performance decreases with increased velocity.

- Performance increases with increased age and filter loading.

- Filter pressure drop increases over time as the filter becomes loaded.

- Increased filter pressure drop may reduce airflow or, if flow is held constant, increase fan energy consumption.

The above may not necessarily be true for certain types of filters like electret filters where filter loading effects are less significant (Tennal et al. 1991). The performance of filters in the range of larger particles (that is, 0.1 to 100 microns) has been studied by various researchers (Ensor et al. 1991; Kemp et al 1995). A review of these and other results confirms that the filter models produce reasonably accurate predictions over this size range (Kowalski et al. 1999).

The actual performance of filters in field applications depends to a large extent on the efficiency with which contaminants are swept across rooms and into return air registers. Air mixing can also be an important factor. Some studies have shown that the actual efficiency with which indoor contaminants are removed can be far less than the efficiency rating of the filter itself (Offerman et al. 1992). Still, substantive reductions

TABLE 8.3 Multifiber Filter Model Parameters

Parameter	Units	MERV 6 20%	MERV 8 40%	MERV 11 60%	MERV 13 80%	MERV 14 90%	HEPA 99.97%
Face area	m^2	0.3530	0.3530	0.3530	0.3530	0.3530	0.3502
	ft^2	3.8	3.8	3.8	3.8	3.8	3.77
Media area	m^2	2.6942	2.6942	5.3884	5.3884	5.3884	14.0284
	ft^2	29	29	58	58	58	151
Media length	m	0.015	0.015	0.015	0.015	0.015	0.017
	ft	0.0492	0.0492	0.0492	0.0492	0.0492	0.0558
Nominal face velocity	m/s	3.81	3.81	3.302	2.54	2.54	1.27
	fpm	750	750	650	500	500	250
Media velocity	m/s	0.5253	0.5253	0.2276	0.1753	0.1753	0.0335
	fpm	103.4	103.4	44.8	34.5	34.5	6.6
Total volume fraction		0.002	0.002	0.002	0.002	0.002	0.0051
Fiber 1							
d_{f_1}	μm	2.6	3.2	1.5	0.65	0.65	0.65
Fraction of total volume		0.15	0.01	0.1	0.1	0.16	0.5
Fiber 2							
d_{f_2}	μm	4	4	3.8	2.8	2.8	2.7
Fraction of total volume		0.3	0.1	0.4	0.5	0.5	0.35
Fiber 3							
d_{f_3}	μm	7	6.5	6.5	6.5	6.5	6.5
Fraction of total volume		0.55	0.89	0.5	0.4	0.34	0.15

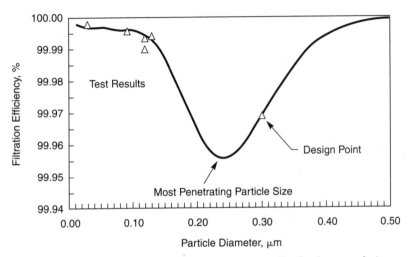

Figure 8.9 Comparison of HEPA filter model with test results for several viruses (Kowalski et al. 1999).

in indoor air particle counts can be achieved with high-efficiency filters (Burroughs 1998; Hicks et al. 1996).

When using a HEPA filter, it is necessary to add a prefilter to preserve filter life; otherwise, maintenance costs can become extreme. For laboratories, ASHRAE recommends placing a 25 percent, a 90 percent, and a HEPA filter in series (ASHRAE 1991). Prefilters are often recommended for 80 percent and 90 percent filters also (AIA 1993). At the very least, dust filters should be included to preserve the performance of high-efficiency filters.

Bacteria intercepted by filters will eventually die from dehydration or natural causes, but spores may last indefinitely on filter media (Maus et al. 1997, 2001). Antimicrobial filters provide an option for preventing microbial growth on filters in humid conditions (Foarde et al. 2000). UVGI exposure of the filter can also prevent microbial growth on the filter surface (Shaughnessy et al. 1999).

Most filter media consist of a random array of fibers with diameters in the 1- to 20-micron size range (Fig. 8.10). A recent development in filtration technology, expanded polytetrafluoroethylene (ePTFE) (Fig. 8.11), consists of a relatively ordered arrangement of polymer fibers (Folmsbee and Ganatra 1996). The denser array of smaller-diameter polymer fibers allows thinner media to be used while producing higher efficiency, although at the cost of higher pressure drops.

This model offers two advantages over typical performance curves. First is the fact that the performance curve can be extended down into the size range of viruses. This is, of course, essentially an extrapolation

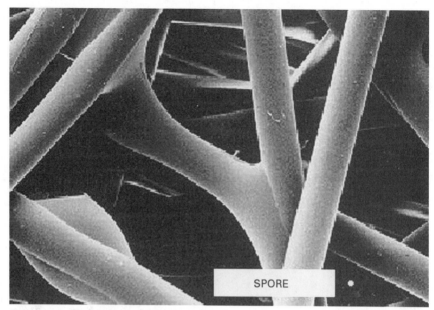

Figure 8.10 The photo above shows typical filter media with fiber diameters of approximately 20 μm. Shown for reference is a 2-μm spore. (*Image courtesy of W. L. Gore & Associates, Inc., Elkton, MD.*)

Figure 8.11 Photomicrograph showing the structure of ePTFE filter media. Shown for reference is a 2-μm spore. (*Image courtesy of W. L. Gore & Associates, Inc., Elkton, MD.*)

and should be used only if actual test data or other performance verification are unavailable. The second advantage is that the model facilitates the study of filters operated above or below their design air velocities. When filters are operated above their design velocities, filter efficiency may decrease below that shown on their performance curves. A HEPA filter operated at 500 feet per minute (fpm) face velocity may allow up to 10 times as many microbial penetrations as one operated at the normal design face velocity of 250 fpm. These penetrations will occur primarily in the most penetrating particle size range. It is important, both for HEPA filters and high-efficiency filters, to account for the actual operating velocity through the filter, and not simply accept the manufacturer's design velocity–based performance curve.

In order to determine how effective filtration can be in comparison with, or in combination with, dilution ventilation, filter performance must be scrutinized in terms of the targeted BW agents.

The size distributions of airborne pathogens form lognormal curves, as detailed in Chap. 2. The most accurate means of determining the filtration rate of a microbe is to use the full size distribution. Unfortunately, these distributions are not known in any detail but an excellent approximation is to use the logmean diameters instead. These have been tabulated in Appendix A for use with filter curves. It is often stated in the literature that microbes, especially viruses, clump together or are attached to larger particles, but such particles are likely to break up on impact with filter fibers, making this a moot point in regard to filtration. Also, weaponized agents often have ingredients to prevent clumping and promote aerosolization.

Mathematical filter models can provide a variety of useful information about filter performance against airborne microbes, and under various conditions like high velocities and recirculation. Figure 8.12 shows the results of such modeling to determine the most penetrating BW agents for a HEPA filter. The number of microbes penetrating a HEPA filter may not be significant since it depends on risk levels and concentrations, but this example illustrates a characteristic of all filters—that certain microbes penetrate more effectively than others.

Figure 8.13 graphically illustrates how the microbes line up against their filtration efficiencies for a 60 percent filter. The largest-sized toxins and fungal spores are removed at a 100 percent rate. The smallest viruses, Marburg and Ebola (GE for genetically engineered), are filtered out at a 30 percent rate. The most penetrating microorganisms exist in the 0.1- to 0.3-micron size range, and these include smallpox, *Francisella tularensis,* and the hemorrhagic fevers.

The logmean diameters given in Appendix A can be used in conjunction with any filter performance curve to determine a similar array of most penetrating microbes. However, as noted previously, most catalog

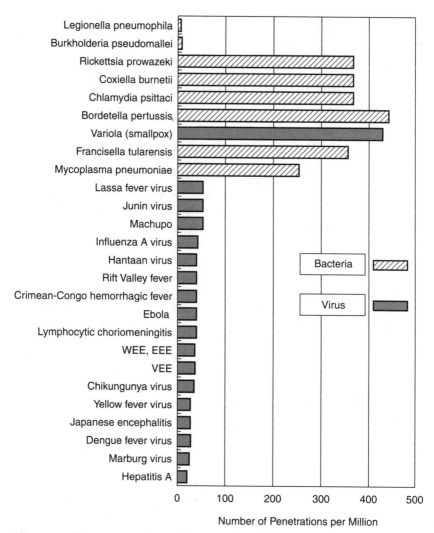

Figure 8.12 The most penetrating BW agents for a HEPA filter.

performance curves do not extend into the virus size range. Appendix B includes the estimated filtration efficiencies for the nominal filter sizes MERV 6, 8, 10, 13, and 15. Interpolation could be used, if necessary, to determine the filtration efficiency of filters in between these sizes, such as the MERV-rated filters from Table 8.2.

Recirculating air through any filter increases the effective filtration rate. The overall filtration efficiency increases with the number of passes through the filter. With a large number of passes, an 80 percent or 90 percent filter can approach the efficiency of a single pass through

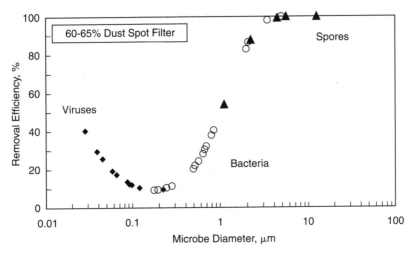

Figure 8.13 BW agents shown superpositioned over the performance curve of a 60 percent DSP filter (approximately MERV 11).

a HEPA filter. Figure 8.14a and b illustrate this effect for a recirculating filter unit in a model room with no outside air, and assuming perfect plug flow. Under normal conditions, airflow would be mixed and these filters would not approach HEPA filter performance so closely, but the comparison illustrates the potential increase in overall performance when filters are used in recirculated air.

The HEPA filter penetrations, being approximately zero, are not visible on the scale in Fig. 8.14. In Fig. 8.14b, note how closely the 90 percent filter simulates a HEPA filter after six passes.

Figure 8.15 shows removal rates of airborne microbial populations for various filters in combination with dilution ventilation in a well-mixed space for a model building with a 100,000-cfm system engaged at time $t = 0$. The microbes include equal proportions of viruses, bacteria, and fungi with an initial concentration of 90,000 cfu/m^3. This model considers both internal generation of viruses and bacteria and fungal spores entering from the outside air. In this well-mixed model, indoor concentrations reach steady-state conditions within a few hours and the final building concentrations are seen to differ in each case. Real-world conditions might not allow cleaning building air so rapidly, but the point is that the filter choice will determine the final level of air quality for any building application.

This analysis suggests that the HEPA filter may provide little improvement over the use of an 80 percent or 90 percent filter in cases when immediate removal of a contaminant is not critical. HEPA filters, however, cost considerably more to own and operate. The analysis

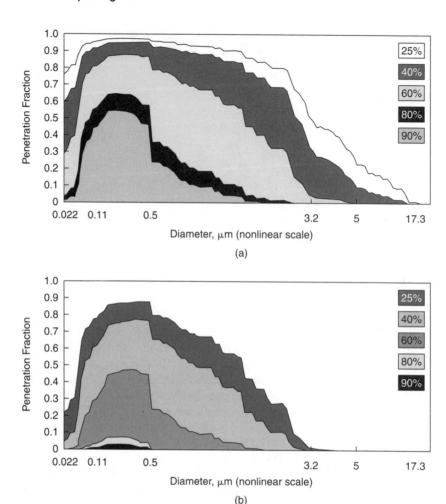

Figure 8.14 Comparison of the penetrations from an extended array of microbes (a) after a single pass and (b) after six passes through the model filters.

applies to a single space or building volume only. In a multizone system, consideration must be given to the fact that microbes may be recirculated to other areas, and a detailed analysis would be required to assess the overall system performance.

8.2.3 Filtration test results for microbes

A number of tests have been performed on the filtration of viruses, and these results were shown previously in Fig. 8.9. The detailed results of these tests are summarized in Table 8.4. These tests all produced results that are essentially in agreement with expectations for the indi-

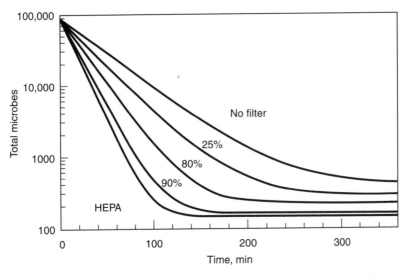

Figure 8.15 Comparison of removal rates of filters located in the supply air duct in combination with dilution ventilation with 25 percent outside air.

cated filters. The viruses that were filtered were removed at rates that suggest they existed as single particles in the test aerosols. These results offer corroboration for expected filter performance against viruses.

Some limited data are available on the effectiveness of high-efficiency filters used in recirculating units from tests done in agricultural animal houses (Carpenter et al. 1986a, 1986b). Figure 8.16 illustrates the reduction in airborne bacteria by recirculation filter units. In the first test, levels were reduced to 31 percent, compared against the unfiltered condition, and in the second test they were reduced to 18 percent of the unfiltered condition. The filter ratings were not specified in these tests but appear to have been 60 to 90 percent DSP filters (MERV 11 to 14). Although the airborne concentrations were significantly reduced, the performance is somewhat less than might be expected for high-efficiency filters. The reason for the reduced performance was apparently due to poor airflow distribution in the areas where these units were placed. This highlights the fact that airflow distribution is an important component of an efficient filtration system, and attention must be given to placement of recirculation units in order to achieve optimum performance.

8.2.4 Filtration of liquid aerosols

Chemical agents are not considered to be filterable by ordinary filters, but chemical agents occur in two forms—gaseous and liquid aerosols. Gaseous chemical agents can only be removed by gas phase filtration

TABLE 8.4   Summary of Virus Filtration Test Results

Researcher	Filter type	Microbe	Removal, %
Roelants et al. (1968)	HEPA	Actinophage *S. virginiae* S1	99.997
Washam (1966)	HEPA	T phage *E. coli* B	99.9915
Jensen (1967)	HEPA	T phage *E. coli* B	99.99
Harstad et al. (1967)	HEPA	T phage *E. coli* B	99.997
Thorne and Burrows (1960)	HEPA	Foot-and-mouth disease virus	99.998
Burmester and Wetter (1972)	40–45%	Marek's disease virus (MDV)	0
	80–85%	Marek's disease virus (MDV)	100
	HEPA	Marek's disease virus (MDV)	100

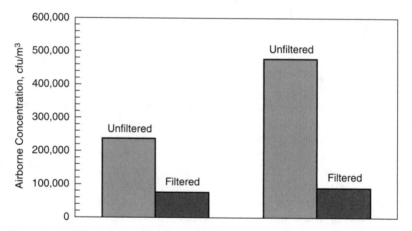

Figure 8.16 Effect of recirculation filter units on reducing airborne bacteria levels. Based on data from Carpenter et al. (1986a, 1986b).

equipment and not by filters. However, most chemical agents do not vaporize easily at normal temperatures, and therefore they exist as droplets when aerosolized (Somani 1992). These droplets will tend to be filtered at a rate that depends on their particle size, but the problem is not quite as simple as the filtration of solid particles. Droplets will also evaporate at a rate that depends mainly on their vapor pressure (Heinsohn 1991). As they evaporate, they become smaller, and if they evaporate completely, they become vapor and in that state they are no longer removable at any significant rate by filters.

The state of a chemical agent, whether gaseous or liquid, depends on its boiling point and vapor pressure. Boiling points above about 24°C will usually indicate the chemical will tend to remain in liquid droplet form when aerosolized at room temperature, depending on how fine the aerosol is to which it has been reduced. The evaporation rate of a chemical, whether in solution or in aerosol form, is primarily a function of its partial pressure and the degree to which the surrounding air is saturated. Needless to say, these factors make the problem too complex to predict, even if the atmospheric conditions and the aerosol generator characteristics are known.

In spite of the difficulty of predicting removal rates, filters can be expected to have some limited effect on liquid aerosols. Unless atomization is complete, liquid aerosols must exist as droplets of varying sizes and may attach or adsorb to filter fibers. However, they may also break into smaller aerosol particles upon impact, a process that depends on their surface tension, density, air velocity, and other factors, and after attachment they may evaporate (Raynor and Leith 1999, 2000). Some test results are available that provide an indication of the degree of their filterability. In fact, some filters are tested with liquid aerosols like dioctyl phthalate (DOP) and methyl salicylate, although most are tested with powders.

Figure 8.17 shows test results for the removal of methyl salicylate aerosol by two commercial recirculation filter units (Janney et al. 2000). These are compared against the removal rate of a filter unit with the filters removed. The effect of the filters is clearly to remove the vapor, although the same test showed much higher removal rates when carbon adsorbers were included.

It can be expected, therefore, that some degree of protection will be offered against chemical aerosols by ordinary filters, although this

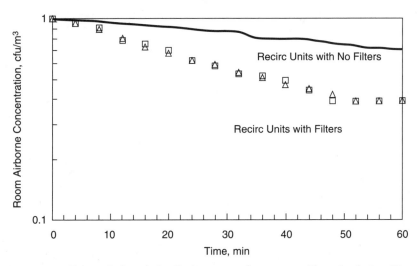

Figure 8.17 Removal of methyl salicylate vapor from room with recirculation filter units. Based on data from Janney et al. (2000).

would be difficult to estimate in advance. It is true that the filters may simultaneously play a part in reducing such aerosols to vapor or gaseous form, but this process takes time. In effect, the filter will behave initially as a collector, and then as a point source as the chemicals evaporate. Depending on the size range of the released aerosol, filters may effectively reduce the release rate of most chemical agents, and thereby mitigate their effects on building occupants.

8.3   Ultraviolet Germicidal Irradiation

When appropriately sized and designed, UVGI systems can be highly effective against a wide range of bacterial and viral pathogens. The recent development of sizing methods (Kowalski and Bahnfleth 2000a) has provided a means of predicting the effectiveness of UVGI systems to a degree not previously possible. These methods, however, are computationally intensive. This section reviews the basic modeling techniques and provides some guidelines for sizing UVGI systems to deal with BW agents.

Two aspects of UVGI system design must be evaluated in order to size a UVGI system: (1) the response of microorganisms to UV exposure and (2) the average UV intensity field inside an enclosure through which the microbes pass. The survival curves of microbes under UV exposure have been addressed in Chap. 2. The subject of the UV intensity field is treated here in as simple a fashion as is reasonable, but more detailed treatment is available in the references (Kowalski 2001; Kowalski and Bahnfleth 2000a, 2000b; Kowalski et al. 2002b).

The dose received by any airborne microorganism is dependent on the lamp intensity field through which it passes. In an in-duct UVGI system such as shown in Fig. 8.18, an intensity field is created by the lamps and the reflective surfaces of the enclosure. Determination of the intensity field is essential for the sizing of any system. The entire intensity field can be reduced to a single value for average intensity by mathematical modeling, as will be shown in Sec. 8.3.1.

The inverse square law (ISL) is often used to compute the intensity of light at any distance from a lamp. This is adequate for lighting purposes but is inaccurate in the near field where much of the germicidal effect occurs. In water-based systems the ISL can be effective, but the attenuation of UVGI by water plays a more significant role than the lamp geometry (Qualls and Johnson 1983, 1985; Severin 1986; Severin et al. 1984). In air, where there is negligible attenuation of UVGI, the ISL fails to account for the lamp diameter and is, therefore, merely an approximation. Even numerical integration of the ISL to create a line source fails to result in agreement with UV photosensor data (Kowalski et al. 2002b).

A radiation view factor represents the fraction of diffuse radiative energy emitted by one surface that is absorbed by another surface. It

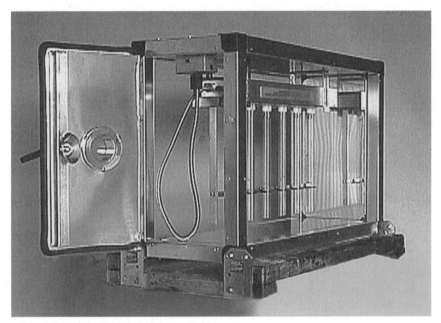

Figure 8.18 In-duct UVGI system with access door. (*Photo courtesy of Lumalier / Commercial Lighting Design.*)

defines the geometric relationship between the emitting and the receiving surface. If the emitting surface is at a constant intensity and the receiving surface is a differential element, the view factor can be used to determine the intensity at the location of the differential element. If the receiving surface is a finite area or the emitting surface is not uniform, the relationship can be integrated over the respective surfaces to find the total radiation absorbed.

8.3.1 Mathematical modeling of UVGI

The intensity at an arbitrary point outside a UVGI lamp can be computed using the radiation view factor from a differential planar element to a finite cylinder when the element is perpendicular to the cylinder axis and located at one end of the cylinder (Modest 1993):

$$F_{d1-2}(x,l,r) = \frac{L}{\pi H} \left[\begin{array}{c} \dfrac{1}{L} \text{ atan}\left(\dfrac{L}{\sqrt{H^2-1}}\right) - \text{atan}\left(\sqrt{\dfrac{H-1}{H+1}}\right) \\ + \dfrac{X-2H}{\sqrt{XY}} \text{ atan}\left(\sqrt{\dfrac{X(H-1)}{Y(H+1)}}\right) \end{array} \right] \quad (8.15)$$

The parameters in the above equation are defined as follows:

$$H = \frac{x}{r}$$

$$L = \frac{l}{r}$$

$$X = (1 + H)^2 + L^2$$

$$Y = (1 - H)^2 + L^2$$

where x = distance from the lamp, cm
l = length of the lamp segment, cm
r = radius of the lamp, cm

The view factor model is compared with photosensor data in Figs. 8.19 and 8.20. Also shown in these figures are two variations of the ISL, and it can be observed that they are relatively inaccurate in either the near field or the far field, depending on how they are applied.

To compute the view factor along the axis of the lamp at any radial distance from the axis, the lamp must be divided into two segments of lengths I_{g1} and $I_{g2} = I_g - I_{g1}$, the sum of which will be the total lamp length l_g, as shown in Fig. 8.21. The total view factor at any point will be

$$F_{tot}(x,l_1,l_g,r) = F_{d1-2}(x,l_1,r) + F_{d1-2}(x,l_g - l_1,r) \qquad (8.16)$$

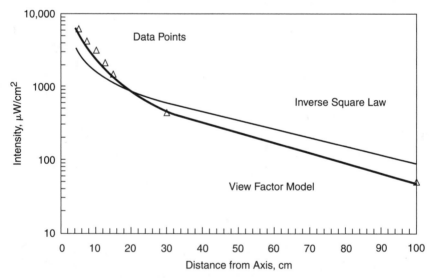

Figure 8.19 Comparison of view factor model with lamp data and inverse square law for a model TUV36W PL-L lamp. Data provided by UVDI (1999).

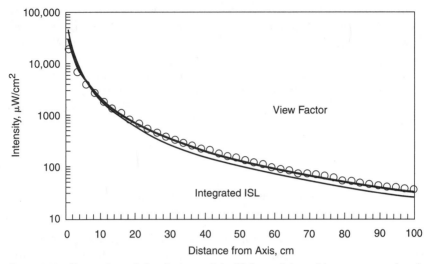

Figure 8.20 Comparison of view factor model with lamp data and inverse square law for a model GHO287T5L lamp. Data provided by UVDI (1999).

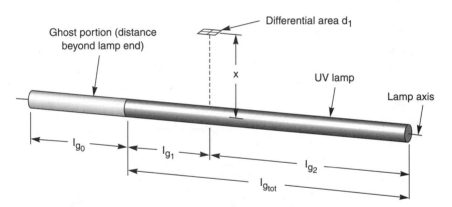

Figure 8.21 View factor model of the lamp showing the lamp modeled in two segments of lengths l_{g1} and l_{g2}, which mark the distance along the lamp axis at which the element is located. The ghost portion represents the distance an element is located beyond the lamp's end.

To compute the view factor for a point beyond the ends of the lamp, a ghost lamp is constructed from the real lamp length l_g and a "ghost" portion of some length equivalent to the distance beyond, or before, the end of the lamp, l_b. Subtracting the view factor of the ghost portion from the total length gives the final actual view factor:

$$F_{\text{tot}}(x, l_b + l_g, l_g, r) = F_{d1-2}(x, l_b + l_g, r) + F_{d1-2}(x, l_b, r) \qquad (8.17)$$

An example of the above model compared with lamp data and with the ISL is shown in Fig. 8.22. Discrepancies that exist here between the view factor model and the data are apparently due to errors in the photosensor readings.

The intensity at any point will be the product of the surface intensity I_s and the total view factor F_{tot}. The lamp surface intensity is the UV power output divided by the lamp surface area. The intensity at any coordinates x, y, and z will be described by

$$I_s = \frac{E_{uv}F_{tot}}{2\pi rl} \tag{8.18}$$

where I_s = UV intensity at any (x, y, z) point, $\mu W/cm^2$
$\quad\quad\;\; E_{uv}$ = UV power output of lamp, μW

The previous equations can be used to model any type of noncylindrical lamp by approximating the shape of the lamp with multiple cylinders. For example, the biaxial and U-tube lamps shown in Fig. 8.23 are in increasingly common use today since they require a connection at one end only. Each of these lamps can be modeled as two cylinders. They could also be modeled as three cylinders by defining a small cylindrical segment at the tip of the lamp. However, modeling either lamp as a single cylinder with an equivalent diameter (one that produces an equal lamp surface area) should be adequate for most purposes.

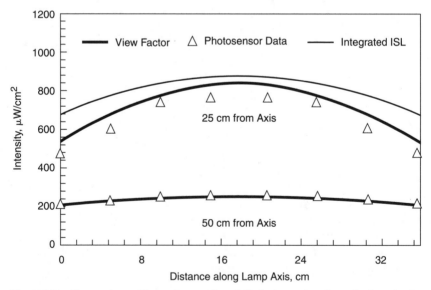

Figure 8.22 Comparison of lamp data and models for intensity along the length of a lamp. At 50 cm the view factor and ISL overlap.

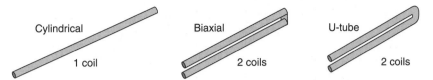

Figure 8.23 Types of UV lamps currently available that can be modeled by the view factor for cylinders. The biaxial and U-tube lamps can be divided into two or more cylinders.

Equations (8.14) through (8.18) can be used to create a three-dimensional matrix defining the intensity field around a lamp in any enclosure, as shown in Fig. 8.24. This matrix can then be used to determine the dose received by a passing airborne microorganism, either by averaging the field intensity or by integrating the dose along airflow streamlines.

Reflective materials such as aluminum or ePTFE can boost the intensity field both from direct reflections and from interreflections. The intensity field due to reflectivity depends on surface geometry and the field of the lamp. Some lamps include fixtures that are themselves reflective surfaces.

Assuming the surfaces to be purely diffuse reflectors, view factors can be used to compute the intensity field for the first reflection. This assumption is not unreasonable, even for specular surfaces, provided the intensity field is relatively uniform (i.e., has no hot spots in the intensity contour).

The direct intensity at each surface is determined with Eq. (8.18). Averaging the surface intensity to simplify computations may be possible provided the intensity distribution on the wall is relatively even. Otherwise, the surface must be subdivided into elements over which the assumption of uniform intensity is valid. The reflected intensity I_{R_n} for each of n surfaces is

$$I_{R_n} = I_{D_n}\rho \qquad (8.19)$$

where I_{D_n} = direct intensity incident on surface n, $\mu W/cm^2$
ρ = the diffuse global spectral reflectivity in the UVGI range

The view factor from a differential element to a rectangular surface, through the corner and located perpendicular to the surface, is given by the following (Modest 1993):

$$F_{d1-2} = \frac{1}{2\pi} \left[\frac{X}{\sqrt{1+X^2}} \operatorname{atan}\left(\frac{Y}{\sqrt{1+X^2}}\right) + \frac{Y}{\sqrt{1+Y^2}} \operatorname{atan}\left(\frac{X}{1+Y^2}\right) \right] \qquad (8.20)$$

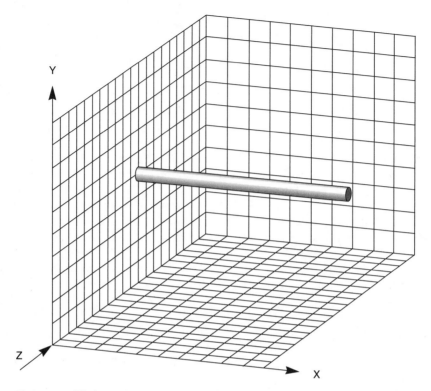

Figure 8.24   Block matrix $10 \times 10 \times 20$ used to define the reflected lamp intensity field.

where X = height / x
  Y = length / x
  x = distance to the corner, perpendicular to the surface, cm

  By dividing a large rectangle into four smaller rectangles, the intensity along any perpendicular line intersecting the large rectangle can be determined. For points that are beyond the edges of the rectangular surface, ghost rectangles representing the nonexistent areas must be included and their intensity contribution subtracted from the total. View factor algebra can be used when surfaces must be subdivided into numerous smaller elements to account for uneven intensity contours. Modest (1993) and Howell (1982) may be consulted for more details on view factor algebra.

  Equation (8.20) can be used to compute a three-dimensional matrix defining the intensity field due to the first reflections. This matrix can then be used to compute both the average and integrated dose, as described later.

  For a rectangular duct such as the one shown in Fig. 8.25, there will be four surfaces (top, bottom, left, and right), but the model can be gen-

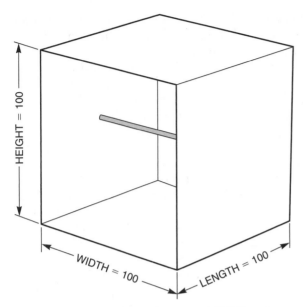

Figure 8.25 Configuration of a typical UVGI enclosure with a crossflow configuration and four reflective surfaces.

eralized to any number of surfaces. For example, a round duct could be approximated by a polygon with a sufficiently large number of sides. The reflected intensity I_R seen by a microbe at any (x, y, z) point will be the algebraic sum of those from all surfaces:

$$I_R = \sum_{i}^{n} I_{R_n} F_{d1 - 2,n} \qquad (8.21)$$

where $F_{d1 - 2}$ = view factor to n facing walls

The above procedure assumes reflectivity is diffuse. Most reflective surfaces in current use for UVGI systems are specular, or mirror-like. Analysis of specular reflections can be complex, but the assumption of diffuse surfaces will produce results that approach those of specular modeling, especially when computing the average intensity (Kowalski and Bahnfleth 2000b).

One method for dealing with specular reflections, for a finite number of surfaces, is to apply Eq. (8.21) for each specular reflection of the lamp. The lamp reflections must be reduced proportionally by the reflectivity of the surface. The interreflections can be computed in an analogous manner to that described for the first reflections, but the computations can become involved. It is, however, conservative to ignore them, especially when reflectivity is low. For additional details on computing the interreflected field, see Kowalski and Bahnfleth (2000b) and Kowalski (2001).

The duration of exposure and UVGI intensity determines the dose. Both depend on the velocity profile and amount of air mixing in the airstream. The velocity profile inside the duct or chamber depends on local conditions and may be impossible to define without CFD modeling. In any event, the design velocity of a typical UVGI unit, approximately 2 m/s (400 fpm), assures that sufficient mixing will occur and the effects of nonuniformity of the velocity profile will be moderated.

For simplicity, the reflected intensity field can be ignored, and this will provide a conservative estimate of the average intensity. The source code for computing the average intensity of the direct intensity field has been provided in Appendix F. Although this is coded in C++, it should be transparent to the engineer since it is primarily math functions. These equations and functions can be translated almost verbatim into Fortran, Basic, or Pascal. The input and output functions (filing and printing) are left to the engineer as these are unique to the programming language used.

The source code performs several necessary sequential functions. These include looping through all lamps, then looping through all points of the $50 \times 50 \times 100$ intensity matrix, computing the distance of each matrix point to the lamp axis, computing the position on the lamp axis (relative the first end of the lamp), deciding if the matrix point is beyond or within the lamp axis, and then computing the intensity at that point. Once the matrix is filled with intensity values at each point, the numerical average can be computed. Figure 8.26 provides a flowchart giving an overview of these functions.

8.3.2 UVGI applications

A limited number of guides and articles are available for assisting designers with various aspects of implementing various types of UVGI systems (Bolton 2001; Kowalski and Bahnfleth 2000a, 2000b; Luciano 1977; Luckiesh 1946; Philips 1985; Sylvania 1981; Westinghouse 1982). In most cases manufacturers can aid the engineer with implementing designs, depending on the application. Figure 8.27 shows just such a system that was sized and installed by Honeywell using their own proprietary software package (Honeywell 2001). At least one other company has a proprietary software package for automated design of UVGI systems (UVDI 2001).

In order to precisely design a system for protection against BW agents, it is necessary to establish either performance criteria or a budget. Often, the engineer must work with both. The performance of any system can be predicted using the methods described in Sec. 8.3.1. These methods will establish the average intensity for any rectangular duct system. The average intensity multiplied by the exposure time

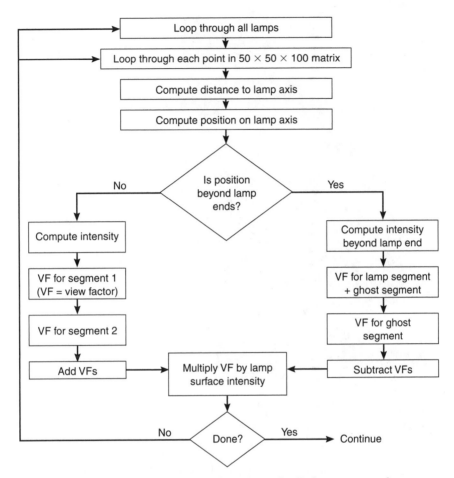

Figure 8.26 Flowchart for the source code in Appendix F that computes the average direct intensity for any number of lamps in a rectangular enclosure.

will yield the dose. The dose can then be used to predict the kill rate of any microbe for which the UVGI rate constant is known, as detailed in Chap. 2. The average intensity is, therefore, the single defining characteristic of a UVGI system.

It is convenient to introduce here the concept of a *UVGI rating value*, or URV, to define the range of average intensities that are typical in UVGI air-disinfection systems. The URV is analogous to the ASHRAE 52.2-1999 MERV system of rating filters and can be used in a complementary fashion to design combined UVGI and filtration systems. Table 8.5 lists some of the proposed URV ratings along with the average intensities that they represent. The remaining URVs are undefined at present since systems like those used for microbial growth control

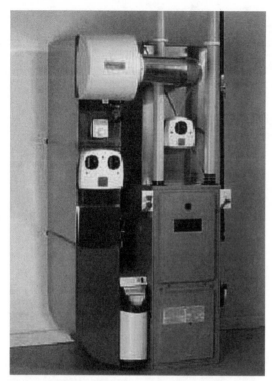

Figure 8.27 Residential ventilation system with UVGI lamps installed on the upstream (double-lamp unit on left riser) and downstream sides (single-lamp unit on right side between vertical pipes) of the coils. (*Image courtesy of Honeywell.*)

(Levetin et al. 2001), or upper air systems (Dumyahn and First 1999), may use much lower values than 100 μW/cm^2, and some very-high-intensity systems exist for which an upper size limit has not yet been identified.

The URV system is designed such that similar URV and MERV ratings can be combined to yield similar reductions in microbial populations across the entire spectrum of airborne pathogens. For example, a MERV 13 filter will reduce anthrax spore concentrations by roughly the same amount that a URV 13 system will reduce smallpox concentrations. This "leveling" characteristic of combined MERV/URV systems will be demonstrated further in Chap. 9.

The average intensity of a UVGI system, coupled with the exposure time, can be used as a sizing criterion. Typically, exposure times are on the order of fractions of a second, and this is defined by the available space in a duct or air-handling unit and the air velocity. Table 8.5 shows an example of the average UVGI intensity and exposure time

TABLE 8.5   URV Average Intensities and TB Kill Rates

URV	Average intensity, $\mu W/cm^2$	Exposure time, s						
		0.1	0.2	0.25	0.35	0.5	0.75	1
		Kill rate of TB bacilli, %						
8	125	3	5	6	9	12	18	23
9	250	5	10	12	17	23	33	41
10	500	10	19	23	31	41	55	66
11	1000	19	35	41	53	66	80	88
12	1500	27	47	55	67	80	91	96
13	2000	35	57	66	78	88	96	99
14	3000	47	72	80	89	96	99	100
15	4000	57	82	88	95	99	100	100

for a variety of typical system sizes. The UVGI rate constant for TB, or any other pathogen for which the rate is known, can be looked up in Appendix A. The survival and kill rates are computed per the methods described in Chap. 2.

It can be observed that the range of average intensity is not extreme, being only one order of magnitude, yet the range of kill rates goes from a few percent to nearly 90 percent (i.e., at 0.25 s). The exposure times shown are typical of what might be found in the space of an air-handling unit. A chart like Table 8.5 can be set up for any pathogen, including anthrax spores and smallpox. The results will clarify what kill rates are possible and what the average intensity field may have to be. Basically, 100 $\mu W/cm^2$ could be considered a low-power system, as opposed to 4000 $\mu W/cm^2$ for a high-power system. This is likely to be true regardless of the actual duct dimensions, and it would be beneficial for engineers to become as familiar with the concepts of URV and average UVGI intensity as they are with DSP or MERV ratings for filters.

Lamps are available in racks that can be mounted inside ductwork, and also as externally mounted units in which only the lamp projects into the duct. Lamp mounting and location are not always critical, although such factors can be used to optimize system performance for any given set of conditions. The use of reflective panels made from polished aluminum or other UV-reflective materials is highly recommended because of the nearly cost-free amplification of the intensity field. Installation of UVGI systems is as straightforward as it appears to be, once the sizing problem is solved. Optimization of the performance of UVGI systems is a topic that is treated elsewhere (Kowalski 2001).

The subject of sizing UVGI systems would not be complete if the filtration component was ignored, and Sec. 8.4 presents this topic in more detail. Like filtration, UVGI can also be used in recirculation units placed inside of rooms. The effectiveness of room recirculation units depends largely on the airflow rate or air change rate, the degree of air mixing, unit location, and local room airflow patterns. These matters should be considered carefully when adding a UVGI recirculation unit to a room. Figure 8.28 shows an example of the reduction of airborne TB bacilli resulting from the use of a recirculation UVGI unit inside an enclosed space (Allegra et al. 1997). Although TB is destroyed at a high rate by UV exposure, only the air passing through the unit is disinfected, and so the room volume is disinfected at a rate that depends on the unit airflow, as well as other factors.

8.4   Combining UVGI and Filtration

UVGI and filtration are mutually complementary technologies. Filtration removes most of the microbes that tend to be resistant to UVGI, and vice versa. This is because the larger microbes, like spores, tend to be hardier. Figure 8.29 shows the passage of a select group of BW agents through a prefilter, an 80 percent filter, and a UVGI system. Notice how the microbes in the most penetrating particle size range of the filters tend to be susceptible to UVGI. Likewise, UVGI destroys small microbes like viruses that may be difficult to filter out, while filtration easily removes spores, which tend to be resistant to UVGI (Kowalski and Bahnfleth 2000a).

Not all the microbes in Appendix A can be used in this example because most UVGI rate constants for microbes remain unknown. It can be reasonably assumed, however, that most viruses will succumb to UVGI exposure and most spores will be taken out by the filters; therefore, combination systems offer an ideal solution.

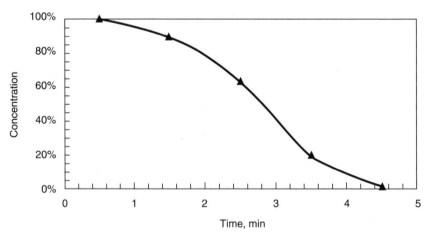

Figure 8.28 Effect of a recirculating UVGI system on airborne *Mycobacterium tuberculosis*. [*Based on data from Allegra et al. (1997).*]

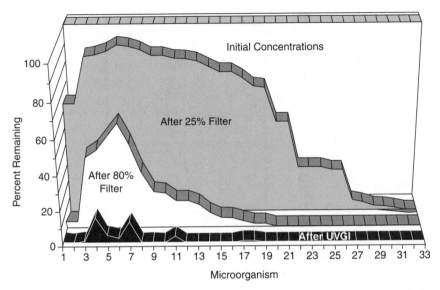

Figure 8.29 Microbial populations before and after filters and a UVGI system. Only microbes with known UVGI rate constants were included here, ordered in size from smallest (1) to largest (33), including many BW agents.

Such a combined filtration and UVGI system can be 'tuned' to target the microbes of concern for any particular facility, and to achieve any desired level of disinfection, up to and including sterilization. In this example, tuning the system to remove all microbes completely could be accomplished by (1) decreasing the airflow rate, (2) increasing UV power, or (3) changing the filter to a more efficient model. Economics would likely dictate the choice of which method or methods to use, and basic techniques of economic optimization can be used to seek the perfect solution.

The combination of UVGI and filtration is as useful against BW agents as it is against pathogens in general. Figure 8.30 shows the reduction of BW agents as they pass successively through a prefilter, an 80 percent filter, and a UVGI system. The pathogens shown are those for which a UVGI rate constant has been established or estimated.

Figure 8.30 suggests that a basic system composed of UVGI and an appropriately sized filter should be sufficient to remove most airborne BW agents. Given any inlet concentration, it is possible to compute the exit concentration. What is not known with certainty, however, is the exact dose of most of these agents that would be lethal or infectious. Approximations of the lethal doses were addressed in Chap. 5 and are modeled in Appendix B.

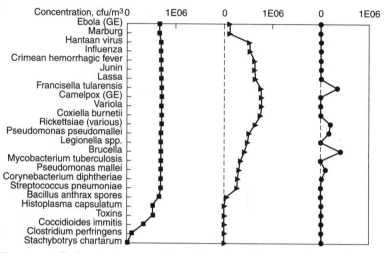

Figure 8.30 Reduction of BW agents passing through filters and a UVGI system.

Combining the filter removal rate with the UVGI kill rate is a simple algebraic process. The population that penetrates the filter is subject to the kill rate of the UVGI system. Figure 8.31 illustrates the process, in which the removal rate or filter efficiency is termed a "kill rate" to conveniently identify it as the same type of process as in the UVGI system. The term S_0 defines the initial population, while S_n denotes the survivors after each step.

The total combined kill rate, KR_T, is then

$$KR_T = \frac{S_0 - S_2}{S_0} = \frac{1000 - 150}{1000} = 85\% \tag{8.22}$$

In mathematical form, we can write the combined total survival for the example in Fig. 8.31 as follows:

$$S_2 = (1 - KR_1)(1 - KR_2)S_0 \tag{8.23}$$

If $S_0 = 1$, then the total survival for three systems in series is

$$S_3 = (1 - KR_1)(1 - KR_2)(1 - KR_3) \tag{8.24}$$

This pattern simply repeats itself for as many systems as there may be, and the total kill rate is

$$KR_T = 1 - S_T \tag{8.25}$$

where S_T = total survival.

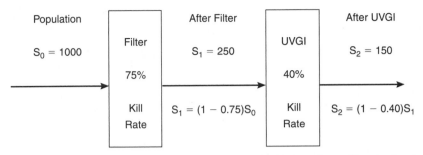

Figure 8.31 Schematic for computing total combined kill rate when a filter is placed in series with a UVGI system.

8.5 Gas Phase Filtration

Gas phase filtration is the removal of gases or vapors by any equipment designed for this purpose. Most gas phase filtration systems involve the use of activated carbon for adsorption, or activated carbon impregnated with various compounds. Some of these compounds enhance the adsorption of gases and vapors, or catalytically destroy them. Although forms of carbon adsorption have been successfully used for decades, a number of gas phase removal systems have recently been developed that provide new options; however, little information is available for most of these technologies in relation to removal of CBW agents. As a result, only those technologies that are reasonably well understood are discussed here.

Figure 8.32 illustrates the breakdown of the more common technologies used for gas phase filtration. They can be subdivided into carbon adsorber–type equipment and noncarbon equipment. This breakdown is by no means comprehensive—a number of technologies are currently under development that have great potential for removing gaseous contaminants, but little is known about their actual performance.

Carbon adsorption is used primarily for the removal of gases and vapors. It is effective against volatile organic compounds (VOCs) but is not used for the control of airborne dust or microorganisms. It is, in fact, not advisable to use carbon adsorption where particulate matter is present and may clog the adsorbent bed. Carbon adsorption depends on the use of materials like activated charcoal that possess an enormous amount of surface area per unit mass. The presence of this large surface area allows gas molecules to adhere to the surface through natural attractive forces at the molecular level (VanOsdell and Sparks 1995).

The attractive force that causes gas molecules to adsorb to the surface of solid carbon, the van der Waal's force, is small, but when the total surface area available for adsorption is large, high rates of removal are possible. The main caveat is that the larger the surface

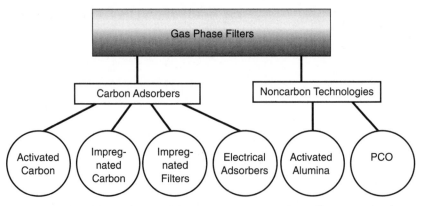

Figure 8.32   Breakdown of some of the more common gas phase filtration technologies.

area needed, the higher will be the pressure loss through the adsorber unit. If necessary, this can be compensated for in the same way as is done for filters—expanding the total face area.

Carbon adsorbers are used in the heating, ventilating, and air conditioning (HVAC) industry for controlling odors and VOCs. The most common form of activated carbon used is 60 percent minimum activity level, for which the internal surface area is approximately 1000 to 1100 m^2/gram. This form of carbon is either 6/8 mesh (3 mm) pellets or 6/12 mesh in granular form. These pellets or granules are typically used for filling perforated metal containers. Some adsorbers are available in which the pellets are processed into a solid media, such as the one shown in Fig. 8.33.

Carbon adsorbers might have some effect on small viruses near 0.01 micron in diameter, but the activated carbon pore size is typically 0.001 micron, or about 10 times too small to accommodate the smallest virus. If the pore size were on the order of 0.05 micron, it might have some effect in the removal of airborne viruses, but this possibility has not been studied.

Carbon adsorption has been reported capable of removing the botulinum toxin in addition to removing gaseous contaminants. It can also be processed to manipulate the pore size and remove large molecules (Tamai et al. 1996). It has a demonstrated ability to remove pesticides, which are chemically similar to number of chemical weapon agents (Cheremisinoff and Ellerbusch 1978).

The several types of gas phase air filters available today include carbon adsorbers or granulated activated carbon (GAC) (Fig. 8.34), activated alumina impregnated with potassium permanganate, and carbon impregnated with other compounds to improve performance against specific contaminants (VanOsdell and Sparks 1995). There are

Figure 8.33 Cutaway model showing carbon media located behind filter media. (*Model provided courtesy of D-Mark, Inc., Chesterfield, MI.*)

also filters impregnated with carbon or related compounds to give them the ability to remove gaseous contaminants. Photocatalytic systems are also capable of gas phase filtration, but insufficient data are available to quantify their performance against any CBW agents.

There are, in addition, types of carbon adsorbers that utilize electrical currents for both the destruction of adsorbed compounds and regeneration purposes. Finally, there are adsorbers that use noncarbon material such as alumina. Although these are not properly classified as carbon adsorbers, they perform the identical function and so are included here.

The bed depths of commercial carbon filters range from 1 to 8 cm with nominal residence times of 0.025 to 0.1 s. Carbon adsorbers are normally tested with contaminants with concentrations from 400 to 4000 mg/m^3. Although such levels are comparatively high for normal indoor contaminants, they are basically appropriate for levels that may be seen following an indoor chemical agent release.

The adsorption of a CW agent onto carbon is a complex function of pore size, CW agent molecule size, relative humidity, temperature, and residence time. Computer models may be available for sizing carbon adsorbers, but catalog sizing methods are commonly used to select units for specific applications. The percentage removal of a chemical agent is approximately an exponential function of residence time, as shown in Fig. 8.35 for chlorine.

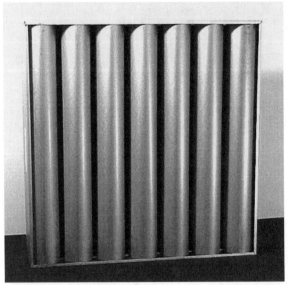

Figure 8.34 Modular carbon adsorber unit with metal framing. Round aluminum sections are filled with activated carbon and perforated to allow air passage. Opposite (exhaust) side has an integral air filter. (*Unit provided courtesy of Donaldson Company, Inc., Minneapolis, MN.*)

The basic procedure for sizing a carbon adsorber is to first define the air velocity and then identify the agents targeted for removal. Air velocities will typically be predefined by the system. The exact type of carbon adsorber selected is generally dependent on the agents to be removed. High levels of performance against CW agents may not be possible yet with GAC systems, but the rates of removal possible by typical carbon adsorbers may be sufficient to mitigate the effects of a chemical release.

The time a carbon adsorber can operate before it becomes overloaded and has to be regenerated or replaced is defined by the breakthrough time. No general computing method can be provided for the various types of carbon adsorbers, but the basic procedures for determination of breakthrough times for standard GAC systems are treated in the *ASHRAE Handbook of Applications* (ASHRAE 1991).

The choice of target agents is limited by the fact that few chemical weapons agents have been tested against carbon filters. Those that have been tested and have known properties are identified in Table 8.6, along with some related compounds and simulants.

The following parameters are necessary for the sizing of a carbon adsorber:

- The air velocity

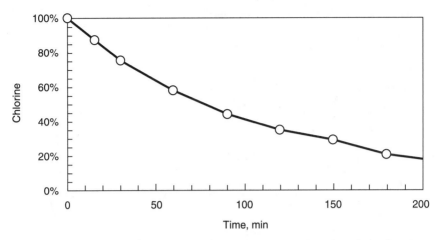

Figure 8.35 Removal of chlorine from water by a carbon fabric. Based on data from Martin and Shackleton (1990).

- The air temperature
- The agent to be adsorbed
- The concentration of the agent
- The maximum allowable downstream concentration

Although the adsorber is presumably designed to deal with a single chemical release incident, additional data on the regeneration of the adsorber are necessary in order to maintain the adsorber's capacity since it will decrease over time from adsorbing pollutants in the air. Carbon adsorbers are normally regenerated by passing a stream of hot air through them to remove moisture or destroy any adsorbed contaminants. In one type of carbon adsorber, an electrical charge is passed through the carbon to regenerate it (BioChem 2002). In the latter system, the carbon is packaged as a solid medium and during regeneration the airflow is reversed as it is charged.

For additional information on carbon adsorption, the References can be consulted. For assistance in sizing a carbon adsorber, or any gas phase filtration equipment, it is recommended that the manufacturer of that equipment be contacted for assistance. Recirculation carbon adsorber units, such as in Fig. 8.36, are presized for room volumes. Until such time as data become available on the performance of GAC systems against CW agents, or until an effective removal technology is developed, there is little more technical information that can be provided here on carbon adsorbers that would be of much assistance. There is clearly a need for more research, and for some major advances, in the field of gas phase filtration.

TABLE 8.6   Some CW Agents Removable by Carbon

CW agent or simulant	Adsorption at 20°C in dry air	Reference
Chlorine	10–25%	ASHRAE 1991
Chloropicrin	20–40%	ASHRAE 1991
Phosgene	10–25%	ASHRAE 1991
Sulfur dioxide	15%	ASHRAE 1991
Sulfur trioxide	15%	Cheremisinoff and Ellerbusch 1978
Nitrogen dioxide	10%	Cheremisinoff and Ellerbusch 1978
Arsine	Low	ASHRAE 1991
Cyanogen chloride	Low	ASHRAE 1991
Hydrogen chloride	Negligible	ASHRAE 1991

Figure 8.36 A carbon adsorber recirculation unit. (*Image courtesy of Electrocorp, California.*)

8.6   Summary

In summary, this chapter has presented detailed methods for sizing filters and UVGI systems for the purpose of removing biological agents from airstreams. The subject of gas phase filtration with carbon adsorbers has been presented, but without a specific sizing methodology due to the limited information on how GAC performs against chemical agents. GAC units can be sized and installed per manufacturers' recommendations, but the prediction of their performance against CW agents will have to await new information as this field of study develops. This is not necessarily a problem because (1) GAC will remove some or most chemical agents to a greater or lesser degree, (2) filtration and dilution ventilation may contribute to removing some degree of liquid aerosols, and (3) chemical agents may be less of a threat to large buildings than biological agents.

9

Simulation of
Building Attack Scenarios

9.1 Introduction

In order to design a CBW defense system for a particular building, it is necessary to establish some criteria for minimum performance in terms of kill rates against BW agents and removal rates against CW agents. Although it might be desired to design systems with 100 percent kill or removal rates, this is neither a practical nor a feasible approach in most cases since reduction of building airborne concentrations is limited by the ventilation system characteristics. Furthermore, designing near the high end of performance often entails prohibitive costs.

Instead of using performance criteria that require 100 percent removal of all CBW agents, it is more practical to seek the level of performance that will provide maximum protection for building occupants. This is a way of balancing the risk versus the cost of a system and can be accomplished through simulating building attack scenarios. The details of system modeling for dilution ventilation, filtration, UVGI, and carbon adsorption have been addressed in the previous chapters. The performance of air-cleaning systems, in terms of their removal rates, can be incorporated into the dilution ventilation model to assess the overall removal rates of a building.

Given the overall removal rates of CBW agents in an integrated building system, the casualties and fatalities can be estimated before and after inclusion of the air-cleaning systems. The baseline condition is the building without air cleaning, or without additional air cleaning if it already has an air-cleaning system. The baseline condition can then be compared with various combinations of the air-cleaning technologies incorporated to determine the reduction of casualties and fatalities. If the reduction appears insufficient, the components can be scaled up until some acceptable level of performance is achieved. This process is iterative and is illustrated in Fig. 9.1.

Once a simulation model is set up, the CBW defense system size, and therefore total cost, becomes dependent on the criteria used to judge the acceptability of the reduction of casualties or fatalities. This is, of course, a judgment call and must be weighed against the perceived risk. As it turns out, however, the system cost is far less sensitive to maximizing reduction of fatalities in the building than it is to maximizing air-cleaning system performance. That is, achieving protection for 90 to 99 percent of building occupants is much easier to do than achieving a 90 to 99 percent reduction of the CBW agents. This is simply because it is not necessary to achieve a 90 to 99 percent reduction of airborne CBW agents to protect building occupants, only to reduce the indoor concentrations to a safe level. The methodology presented

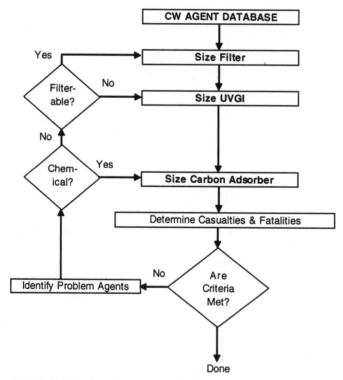

Figure 9.1 Flowchart for sizing and optimizing a CBW defense system.

here for simulating building attack scenarios will allow designers and engineers to seek system component sizes that will reduce overall costs and provide almost any level of desired protection.

9.2 Baseline Building Attack Scenarios

A variety of buildings and ventilation systems have been identified in Chap. 6, along with the basic attack scenarios. In this section, a baseline building and attack scenario is defined and the techniques of simulating system performance are detailed. This model can then be adapted to address other buildings and scenarios.

The primary building type used in this analysis is the multistory office building with forced ventilation, as shown in Fig. 9.2. The model building shown here has 50 stories with 1524 m² of floor area, but can be scaled to any number of floors and any floor area. The ventilation system consists of an air-handling unit with a supply air duct, return air duct, and an outside air intake. The outside air flowrate is 15 percent of the total volumetric airflow.

The design basis attack scenario is the worst-case scenario discussed in Chap. 6 for office buildings—the covert gradual release of CBW agents in the air-handling unit downstream of the filters. The release

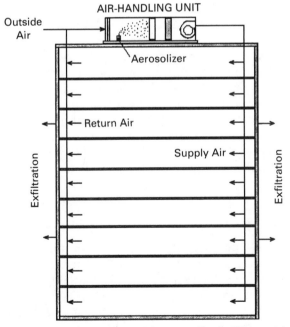

Figure 9.2 Schematic of a multistory office building used in the attack simulation.

occurs gradually for 8 h. Releases in other building or duct locations will produce fewer fatalities for the reasons discussed previously. This will be true for most buildings except for stadiums, malls, and similar large facilities with high occupancies in general areas, in which case a general area release might produce equal or higher fatality rates. In this building, the model can be adapted to examine all cases to ensure that the worst-case scenario for the particular facility is addressed.

The design basis microbes used in this analysis are the three described previously as being the most representative of the full array of potential biological agents—anthrax spores, smallpox, and TB bacilli. If it is desired to focus on other specific pathogens, or to examine all possible pathogens, it is a simple matter to include their properties when determining the kill rates for the air-cleaning components.

Chemical releases are not evaluated here in terms of the performance of gas phase filtration systems since this is not generally quantifiable. An example is provided, however, of the removal rates of chemical agents by 100 percent outside air systems. In the event information becomes available on the removal rates of CW agents by any particular gas phase filter, the model is easy to adapt to such analyses.

The design basis chemicals are more difficult to assign than BW agents for a number of reasons. The chemicals may be removed by the carbon adsorber at rates that vary considerably. In at least one case, the carbon may even convert the CW agent to another dangerous CW agent, depending on conditions (Lewis 1993). Three design basis CW agents are suggested on the basis of their availability and a knowledge of their properties and lethality—chlorine, sarin, and phosgene.

In order to begin a simulation, the quantity of agent to be released must be established. It is assumed the agent will be released gradually over an 8-h period, as this will maximize casualties. It is further assumed that the fatality rate achieved will be 99 percent of building occupants at the end of this 8-h period. For any given agent, the quantity released can be manually adjusted to achieve this fatality level for the baseline condition. This procedure will become clear after the model is developed.

9.3   Simulation of CBW Attack Scenarios

Two approaches can be used to develop the simulation—a minute-by-minute computation using a spreadsheet, or a programmed approach using software. Some software packages are currently available to perform such simulations, the most notable being the CONTAMW program (NIST 2002). The CONTAMW program is capable of modeling a building and its ventilation system in considerable detail, including modeling of wall and door leakage and wind conditions, while the spreadsheet method is simplistic, but adequate for buildings that can be approximated by single zones. Both methods will be presented here, and it will be shown that the differences between these modeling techniques can be relatively minor for simple systems in terms of the resulting CBW system sizes. For complex systems in which factors like the leakage between zones, outside weather conditions, elevator shafts, etc., are included, CONTAMW or similar advanced computer models should be used.

For single-zone modeling, the spreadsheet approach offers considerable flexibility and simplicity, while the CONTAMW model has the ability to evaluate building contaminant processes in considerable detail and has been extensively verified. After the following development of the spreadsheet model, the results will be compared with the CONTAMW model for verification.

The spreadsheet method previously detailed in Chap. 7 for dilution ventilation modeling provides the basic model of the building and ventilation system. This model can be expanded to include first a filter and a UVGI system, and later the carbon adsorber.

Table 9.1 summarizes the input data for a building attack scenario. The essential parameters defining the building are the floor area, the floor height, and the number of floors. The essential parameters defining the ventilation system are the total flowrate and the outside air percentage. The design basis BW agent used in the simulation can be anthrax, smallpox, or TB bacilli. The air-cleaning technologies are filters and ultraviolet germicidal irradiation (UVGI), and they are defined by a nominal DSP filtration rate and the average UV intensity and exposure time, respectively.

Table 9.1 reflects the options for inputting filter types and UVGI system size. In this particular case, filter size 2 is selected, which represents a nominal 40 percent DSP filter. For the UVGI system, an average intensity of 2000 $\mu W/cm^2$ with an exposure time of 0.25 s is selected.

For each agent, the filter must have a defined filtration rate, and the UVGI system must have a defined kill rate. The latter is determined by the methods previously detailed in Chaps. 2 and 8. Table 9.2 summarizes the UVGI rate constants and the filter removal rates for each of the design basis pathogens.

Based on the input parameters of Tables 9.1 and 9.2, the remaining simulation parameters can be computed. Table 9.3 shows the computed values for return air flowrate, UVGI kill rate, baseline fatalities, and other values of importance.

The release rate of the BW agent needs to be specified as an input parameter. This cannot be known in advance since it is necessary to establish the baseline fatality rate for each agent. That is, the baseline release rate is that which produces a fatality rate of approximately 99

TABLE 9.1   Input Parameters for CBW Attack Simulation

	Input data		
Dilution	OA*	15	%
	Total airflow	8,216	m³/min
Filtration	Filter type (0–6)	2	0 = none
	1 = 25%	4 = 80%	
	2 = 40%	5 = 90%	
	3 = 60%	6 = 99.97% (HEPA)	
UVGI	Intensity	2,000	μW/cm²
	Exposure time	0.25	s
Building parameters	Floor area	465	m²
	Floor height	2.7	m
	Number of floors	50	
BW agent	BW number (1–3)	1	(Anthrax)
	Filter removal rate	13.6	%
	Release rate of agent	18,000,000	cfu/min

*OA = outside air.

TABLE 9.2   Pathogen Input Data for BW Attack Simulation

Pathogen	Bacillus anthracis	M. tuberculosis	Smallpox
Mean diameter, μm	1.118	0.637	0.224
UVGI rate constant, cm²/μW·s	0.000424	0.002132	0.001528
Lethal dose (LD_{50})	28,000	5	100,000
40% filter removal rate	0.136	0.057	0.024
60% filter removal rate	0.543	0.286	0.120
80% filter removal rate	0.955	0.765	0.370
90% filter removal rate	0.986	0.864	0.470
HEPA filter removal rate	1.000	1.000	0.999

Note: Anthrax airborne rate constant assumed as 4× the surface rate constant from Appendix A.

percent without filtration or UVGI. The release rate, in cfu/min, must be adjusted until the baseline fatality rate is achieved. Once this is done, the effect of filtration and UVGI can then be compared in terms of reducing the fatality rate. Obviously, the baseline release rate cannot be known in advance and depends on the model of the building and ventilation system being complete.

Table 9.4 shows the minute-by-minute calculations of the first few parameters used in the simulation for the first 10 min. Note that the first minute shows zero concentration throughout. The table shows the release rate, 18,000,000 cfu/min, which has been established by trial and error to produce approximately 99 percent fatalities without fil-

TABLE 9.3   Computed Data for BW Attack Simulation

OA* fraction	0.15	
Total airflow	290,148	cfm
Outside airflow	43,522	cfm
	1,232	m^3/min
Return airflow	6,984	m^3/min
UVGI dose	500	$\mu W \cdot s/cm^2$
BW agent	*Bacillus anthracis* spores	
BW agent diameter	1.118	μm
BW agent UVGI k*	0.000292	$cm^2/\mu W \cdot s$
UVGI kill rate	0.1358	Fraction
Lethal dose (LD_{50})	28,000	cfu
4-h dose	7,151	
Release point	AHU*	
Floor area	5,005	ft^2
Floor height	9	ft
Floor volume	1,256	m^3
	44,336	ft^3
Estimated occupancy	2,176	Persons
Breathing rate	0.012	m^3/min
Suggested airflow	8,216	m^3/min
(function of floor area)	43,523	cfm
ACH*	1.18	
Filter removal rate	13.6	Fraction
Total building volume	62,775	m^3
	2,216,797	ft^3
Baseline fatalities	99	%
Fatalities with BW	12.30	
Filter efficiency	40	%
Fractional filter efficiency	0.4	

*OA = outside air; k = UVGI rate constant; AHU = air-handling unit; ACH = air change rate.

tration or UVGI. The next two columns show the number of microbes and the concentration inside the building. The next column, Microbes Exhausted, indicates how many microbes have been removed each minute by the purging action of the outside air under complete mixing. The remaining columns show the computations of the concentration and the number of microbes that enter the air-handling unit. This simulation assumes the microbes are either exfiltrated from the rooms or are forcibly exhausted in the air-handling unit inlet. The exact location of microbe exfiltration is not significant in this model.

TABLE 9.4 Simulation of Baseline Scenario—Part 1

Time, h	Time, min	Microbes released, cfu	Microbes in building, cfu	Building concentration, cfu/m³	Microbes exhausted, cfu	Microbes remaining, cfu	RA microbes, cfu	AHU in microbes, cfu/m³	AHU in microbes, cfu
0.00	0	0	0	0.00	0	0	0	0	0
0.02	1	18,000,000	18,000,000	286.74	353,376	17,646,624	2,002,466	244	2,002,466
0.03	2	18,000,000	35,646,624	567.85	699,815	34,946,808	3,965,619	483	3,965,619
0.05	3	18,000,000	52,946,808	843.44	1,039,453	51,907,356	5,890,232	717	5,890,232
0.07	4	18,000,000	69,907,356	1113.62	1,372,423	68,534,933	7,777,061	947	7,777,061
0.08	5	18,000,000	86,534,933	1378.49	1,698,855	84,836,078	9,626,848	1172	9,626,848
0.10	6	18,000,000	102,836,078	1638.17	2,018,880	100,817,198	11,440,319	1392	11,440,319
0.12	7	18,000,000	118,817,198	1892.75	2,332,621	116,484,576	13,218,188	1609	13,218,188
0.13	8	18,000,000	134,484,576	2142.33	2,640,204	131,844,373	14,961,155	1821	14,961,155
0.15	9	18,000,000	149,844,373	2387.01	2,941,748	146,902,625	16,669,903	2029	16,669,903
0.17	10	18,000,000	164,902,625	2626.88	3,237,371	161,665,254	18,345,105	2233	18,345,105

RA = return airflow; AHU = air-handling unit.

Figure 9.3 shows the total number of microbes in the building, the number of microbes exhausted, and the number of microbes in the return air for each of the first 10 min. The totals increase linearly, as would be expected for a constant release rate.

The total microbes in the building at any time, MIB_i, can be written as follows:

$$MIB_{i+1} = MIB_i + MR_i - ME_i - MF_i - MU_i \qquad (9.1)$$

In Eq. (9.1), the quantities defined at each minute or time period i are as follows:

MR = microbes released each minute (release rate)

ME = microbes exhausted by ventilation purging

MF = microbes filtered out

MU = microbes killed by UVGI

The units used in Eq. (9.1) are arbitrary and can be either cfu (spores, bacteria, or virions) or concentrations of agents. Spores like anthrax can be equally well modeled with units of $\mu g/m^3$ as with units of spores/m^3 or cfu/m^3. For chemical agents the units $\mu g/m^3$ should be used.

The dose inhaled each minute, or time period i, as described previously in Chap. 5, is a function of the breathing rate as follows:

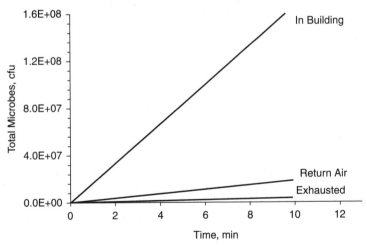

Figure 9.3 Total microbes at each point in the system. From the data in Table 9.4.

$$\text{Dose}_i = \sum_{n=1}^{i} E_i B_r C_i \qquad (9.2)$$

In Eq. (9.2), the parameters are defined as follows:

E_i = exposure time, or time period i, s or min

B_r = breathing rate (typically 0.005–0.032 m³/min)

C_i = concentration in air at time i, cfu/m³ or μg/m³

The remaining columns of the spreadsheet model are shown in Table 9.5. These columns show the effect of the air-cleaning components, compute the supply air concentration of microbes, and estimate the building fatalities. In this example, the baseline condition, there is no filter or UVGI system and so the supply air (SA) concentration is identical to that of the air-handling unit inlet shown in Table 9.4. The sum total of microbes removed by the purging effect of the outside air is then subtracted from the total number of microbes in the building (the fourth column in Table 9.5). The number of microbes in the building is then used to compute the building airborne concentration (the fifth column in Table 9.4). The calculational process is repeated for each subsequent minute using the previous minute's concentration.

The total number of spores inhaled each minute, shown in Table 9.5, is computed from the breathing rate at the given concentration. The dose of total inhaled spores is then computed by summing the total spores inhaled. The percent fatalities are then computed using the lethal dose equation presented in Chap. 5.

Tables 9.4 and 9.5 represent the complete calculations for simulating the baseline of a building attack scenario. The results of this scenario can be checked by computing the steady-state concentration, called SSC, as follows:

$$\text{SSC} = \frac{\text{RR}}{\text{OA}} \qquad (9.3)$$

In Eq. (9.3), RR is the release rate of the microbes (in cfu/min or μg/min) and OA is the outside air flowrate in m³/min or cfm. If steady state is reached, and this should be obvious from a graph of the data, then the steady-state concentration must equal the value computed by Eq. (9.3).

Given the baseline spreadsheet model as detailed above, the release rate that produces 99 percent fatalities can be sought by trial and error. It can also be determined easily by use of goal seeking or other iterative spreadsheet functions. The release rate is simply adjusted until

TABLE 9.5 Simulation of Air-Cleaning Components and Fatalities—Part 2

Filtered out, cfu	UVGI in microbes, cfu	UVGI killed, cfu	UVGI out microbes, cfu	Microbes removed, cfu	SA microbes, cfu/m³	Total spores inhaled	Dose total inhaled	% total fatalities
0	0	0	0	0	0	0	0	0.1
0	2,002,466	0	2,002,466	353,376	244	3	3	0.1
0	3,965,619	0	3,965,619	699,815	483	7	10	0.1
0	5,890,232	0	5,890,232	1,039,453	717	10	20	0.1
0	7,777,061	0	7,777,061	1,372,423	947	13	34	0.1
0	9,626,848	0	9,626,848	1,698,855	1172	17	50	0.1
0	11,440,319	0	11,440,319	2,018,880	1392	20	70	0.1
0	13,218,188	0	13,218,188	2,332,621	1609	23	93	0.1
0	14,961,155	0	14,961,155	2,640,204	1821	26	118	0.1
0	16,669,903	0	16,669,903	2,941,748	2029	29	147	0.1
0	18,345,105	0	18,345,105	3,237,371	2233	32	179	0.1
0	19,987,420	0	19,987,420	3,527,192	2433	34	213	0.1

SA = supply air.

fatalities in this range are achieved. In a goal-seeking routine, the variable to be adjusted is the fatality rate, the value it is to be adjusted to is 99 percent, and the variable to be manipulated is the release rate of the agent.

The reason that a 99 percent fatality rate is used in place of a 100 percent fatality rate is that an asymptotic approach of the fatality rate toward 100 percent may greatly skew the results, and, furthermore, the release rate might be unintentionally estimated many times higher than necessary. The actual value chosen for the upper limit of lethality, whether 98, 99, or even 99.9 percent, is arbitrary. This does not preclude performing an analysis in which much more of the agent is used than is necessary; it merely provides a baseline for understanding the results.

Once the release rate is determined, the simulation of filter and UVGI components can be accomplished by duplicating the entire spreadsheet and adding columns for the component removal rates. It is necessary to duplicate the spreadsheet so that the results can be compared with the baseline condition. There are other ways of handling this, but creating two separate spreadsheets allows for simulations that are fully automatic. That is, with two spreadsheets linked together, the change of a single variable will automatically produce a full analysis and allow an instant comparison of results.

Table 9.6 shows the results when a filtration rate of 80 percent and a UVGI kill rate of 8 percent (for anthrax spores) are used in the baseline condition shown in the previous tables. The number of microbes filtered out is computed based on the filtration rate, and this produces the number of microbes (in cfu per each minute) that enter the UVGI system, assuming the UVGI system is downstream of the filter. Next, the UVGI kill rate is used to determine the number of microbes that remain after the UVGI system. The new building concentration (for the next minute) can then be computed by subtracting the microbes purged, filtered, or destroyed by UVGI from the total building load of microbes. As before, the process is repeated each minute.

The results of the previous example are illustrated in the following graphs. Figure 9.4 shows the airborne concentration of anthrax spores in the baseline condition compared with the concentrations after the filter and UVGI system have been included. Obviously, there is a considerable reduction in building levels as a result of adding the air-cleaning components.

Figure 9.5 shows a comparison of the predicted fatalities in the baseline condition and those predicted after air-cleaning components have been added. Again, there is a significant difference between these scenarios. It would appear that these relatively modest components, a 40 percent filter and a low-power UVGI system, have a marked effect on

TABLE 9.6   Air-Cleaning Component Simulation—Part 2

Filtered out, cfu	UVGI in microbes, cfu	UVGI killed, cfu	UVGI out microbes, cfu	Microbes removed, cfu	SA microbes, cfu/m³	Total spores inhaled	Dose total inhaled	% total fatalities
0	0	0	0	0	0	0	0	0.1
1,601,973	400,493	54,404	346,089	2,009,753	42	3	3	0.1
3,025,081	756,270	102,733	653,537	3,795,111	80	6	10	0.1
4,289,294	1,072,324	145,667	926,657	5,381,129	113	9	19	0.1
5,412,355	1,353,089	183,807	1,169,282	6,790,063	142	12	31	0.1
6,410,022	1,602,506	217,688	1,384,817	8,041,685	169	14	45	0.1
7,296,297	1,824,074	247,786	1,576,288	9,153,561	192	16	60	0.1
8,083,617	2,020,904	274,524	1,746,380	10,141,292	213	17	78	0.1
8,783,030	2,195,757	298,277	1,897,481	11,018,740	231	19	96	0.1

SA = supply air.

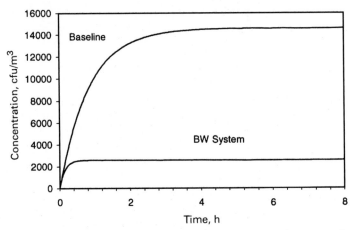

Figure 9.4 Comparison of results of building attack simulation example before (the baseline) and after (the BW system) addition of air-cleaning components.

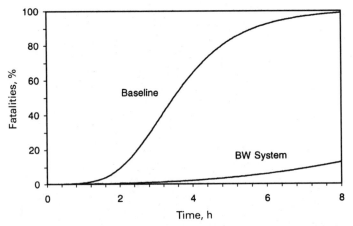

Figure 9.5 Comparison of predicted fatalities before (baseline) and after (BW system) addition of air-cleaning components.

the survivability of building occupants under this attack scenario. In fact, this seems to be generally the case—relatively low-cost components offer a high degree of protection when combined with an ordinary building ventilation system—but more will be said about this later.

In the above example, the attack scenario consisted of a quantity of anthrax spores released over an 8-h period into the return air duct of a 50-story building. The outside air exchange rate is 15 percent. The number of occupants is not relevant since we are predicting the percentage of infections. The breathing rate for office workers is taken as 0.032 m^3/min.

It should be reiterated here that treatment of the building as a single zone is an idealized analysis. The predicted fatalities could vary considerably for buildings that have high-occupancy zones in which a release occurs in a single zone only. Such building scenarios are best analyzed using a detailed CONTAMW model.

Table 9.7 summarizes the results of the attack simulation. Without air treatment, some 99 percent of building occupants would be likely to receive a fatal dose, while 100 percent would be infected. The presence of a 40 percent filter reduces the fatalities to 13 percent and the total infections drop to 80 percent. The combination of the filter and the UVGI system cuts the fatalities down to 12 percent and the infections to 78 percent. Although the UVGI system seems to make a negligible contribution when used in conjunction with the filter, this is partly due to the position of the UVGI system downstream of the filter and applies only to anthrax spores, which are UV-resistant. As will be seen later, the UVGI contribution to the reduction of fatalities in regard to agents like smallpox and TB can be significant. All three baseline microbes should be used to design a BW defense system so as not to overlook the advantages of either component, filtration or UVGI, in protecting against the entire array of pathogens.

Obviously the system could be improved by the addition of an 80 percent filter, or a 90 percent filter, or a higher-powered UVGI system. The choice of which components to add depends on many factors, but for a commercial office building both performance and economics will be factors in the selection of components. A high-efficiency particulate air (HEPA) filter, for example, would remove almost all anthrax spores in the supply air, but the cost of operating the air-handling unit with HEPAs in place might become prohibitive. These factors must be weighed in the consideration of designing building defense systems.

Building immunity could also be improved by increasing the outside airflow. This is another question that must be resolved in each case by evaluating the economics of system operation. There are a variety of high-efficiency air-to-air heat exchangers and energy recovery systems available today that could make even 100 percent outdoor air systems feasible. It is basically a matter of engineering to select the most appropriate system for any given budget.

TABLE 9.7 Building Model Predictions of Anthrax Fatalities

System	Inhaled spores	Infections, %	Fatalities, %
No filter	75,377	100	99
UVGI	44,785	100	84
40% filter	14,944	80	13
40% filter + UVGI	14,547	78	12

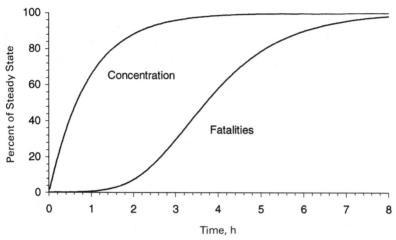

Figure 9.6 Combined graph of the predicted fatalities and the percent of steady-state concentration versus time. Note that the lag in fatalities in relation to concentration is a general characteristic of all exposure scenarios.

A convenient means of depicting the relation between the lethality curve and the concentration is to plot them together as a percentage of their steady-state values. The lethality is already given in terms of percent, and the concentration under steady state is computed from Eq. (9.3). Figure 9.6 shows these curves graphed together for the previous example. The nature of the curves is such that there will always be a lag between the predicted fatalities and the airborne concentration of the agent. The degree of this lag will depend on several factors, primary among which will be the operating parameters of the building ventilation system and the toxicity of the agent. The time lag also provides a convenient window within which to design detection systems.

9.4   Simulation with CONTAMW

The CONTAMW program, which was introduced in Chap. 8, is used here to analyze a 10-story building as both an example of the program and a corroboration of the spreadsheet model.

Figure 9.7 shows the layout of one floor of the building model. The symbol © indicates the contaminant source. This source must be located inside the rooms since it cannot be placed inside the air-handling unit in CONTAMW. In order to simulate a release in the air-handling unit, each room in a multizone building must have a source proportional to its supply airflow. The center zone in Fig. 9.7 represents an elevator shaft. The symbol with the arrows represents an air-handling unit—only one of these is used, and it is placed on any arbitrary

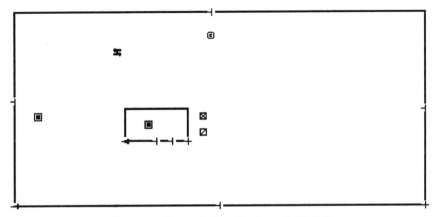

Figure 9.7   CONTAMW layout of a single floor for the model building.

floor. The remaining symbols represent either zone numbers (room numbers) or the supply and exhaust airflow registers.

The ability of CONTAMW to model elevator shafts can be of great value in achieving realism and in modeling certain scenarios. Elevator shafts provide a direct connection between all floors, and a leaky shaft can cause a wider distribution of contaminants from a single contaminated zone. Furthermore, a release that occurs in the elevator shaft is a scenario that may be worth investigating for any tall building.

All the parameters in the CONTAMW building model are as stated in Table 9.8, which is the control panel for the spreadsheet model. The differences between the spreadsheet model and the CONTAMW model are relatively minor, such as the leakage paths through the walls and the elevator shaft, but these do not have a significant effect on these results since it is effectively a single-zone model and no infiltration is considered. The baseline model analyzed here does not contain any filters or any UVGI system.

Operation of the CONTAMW program is beyond the scope of this chapter and information on this subject can be found elsewhere, especially in the CONTAMW help files. It is sufficient, for the purposes of this example, to describe the input and the output of the program. Readers should refer to the extensive body of information and publications for further information on this public domain software package (NIST 2002).

Output of the CONTAMW program consists of various charts and data that may be selectively printed or sent to file. The most useful data are obtained by exporting a data file of contaminant concentrations and reading this into a spreadsheet. The output is configured for units of mass concentration, and therefore they may have to be converted for biological agents. As an example, the mass of anthrax is

TABLE 9.8   Building Attack Simulation Control Panel

	Input			Data		
Dilution	OA	15	%	OA fraction	0.15	
	Total airflow	7,068	m³/min	Total airflow	249,606	cfm
		117.8	m³/s	...per floor	11.78	m³/s
	Suggested airflow (function of floor area)	7,068	m³/min	Outside airflow	37,441	cfm
		37,439	cfm		1,060	m³/min
	ACH	1.06		Return airflow	6,008	m³/min
Filtration	Filter type (0–6)	3		Anthrax removal rate	0.544	Fraction
	Nominal filter efficiency	60	%	Smallpox removal rate	0.287	Fraction
	Filter removal rate	0.54351	Fraction	TB removal rate	0.120	Fraction
UVGI	UVGI size (0–6)	4		Intensity	1,000	μW/cm²
	Exposure time	0.5	s	UVGI dose	500	μW·s/cm²
BW agent	BW number (1–8)	1		BW agent		*Bacillus anthracis*
	Release rate of agent	55,000,000	cfu/min	BW agent diameter	1.118	μm
		44.267	μg/min	Mass of microbe	8.04853E-07	μg
	Breathing rate	0.012	m³/min	BW agent UVGI k	0.000424	cm²/μW·s
	Baseline fatalities	99.00	%	UVGI kill rate	0.1910	Fraction
	Fatalities with CBW	23.91	%	Steady state	51,877	cfu/m³
	Lethal dose, LD$_{50}$	28,000	cfu	Steady state	0.041753	μg/min
Building parameters	Floor area	2,000	m²	Floor area	21,527	ft²
	Floor height	3	m	Floor height	10	ft
	Number of floors	10		Floor volume	6,000	m³
	Total building volume	60,000	m³		211,880	ft³
		2,118,803	ft³	Estimated occupancy	1,872	persons

OA = outside air; ACH = air change rate; k = UVGI rate constant.

computed in Table 9.9 based on a spore density 1.1 times that of water
(Bakken and Olsen 1983; Bratbak and Dundas 1984; Poindexter and
Leadbetter 1985; van Veen and Paul 1979), and by assuming a perfectly
spherical spore (see Fig. 9.8). Similar calculations can be performed for
other microbes even when they are nonspherical. Bacterial density can
vary slightly from spore density, but a value of 1.1 times that of water
should be representative.

Figure 9.9 shows graphic output from the CONTAMW model for a
10-story building. It can be observed that the steady-state concentration of approximately 0.000335 $\mu g/m^3$ is achieved after a few hours of
contaminant release, beginning at hour 6 (or 6 a.m.). This graph also
shows the decay of the contaminant concentration for several hours
after the release ends.

Table 9.10 shows the output of the CONTAMW simulation for the
first 10 min. The first four columns are the exported results of CON-
TAMW. The remaining columns represent the computations of fatalities, including conversion of the anthrax concentrations from units of
kg/kg (kg of anthrax spores per kg of dry air) into airborne concentrations in cfu/m³. The fatalities are computed by the same lethal dose
equations used in the previous example, and which were detailed in
Chap. 4.

Figure 9.10 shows a graphic comparison of the spreadsheet model
with results of the CONTAMW simulation of an anthrax attack on a 10-
story building. The results are virtually identical, although minor differences exist in the levels in the first hour. The two lines cannot be
distinguished except on a larger scale. The reason the results are so
similar is that both models are essentially identical single-zone models,
with the CONTAMW model including an elevator shaft and leakage
across walls, which did not contribute significantly to changing the
results. These results corroborate the spreadsheet model, which can
now be used for further simulations.

TABLE 9.9 Mass of Anthrax Spores

Mean diameter of spore	1.118	μm
	0.000001118	m
Volume of spherical spore	7.31685E-19	m³
Spore density	1100	kg/m³
	8.04853E-16	kg
Mass of spore	8.04853E-13	g
	8.04853E-07	μg

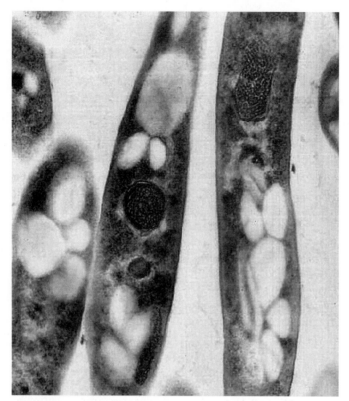

Figure 9.8 Electron micrograph of *Bacillus anthracis* showing a near spherical spore (dark sphere in center of bacilli). [*Image reprinted courtesy of Centers for Disease Control (PHIL 1824, provided by Dr. Sherif Zaki and Elizabeth White).*]

9.5   Simulation of Several Model Buildings

The spreadsheet model developed in the previous sections and verified by comparing results with those of the CONTAMW program can now be used to analyze a variety of buildings from small to large. Six basic building types have been selected for analysis, ranging from a single-room auditorium (building 1) to a 50-story office building (building 6). For each building, a baseline release rate is developed for anthrax spores and for smallpox.

The reduction in fatalities from the baseline rate, 99 percent, is then computed for each of four air-cleaning systems. Table 9.11 shows the results for the first three building types and lists the air-cleaning systems that are being tested against the building baseline attack scenario. The smallest system consists of a MERV 9 (40 percent DSP) filter aligned with an URV 10 UVGI system that produces a dose of 250 μW·s/cm². The latter URV rating produces an average intensity of 500

CONTAM Project: 10story.prj
Date: Jan1

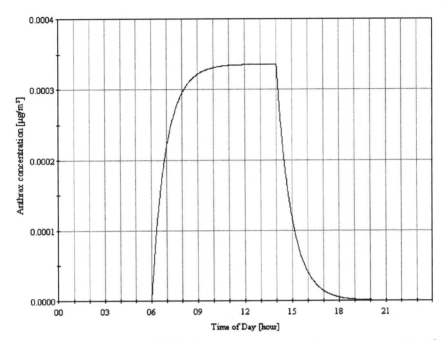

Figure 9.9 Output from CONTAMW program showing airborne concentrations of anthrax spores in a 10-story building.

μW/cm^2 for an exposure time of 0.5 s. Any UVGI system that produces the same dose will have the same kill rate regardless of the exact intensity or exposure time. This may not be true in extreme cases, but it is a good approximation for typical systems.

The next three air-cleaning systems are scaled up accordingly, with the most powerful system consisting of a MERV 15 (90 percent DSP) filter and an URV 15 UVGI system. Table 9.11 shows the fatalities that would result if each of these systems were in place and used against either anthrax spores or smallpox. The lower the predicted fatalities are, the better is the system performance.

The types of buildings shown in these simulations include an auditorium, an apartment building, a government building, and commercial buildings. The occupancy levels differ between these buildings, and this is what determines the required total airflow rate, which is based on approximately 0.566 m^3/min (20 cfm) per person.

Table 9.12 shows results for the same simulations involving large buildings. In both this and the previous table, it can be observed that the reduction in fatalities levels off between the third system and the

TABLE 9.10 CONTAMW Simulation Results

	CONTAMW output			Converted results and fatalities				
Level: date	Time of day, a.m.	Zone: time, s	W anthrax, kg/kg	Inhaled spores	Total inhaled spores	Total fatalities, %	Building concentration, μg/m³	Building concentration, cfu/m³
1-Jan	6:00:00	21600	0	0	0	0.10	0	0
1-Jan	6:01:00	21660	4.89E-15	0	0	0.10	5.87E-06	907.4925
1-Jan	6:02:00	21720	9.7E-15	0	0	0.10	1.17E-05	1800.139
1-Jan	6:03:00	21780	1.44E-14	0	1	0.10	1.73E-05	2672.371
1-Jan	6:04:00	21840	1.91E-14	0	1	0.10	2.29E-05	3544.603
1-Jan	6:05:00	21900	2.36E-14	0	1	0.10	2.84E-05	4379.718
1-Jan	6:06:00	21960	2.81E-14	1	2	0.10	3.38E-05	5214.834
1-Jan	6:07:00	22020	3.25E-14	1	2	0.10	3.9E-05	6031.392
1-Jan	6:08:00	22080	3.68E-14	1	3	0.10	4.42E-05	6829.391
1-Jan	6:09:00	22140	4.11E-14	1	4	0.10	4.94E-05	7627.391
1-Jan	6:10:00	22200	4.52E-14	1	5	0.10	5.43E-05	8388.274

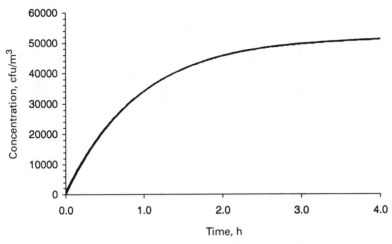

Figure 9.10 Comparison of CONTAMW results with the spreadsheet model. Deviations are so small that the two lines are practically indistinguishable.

fourth. That is, no significant increase in the reduction of fatalities occurs with the MERV 15 (90 percent DSP) filter combined with the URV 15 UVGI system over the next smallest system. Furthermore, it can be observed that the predicted fatality rates for any air-cleaning system converge to similar results for all three buildings. This should not be too surprising since all the systems are assumed to be operating under design air velocity, and all the buildings are assumed to have the same outside air flowrate and occupancies proportional to floor area. In real-world buildings, these parameters may not be proportional, and it is possible that kill rates and removal rates could vary widely from building to building.

The summary of predicted fatalities in the previous tables should provide the engineer with a reasonable range of system sizes to start with when designing a building air-cleaning system. These system sizes, in terms of filter MERV and UVGI URV ratings, may be sufficient to provide a reasonable level of protection against BW agent releases, but a detailed analysis should always be undertaken to verify system performance, since many variables can affect the way a system may operate in any actual applications. The addition of a filter, for example, may decrease the total system airflow and even affect the quantity of outside air. This change in operating system parameters may, in turn, affect the predicted fatality rates of the integrated system.

Figure 9.11 illustrates the trend that is apparent in the previous tables. This chart combines all simulation results for both anthrax and smallpox and shows the reduction in predicted fatalities as increased

TABLE 9.11   Predicted Fatalities for Medium Building Sizes

Building type	Auditorium		Apartment		Government	
Floors	1		5		10	
Floor area, m^2	8,000		1,000		2,000	
Total airflow, m^3/min	32,000		410		16,250	
Outside air	15%		15%		15%	
Occupancy	8,611		538		4,306	
Baseline fatalities	99%		99%		99%	
Predicted fatalities	Anthrax	Smallpox	Anthrax	Smallpox	Anthrax	Smallpox
MERV 9 (40% DSP)	84	95	87	96	85	95
URV 10—500 μW/cm^2	97	52	98	57	98	54
Combined	80	46	83	51	81	48
MERV 11 (60% DSP)	25	57	30	62	27	59
URV 11—1000 μW/cm^2	95	26	96	31	95	28
Combined	23	17	27	21	25	19
MERV 13 (80% DSP)	8	13	10	16	9	15
URV 13—2000 μW/cm^2	88	13	89	16	88	14
Combined	8	8	10	11	9	9
MERV 15 (90% DSP)	8	10	10	13	9	11
URV 15—4000 μW/cm^2	69	8	81	12	71	9
Combined	8	8	10	10	9	8

levels of filtration and UVGI power are added to the buildings. This chart makes it clear that the advantages of adding air-cleaning equipment level off after the third increment in size, which has a MERV 13 (80 percent DSP) filter and an URV 13 (2000 μW/cm^2) UVGI system. The conclusion can be drawn that there is no real advantage in using extreme levels of filtration or UVGI power for the types of buildings evaluated. This is called the "diminishing returns" principle. The reason for these diminishing returns is simply that only a fraction of the building air volume is being filtered or disinfected at any given time. The diminishing returns effect can be viewed as a property of building ventilation systems, and is not the result of any limitations of air-cleaning technologies.

The diminishing returns effect leads to the corollary that removal rates can be increased by increasing total system airflow. Such a system modification would involve increased costs in terms of fan power

TABLE 9.12   Predicted Fatalities for Tall Buildings

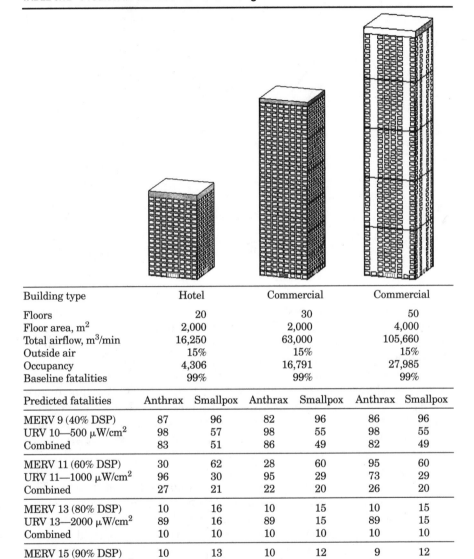

Building type	Hotel		Commercial		Commercial	
Floors	20		30		50	
Floor area, m^2	2,000		2,000		4,000	
Total airflow, m^3/min	16,250		63,000		105,660	
Outside air	15%		15%		15%	
Occupancy	4,306		16,791		27,985	
Baseline fatalities	99%		99%		99%	
Predicted fatalities	Anthrax	Smallpox	Anthrax	Smallpox	Anthrax	Smallpox
MERV 9 (40% DSP)	87	96	82	96	86	96
URV 10—500 μW/cm^2	98	57	98	55	98	55
Combined	83	51	86	49	82	49
MERV 11 (60% DSP)	30	62	28	60	95	60
URV 11—1000 μW/cm^2	96	30	95	29	73	29
Combined	27	21	22	20	26	20
MERV 13 (80% DSP)	10	16	10	15	10	15
URV 13—2000 μW/cm^2	89	16	89	15	89	15
Combined	10	10	10	10	10	10
MERV 15 (90% DSP)	10	13	10	12	9	12
URV 15—4000 μW/cm^2	81	12	72	11	72	11
Combined	10	10	9	9	9	9

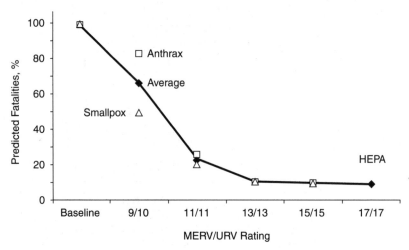

Figure 9.11 The diminishing returns effect for a release in the AHU: Increasing filter efficiency and UVGI power reduces predicted fatalities only up to a point.

and other effects on system performance. The performance of both the filters and the UVGI system would decrease, on a per unit airflow basis, if the air velocity through the air-handling unit were increased without increasing face area. However, these are factors that can be considered in an economic evaluation, since there must be some optimum level of performance. This is somewhat beyond the scope of current knowledge, however, and will be left as a matter for future research. It is potentially useful to note, however, that an emergency system with a high airflow rate and that uses filters and UVGI could be designed to increase removal rates far beyond those that are possible with normal building ventilation systems.

The previous scenario involved a release inside the building AHU, which is the most severe scenario but may not be the most likely. The outside air intake release scenario, with or without the use of an aerosolization device, is probably the most likely scenario due to ease of access. The analysis of this scenario uses the same methods as previously, and assumes an 8-h continuous release, but in this case two additional BW agents are analyzed, and the filter models used are based on actual MERV test data and have slightly different performance than the previous DSP-rated filters (for which the MERV ratings were only estimates).

Table 9.13 summarizes the MERV/URV performance against the four design basis BW agents: anthrax, TB bacilli, smallpox, and botulinum toxin. The performance of the MERV/URV combination is computed by the algebraic method shown previously in Chap. 8.

TABLE 9.13 Removal Rates for Design Basis BW Agents

Filter rating	MERV 6	MERV 8	MERV 11	MERV 13	MERV 15	MERV 16
Bacillus anthrax spores	21.7	39.2	56.7	95.6	98.6	100.0
TB bacilli	3.8	7.8	14.1	76.6	86.5	100.0
Smallpox virus	7.5	19.5	32.3	37.1	47.0	99.96
Botulinum toxin	43.9	67.5	85.8	99.9	99.997	100.0

UVGI system rating	URV 9	URV 10	URV 11	URV 13	URV 15	URV 17
Average intensity, μW/cm^2	250	500	1000	2000	4000	6000
Dose ($t = 0.5$ s), μW·s/cm^2	125	250	500	1000	2000	3000
Bacillus anthrax spores	2.1	4.1	8.0	15.4	28.4	39.4
TB bacilli	23.4	41.3	65.6	88.1	98.6	99.8
Smallpox virus	17.4	31.8	53.4	78.3	95.3	99.0
Botulinum toxin	Unknown	Unknown	Unknown	Unknown	Unknown	Unknown

Combined removal rate, %						
Bacillus anthrax spores	23.3	41.7	60.2	96.3	99.0	100.0000
TB bacilli	26.3	45.9	70.4	97.2	99.8	100.0000
Smallpox virus	23.6	45.1	68.5	86.4	97.5	99.9996
Botulinum toxin	43.9	67.5	85.8	99.9	99.997	100.0000

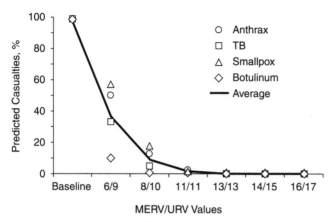

Figure 9.12   Predicted fatalities for the outside air intake release scenario in a 50-story building.

In the case of TB bacilli, only the infections can be predicted, and so Tables 9.11 and 9.12 refer to predicted casualties instead of fatalities. In the case of botulinum toxin, the UVGI rate constant is unknown and is assumed to be zero.

Results of the analysis of the continuous release in the air intakes in a commercial 50-story building are summarized graphically in Fig. 9.12. It can be observed that the diminishing returns effect occurs in this case also, except that the predicted fatalities converge to approximately zero. Again, it is evident that moderate levels of air cleaning, or about a MERV 13/URV 13 system, are sufficient to provide near total protection for building occupants. The other release scenarios, including the sudden release in the air intakes, and releases in general areas, sudden or continuous, produce similar results, but they are not as severe.

At this point it is of interest to examine the comparative performance of the MERV component versus the URV component in the above analysis. Figure 9.13 summarizes the reduction in casualties that can be ascribed to each component if operated in isolation for the previous scenario. In this case a MERV 11 filter is compared to an URV 11 UVGI system. It can be seen that across this range of three pathogens, the effect of each component counterbalances the deficiencies of the other, producing a "leveling" effect.

9.6   Multizone Simulation with CONTAMW

Considerable power and flexibility of analysis in the modeling of buildings and the simulation of attack scenarios are offered by programs

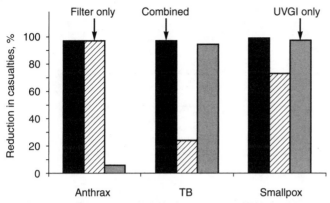

Figure 9.13 Comparison of the performance of a MERV 11 filter versus an URV 11 UVGI system when operated in isolation, for the previous analysis.

such as CONTAMW. With this program, it is possible to create a detailed building model that reflects actual operating data or test data on building performance, including factors like doors, windows, leakage through walls, buoyancy effects, and pressure imbalances between areas and zones.

By using actual operating data for a building, more confidence can be given to the results of a simulation, and a more detailed study of design remedies may be possible. For example, most tall buildings have a stairwell and an elevator shaft. Both of these passageways can autonomously distribute contaminants to all floors. As a result, contaminants that are released on the main floor, which normally distribute more heavily to the immediate surrounding areas, may be released more equally to other floors, or even be directed in preference toward the uppermost areas. Such phenomena can occur during fires, where flames and smoke can funnel through shafts to upper areas.

Stairwell effects and other factors can be modeled automatically with the basic building blocks of the CONTAMW program. Figure 9.14 shows a CONTAMW layout of the main floor of a 10-story building. Some of the symbols have been described previously. Note the contaminant symbol in the north (N) zone—this is the only contaminant release point. The BW agent used in this example, anthrax spores, is being released continuously in this one room for 8 h.

The model building in this example has a floor area of 2650 m²/floor and a total airflow rate of 10,000 m³/min. Other relevant data are given in Table 9.14. This is the same basic input data form used in the spreadsheet model. The spreadsheet provides a convenient way of estimating the release rate that will produce 99 percent fatalities. This 99

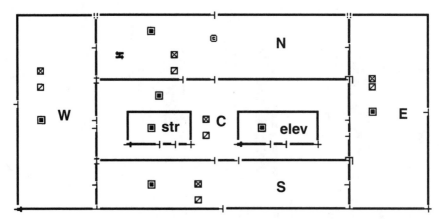

Figure 9.14 CONTAMW model layout for the main floor of a 10-story building with a stairway (str) and an elevator (elev) shaft running through all floors. Letters designate the separate zones or areas.

percent value will be used to compare reductions in fatalities when air-cleaning systems are added, as before. In this case the release rate that will produce 99 percent fatalities is estimated to be 6.46 µg/min. All other data in Table 9.14 are similar to values that have been used in the previous analyses, except the breathing rate, which is 0.01 m³/min (10 liters/min).

The CONTAMW model was run with 1-min time steps for 12 h. The anthrax spores were released in the north room on the main floor. Figure 9.15 shows the results in terms of the concentrations reached on every floor, and each of the surrounding rooms. The highest concentrations occurred in the room subject to the release, as would be expected. Most of the other floors, and the adjacent rooms, received approximately the same dose. This result confirms the earlier statements that the release of an agent in a local room, instead of in the ductwork, is a less severe scenario. That is to say, the release of an agent in a single zone usually affects that zone the most unless some external influences affect the pressurization.

The release is stopped at the eighth hour, and the dropoff in concentration can be observed to begin at that point. The exponential decrease is the result of the combined purging effects of the ventilation system and the air-cleaning effect of the filter and UVGI system. This latter process can also be modeled with CONTAMW or with the previously described spreadsheet model.

Figure 9.16 shows the predictions of fatalities for each floor in the model building and each of the adjacent areas. The north zone reaches 100 percent fatalities within a short time, but the fatalities in the adjacent areas and on the other floors are considerably lower. It could be

TABLE 9.14 BW Attack Simulation Input and Data for CONTAMW Model

	Input			Data		
Dilution	OA	15	%	OA fraction	0.15	
	Total airflow	10,000	m³/min	Total airflow	35,150	cfm
		166.6666667	m³/s	...per floor	16.66666667	m³/s
	Suggested airflow	9,365	m³/min	Outside airflow	52,973	cfm
	(function of floor area)	49,606	cfm		1,500	m³/min
	ACH	1.13		Return airflow	8,500	m³/min
Filtration	Filter type (0–6)	5		Anthrax removal rate	0.986	Fraction
	Nominal filter efficiency	90	%	Smallpox removal rate	0.865	Fraction
	Filter removal rate	0.98646	Fraction	TB removal rate	0.470	Fraction
UVGI	UVGI size (0–6)	6		Intensity	3,000	μW/cm²
	Exposure time	0.5	s	UVGI dose	1,500	μW·s/cm²
BW agent	BW number (1–3)	1		BW agent		*Bacillus anthracis*
	Release rate of agent	32,207,569	cfu/min	BW agent diameter	1.118	μm
		25.92236754	μg/min	Mass of microbe	8.04853E-07	μg
	Breathing rate	0.01	m³/min	BW agent UVGI k	0.000424	cm²/μW·s
	Baseline fatalities	35.07	~98–99%	UVGI kill rate	0.4706	Fraction
	Fatalities with CBW	0.64		Steady state	21,472	cfu/m³
	Lethal dose, LD_{50}	28,000	cfu	Steady state	0.017281578	μg/min
Building parameters	Floor area	2,650	m²	Floor area	28,524	ft²
	Floor height	3	m	Floor height	10	ft
	Number of floors	10		Floor volume	7950	m³
	Total building volume	79,500	m³		280,741	ft³
		2,807,414	ft³	Estimated occupancy	2,480	persons

OA = outside air; ACH = air change rate; k = UVGI rate constant.

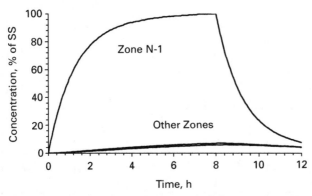

Figure 9.15 Predicted concentrations of all floors in the CON-TAMW model. Zone N-1 is the first floor north zone. All other zones are nearly overlapped below.

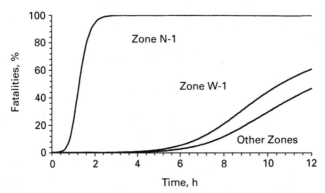

Figure 9.16 Predicted fatalities of all floors in the CONTAMW model. Zone N-1 is the first floor north (N) zone and Zone W-1 is the first floor west (W) zone. All other zones are overlapped below.

concluded, as stated before, that this particular scenario, the release of anthrax spores in a single room, is not as severe as the kind of effects seen in all the previous examples. That would generally be the case; however, the quantity of agent released and the airflow characteristics of specific buildings may alter this state of affairs.

9.7   Overload Attack Scenarios

Another aspect that may be of interest for simulation purposes is the scenario in which the release rate of anthrax spores is much higher than would be needed to cause 99 percent fatalities. Consider the same model 10-story building from the previous example with a BW release

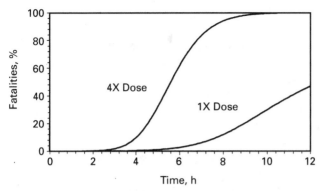

Figure 9.17 Effect of a fourfold increase of the BW release rate on predicted fatalities.

rate four times the design basis scenario. Figure 9.17 shows the predicted fatalities for all zones except the main floor north zone. The effect is clearly to shift the lethality curve to the left, reducing the window for treatment.

Attempting to overload immune building systems with larger quantities of an agent than are necessary is an unlikely scenario because of several factors. First of all, it is not a simple matter to create biological agents. The process is time-consuming, expensive, and dangerous. Industrial bioweapons programs like those of Iraq and the former Soviet Union may have been able to produce kilogram quantities of anthrax, but these facilities have been shut down for many years. Weapons-grade anthrax will decay over time regardless of how it is stored, and so the quantities of anthrax from these operations that have been purported to be missing may eventually become useless.

Terrorists are more likely to attempt to culture fresh quantities from samples they have obtained, and this will require painstaking work and equipment that might not go unnoticed if it is purchased outside of research circles. Such attempts to grow any biological agent must of necessity be done on a small scale with limited equipment. It could take weeks, months, or even years to develop a sufficient quantity to launch an overload attack against a large building. More likely, they would use whatever quantities they had when they had enough, and therefore it is more probable that a building would be "underloaded" rather than overloaded.

In order to defend against an overload attack, it is necessary to go beyond the design of air-cleaning components, since their performance is limited by the ventilation system characteristics, as explained previously. The building can be protected against an overload attack by boosting either the outside air flowrate or boosting the total system flowrate and then sizing the air-cleaning components accordingly. Any level of protection is possible, but the costs may become prohibitive. The problem of overload attacks highlights the fact that installed immune building systems, and the level of protection that they can provide, should not be advertised. All details of the performance and size of such systems, as well as their very existence, should be treated as sensitive information.

Consider again the example from Fig. 9.17 of an overload attack with four times the amount of agent that would be needed to cause 99 percent fatalities. Increasing the total system airflow is one approach. Assume the total airflow is doubled, for example, and the filter and UVGI system size remain the same. Figure 9.18 shows the results. The fatalities have been reduced, although only to 44 percent. This illustrates the fact that manipulating total airflow can be used to counteract the effects of an overload attack scenario. Other alternatives include increasing the outside airflow, and further increasing the filter and UVGI systems' sizes. Although increasing total system airflow may not always be an option, the addition of recirculation units in various parts of the building would have a similar effect. That is, the total quantity of air disinfected by the added recirculation units would have the equivalent effect of increasing the airflow without altering the ventilation system characteristics.

9.8   Sudden-Release Scenarios

An additional scenario that may be of interest for modeling purposes involves the sudden release of a large quantity of an agent at the outside air intakes. It has previously been stated that the continuous-release scenario is the worst-case scenario for most types of buildings, and it will now be shown why this tends to be true.

In this case it is assumed that the same quantity of agent, say smallpox virus, released over 8 h in the previous models is now released into the air intakes in the space of 1 min. It can be assumed that this is dumped into the air intakes or released from a pressure bomb in such a way that the entire quantity becomes aerosolized. Upon entering the outside air intakes and mixing with the return air, the aerosol will enter the air-handling unit. In the design basis scenario, there is no air cleaning and the entire aerosol quantity disperses into the building volume. For this example, only the 60 percent filter and 1000-μW/cm^2 UVGI system are considered. With the filter and UVGI system present, the entire aerosol will make an initial single pass through these components before entering the building.

Figure 9.19 shows the results of the sudden air intake release scenario, showing the baseline condition and the filtered condition. The

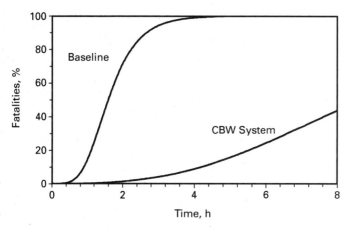

Figure 9.18 Effect of doubling the total airflow for the 4× overload scenario shown in Fig. 9.17.

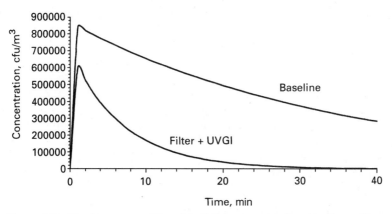

Figure 9.19 Baseline versus filtered building concentrations for the sudden release at air intake scenario. Filter DSP is 60 percent and UVGI system is 1000 μW/cm^2.

building concentrations peak initially and then decrease exponentially. Only the first 40 min are shown in order to highlight the initial transient effects, but it is clear that this scenario produces a lower concentration after the first hour than the slow- or continuous-release scenario.

Figure 9.20 shows the predicted fatalities for the same sudden-release scenario compared against the continuous-release case. Note that although the predicted fatalities peak earlier for the sudden-release case, they are ultimately slightly less than for the continuous-release case. It should be noted that the details of the simulation can have a major effect on the final results. It was assumed that the sudden release

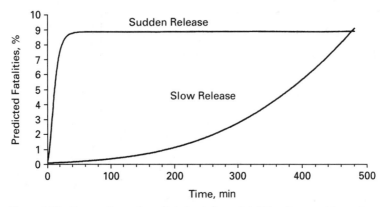

Figure 9.20 Comparison of predicted smallpox fatalities for a sudden-release versus slow-release scenario, for a MERV 11 filter and an URV 11 system.

occurred at the outside air intakes, but a sudden release could be postulated at any of the points previously discussed, including inside the air-handling unit and upstream or downstream of the filters or fans. All of these could be analyzed for any particular building; however, the worst-case results of most scenarios will tend to converge to the approximate results of the slow-release scenario. In any event, the postulated release scenarios should be well thought out for any building or facility under consideration and identified prior to performing simulations, since some buildings may have unique vulnerabilities that cannot be generalized with a simple single-zone model. If there is any question about the validity of a single-zone building model, programs like CONTAMW should be used in place of, or to verify, the modeling results.

9.9   CW Removal by Outside Air Purging

Although no modeling information is available for gas phase filtration, dilution ventilation can be used to purge a building of chemicals in gaseous or vapor form. In this attack scenario, we consider the chemical agent sarin to be released at a constant rate into the same 50-story building previously modeled. The LD_{50} for sarin is 100 mg·min/m³. After establishing the release rate to obtain 99 percent predicted fatalities under the normal operating condition of 15 percent outside air, with no air cleaning, we examine the cases of 50 percent outside air and 100 percent outside air.

Figure 9.21 shows the results of the analysis for the three outside air flowrates. The 50 percent outside air flowrate reduces predicted fatalities to 42 percent at the end of 8 h, while the 100 percent outside air flowrate drops them to only 9 percent. Obviously, the 100 percent outside air flowrate can have as much effect on the CW agents as the high-

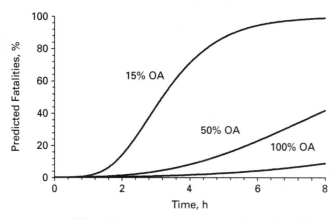

Figure 9.21 Effect of increasing purge rates on the predicted fatalities from a sarin gas attack on a 50-story building.

est filtration and UVGI rates have on BW agents. Of curse, purging continuously with outside air is not an economically viable alternative, but if chemical detection systems were installed, the use of a purge mode would become a feasible means of dealing with the potential threat of chemical agent releases.

9.10 Summary

In this chapter a number of building attack scenarios have been simulated, and the details of modeling have been presented that will enable the engineer to develop further simulations. These models can be adapted to any building and ventilation system, and taken to any level of detail that is desired. The CONTAMW model is particularly well suited for complex simulations, and further references have been provided, including a download website (NIST 2002). The simulations of generalized buildings performed in this chapter have led to the useful conclusion that the performance of air-cleaning systems is limited by building and ventilation system characteristics, and that only moderately sized filtration and UVGI systems may be necessary to provide a high degree of protection to building occupants. The primary advantage of this conclusion is in the economics of providing such protection, which will be presented in Chap. 16, and in the fact that the sizing of such systems becomes academic. It has also been shown that some attack scenarios are more dangerous than others, but that increased levels of protection, if desired, can be achieved through increasing the total system airflow, or by the effectively equivalent addition of internal recirculation units. It has also been shown that outdoor air purging may be a viable method for dealing with chemical agent releases.

10

Detection of CBW Agents

10.1 Introduction

In a completely integrated immune building system, the detection of chemical and biological agents may perform important control system functions. Rapid detection of hazardous agents would allow alarms to be triggered, ventilation systems to be shut down, areas to be isolated, and emergency systems to be engaged to minimize the danger to any building occupants.

Although the detection of chemicals and aerosols composed of chemical and biological agents has been under intense study for decades, the present state of chemical and biological sensing technology does not allow for all-encompassing solutions and is limited to the detection of specific agents. Several types of detectors may have to be combined in order to provide detection of a broad range of threats (see Fig. 10.1). Such detection systems, furthermore, are often subject to delays in response time, limiting their usefulness in applications that require rapid responses. Finally, the costs associated with many types of detection systems may be prohibitive for normal commercial building applications. In spite of these limitations, there are applications where CBW detection is desirable, and new developments may yet make applications of such technology feasible.

This chapter reviews the state of the art of biodetection and chemical sensing and describes the various ways in which such technologies would be applied in an integrated immune building system. Air-sampling methods are also discussed as a means of developing low-cost alternatives to biological sensing. As this section is mainly intended to familiarize engineers and designers with the options, reliance

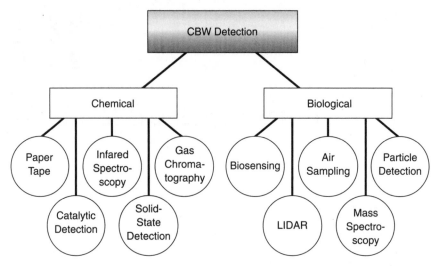

Figure 10.1 Breakdown of the most common types of chemical and biological detection systems (LIDAR = light detection and ranging, a developmental military technology). Additional sensing technologies are described in the text.

on the details of chemistry and microbiology is avoided. An extensive amount of literature is available on these subjects, and readers may consult the references for more detailed information on these detection technologies.

10.2   Chemical Detection

The science of detecting airborne chemicals has a long history and currently exists in a relatively advanced state. Much of the progress in this field has been driven by a desire to protect workers and individuals from toxic or flammable vapors, or to measure carbon dioxide levels as an indicator of oxygen deficiency in enclosed spaces (Cullis and Firth 1981). Industrial environments, including mines, are subject to the release of all sorts of hazardous vapors, and legislation going back to the 1800s has spawned research into methods of detecting extremely low levels of toxic substances, as well as establishing acceptable limits for human exposure.

Because of the high toxicity of chemical warfare agents, and the fact that many of these agents often have dual uses in various industries, considerable information exists on their limits of exposure. A variety of chemical sensors like those shown in Figs. 10.2, 10.3, and 10.4 are available to detect harmful gases and vapors, but no general-purpose sensing technology exists (Leonelli and Althouse 1998; Paddle 1996). As a result, many chemical detection systems exist that are designed to detect a specific chemical or group of related chemicals, and systems designed to detect a broad range of chemicals often include several separate sensing systems combined into one unit. Such systems are invariably complex and expensive.

The problem with detecting the broad range of CW agents listed in Appendix D by a single method is that they have no universal physical property that would lend itself to detection by any single method. Electrochemical techniques are probably the most common method for detecting hazardous gases and vapors. Although these can detect a number of CW agents, they can also produce false alarms from relatively innocuous chemicals (Cullis and Firth 1981).

Another method is the use of chemically impregnated paper tape devices. Such systems use various chemicals that change color when exposed to specific chemical agents, allowing colorimetric sensing to ascertain the presence of a toxic chemical agent (Brletich et al. 1995).

Gas chromatography systems employ methods in which a gas is sampled through a tube and analyzed by instruments. In some systems, the conductivity changes measured in a material to which the gas has adsorbed are measured as an indicator of the presence or concentration of the agent.

Perhaps the most successful and widely used technologies are the various spectroscopic devices that use techniques like infrared spectroscopy to determine the presence of a toxic gas in a sample (Gardner 2000; McLoughlin et al. 1999; Morgan et al. 1999).

The measurement of flammable gases and vapors is in an advanced state, and these systems often test small samples of a gas

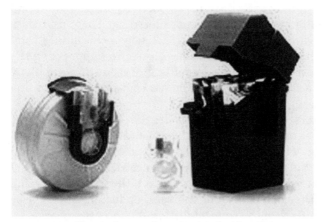

Figure 10.2 Chemical agent detection kit for identifying nerve and blister agents. (*Image courtesy of Dr. Amram Golombek, Israel Institute for Biological Research.*)

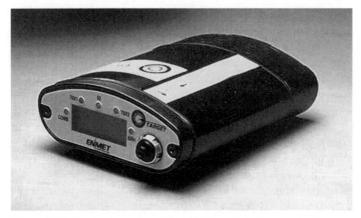

Figure 10.3 Compact gas detector capable of detecting up to six separate gases, and which includes a broad range toxic gas sensors. (*Image courtesy of ENMET Corporation, Ann Arbor, MI.*)

for flammability using a spark or other ignition method (Cullis and Firth 1981). Other systems use a catalytic detector that oxidizes the vapor and senses any increase in sensing element temperature. These detectors are of limited use since most CW agents are not flammable, but they may, however, form one component of a multiagent detection system.

Photoionization detectors can be used to measure a wide range of toxic agents. These devices expose a sample of a gas inside a chamber to ultraviolet light, which can ionize many chemical compounds. The ions generated are drawn to a collector electrode, which then amplifies the electrical signal.

Electron capture detectors contain a radioactive source, usually an isotope of hydrogen or nickel, which emits electrons. The output is a continuous current unless a gas, usually a halogen-containing compound, enters the chamber and absorbs the electrons, suppressing the current. However, few of the chemical agents in Appendix D are likely to trigger this device since most of them do not contain halogen or any similar compounds.

Solid-state detectors exist in two types. In the first type, the heat released when a gas is oxidized on the surface of a catalytic solid is measured as an indicator of the presence of a hazardous vapor. In the second type, a measurement is made of the change in electrical conductivity of a semiconductor material exposed to a toxic gas. If the gas is a reducing agent, it will alter the conductivity of the semiconductor. Other gases may adsorb to the surface of the semiconductor and impart a charge.

Many portable gas detectors such as the one shown in Fig. 10.5 are available that are capable of detecting several gases simultaneously.

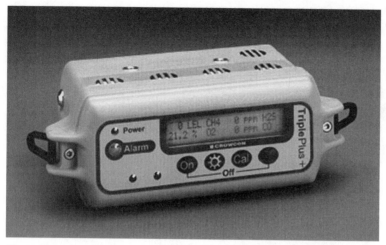

Figure 10.4 A multisensor portable gas detector capable of detecting a broad range of toxic gases. (*Image courtesy of Crowcon Detection Instruments, Ltd., Abingdon, U.K.*)

Most of the gases they are designed to detect are typical hazardous gases such as carbon monoxide or hydrogen sulfide. These units are usually capable of detecting certain chemical weapon agents also, such as chlorine, phosgene, and nitrogen oxide.

Mass spectrometry (MS), also called "rotational-vibrational spectroscopy," can be used to detect chemical agents by identifying their spectral signature in air. All mass spectrometers include a sampling system, an ion source, a mass analyzer, a detector, and a signal processing system. A charge is imparted to the air sample and the resulting ions are separated by the mass analyzer. A mass spectrum is produced showing a specific number of peaks on a graph of mass-to-charge ratio versus intensity. This mass spectrum is unique to each chemical or organic compound. Mass spectrometry is often used in conjunction with gas chromatography (GC). In combination, this method (GC/MS) is one of the most sensitive and discriminating tools for identifying chemical and biological compounds, although it requires considerable skill to operate the equipment and interpret the results (Brletich et al. 1995).

Ion mobility spectrometry (IMS) is used in point detectors and alarms by military forces worldwide. In IMS technology, chemical substances are characterized through gas phase ion mobilities that are triggered by a radioactive source. IMS is a reliable technology that has high sensitivity (Brletich et al. 1995).

In Raman spectroscopy, a laser is directed to an air sample and the phenomenon known as "Raman scattering" provides a means of identi-

Figure 10.5 Portable gas detector capable of monitoring for chlorine and nitrogen oxide. (*Image courtesy of Biosystems, Middletown, CT.*)

fying the particular chemical agent through spectroscopic analysis (Hobson et al. 1996).

Flame photometry involves the burning of an air sample in a hydrogen-rich flame. The wavelengths of the light emitted are passed through an optical filter that isolates a characteristic wavelength for each substance burned. A photomultiplier tube produces an analog signal that will register the presence of specific chemical agents (Brletich et al. 1995).

Surface acoustic wave (SAW) detection involves the use of a piezoelectric crystal substrate on which electrodes are placed to create an electric field. A surface acoustic Rayleigh wave is generated between the electrodes that becomes perturbed by any substance making contact with the surface. A chemical coating on the surface absorbs any vapors, and this produces changes in the wave as it passes through the surface. Different chemical coatings can be used for the detection of specific chemical agents (Brletich et al. 1995).

Biosensors have also been developed that can identify chemical weapon agents such as DFP (Rogers et al. 1991), hydrogen cyanide (Smit and Cass 1990), and soman (Lee and Hall 1992). Detection times, however, vary from 1 to 30 min, which does not necessarily improve upon conventional chemical detection methods.

For a detection system to be integrated with a building ventilation system for purposes of shutdown or isolation, fixed-station monitoring is necessary. Monitoring can be of a single point—monitoring of the air-handling unit (AHU) only, for example. This is because all air in the building ultimately circulates through the air-handling unit. Also, as described in previous chapters, the air-handling unit is perhaps the most sensitive location for most buildings, since a chemical release in general areas would likely produce fewer casualties.

A fixed-point gas monitoring system would draw a continuous sample from the airstream and cycle through all available tests to check the presence of any one of a number of chemical agents. A positive signal would then be used to alarm and/or isolate the ventilation system. It is important that the air-handling unit be immediately shut down since it would invariably disseminate any chemical agent throughout the building. This response may cause higher concentrations in some areas than in other areas, but the principle of protecting as many occupants of the building as possible precludes favoring any one area over another, unless this is done by design.

Most chemical detection systems, such as the one shown in Fig. 10.6, require some finite amount of time to produce a positive signal. This can be as long as tens of seconds, or even minutes, depending on the chemical and detection levels. If an agent is detected in the ductwork at the air-handling unit, the system should be isolated and shut down before any significant amount of the CW agent can pass into the building's general areas. This control strategy is known as a "detect-to-isolate" architecture. Simply put, any detection signal will automatically shut down the ventilation system and produce an alarm. Additional

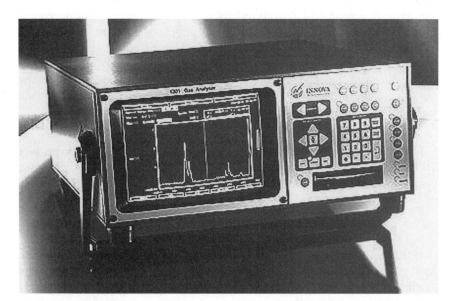

Figure 10.6   A photoacoustic gas analyzer that uses infrared interferometry to identify up to seven gases simultaneously. (*Image courtesy of Innova AirTech Instruments.*)

functions may include engaging auxiliary systems, purging air systems, or isolating sheltering areas.

More will be said about detect-to-isolate control systems in Chap. 11, but one physical design consideration for incorporating such a detection system may include the use of extended ductwork to provide sufficient time for complete isolation.

10.2.1   Chemical detector response time

The most rapid-acting, broad-range chemical detectors may require anywhere from 10 to 20 s or more to respond to the presence of a chemical agent in the supply airstream. The delay in response time might not necessarily be critical, but the standard approach for dealing with this situation is to provide extended ductwork with an isolation damper at the end. The ductwork is sized such that the travel time of the airstream from the point of detection to the nearest exhaust register is greater than or equal to the detector response time plus the damper isolation time. The damper isolation time is how long it takes for the damper to completely close, which can be as little as a fraction of a second or as long as several seconds.

Figure 10.7 shows an example of a system designed to allow a specific travel time from the detection point to the isolation point. The system has a toxic gas detector with a sample point in the AHU. Upon detection of a toxic gas at point A, an isolation signal is sent to close the damper at point B. The signal should also be sent to shut down the fan. The duct must be physically extended so that the travel time between point A and point B is greater than or equal to the system response time.

The ductwork length required to provide isolation is computed from the air velocity, V, in the duct and the system response time, RT, as follows:

$$\text{Length} = V \cdot \text{RT} \qquad (10.1)$$

For example, if the air velocity in the duct is 5 m/s (approximately 1000 fpm), and the system response time (including detector response and damper isolation), RT, is 15 s, then the required length of the duct is computed as follows:

$$\text{Length} = (5)\,(15) = 75 \text{ m} \qquad (10.2)$$

If the air velocity in the duct varies from duct section to duct section, as is often the case, the individual travel times can be computed and summed to determine the total duct length.

In situations where the detection system is being retrofitted and the ductwork cannot be extended to accommodate the detector response

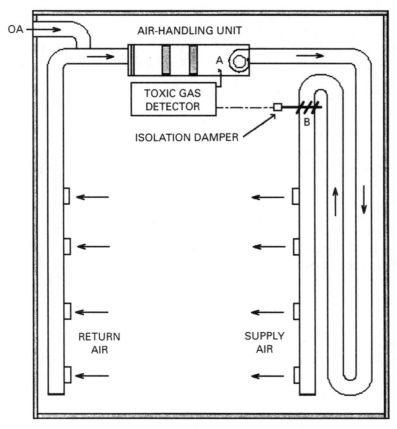

Figure 10.7 Ventilation system with extended supply ductwork to allow toxic gas detection at point A to isolate the system before gas reaches point B.

time, then it must be accepted that some amount of contaminant will enter the building. If the detector response is reasonably rapid, the actual dose received by any occupants may not be hazardous. In any event, isolating the system and/or shutting down the fan will usually be the best response, but this may depend on the characteristics of the building and the release point (see Chap. 7 for further discussion on this matter).

10.3 Biological Detection

Various technologies for the detection and identification of biological agents are available today, but the number of agents that can be detected are limited. Methods range from manual sampling and culturing of microorganisms to developmental military technologies such as light detection and ranging (LIDAR) for detecting bioaerosols in the atmosphere. The costs associated with advanced detection technologies can be prohibitive. This section summarizes some of the technologies that can be used or adapted for detection of biological agents, including toxins and pathogens, in indoor environments and especially in ductwork.

10.3.1   Air sampling

Sampling of microorganisms in the air or on surfaces has been in use for over a century and is ultimately the most reliable means for identification of pathogens. Air sampling is in common use for identifying airborne microbial contaminants such as fungi for the purpose of remediating indoor air quality (IAQ) problems (Boss and Day 2001). Surface sampling is also used to detect contaminants in buildings, but this is mainly relevant to remediation or cleanup activities. Air sampling can take anywhere from hours to days to identify the presence of an airborne contaminant, but this is not necessarily a problem because detection may still occur in sufficient time to treat any exposed occupants. When this is used in such a fashion, it is known as "detect-to-treat" control architecture, as opposed to "detect-to-isolate" or "detect-to-alarm" architecture.

Air sampling normally involves drawing a sample from an airstream or room by means of equipment like that shown in Figs. 10.8 and 10.9 and directing it to impinge upon a petri dish or sample plate such as

Figure 10.8 Air sampler. (*Image courtesy of Aerotech Laboratories, Inc., Phoenix, AZ.*)

shown in Fig. 10.10. The gelatinous media is a growth medium that is usually selected for the particular microbes that are expected. The plate is exposed to a measured flow of air for a specific period of time, usually about 20 min. The plate is then incubated for a period of 12 to 48 h, during which colonies grow and become visible. The plate culture can often be visually identified, but microscopy or other tests are available to make any final determination of what microbes are present (Aerotech 2001).

Some progress has been made in the development of automated sampling systems and samplers geared to identify a wide range of possible microbes (Hobson et al. 1996). The use of strips containing selective growth media, accelerated incubation times, and reduction of colonies to droplet size capable of being evaluated under a microscope are some of the methods being used to provide rapid detection times.

Air samplers typically house a petri dish and direct an airstream toward the surface of the dish containing the growth media. A variety of air samplers are available today, and they have diverse characteristics, applications, and varying levels of accuracy (Jensen et al. 1992). Details on the use and performance of aerosol samplers can be found in the references (Aerotech 2001; Bradley et al. 1992, Griffiths et al. 1993; Han et al. 1993; Li et al. 1999; Straja and Leonard 1996).

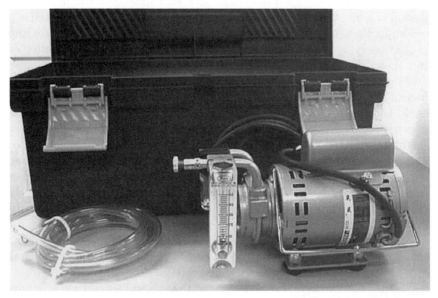

Figure 10.9 Air compressor for drawing air samples. (*Image courtesy of Aerotech Laboratories, Inc., Phoenix, AZ.*)

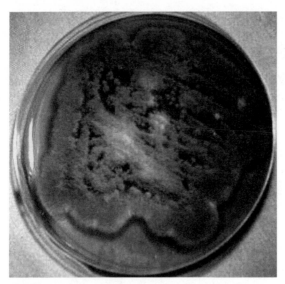

Figure 10.10 Petri dish on which colonies of *Penicillium* have been cultured. (*Image courtesy of BioChem Technologies, LLC, Canada.*)

Air samplers may be driven by an air pump or an integral fan. Figure 10.11 shows a self-contained air sampler in which a small fan draws air over the surface of the plate near the entrance. After a prescribed period of time, the petri dish is removed and the culture incubated.

The implementation of an air sampler as part of an immune building protection system would involve routine sampling of air in the building or in a duct. The sample time might be extended to a period of 8 h or more and the sample could be removed daily and sent to a lab for culturing and identification. In the event any pathogens were detected, the building could be quarantined and all occupants given medical treatment. This approach embodies the principle of detect-to-treat architecture. The alarm might be sounded too late to stop the release of a biological agent but in sufficient time to treat any infections before they became life-threatening.

A detect-to-treat system could be implemented economically in a commercial building. A low-cost air sampler could be permanently installed in the air-handling unit and designated personnel would remove the plates on a regular schedule. These would be delivered for analysis to a local laboratory, which would return results in an expedient manner. Procedures would dictate the response in the event that any dangerous pathogens were detected.

One advantage of the air-sampling approach is the fact that any normally occurring airborne microbes would also be detected, and this would enable a record of building air quality or microbial air loading to

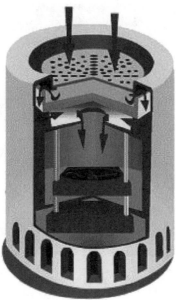

Figure 10.11   Air sampler with an integral fan. Cutaway diagram at right shows airflow direction. Petri dish is at the top under the grated inlet. (*Images courtesy of BioChem Technologies, LLC, Canada.*)

be created. Variations of airborne fungi or bacteria beyond normal levels would alert building managers to the possible need for some remedial action, especially if air-cleaning systems were in place.

10.3.2 Biosensors

Biosensors may be any type of detection system that provides reasonably rapid automatic identification of biological agents. Some biosensors are solid-state devices that provide an electrical output signal on detection. Many of these devices involve the fixing of a biologically sensitive material or compound onto a hard surface. In general, biosensors can detect only one agent and can only be used once, since the process of detection consumes the sensing material. Some biosensors are in use today for the detection of specific chemical compounds (Gardner 2000; Leonelli and Althouse 1998; Paula 1998). Some biosensors are currently in use in the food industry for detecting food pathogens (Hobson et al. 1996). Piezoelectric biosensors have been developed that can identify *Salmonella typhi* in water (Prusak-Sochaczewski et al. 1990).

Some progress is being made in the development of combined sensors (see Fig. 10.12) that can automatically detect a wide range of agents

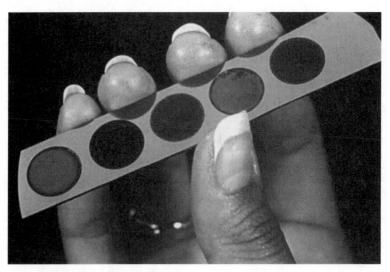

Figure 10.12   Biosensor strip with five separate sensors for detecting specific compounds. (*Image provided courtesy of Michael Sailor, University of California.*)

(Belgrader et al. 1998; Donlon and Jackman 1999), but such technology is still in a developmental stage.

When a pathogen invades a host, the immune system often recognizes the invader by the presence of antigens, or surface molecules that are unique to a particular species. The immune system may then attempt to release antibodies to neutralize or otherwise limit the damage done by the pathogen. These antibodies have a specific affinity for individual antigens and can be used to identify the presence of a specific pathogen. Biosensors have been developed that immobilize antibodies on an inert surface and use them to generate a signal through electrochemical reactions. Biosensors using this technique have been created that can detect *Clostridium perfingens* (Cardosi et al. 1991), *Salmonella* (Plomer et al. 1992), *Francisella tularensis* (Thompson and Lee 1992), and *Yersinia pestis* (Cao et al. 1995). The detection of toxins like ricin with biosensors has also been demonstrated (Cater et al. 1995), and fiberoptic sensing has been used to detect botulinum A toxin (Ogert et al. 1992).

Many other approaches are possible that can identify the reaction of an antigen to an antibody. Some of these methods use optics, such as surface plasmon resonance (SPR), evanescent waves (EW), luminescence, and fluorescence to identify an antigen-antibody reaction. Electrochemical methods can also be used to detect such reactions, such as potentiometric, amperometric, and conductimetric techniques. Piezoelectric techniques can be used to detect viruses and bacteria (Konig and Gratzel 1993). In addition, there are acoustic techniques that use piezoelectric devices and calorimetric techniques that use thermistors or thermopiles to detect reactions (Paddle 1996). The photoelectric detection of agents is also possible (Arrieta and Huebner 2000). These techniques are currently under development, and the number of pathogens that can be detected is small. In time, biosensors may provide an ideal and economic solution to the problem of detecting airborne pathogens, but at present such technology is not quite feasible for incorporation into a commercial immune building system.

One of the main limitations today on the development of biosensors for BW defense is a lack of information about the target molecules on the surfaces of microbes. The molecules necessary for use in biosensing have either not yet been identified or cannot be isolated in a stable form.

Biosensors depend on the selective recognition of the biological molecule that is being targeted. Two distinct types of biosensors can be identified—catalytic and noncatalytic. Catalytic sensors include enzymes, bacterial cells, and tissue cells. Noncatalytic biosensors are also known as "affinity class biological sensors" and these include antibodies, lectins, receptors, and nucleic acids. Noncatalytic sensors can be used to identify microbial toxins.

Table 10.1 lists the various biological agents for which biosensors have been developed or reported. This list is summarized from Paddle (1996), where the source documents may be found for each biosensor type. The column showing the medium refers to the medium in which the biological agent was detected in the test, whether liquid solution or dry air. Although some tests were done in the medium of dry air, no results were available. The possible dependence of a biosensor on liquid media is not necessarily a problem since air can be sampled into a liquid solution.

Systems are under development for military applications that combine mass spectrometry and biosensors for both toxic gas and biological detection (Donlon and Jackman 1999). These systems use mass spectrometers to identify the presence of any aerosolized agents (Cornish and Bryden 1999; McLoughlin et al. 1999; Scholl et al. 1999). In this integrated system, toxic gases can be identified purely through mass spectrometry (Hayek et al. 1999), while toxins can be detected by the use of a fluorometric biosensor (Carlson et al. 1999).

10.3.3   Particle detectors

A feasible alternative to detecting airborne pathogens is the use of particle detectors to identify the presence of microbial-sized particles in the air. In theory, all pathogens have a characteristic size range and shape, and it may be possible, although not essential, to identify many

TABLE 10.1 Biosensors for BW Agents and Assay Times

BW agent	Biosensor type	Medium	Assay time	Reference
α-Bungarotoxin	Optical fiber, evanescent wave, fluorescence (receptor)	Liquid	Minutes	Rogers et al. (1989)
α-Bungarotoxin	LAPS* (receptor-enzyme amplification)	Liquid	20–30 min	Rogers et al. (1992)
Botulinum toxin A	Optical fiber, evanescent wave, fluorescence (receptor)	Liquid	1 min	Ogert et al. (1992)
Botulinum toxin A	LAPS botinylated membrane capture of enzyme-labeled antibody-antigen complex	Liquid	15 min	Menking and Goode (1993)
Staphylococcal enterotoxin B	Impedance (Pt film-antibody layer)	Liquid	—	DiSilva et al. (1995)
Ricin	Piezoelectric (QCM)*	Liquid	1 h	Carter et al. (1995)
Ricin	Optical fiber (evanescent wave, fluorescent antibody)	Liquid	15 min	Ogert et al. (1994)
Ricin	LAPS (antibody-capture, enzyme label)	Liquid	—	Dill et al. (1994)
Bacillus anthracis spores	Optical fiber (evanescent wave, fluorescent dye)	Liquid	35 s	Wijesuriya et al. (1994)
Yersinia pestis	Optical fiber (evanescent wave, fluorescent antibody)	Liquid	30 min	Cao et al. (1995)
Yersinia pestis	Acoustic (QCM) (antibody capture)	Liquid	45 min	Konig and Gratzel (1993)
Shigella dysenteriae	Acoustic (QCM) (antibody capture)	Liquid	46 min	Konig and Gratzel (1993)
Francisella tularensis	LAPS (antigen-antibody, enzyme-labeled urease)	Liquid	65 min	Thompson and Lee (1993)
Salmonella typhimurium	Acoustic-piezoelectric (antibody-antigen)	Liquid	5 h	Prusak-Sochaczewski et al. (1990)
Salmonella typhimurium	Optical (optride-luminescence) (antibody-capture, membrane-enzyme label)	Liquid	30 min	Downs (1991)
Brucella melitensis	LAPS	Liquid	—	Lee et al. (1993)
Coxiella burnetii	LAPS	Liquid	—	Menking and Goode (1993)

*LAPS = light addressable potentiometric sensor, QCM = quartz crystal microbalance.

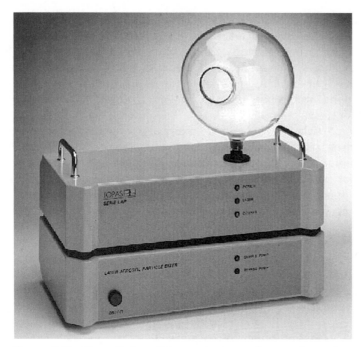

Figure 10.13 The optical particle counter shown here can be used to detect particles in the BW agent size range of 0.3 to 20 microns. (*Image courtesy of Topas GmbH, Dresden, Germany.*)

pathogens on this basis alone. Particle sizers and particle counters are scientific instruments that use a variety of techniques. The ones that are perhaps the most adaptable to real-time air sampling are the optical particle sizers such as the one shown in Fig. 10.13. This unit uses laser spectrometry to measure the size distribution of particles in the 0.3- to 20-μm size range, which is the range that includes most bacteria and all spores. Particle sizers are available that can measure particles down to 0.005 μm in diameter, which is below the size range of the smallest viruses (approximately 0.03 μm).

The actual identification of microbes based on particle size distribution has certain limitations. Airborne microbes do not always occur in singlets, multiple species and particles may be present in any sample, and equipment may not be sensitive enough to establish a complete size range; thus, the identification of airborne microbes on the basis of size alone may be fraught with difficulties.

However, if it is not critical that airborne contaminants be specifically identified, particle detectors can be used to determine if any unusual concentrations of particles are present. For example, if particle concentrations in the microbial size range were monitored downstream of the air filters, any high concentrations would represent either a filter breakthrough or a massive concentration of particles upstream. That is, if the filter is functioning normally, then no high concentrations of particles should exist downstream unless there was a major release of them upstream of the filter or a release downstream of the filter. In any event, a prudent response would be to alarm or isolate the system pending an investigation of the cause.

Many of the particle sizers available today are somewhat expensive, but in comparison with biosensing technology they may be quite feasible for high-end immune building applications. Some low-end models may be adequate for applications were extreme sensitivity is not required, and it may be possible to develop low-cost particle sizers devoted to sensing those BW agents that can penetrate medium-efficiency filters. Even basic optical sensors might be adaptable to the detection of high concentrations of particles in ductwork, especially over long spans.

Many lower-cost instruments are available that are designed to size particles in fluid suspension, like Coulter counters, and these could analyze particles after sampling air into a solution. Particle sedimentation of centrifugation can be used after the capture of an air sample to identify the presence and size range of particles. This method, however, may take 30 min or longer to complete.

Nephelometry is a method for measuring the light scattered by microorganisms in fluid suspension under controlled conditions. The technique produces signals that are directly proportional to the cell mass (Koch 1961).

Dynamic light scattering is one of the primary methods of detecting or sizing particles in air. Submicron particles subject to Brownian motion are analyzed in terms of the random light intensity fluctuations of scattered light. Algorithms are then used to interpret the fluctuations to establish particle size and particle density. These methods can be accurate and can rapidly determine the mean size and size distribution of particles.

Particle sizers have the ability to distinguish some species based on their size distribution, as shown in Fig. 10.14. In this figure the distribution of aerodynamic diameters measured by a particle sizer is characteristic of each bacterium. Although not every bacterial species may be identifiable or differentiable by this method, it does offer the possibility of classifying microbes and thereby narrowing the potential threat.

Other methods using light scattering techniques have also been applied to the identification of biological aerosols. The use of high-resolution two-dimensional angular optical scattering (TAOS) has been shown to be capable of characterizing clusters of particles, including spores (Holler et al. 1998).

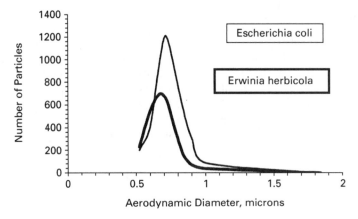

Figure 10.14 Particle size distributions as measured by a particle sizer. Based on data from Eversole et al. (1998).

10.3.4 Mass spectrometry and LIDAR

Mass spectrometry uses light of varying wavelengths to determine the spectra of chemical and organic compounds. Although microbes may contain hundreds or thousands of compounds, certain compounds may be dominant, or certain groups of compounds may occur uniquely with specific microorganisms. Rapid spectroscopic analysis of the various spectra provides a possible means of identifying biological agents. This approach is complex and still under development, but it holds the promise of producing a broad-range biological detector.

Short-wavelength near-infrared spectrometry can be used to rapidly measure hundreds of spectra and can estimate biomass (Singh et al. 1994). A related method, UV resonance Raman spectroscopy, has also been reported to be capable of identifying bacteria and spores (Farquharson and Smith 1998). Ultraviolet laser-induced fluorescence (UV LIF) is also under development as a means of monitoring the number, density, and fluorescent emissions of airborne particles (Eversole et al. 1998).

LIDAR (light detection and ranging) systems are developmental military technologies intended for standoff chemical and biological detection of atmospheric aerosols. These systems use lasers of various wavelengths, including ultraviolet, to detect agents at ranges of several kilometers and can discriminate between biological and nonbiological aerosols (Cannaliato et al. 2000). Such systems are designed for outdoor use and have limited effectiveness, but they may one day be perfected and adapted to immune building applications.

10.4 Summary

This chapter has presented a glimpse of the complex field of gas detection and biodetection as it relates to CBW agents. Currently available equipment has been described that can be adapted for this purpose, and the limitations of these technologies have been discussed. Air samplers and particle detectors have both been suggested as potential cost-effective components of an integrated immune building system. In Chap. 11 it will be shown how detection systems might form part of automatic control systems for immune buildings.

11

Immune Building Control Systems

11.1 Introduction

Control systems for immune buildings that have CBW agent detection capability may perform critical functions like the shutdown of the air-handling unit (AHU), isolation of the building envelope, and alarming for the purpose of building evacuation. In addition, immune buildings may have emergency systems that are designed to engage on detection of hazardous agents. Emergency systems might include 100 percent outside air purging, the use of a secondary train of air-cleaning equipment, or the isolation of sheltering zones. The coordination and activation of the various components of an immune building system can be controlled through basic analog or digital control systems. Facilities with building automation systems (BAS) or with Internet-based building automatic controls (BACnet) can be programmed for control of any air-cleaning systems or CBW detection systems that are retrofitted.

The basic control system architectures necessary for responding to the detection of chemical and biological agents are addressed here in sufficient detail to guide engineers in the implementation of successful control strategies. It is assumed for the purposes of this chapter that detection systems are capable of identifying all threats, although, as explained in Chap. 10, only a limited number of CBW agents can be accurately identified by current detection technology. This may change in the future, and the control system architectures presented here should be equally well applicable regardless of the type and number of detection systems that are implemented.

11.2   Control Systems

Many types of control systems are available that can be connected to CBW agent sensors or detection systems. Some detection systems will output a signal on detection that can be directly routed to an alarm panel or otherwise used to control equipment operation. The output signal from a gas detector, for example, could be used to shut down the ventilation fan. More sophisticated systems, especially those using multiple sensors and sample points at multiple locations, may need a control unit to rout signals to the appropriate equipment or to computers that are programmed to respond to the signals and perform the appropriate functions such as building isolation or smoke purging.

Multipoint detection will be necessary if more than one sample point is used in a control system. Sample points may include the air-handling unit, points in the supply duct, points in the return duct, points in general areas, and the outside air intake. Multipoint sampling can be implemented through the use of multiplexing equipment such as that shown in Fig. 11.1. A multipoint sampler will draw samples from separate locations in timed sequences, and will allow the control system to identify where the release of an agent has occurred.

A number of references are available for more advanced information and more specific details on building controls, digital controls, building automation systems, and BACnet, and these should be consulted if fur-

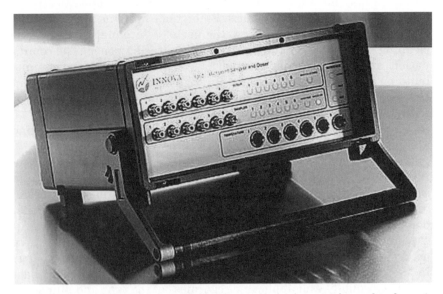

Figure 11.1   A multipoint air sampler like the one shown here can be used to draw air samples from several different duct or building locations. (*Image courtesy of Innova AirTech Instruments.*)

ther information is required (ASHRAE 1995; Bushby 1999; Hartman 1993; McGowan 1995; Newton 1994; Stoecker and Stoecker 1989). Considerable support and information is also available from manufacturers of detectors, automated control systems, and BACnet tools and software. These suppliers should be consulted concerning installation of CBW detectors into existing facilities that use their systems.

11.3 Control System Architectures

Control systems for the protection of immune buildings can be classified into at least three categories—detect-to-alarm, detect-to-isolate, and detect-to-treat. These architectures are depicted in Fig. 11.2 along with the functions they are intended to perform. These control systems may be incorporated regardless of the presence of any air-cleaning technologies, or even in place of such technologies. Ideally, such control systems would be fully integrated into any building or ventilation control system and would function automatically without the need for action by building maintenance, security, or operations personnel, but in actuality these systems may require some degree of personnel involvement, especially the detect-to-treat system.

Detect-to-alarm systems are primarily for evacuation and emergency response. Detect-to-isolate systems are primarily for shutting down the ventilation system or isolating the building envelope. Detect-to-treat systems are primarily for determining if a biological attack has already occurred and will result in building occupants being sent for medical

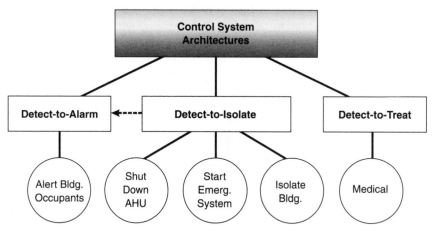

Figure 11.2 Breakdown of control system architectures that would serve to perform specific functions in the event of detection of a CBW agent.

treatment in sufficient time to prevent fatalities. These systems are addressed in detail in the following sections.

11.3.1 Detect-to-alarm

Detect-to-alarm systems serve the purpose of alerting building occupants to the potential presence of hazardous agents in a building's general areas or ventilation system. The alarm signal could be triggered by toxic gas detectors, biosensors, particle detectors, or even manual initiation in the event a CBW agent attack is suspected or imminent. The purpose of the alarm would, in general, be to alert building occupants and initiate a building evacuation. Evacuation out of the building, or to sheltering zones, would be per emergency procedures that should be in place, and for which personnel would be trained to respond in an appropriate manner.

A detect-to-alarm system may distinguish between the type of agent (i.e., biological or chemical) and the degree of the threat (i.e., concentration). Therefore, the type of alarm may dictate a different procedural response. Emergency protocols will be addressed in more detail in Chap. 16. In the event a detect-to-isolate system is installed, the detect-to-alarm function is also performed by this system, and hence the link shown in Fig. 11.2. Detect-to-treat architectures would likely involve a manual initiation of any alarm, since such detection, as explained later, may not occur until the CBW attack is over.

Figure 11.3 illustrates a CBW agent detector with a single sample point inside the air-handling unit. This air-handling unit is shown with filters, an ultraviolet germicidal irradiation (UVGI) system, and carbon adsorbers, but it could just as well have no air-cleaning components at all. The alarm will be triggered if any significant quantity of the agent penetrates the air-cleaning system, which would indicate either a release or equipment failure.

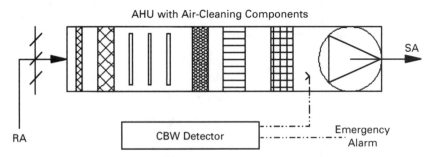

Figure 11.3 Detect-to-alarm system with a single sample point in the AHU. RA and SA indicate the return air and the supply air, respectively.

A detection system could have its sample point located upstream of the air-handling unit so that the alarm would be sounded even when the air-cleaning components were capable of removing the hazard. Of course, this makes the AHU an even more vulnerable target for a downstream release of CBW agents. If a single sample point is used, and it is not unreasonable to use only a single sample point, the best location may be just before or after the fan.

Figure 11.4 shows the basic control logic for a detect-to-alarm system. A caveat of this type of control system is that there is no general-purpose CBW agent detector available today. At least two separate systems may have to be used—one for chemical agents and one for biological agents. In any event, the wiring of two or more detectors to one alarm unit can be effected with a multiplexer.

11.3.2 Detect-to-isolate

Detect-to-isolate systems may serve a critical purpose in protecting building occupants and can perform various control functions. Such a system would likely have a toxic gas detector operating alongside a biological agent detection system, as depicted in Fig. 11.5. Sensors would be located after the filters in the AHU, and possibly also at various other locations, like the supply duct, the return duct, the outside air intakes, and the general areas. A positive detection signal from either the gas detector or the biological sensor would initiate the various control functions depending on what was detected and in what location any agent was sensed. These functions would include alarms and isolation signals, and initiation of any purging or secondary equipment trains if they were present.

In Fig. 11.5 a particle detector is used in place of a biological detector because of its ability to indiscriminately detect any agents in the microbial size range in sufficiently short time to make the system viable.

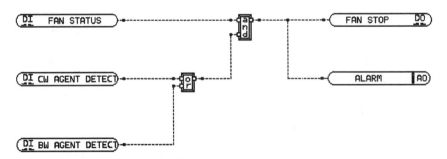

Figure 11.4 Control logic for a detect-to-alarm system with fan shutdown. DI = digital input, DO = digital output, AO = analog output. Prepared using Eikon™ from Automated Logic Corporation (AL 2002).

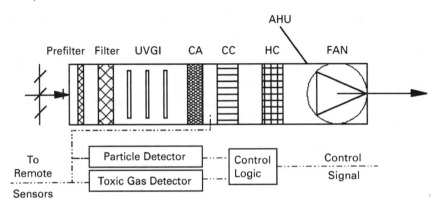

Figure 11.5 Basic components of a CBW detection system. CA = charcoal adsorber, CC = cooling coils, HC = heating coils.

Biological sensors could be used in its place if such were available and were capable of detecting all BW agents, but such technologies are not fully developed at present. The primary sensing location is depicted downstream of the air-cleaning components such that detection of any particles by the particle detector would indicate a massive release upstream. This is merely one possible configuration, and several others may be possible depending on what sensors are available and the building and ventilation system characteristics.

Upon detection of a CBW agent, the first function to be performed is a shutdown of the AHU to prevent the spread of contaminants if this is warranted by the nature of the threat. This may not always be the best choice, depending on the building and the situation, and so the selection of the appropriate control logic should be given careful consideration. Some sources recommend leaving the fan operating (ASHRAE 2002), but this would only be feasible in certain situations. In auditoriums, for example, the fans should probably never be shut down unless they are the source of the contaminant dispersal. In office buildings, with an internal general area release, stopping the fans may protect other areas but fail to clear the air in the area subject to the release. In such a situation, shutting down the fans may be an unfortunate compromise for those in the area of the release, yet it may also protect the largest number of building occupants. Each building and attack scenario should be considered on a case-by-case basis to determine the best response to any CBW agent release.

If the fan is shut down automatically on detection of any internal agent release, it should be done in parallel with the issuance of an alarm signal. The reason the AHU system should be shut down is that most CBW attack scenarios depend on the ventilation system to spread the agent. This is particularly true in buildings with forced ventilation. It may not necessarily be true for large auditoriums with 100 percent out-side air systems, or for buildings with 100 percent outside air. In these latter cases, the shutdown of the AHU would probably be counterpro-ductive, but to reiterate, any system design should be reviewed to deter-mine whether shutdown of the AHU is an appropriate response.

Figure 11.6 shows the basic control logic for a detect-to-isolate sys-tem that has CBW sensors in the return air (RA) duct or AHU, the sup-ply air (SA) duct or general areas, and the outside air (OA) intakes. Upon detection of either chemical or biological agents when the system is running, the system will shut down the fan and close all SA, RA, and OA intake dampers. Alarms will also be triggered.

The shutdown of a ventilation system should include isolating the system by closing all the main dampers. The isolation of the dampers prevents the recirculation of any contaminants. The question of whether the outside air damper should be closed depends on whether the agent is being released outdoors, inside the outside air intakes, or in some other location.

Isolation systems require the use of dampers that can close rapidly. Since most dampers close slowly, the hazard due to any contami-nants that enter as the result of slow-closing dampers may need some evaluation.

11.3.3 Detect-to-treat

Biological detection systems cannot, at present, detect all possible bio-logical agents in sufficient time to issue an isolation or alarm signal.

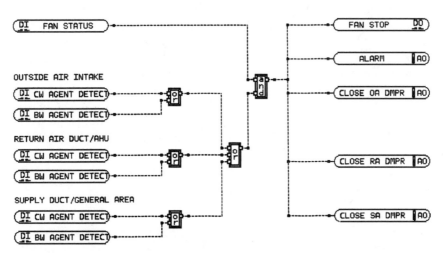

Figure 11.6 Control logic for a detect-to-isolate system with sensors in OA, RA, and SA ducts, and isolation dampers. DI = digital input, DO = digital output, AO = analog out-put. Prepared using Eikon™ from Automated Logic Corporation (AL 2002).

However, because of the delay in symptoms and the incubation period of most pathogens, instant or rapid detection is not an absolute necessity. What is important is to identify the agents in sufficient time to treat the exposed occupants of a building. For this purpose, detect-to-treat control architectures may suffice even if it takes days to identify the agent, and even if this identification is performed manually.

Some biosensors are capable of detecting certain BW agents within minutes or hours. Automatic operation is even possible with biosensors, and a signal could be issued that would perform alarm and isolation functions. Particle detectors can also be used to rapidly identify the presence of microbial-sized particles and issue alarm and isolation signals, but they would not identify the specific pathogen or agent. An alternative to automatic control systems for the detection of bioaerosols is the use of air sampling, as discussed in Chap. 10.

In a detect-to-treat system using air sampling, air samples would be taken at some regular interval, from daily to weekly, and delivered for culturing and identification to a lab or a third party. Upon identification of a dangerous pathogen, emergency procedures would be put into effect. These might include evacuation and isolation of the building, and treatment of all occupants who had been in the building during the previous sampling period. For some agents, like smallpox, regulations may even require quarantining the building.

The amount of time available to treat occupants who have been exposed varies with the pathogen. The incubation period of each pathogen is often approximately representative of the time available to provide treatment. These incubation periods have been estimated and are summarized in Appendix A and graphed in Appendix B.

An alternative to using the disease progression curves and equations is to use a chart of symptoms and incubation times like the one shown for anthrax in Fig. 11.7. Such charts are used by medical and other personnel to assess the time available to treat a disease, and can also be used to establish the time for detection to be effective. Sources on medical treatment should be consulted for more detailed information about symptoms and diagnosis (CDC 2000a, 2000b; DOT 2000; Dept. of the Army 1989; ERG 2000; NRC 1999; Sidell et al 1997; USAMRIID 2001).

11.4 Emergency Systems

Emergency systems can be put into operation by a CBW agent detection signal. An emergency system might be the main ventilation system operated in a different mode, such as a variable-air-volume (VAV) system operated with outside air dampers wide open and no recirculation. An emergency system could also be a separate ventilation system designed for high-volume purging with outdoor air, as in smoke removal

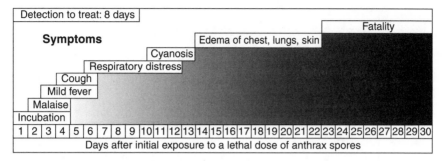

Figure 11.7 Chart of disease symptoms showing incubation time for anthrax.

systems. It could also be a secondary train of equipment designed for high-volume air cleaning. All such systems would be triggered by a detection signal or by a manual switch, and their controls are straightforward, as detailed in the following sections.

11.4.1 Outside air purging

If the building has the capability to purge with a large volume of outside air, such as in 100 percent outside air systems, or in some VAV systems, it would generally be prudent to initiate a purge if it could be determined that the release location was (1) not in the outside air intake, (2) not in the AHU, or (3) not in the supply ductwork. The latter information may or may not be provided by the detection system, but a system could be designed to accommodate this function and determine the location of the release. Alternatively, if the release point of the CBW agent was determined to be in a general area or in the return ductwork, the detect-to-isolate system could isolate the return duct and initiate an immediate purge of the building with maximum outside air.

Outside air purge systems could exist as separate sets of fans, as is sometimes done in the industrial sector for emergency purging of smoke or hazardous agents. Basically, such systems consist of high-volume supply fans and exhaust fans that are started in response to an emergency signal, as shown in Fig. 11.8. In a detect-to-isolate system, the main ventilation system would be shut down and the emergency purge system would be initiated.

Figure 11.9 shows a proposed control logic diagram for a secondary purge system which uses the existing ductwork and outside air intakes to purge a building upon detection of a chemical or biological agent internally. The outside air damper may isolate separately from the remaining system. In the event the outside air damper closes on a CBW agent detection signal, this system would recirculate and process the

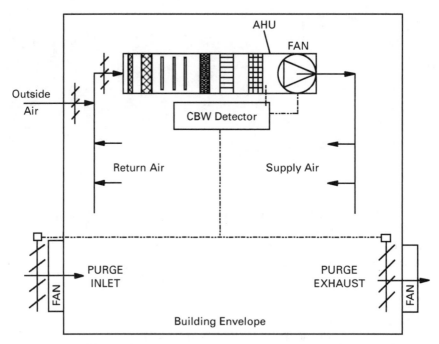

Figure 11.8 Building with a separate 100 percent outside air purging system. The main fan is shut down and the purge system is initiated on detection of CBW agents in the main ventilation system or general areas.

indoor air. This is only one suggested scheme, and several other variations are possible.

11.4.2 Secondary system operation

Immune buildings that have secondary trains of air-cleaning equipment, sometimes called "emergency systems," require controls to initiate a start upon detection and shutdown of the main ventilation system. Such systems may include high-power UVGI, high-efficiency particulate air (HEPA) filters, and combinations of carbon adsorbers designed to deal with a variety of toxic gases. Secondary trains of emergency air-cleaning equipment would presumably save the costs of full-time operation, but also require reliable, fast-acting, detection systems, which may not be entirely feasible at present.

Secondary trains of air-cleaning equipment might also be designed into systems that have primary air-cleaning systems. An example is shown in Fig. 11.10. The primary air-cleaning system might be designed for full-time operation and consist of modest levels of filtration and UVGI. The secondary system would consist of a high-power

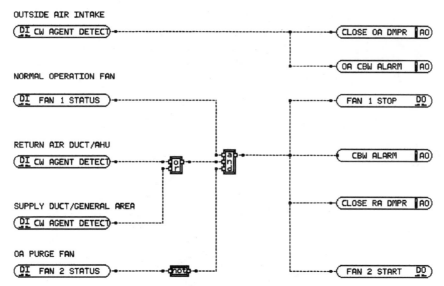

Figure 11.9 Control logic for a secondary purge system with sensors in OA, RA, and SA ducts, and isolation dampers. Prepared using Eikon™ from Automated Logic Corporation (AL 2002).

UVGI system, high-efficiency or HEPA filters, and carbon adsorbers. The purpose of such systems would be to provide extra levels of protection when needed, without the expense of operating them full time. The primary system would offer a reasonable amount of protection by itself without high operating costs.

The control logic for a secondary air-cleaning system would be similar to that shown in Fig. 11.9 except that the return air damper would not be shut. This is because the secondary air-cleaning system would process the indoor air rather than exhaust it all to the outside. The advantage of this system over the secondary purge system is that it would not be as vulnerable to an outside air intake release. Of course, multiple outside air intakes like those shown in Fig. 11.8 might also resolve this problem.

11.4.3   Sheltering zone isolation

The concept of sheltering zones involves defining or isolating particular rooms or zones of a building to which occupants can flee in the event of an emergency (Ellis 1999). Such might be the case in high-rise buildings where evacuation to the outdoors might not always be feasible.

A sheltering zone, in its simplest form, is simply a room without any ventilation supply or exhaust, and that is airtight to the surrounding areas. It would be a neutral pressure zone in this form. If the room,

Primary System

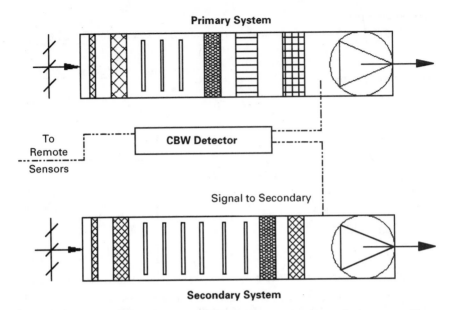

Secondary System

Figure 11.10 Example of a secondary AHU train, or an emergency train, engaged by a CBW detection signal that shuts down the primary train.

including the doors, were reasonably airtight, it would not be subject to as much contamination as other areas of the building. Individuals would have to enter the room in a manner that prevented any of the contaminant from entering at the same time. A double-door entry way, a ventilated hall outside the room, plastic drapes, or even a shower might provide such a barrier. Existing rooms in tall buildings could be converted to such applications by sealing the supply and exhaust ducts and then sealing all penetrations, including those around the doors. Without ventilation, however, such a room would only provide temporary shelter.

By positively pressurizing a room or zone with a separate ventilation system, a sheltering zone could be created to provide a high degree of protection against releases of CBW agents in other parts of the building. Figure 11.11 shows an isolation room, which is common in hospitals, in which a supply duct and an exhaust duct, each with control dampers, are used to regulate the pressure in the room relative to its surrounding areas. The exhaust damper is automatically adjusted to maintain positive pressure in the room by reducing the exhaust airflow below that of the supply. This causes excess air to exfiltrate through the walls, door gaps, and other openings, and so prevents the entry of CBW agents.

In a sheltering zone, the supply and exhaust airflow may have to come from a separate ventilation system, perhaps one that supplies

100 percent outdoor air. Since it is for emergency use only, the air conditions are not critical. Heaters could also be installed, as well as modular air-cleaning systems for added protection.

11.5 Building Automation for Immune Buildings

Building automation has made great strides in recent years, with standardization of communication protocols and equipment interfaces. Much of the building ventilation system control equipment available today is equipped with standard interfaces that respond to common control commands. The most innovative developments involve the use of the Internet to monitor and communicate with building control systems.

One standard for such interfacing that has been widely adopted is BACnet, an acronym for "building automation and controls via networks." BACnet allows control of building automation systems through networks that may include local area networks (LANs) and the Internet. Remote operation of building systems from computers connected to the Internet is possible and provides for centralized monitoring and control of multiple buildings. It also carries the danger that a terrorist CBW attack may be preceded by Internet-based sabotage of an automated building ventilation system.

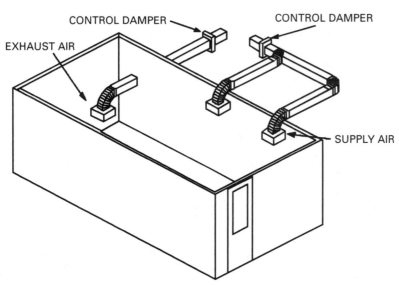

Figure 11.11   Isolation room or sheltering zone using control dampers to pressurize an area.

Building automation systems are normally programmed for automatic response to fires and for smoke control. These systems can be reprogrammed to respond to BW agent threats. If detectors are added, they can be wired into the system such that a detection signal would activate the smoke control or fire systems. Even without detectors, a manual emergency alarm could be used to activate the same smoke control systems or emergency mode of operation. If the building system includes fire dampers, building zones could be preferentially isolated to protect most building occupants, or to create sheltering zones.

One of the problems with converting a smoke removal system to a CBW removal system is that the CBW agents are almost all heavier than air and will tend to sink instead of rise under normal conditions. Smoke control systems often rely on rooftop vent fans for smoke removal, but this may not be effective against many CBW agents where internal building areas are large and open. Figure 11.12 illustrates this process comparatively for a large open atrium. The degree to which heavy gases will hug the lower elevations depends on the molecular weights of the compounds and air currents. Gases that are lighter than air, or only marginally heavier than air, may behave sufficiently like smoke to be purged by existing smoke removal systems. Most biological pathogens tend to have densities only slightly greater than that of water and may behave much like vapor clouds. Chemical agents, however, are mostly much heavier than air and may sink rather than rise.

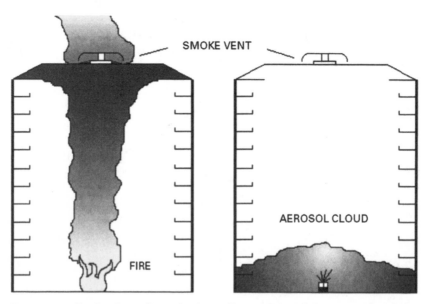

Figure 11.12   Smoke plume dispersion in a tall open atrium (left) compared with dispersion of a heavy gas in the same area (right).

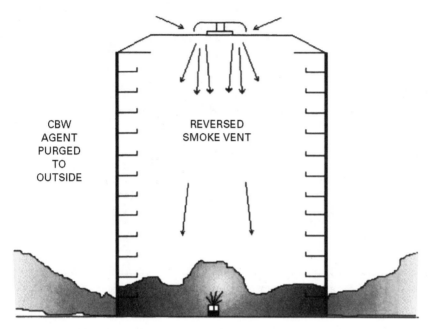

Figure 11.13   Use of a smoke vent operated in reverse to purge heavy gases from a tall open atrium.

In situations where rooftop smoke removal units are to be adapted to CBW agent removal, it may be feasible to wire the exhaust fans to run in an optional reverse mode as well as a forward mode. In this configuration, the smoke vent fans would drive the air downward into the building and it would exhaust through the inlet air paths (i.e., through doors and windows), or any designated flow paths. The logic for deciding whether the function is to purge smoke or gases can be programmed into these systems, or it could be decided by an operator who chooses the appropriate mode of operation with a single control switch or Internet command. Figure 11.13 shows the previous atrium in a CBW agent purge configuration. Forcing the CBW aerosol outdoors at ground level could pose a hazard to pedestrians, and such options should be carefully considered.

Building automation systems also have the potential to divert airflow and pressurize certain areas in relation to others. This process could be implemented in an emergency to contain the area subject to a CBW agent release, or to protect other areas where occupants may not have an escape route. The nature and location of the release would determine the appropriate response. In most cases, these could be preprogrammed, but there is also value in having a skilled operator at the controls.

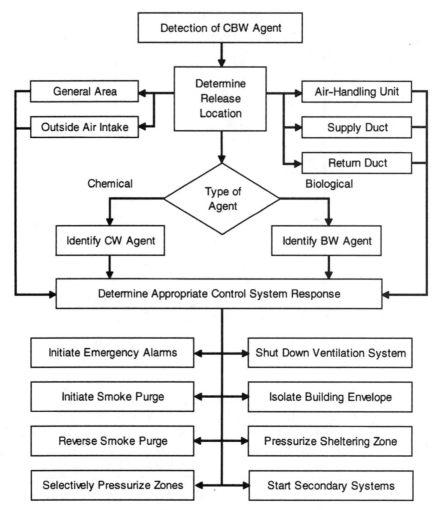

Figure 11.14   Control logic flow diagram for identification and response to a CBW attack in a high-rise building.

In a fully integrated building automation system for an immune building with CBW detection systems, the logic of the response would be, in general, as depicted in the flowchart in Fig. 11.14 for an ordinary high-rise building. Although the detection of every possible agent might not be entirely feasible or affordable today, future developments may allow such a broad-range detection system to be integrated with intelligent building control systems. The default response in the event the agent could not be identified would be building isolation and ventilation system shutdown, depending on the building. The same would be true if the location could not be identified.

The response control logic for certain facilities like large auditoriums may involve alternative functions, like selective ventilation of exit zones, and such buildings should be studied to determine how their unique characteristics might affect control system logic.

11.6 Summary

In summary, this chapter has presented several basic types of integrated immune building systems and how they might be controlled to provide maximum protection to occupants. The actual controls of such systems might be considerably more complicated since they would involve multiple detection points, multiple isolation dampers, decision-making logic based on detection levels and detection location, sheltering zone isolation and pressurization, and BACnet connections. Just as with air-cleaning systems, the ideal choice for a control system may depend on the unique characteristics of the building under study. In essence, however, all these systems will perform a simple function—they will always default to an appropriate operating condition or fail safe for the protection of building occupants. Although detection technologies may not yet be sufficiently developed to allow rapid identification of CBW agents, at least one of the suggested alternatives, particle detectors, is available and may be cost-effective as a means of initiating alarm and isolation of immune building systems.

12

Security and Emergency Procedures

12.1 Introduction

Immune building technologies may provide defense against CBW releases, but no solution is complete without addressing security measures and emergency procedures. Large buildings and high-profile targets often have such measures already in place. This chapter addresses aspects of CBW agents that may be useful for augmenting existing security measures or for developing new procedures.

Ideally, no individual intent on releasing CBW agents would gain access to sensitive areas of a building, and any release would be contained and controlled by building systems. In actuality, neither security nor engineering can guarantee absolute protection. In order to protect the building against would-be terrorists, various security measures may be necessary, since such individuals will attempt to bring in substances or equipment, and will attempt to bypass any systems that are in place. They will also attempt to outsmart any security personnel who may be on duty. The options for defending against a perpetrator include securing equipment and equipment rooms, restricting access to sensitive areas, and using metal detectors and bomb detectors.

Figure 12.1 is a simplified decision tree for managing the problem of CBW defense for any building. The two-pronged approach to the problem includes measures to prevent possible attacks, and measures to deal with any actual attacks. Note that the means for managing biological attacks versus chemical attacks are distinctly different.

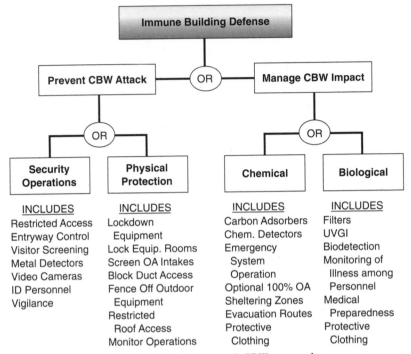

Figure 12.1   Decision flowchart for dealing with CBW agent releases.

In dealing with the human element, engineered systems for immune buildings must rely largely on human security personnel. Security operations cannot necessarily be fully automatic since any would-be perpetrator may attempt to bypass protective systems. The vigilance of security personnel is the one factor that a terrorist may not be able to circumvent. A classic example is the 1996 Centennial Olympic Park bombing in Atlanta, in which a single alert security officer, Richard Jewell, saved numerous lives by his recognition of a suspiciously abandoned backpack.

12.2   Physical Security Measures

Physical security measures can include any devices or methods that protect heating, ventilating, and air conditioning (HVAC) equipment, equipment rooms, building access, potable water supplies, or building control systems. Physical security measures can also include metal detectors, video cameras, and other equipment used by security personnel.

HVAC equipment room access should be restricted (Fig. 12.2). The air-handling equipment for any building is a sensitive area since placement of a device inside the air-handling unit (AHU) could cause maxi-

Figure 12.2 Air-handling units located in equipment rooms, like the one shown above, should be locked and access restricted.

mum fatalities in a building. Rooms should be locked, with key access limited to those personnel that need them, including maintenance personnel, building managers, and security personnel.

Roof access should be restricted if rooftop air-handling units are accessible (Fig. 12.3). Rooftop units are vulnerable to placement of CBW agent release devices and should be locked and secured. Only personnel normally requiring access should have keys to such equipment. If maintenance is handled by outside contractors, arrangements should be made to either limit key access by the contractors or require access only in the presence of building personnel.

Duct access doors should be locked or sealed. Access doors typically exist wherever ventilation equipment is present. Such equipment may include all components in the air-handling unit train, control dampers, fire dampers, and zone reheat equipment.

Air return grilles or supply diffusers larger than approximately 18 in × 12 in should be secured and/or locked to prevent possible human entry (Figs. 12.4 and 12.5). All supply and return grilles should, in fact, be secured in a manner that will prevent tampering or possible placement of a device inside the duct. Various methods can be used to secure grilles, including locks, wire tabs that are cut and replaced after normal maintenance, or alarm connections. Welding grilles shut is one option, as is the use of video cameras.

Outdoor air (OA) intakes should be reviewed for vulnerability. These provide uniquely vulnerable areas where a CBW agent could be disseminated, either inside the duct or just outside the air intakes. The simplest form of attack might be the tossing of a substance, or a jar con-

Figure 12.3 Rooftop air-handling units, like the one above, are vulnerable if access to the roof is uncontrolled.

Figure 12.4 Room exhaust grilles like the one shown here may provide easy placement of an aerosolization device, or may even permit entry to more critical locations in a ventilation system.

Figure 12.5 Ceiling supply air diffusers, like the one shown here, are unlikely placement points for CBW agents, but may provide access to critical areas of a ventilation system.

taining a chemical agent, directly into the outside air intake of a building. If the outside air intake is located on an upper floor or on the roof, then protection is offered by physically securing these areas and denying access.

When outside air intakes are located at ground level (Fig. 12.6), there is the potential for malicious acts. One alternative is to add ductwork and raise the inlet to a height (i.e., 10 to 15 ft) that will limit access. The inlet itself can be blocked with a wire mesh screen and raised at an angle so that no objects can be tossed into it. Any such modification might cause a pressure loss in the system, however, and that effect should be evaluated since it may reduce airflow or boost fan energy consumption, depending on the type of fan.

Video cameras are one option for enhancing building security. Cameras can be used to monitor all sensitive equipment areas as well as all ingress and egress locations. In addition to functioning as a deterrent when they are visible, video cameras can also create a record of people or activities as an investigation resource in the event any CBW incidents occur.

External piping for potable water should be secured. Although not part of the ventilation system, the possibility exists that the potable water piping entering a building could be contaminated with a waterborne pathogen, chemical, or toxic agent. Connections and valved openings should be secured, and such equipment should ideally be screened or fenced off to prevent access. For information and recommendations for protecting municipal water supplies, see Lancaster-Brooks (2000).

Figure 12.6 Protecting or controlling access to outdoor air intakes that are located at ground level, like the one shown here, may require fencing or video cameras.

Metal detectors at main entrances to buildings such as the one shown in Fig. 12.7 can serve a function in protecting against CBW agents being smuggled into a building. The fact is that almost any agent being transported needs to be kept in an airtight container, and this will often involve the use of some kind of metal. In addition, aerosolization units that are equipped with battery-powered compressors, or compressed air tanks, are likely to set off any metal detectors.

Access to any building control systems, including Internet-based building automation systems, should be reviewed for vulnerability. The possibility exists that a building with building automation systems or BACnet controls could be subject to a coordinated attack that would alter critical functions of the ventilation system at the same time a CBW agent was released. Evidence of tampering, reprogramming, or Internet hacking should be taken seriously in light of such possibilities.

Information relating to building mechanical and electrical systems should be controlled. Building drawings showing ventilation system components and ductwork should be classified as sensitive information and access to them should be restricted only to those personnel that normally need to use them. This may include the building engineers, facilities managers, or maintenance personnel. Requests for such information by outsiders should be investigated if the basis of such a request proves invalid or groundless. Information regarding the pres-

Figure 12.7 Walk-through metal detectors and x-ray scanning equipment may be used to detect or identify aerosolization equipment or containers for CBW agents.

ence and the capabilities of immune building systems should be considered as especially sensitive material.

Security bars can be added to ducts that are approximately 18 in wide or larger. This measure is standard in most high-security or sensitive facilities and is intended to prevent anyone from moving freely through the ventilation ductwork.

12.3 Incident Recognition

Rapid response to any CBW incident is essential to savings lives. Detection and alarm systems may not be available or functional, and so recognition of an attack in progress may be the only means of preventing casualties or fatalities. Some agents can incapacitate so rapidly that it may be impossible for anyone to sound an alarm.

Explosive CBW release devices may pop instead of explode, or may explode with a less powerful sound than would normally be expected.

Pressurized devices will make a hissing sound. The use of sprayers around buildings, especially at night, should be viewed with suspicion.

Some factors may give preliminary evidence of a chemical attack, but no immediate symptoms are likely for pathogens or for most toxins. Figure 12.8 shows a summary of possible symptoms for the various types of chemical agents. Not all symptoms will be manifest for each type of agent, and symptoms may even vary from person to person, but observation of the responses of victims may offer strong clues as to which type of agent is present. Recognition of these symptoms, in combination with any odor or the visual appearance of a gas, may also allow some remedial life-saving actions to be taken.

Other indicators of a possible chemical agent release include low-lying fog or mists in indoor areas, especially at the lowest elevations of a building. The presence of oily droplets, an oily film, or oily surfaces on water may indicate a chemical agent release. In outdoor areas, dead insects or discolored or withered vegetation may indicate the same. Unusual metal debris, exploded munitions, discarded liquid containers with oily residue or liquids, abandoned spray equipment, or spraying at night from vehicles may be indicative of a chemical release.

The covering of all exposed parts of the body would be one option during an evacuation. The use of towels and clothes soaked in bleach may provide limited protection to individuals while they are exiting an area, or if they are trapped on an upper floor.

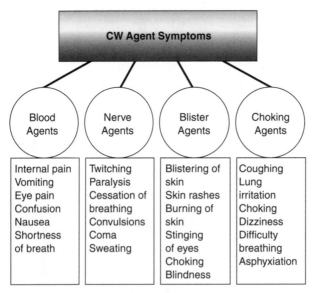

Figure 12.8 Breakdown of symptoms often associated with each of the four major categories of chemical weapons.

The symptoms due to toxins and microbial pathogens are unlikely to occur so rapidly that a building evacuation could occur before the attack was all over. The symptoms of toxins are likely to be severe illness or death within days, hours, or even minutes, depending on concentrations. The symptoms of pathogenic infections could take days to become manifest. The first warning that security or building personnel might have would be an unusual number of illnesses or absences.

A faint smell of anything unusual coming from the air supply ducts should be reported or investigated. Most CBW agents have no smell, but some may have a faint smell. Often such smells are the results of normal processes, construction work, and maintenance procedures, including cleaning activities. Security personnel should be informed of normal maintenance and cleaning schedules so that they can distinguish unusual odors occurring at off-schedule periods.

Visible vapor, powder, or even dust coming from the supply ducts may indicate the release of a CBW agent is in progress. Although most CBW agents are invisible, some chemical agents may condense in cold air and form a visible vapor. The quantities of powdered toxins that might be used are likely to be invisible, but if a sudden release occurs, it may be possible to observe powders or particles highlighted against a dark background. The sudden release of dust or other material from the supply air registers may be one indication that the system has been tampered with or that a device has been placed in the ductwork.

Sudden changes in system performance may indicate equipment tampering. The use of an aerosol generator with a solution containing toxins, pathogens, or a liquid chemical may have some effect on the humidity or temperature of the air. Changes in cooling coil or heating coil performance may indicate liquids have entered or coated the coils. Chemical agents in particular may condense on cooling coils, altering the rate of heat transfer. Chemical agents entering heating coils may combust and produce excess heat. These effects would only be likely to occur if the release point was upstream of the coils.

Casualties in a building may be distributed in relation to the ventilation system if it was used as a delivery mechanism. Casualties may be distributed around a general area if a release occurred in that area. People near supply air ducts in large office areas or other open areas may succumb to chemical agents before others located further away. Most chemical agents are fast acting, and people near supply registers may receive harmful doses before those who are at some distance from any direct supply air.

Weak or susceptible people will tend to succumb to chemical agents before anyone else even becomes aware of a problem. Asthmatics or the elderly, for example, may begin having difficulty breathing before airborne concentrations of a chemical agent become high enough to affect healthy people.

Animals and pets may be "sentinels" because they will succumb to low levels of CBW agents before humans. The Germans used sentinel dogs in World War I to warn of gas attacks. Some pets may die from chemical exposures that are not even noticeable by humans. Humans have inherent natural defenses against a variety of inhalation hazards that animals do not. Although animals may have a heightened sense of smell and may even smell agents that are odorless to us, they may not recognize a chemical smell as a hazard, at least not without training. Animals also tend to inhabit lower areas near the floor where heavier-than-air chemical agents may first drift.

In the case of bioweapon agents, pets may even die before humans become ill. The Aum Shinryko cult released anthrax spores in Tokyo in 1993 and caused no human deaths, but there were reportedly some pet deaths (Olson 1999). Observing animals or pets that are severely ill or dead in close proximity should be considered indicative of a possible chemical release.

Large numbers of people reporting ill within a day or two of each other, and within 2 to 3 days of a work day, may suggest that a biological agent has previously been released. Because the incubation period of biological agents may be a few days or more, a release may not become apparent until long after it is over, and it is essential to identify such trends as quickly as possible before any treatment becomes ineffective.

Any personnel who become suddenly and inexplicably ill should contact security or building managers immediately and not by the normal method of calling in sick the next work day. If there has been widespread contamination of a building with a toxin or pathogen, a difference of a few hours could be critical in saving lives. Waiting until the next business day to call in sick might put many victims of a CBW release beyond medical help. If, however, a large number of personnel reported an illness the night before, instead of the next day, suspicions would be aroused and emergency plans could be put into action. It is essential, therefore, that an illness-reporting protocol be in place to allow employees to report an illness at any hour of the day or night. It is not sufficient to allow supervisors to receive these calls since they may be unaware of company-wide trends. A responsible individual, such as a resident nurse, security head, or perhaps a "designated epidemiologist," should be on call 24/7 to receive illness reports from all company employees and should be trained to recognize trends and symptoms.

The employees themselves can form a critical part of any security plans for dealing with emergency situations in which a CBW release is suspected. Training plans that address incident recognition should be developed and incorporated into normal employee training programs. Employees should become familiar with the basic types of CBW agents and the symptoms they cause so that they are able to make intelligent decisions in the event they become inexplicably ill.

12.4 Emergency Response

It is the responsibility of the police to handle evacuations, and of firefighters or HAZMAT teams (Fig. 12.9) to control hazardous chemicals. However, in the event of a CBW attack on a building, the police may not have the right protective gear that would allow them to enter a contaminated area to assist evacuation. Furthermore, building personnel may find themselves in a position to take life-saving action before any emergency help arrives. Such action is not recommended for anyone

Figure 12.9 HAZMAT teams use chemical protective clothing and are equipped to deal with chemical releases. ("HAZMAT" is a contraction of "hazardous materials.") (*Photo courtesy of DuPont Tyvek.*)

but emergency response teams, but there are some things that may be done to mitigate the effects of a CBW release while it is in progress. In most cases, this means a chemical attack, since a biological agent release might not be recognized for days.

In a CBW attack, the ventilation system itself may be used as a delivery weapon. Shutting down the fan is an option that will limit the further spread of the agent and, in most cases, will be the proper first step to mitigating the consequences. But it may not always be the right choice since, for example, buildings with 100 percent outside air should leave the fans running if the release is internal. Also, variable-air-volume (VAV) systems can be switched to 100 percent outside air to purge the building in the event of an internal release.

Figure 12.10 shows a possible decision-making sequence for emergency response upon identification of a CBW attack in progress. Familiarity with building systems is essential to providing an appropriate response. Most buildings do not have 100 percent outside air systems and shutting down the air-handling unit or fan will usually be the appropriate action, depending on circumstances and the building. In fact, shutting down the air-handling unit is the traditional action recommended to prevent smoke spread in a building during a fire (ASHRAE 1999b).

If the agent is being released externally at the outside air intakes, building evacuation must be away, or upwind, from the release point. If the release has occurred outdoors and simultaneously contaminated the building air, evacuation may be an option if the gas cloud can be avoided. Otherwise, it may be best to remain indoors and seek shelter in the least contaminated area. The outdoor release scenario is, as discussed previously, the least dangerous, and probably the least likely, of all the building attack scenarios.

Emergency teams are trained and prepared to deal with hazardous material releases and a variety of guidelines and procedures are at their disposal, including the U.S. Department of Transportation's *Emergency Response Guidebook* (DOT 2000). This guidebook provides detailed information on both first aid and remediation or decontamination measures for most of the chemical weapon agents listed in Appendix D, and it is an excellent training resource for security personnel.

Specific requirements for protective clothing are defined by federal standards such as the Occupational Safety and Health Administration's (OSHA's) 29 CFR 1910.134 and/or 29 CFR 1910.156 (f) (OSHA 1999). Level A is the highest level of respiratory and physical protection. Level A protective suits are called "Vapor Protective Suits" (NFPA 1991) or "Totally-Encapsulating Chemical Protective Suits" (TECP), and consist of an SCBA (self-contained breathing apparatus) and a full-body, nonpermeable suit. This might be the choice if the hazard is unknown or biological, but Level B and Level C clothing would also offer considerable protection for most CBW agents since the danger is primarily one of inhalation, except for blister agents. Level B suits are known as "Liquid-Splash Protective" suits (OSHA 29 CFR 1910.120, Appendices A and B; OSHA 1999).

In most cases, building managers and security personnel may be aware of the type of ventilation system installed, and the sequence of steps shown in Fig. 12.10 can be simplified or adapted to particular building configurations. There is a need for personnel to be familiar with the type of ventilation system used in the buildings where they work so that they may make the correct decisions about what to do in an emergency. It is also important for each building to draw up an emergency plan and train personnel to take appropriate action.

The most important functions depicted in Fig. 12.10 are the evacuation and the shutdown of the AHU. Some options may exist in the event evacuation is not possible, such as initiating a smoke purge where such systems are present, or even disabling the CBW device. The latter function is best left to emergency response teams who will be equipped to approach such a device.

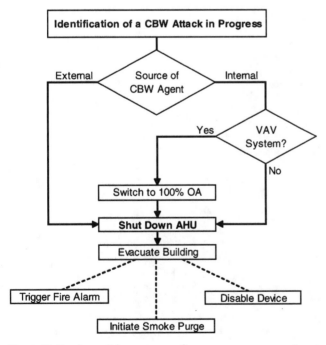

Figure 12.10   A possible sequence of emergency responses to an identified CBW attack in progress.

A variety of unorthodox responses may be possible in situations where people become trapped during a CBW release in a building. If a release occurs on the main floor of a high-rise building, evacuation may be more hazardous than staying put. Knocking out windows is one option, but it should be done on the windward side to have hope of bringing in fresh air. In high-rise buildings, the air may tend to be driven out of the upper stories by the buoyancy effect, so this approach may not necessarily improve the situation.

The use of gas masks like that shown in Fig. 12.11 is an option for security personnel, who could maintain such equipment in case of emergencies. It would not, however, necessarily be practical to stockpile this equipment for all building occupants.

Another unorthodox approach, but one that may save lives if people become trapped in a building during a chemical agent release, is to use the World War I technique of soaking towels in bleach. It is possible to neutralize some CW agents by soaking towels in cleaning agents like bleach and throwing them over the head as a crude form of gas mask. In the case of blister agents, all exposed skin surfaces should be covered with airtight materials, or even wrapped with bleach-soaked towels.

Figure 12.11 The use of gas masks or protective clothing may not be a practical approach for everyone.

Jumping into a shower may offer some limited protection against many agents, except perhaps those that are highly soluble in, or react with, water. If the building has a sprinkler system, engaging it may have a small effect in reducing the airborne concentrations of both chemical and biological agents.

One technique used by HAZMAT teams is to direct a fine water spray against a vapor cloud. The spray can actually move the vapor by affecting the air currents around the water spray. It must be emphasized here that these are unorthodox measures and should not be considered a first line of defense, but desperation measures to be resorted to in the event that all other procedural options are exhausted. The only action recommended for those who are not trained members of an emergency response team is to evacuate and await help. Additional information is available from a variety of sources on emergency response to CBW incidents (CDC 2002b; Dept. of the Army 1996; DOT 2000; Drielak and Brandon 2000; Fatah et al. 2001; Irvine 1998; NATO 1996; SBCCOM 2000).

12.5   Disabling Devices

Emergency response teams, including first responders, firefighters, and police, may discover an aerosolization or gas release device, such as shown in Fig. 12.12, operating during a release event and may need to disable it. Building security and maintenance personnel may also find themselves faced with the need to disable a device prior to the arrival of emergency teams. Since such devices are basically simple in construction and operation, it is not difficult to disable them, unless, of course, explosives or other antitampering mechanisms appear to be present. In all cases, aerosolization devices should not be approached without the proper protective gear.

If a pressurized tank is releasing gas, the valve should be closed. If the valve operator has been disabled or broken off, tools should be used to close the valve. If it is impossible to close the valve, the release nozzle should be sealed shut. This may be difficult to accomplish by denting or welding, but the normal technique in the petroleum industry is to add a connection to the outlet nozzle that includes a valve. After fitting the connection, the new valve is closed. This is a type of skill that should be practiced by CBW emergency response teams.

If a device is being driven by a compressed air source, the source should be disconnected or shut down. Compressed air tanks have valves that can be closed, or else the line running from the tank to the aerosolizer could be disconnected or broken. Figure 12.13 illustrates a device powered by a compressed air tank that could be disabled by such means.

Figure 12.12 Pressurized metal tanks with valves removed and no external source can be disabled by trained personnel using appropriate tools.

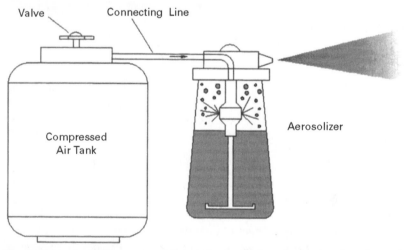

Figure 12.13 Aerosolizer powered by a compressed air tank showing the valve and connecting line, which can be closed and disconnected, respectively, to disable the device.

If a compressor is used instead of a tank, the compressor should be turned off, or disconnected from its power source. Gas-powered compressors can be turned off like any ordinary lawnmower engine. Battery-powered compressors should have the batteries or wires pulled to disengage them.

With high-tech or scientific equipment–style aerosolization equipment, electric power is almost always required. The power should be disconnected to disable these devices. The container holding the agent can also be removed from the device to disable it.

It is possible such aerosolization devices may have explosives attached to prevent anyone from disabling them. The detonation of the explosive would likely spread the CBW agent locally. The explosives should be defused first, by such methods as are normally used by bomb squads.

If a device is releasing a vapor, and not a gas, and it cannot be disabled, it may be possible to throw adsorbent towels or materials over it to limit the aerosol or even cause it to condense. Burying a device with blankets, boxes, or even sand may not stop the release, but may reduce the amount of agent that becomes aerosolized.

Some devices may be designed to disseminate the CBW agent with explosives, and not with an aerosolizer. A timer would likely be used for such a device. Timers may also be used on aerosolization devices to enable the perpetrators to escape without risk. Timers should be disabled by the methods normally used for such devices.

Binary chemicals might be used in conjunction with a mixing device or even with explosives. Binary chemicals are typically a pair of precursor agents that, when combined, will form a chemical weapon agent. These chemicals should be separated or otherwise kept from combining.

If chemical agents are spilled (i.e., in the ductwork), these agents should be contained, ideally by the HAZMAT teams, fire department personnel, or other trained specialists. In an emergency, various bleaching agents, powders, or even water can be poured on them to limit their evaporation and spread. If open containers are found placed in the ventilation duct, they should be sealed to prevent further evaporation.

12.6 Emergency Evacuation

Emergency evacuation procedures should be essentially the same as those for fire or other incidents. Designated evacuation routes and emergency exits should be clearly identified as per OSHA guidelines and other standard fire safety procedures. The distances to evacuate outdoors from an indoor CW release depend on many factors, but the recommended distance for the isolation zone is at least 60 m, based on a small release of hydrogen cyanide (DOT 2000). The area to protect persons in, the protective action distance, for this same scenario, is 0.2 km during the day, and 0.5 km during the night (DOT 2000).

If the release is indoors in a room, the evacuation should be away from that area. In a high-rise building, an evacuation may not be possible if an agent has been released on the lower floors. Once the ventilation system has been shut down, however, it may be possible to find safety by moving to the upper floors. Almost all chemical agents are heavier than air and may not spread easily to upper floors in the absence of forced ventilation.

If the release of a chemical agent has occurred on an upper floor, remaining on a lower floor is not the safest option even if the fan has been shut down, since the heavier gases and vapors will tend to sink through the building and may even collect in the basement.

If the release is outdoors, remaining indoors may be the safest short-term option as long as the ventilation system fan has been shut down and the building isolated. For additional details on evacuation procedures, see the references SBCCOM (2000), DOT (2000), and Ellis (1999). For information on securing a site for investigation, see Drielak and Brandon (2000).

In the event a released BW agent is identified as a dangerous and contagious pathogen, quarantine may be required by federal regulations. Since the building in which the release occurred must be evacuated, the occupants may have to be placed in quarantine in another location. Portable facilities are available for such incidents (Fatah et al. 2001), but evacuation to another building for quarantine purposes is also possible.

12.7   Sheltering in Place

The concept of "sheltering in place" normally involves evacuating to a designated area that has been prepared to act as a temporary shelter against any CBW release. Sheltering in place within an isolated and pressurized zone or building has been described as an option for outdoor toxic gas releases (ASHRAE 2002; Ellis 1999; Somani 1992). However, if the outdoor air is contaminated, no pressurization is possible and it would be best, under such scenarios, to simply shut down the ventilation system and seal the doors and windows.

Ideally, a sheltering area will be relatively airtight, have a separate ventilation system capable of supplying a high volume of outside air, and be capable of maintaining habitable conditions indefinitely (U.S. Army Corps of Engineers 2001). Such a design could be implemented on an upper floor of a high-rise building by providing isolation dampers to a designated area, adding airtight doors and penetration seals, and then installing fans (i.e., in the windows) to bring in outside air. Such a system could be started with a manual switch or automatic signal, which would isolate the room and start the fans. Figure 12.14 illustrates this concept with the middle floor of a high-rise building isolated from the main ventilation system.

One recommendation for sheltering in place is that these areas should be pressurized against both internal and external releases (ASHRAE 2002). A problem with this approach is that the air used for pressurization must come from somewhere. During an internal release, outdoor air can be used to pressurize a sheltering zone. During an external release, no air can be taken from the outdoors unless it is disinfected or decontaminated, and taking air from other parts of the building is

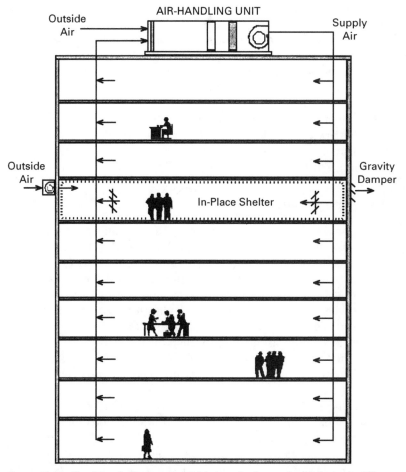

Figure 12.14 In-place sheltering against internal releases for high-rise buildings. One zone is isolated from the main system and pressurized with outside air.

possible but may not always be an option. Simply shutting the ventilation system down may be the best approach for an external release.

A less ideal, but simpler, alternative is to isolate a room without supply air and attempt to maintain it at neutral pressure relative to the surrounding areas, creating a neutral pressure zone. This would be accomplished by simply making it airtight, including sealing all possible leak paths such as ductwork dampers, doors, windows between areas, electrical penetrations, piping penetrations, etc.

An emergency option is to seal off a room to create a sheltering area during a release. If the supply and exhaust air to a room can be blocked, and the doors and other penetrations sealed, no additional agents should enter the area. Supply and exhaust registers may have local dampers that can be shut. Ducts can be blocked with pillows and blankets, duct tape, or other materials. Doors can be sealed with duct tape or wet towels.

Sheltering areas should be equipped with some basic medical supplies, plenty of water, and perhaps gas masks. Nonpermeable protective suits and gloves may be an inexpensive option. Stocking such a room with towels and bleaching compounds, especially sodium hydroxide bleaches, is another inexpensive option. A large number of commercial products for emergency use, including decontamination solutions and protective clothing, are available from various sources that also provide the military with such equipment (Drielak and Brandon 2000; Fatah et al. 2001, vol. 2; NBC 2002).

12.8 Medical Response

Medical triage is the primary responsibility of emergency medical teams, and providing first aid to victims should be left to their expertise. These personnel are trained to deal with hazardous chemical exposures and usually carry medical kits with antidotes like amyl nitrite, anti-Lewisite, atropine, diazepam, and others. Without these antidotes and the skills to recognize the symptoms, there is little that nonprofessionals could do to help any victims of a chemical or biological attack other than getting them to fresh air.

Some building managers may wish to obtain medical kits and maintain them in sheltering areas or first aid rooms. However, without proper training or on-site medical experts, such kits may not be very useful. Training in the medical arts is no simple matter, and the specialized skill of dealing with CBW scenarios may not be one that could be acquired by anyone without a medical background of some sort. There are, however, some texts on this subject that provide comprehensive treatment, such as *Medical Aspects of Chemical and Biological Warfare* (OMH 1997), the *Emergency Response Guidebook* (DOT 2000), and the NATO Handbook FM8-9 (NATO 1996), which can be consulted for more detailed information.

Medical kits for CBW protection typically include a Mark I Nerve Agent Antidote Injector Kit like that shown in Fig. 12.15. Military decontamination kits often include skin decontamination packets and hand-held spray cans of DS2 decontamination solution. Military kits also include a nerve agent pretreatment medication that consists of tablets of pyridostigmine bromide which increase the effectiveness of Mark I injectors.

Guidelines for first aid for most of the chemical agents in Appendix D are provided in the *Emergency Response Guidebook* (DOT 2000). Many of the recommendations are similar and involve moving the victims to fresh air, washing them off with water, soap, or decontaminants, removing their clothing, and rinsing them with more water. Mouth-to-mouth resuscitation can be given only if the person's face has not been contaminated. Other procedures require medical equipment, such as providing oxygen to assist those with breathing difficulties, and the use of antidotes for nerve agents. First aid measures in response to CW agent releases should always be left to medical professionals or trained personnel, but for additional information see the references CDC (2000a, 2000b), Dept. of the Army (1989), DOT (2000), Drielak and Brandon (2000), NRC (1999), Sidell et al. (1997), and USAMRIID (2001).

12.9 Security Protocol

Based on the previous discussions of release scenarios, building vulnerabilities, types of agents, types of devices, and possible placement

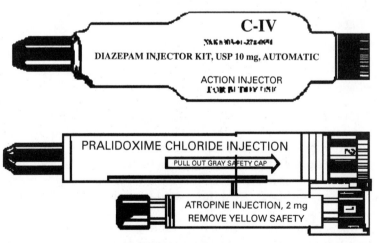

Figure 12.15 Components of the Mark I Nerve Agent Antidote Kit available to medical and military personnel.

locations, the responsibilities in regard to CBW defense of a building and related matters of concern to security personnel can be summarized and tabulated.

The following activities should be incorporated or emphasized in the normal responsibilities of security personnel:

- Monitoring building ingress and egress for suspicious individuals
- Use of metal detectors in entryways
- Video surveillance of critical or sensitive building areas
- Routine inspection of air-handling equipment for tampering
- Routine checking of locks on equipment room doors and AHU equipment

The following types of equipment should be recognized and treated with suspicion if they are used around buildings or brought into a building without a verifiable purpose:

- Foggers or aerosol generators
- Sprayers or spraying devices
- Paint sprayers
- Asthma inhaler equipment
- Devices with nozzles
- Compressed air tanks
- Pressurized gas tanks
- Compressors
- Airtight containers with powder or liquid
- Portable gas-powered spray equipment
- Briefcases or suitcases with apparent nozzles on them
- Protective equipment like face masks and rubber gloves

In addition, suspicious vehicles with spraying equipment or with external modifications, including nozzles, vapor-tight sealing, and remote-operated or computer-operated systems, should be reported to the police. The Aum Shinryko cult used both modified vehicles with computer-operated sprayers and modified briefcases with spray devices in their anthrax, botulinum, and chemical attacks (Olson 1999).

Regular inspection of ventilation or potable water equipment should enable recognition of any equipment tampering. Equipment that should be inspected or monitored includes the following:

- Air-handling units
- Ductwork
- Duct access doors and hatches
- Potable water supply piping or pumps
- Fire extinguishers

Any inspection of ventilation equipment should take special note of the following:

- Holes drilled upstream or downstream of the fan
- Removal of inlet grilles or bars
- Removal and replacement of components like filters
- Tampering or opening of valves in potable water piping systems
- Disabling of ultraviolet germicidal irradiation (UVGI) systems
- Disabling of detector systems
- System malfunctions from control panel tampering
- System malfunctions from Web-based intrusions

Buildings with intelligent controls, especially those with Internet-based control systems, may have unique vulnerabilities. Coordinated attacks on an automated building would begin with building controls altered to facilitate a CBW agent release. Outdoor air dampers might be shut and outside air cut to zero or a minimum. Control dampers for multizone systems might be adjusted via remote Internet access to attack a particular area of the building. Aerosol devices might be strategically located and engaged in concert with an altered building ventilation system. Security or building management personnel should keep a vigilant eye for any such attempts to manipulate a building's controls for an imminent CBW release.

12.10   Personnel Training

The complex nature of the threat posed by CBW agents requires heightened awareness and knowledge about various aspects of the problem. Essentially every employee in any commercial building needs to have some level of training and understanding of the danger. Building or business managers should assemble a training program to educate all employees about those matters that may be directly relevant to their safety in the event of a CBW agent release.

At the minimum, all employees or residents of a building should be familiar with evacuation routes and should recognize the symptoms of a chemical attack. In addition, employees should be advised to report any illness at the time they become ill and not the next business day. An individual, perhaps a security officer or nurse, should be designated to receive these reports at any hour of the day or night so that trends or unusual patterns can be recognized. Educating personnel with some basic BW agent epidemiology would enhance the understanding of the threats and symptoms of biological agents.

Training plans for security personnel may differ considerably from training plans for other employees, depending on the type of building. Administrators charged with responsibility for training should be able to draw up specific training plans for all personnel based on the information presented here, and in some of the references (ASD 2002; CDC 2000a; DOT 2000; Drielak and Brandon 2000; Harris 2002).

12.11 Summary

This chapter has provided information and suggestions about security and emergency procedures relating to CBW attacks, and outlined some basic procedures that may assist security or medical personnel that have been designated with responsibility in the event their building suffers a CBW attack. This information should in no way be considered sufficient for the training and education of professionals who may have to respond to emergencies—it is merely intended to supplement the information they should obtain through official training programs sponsored by accredited sources or government and municipal agencies. The information presented here on building equipment, possible terrorist devices, and emergency equipment shutdown procedures can be used to augment these training programs but is not intended to replace them.

Decontamination and Remediation

13.1 Introduction

Remediation of contaminated buildings is a specialized field that, prior to the events following the September 11 attacks, was almost strictly limited to the decontamination of buildings that had experienced excessive mold growth. In one singular case in Reston, Virginia, in 1989, an entire building was sealed off with plastic and tape when some monkeys being housed there were discovered to have been spreading the virus now known as the Reston strain of Ebola. At the time, this strain was thought to be contagious to humans and so the building was isolated and all the monkeys were destroyed. The building was subsequently disinfected with chemicals.

Remediation of buildings with mold contamination usually involves scrubbing the surfaces with bleach or other disinfectants, or else tearing out the walls and replacing them. In some cases mold contamination has been so bad or so expensive to remediate that buildings have been burned or torn down and replaced (Kowalski and Burnett 2001).

Considerable interest has been generated in the nondestructive remediation of contaminated buildings since the anthrax mailing incidents after the September 11 attacks. There are a great many liquid, vapor, and gaseous disinfectants available, but few are suitable for use inside buildings due to the damage they may cause to building materials or their inherent toxicity to humans. A variety of disinfecting agents have been proposed or marketed for such applications, including chlorine dioxide, SNL foam, and ozone.

The remediation methods discussed here are, at best, developmental technologies, and only limited engineering guidelines or design infor-

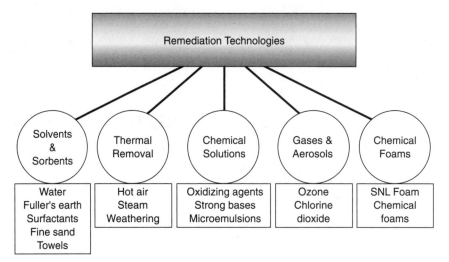

Figure 13.1 Breakdown of common decontamination and remediation technologies.

mation can be provided at present. Therefore, these methods are summarized only in terms of their present or hypothetical use. If new disinfectants are developed in the future for such applications, it is likely they would be applied in a manner similar to the agents described here, and so these techniques can be considered to be generally applicable regardless of the agent used. These methods apply generally to all CBW agents, regardless of whether they are considered persistent agents or nonpersistent agents. The use of ultraviolet germicidal irradiation (UVGI) as a surface disinfectant is not addressed here since it has been adequately covered in Chaps. 2, 8, and 16.

Figure 13.1 provides a breakdown of most of the basic technologies used for decontaminating chemical and biological agents on surfaces, for cleaning up chemical spills, or for remediating contaminated buildings. Each of these is addressed in the following sections.

13.2 Decontamination by Physical Means

Chemical and biological contamination is often dealt with by first washing the area with water. Many agents have some degree of solubility in water, but water with detergents added can be effective for removing most CW agents. Some agents are neutralized by water through hydrolysis. One of the problems with using water is that it will become a contaminated material when so used and must be disposed of properly.

Sorbents, such as diatomaceous earth, fine sand, or absorbent towels and cloth, can be used to soak up spills or scrub surfaces. Fuller's earth is one commercially available product used for this purpose. It is a form of kaolin that contains an aluminum-magnesium silicate (Fatah et al. 2001). Any material used as a sorbent remains a potentially hazardous material and must be disposed of properly.

Weathering is a form of passive decontamination in which natural exposure to temperature or sunlight will slowly break down any contaminants, including biological agents. Sunlight consists of broad-spectrum white light, with some UV and infrared light, and tends to break down many chemical compounds after prolonged exposure (Koller 1952). Sunlight also has biocidal effects and will even destroy spores in time (Beebe 1958; El-Adhami et al. 1994; Futter and Richardson 1967). In the natural environment, spores usually survive in the soil or in shade, but open exposure to sunlight will eventually destroy them (Austin 1991).

Exposing a building interior to sunlight for the purpose of destroying contaminants may be a time-consuming but nontoxic alternative to using disinfectants. One problem is that windows screen out UV, so the biocidal effect of sunlight can be enhanced by opening windows, and also by the use of reflectors to direct sunlight to shadowed areas. Open windows will also allow outside air to enter a building, and the effects of temperature extremes and variations in relative humidity also have a certain biocidal effect (Wilkinson 1966). All pathogens have a natural decay rate that is species-dependent. The natural rate of die-off can be accelerated due to exposure to the elements, and especially from temperature extremes and dehydration. Even spores will die off at some natural rate, although they may persist for decades in soil.

One alternative means of destroying spores is to cause them to germinate or grow. Once spores have germinated, they become as vulnerable to biocidal agents as any bacteria. The way to induce germination is to provide moisture, a substrate, and warm temperatures (Sakuma and Abe 1996). Under ideal conditions, spores can be induced to germinate within a matter of hours, after which they can be disinfected in minutes (i.e., by UVGI) when in their vulnerable growth state.

Hot air or steam can be used to decontaminate surfaces, but the degree to which the agent is destroyed depends on its characteristics. Steam is generally considered more effective for sterilizing equipment and surfaces than hot air and is the method of choice for sterilizing most laboratory equipment (Morrissey and Phillips 1993).

Surfactants solubilize chemical agents and convert them into a solution that can be detoxified. There are basically three types of surfactants in use today—anionic surfactants, cationic surfactants, and nonionic surfactants. Anionic surfactants are generally more effective against CW agents (Fatah et al. 2001).

13.3   Decontamination by Chemical Means

A number of chemical decontaminants are available that can be effective against CW and BW agents. These chemicals are available in various forms, including liquids, powders, foams, and vapors or gases. Some of these decontamination compounds are used for specific chemical agents. The compounds discussed here are representative but not comprehensive, since there are at least as many decontamination agents as there are CBW agents. Chemicals specifically used for the disinfection of biological agents are known as "disinfectants," and only the most common or representative disinfectants are covered here, since there are a large number of them.

Strong oxidizing agents such as calcium hypochlorite and sodium hypochlorite are effective against many CW agents. When these chemicals dissolve in water, they produce hypochlorite ions, which are effective in breaking down chemical agents (Fatah et al. 2001).

Strong bases such as CaO, Ca(OH)$_2$, sodium hydroxide, and potassium hydroxide produce high concentrations of hydroxide ions in aqueous solution. These solutions are effective in hydrolyzing CW agents.

Bleaching powder was first developed in 1798 and is an effective means of cleaning up certain hazardous chemical spills. It was used by the Germans in World War I to neutralize mustard gas. An improvement, supertropical bleach (STB), was developed in the 1950s. It is a mixture of 93 percent calcium hypochlorite and 7 percent sodium hydroxide and can also be used to neutralize nerve agents like soman, tabun, and sarin, as well as HD and VX.

Decontamination Solution 2 (DS2) was developed in 1960 and is highly effective for decontaminating CW agents. It is a nonaqueous liquid made from 70 percent diethylenetriamine, 28 percent ethylene glycol monomethyl ether, and 2 percent sodium hydroxide. DS2 is toxic to humans, and respirators and other protective equipment must be worn when handling it. A less toxic version of DS2 is available and is known as DS2P. These compounds may damage paints, plastics, and leather materials. To avoid such damage, DS2 is used for a timed period of 30 min, followed by flushing with water. The total time for 100 percent decontamination with DS2 is between 10 min and 1 h for agents like HD (Mustard), VX, or GD (Modec 2001). DS2 is available in spray canisters, portable powered brush washing systems, and large liquid containers.

Chloramine-B is an oxidant that is commonly used as an antibacterial agent. It can be used to detoxify chemical agents by soaking towelettes that have been wetted with an aqueous solution of 5 percent zinc chloride, 45 percent ethanol, and 50 percent water. It is effective against HD and VX but is not very effective against nerve agents (Fatah et al. 2001).

Microemulsions are mixtures of water, oil, surfactants, and cosurfactants that behave as a homogenous mixture. Chemical agents dissolve in microemulsions and begin to break down at a rate that depends on the size of the microemulsion particles. The smaller the particles, the faster the chemical breaks down. One such microemulsion, called C8, is commercially available as a multipurpose decontaminant agent (Fatah et al. 2001).

Table 13.1 summarizes the CW agents for which decontaminants have been identified and are effective. Also provided are liquids in which the chemical agents are soluble or miscible. To be "miscible" means the agents are partially soluble in the indicated liquid. If a chemical agent is soluble in a liquid, for example in water, then the liquid may be useful in suppressing evaporation or in containing a chemical spill.

13.4   Ozone

The use of ozone as a method of disinfecting entire rooms was first suggested by Masaoka et al. (1982), but it was studied as a disinfectant of air and surfaces by several researchers much earlier (Elford and van den Eude 1942; Franklin 1914; Jordan and Carlson 1913; Olsen and Ulrich 1914).

Ozone is a corrosive form of oxygen that is harmful to humans in concentrations above about 1 to 20 ppm. At these levels, it can sterilize spores in a matter of hours. It can be used at much higher levels inside enclosed rooms or buildings, with the only major side effect being the destruction of organic rubber and a few other susceptible materials.

Ozone tends to break down rapidly in air, especially in sunlight or in high humidity, and so it can be vented to atmosphere after its use. It can also be removed with carbon adsorbers to some degree or it can be catalytically converted to oxygen by materials like Carulite (Carus 1998).

Ozone can be generated fairly easily with various types of ozone generators like the one shown in Fig. 13.2. Some of these units, like those used for industrial water disinfection, can produce over 1500 ppm. Studies have shown that lower levels of ozone in the range of 1 to 20 ppm may be sufficient for disinfection purposes (Kowalski et al. 1998 and 2002a; Lee and Deininger 2000).

The procedure for ozonating a building is as follows:

1. Isolate the building envelope (seal exits and leak points).

2. Locate ozone sensors throughout the building with external readouts.

3. Configure the ventilation system to run in full recirculation mode.

4. Connect a blower to an ozone generator and run a tube or connection to the ventilation ductwork.

TABLE 13.1 CW Agent Decontaminants and Solubilities

CW agent	Decontaminants	Solubility
Amiton	SNL Foam	Soluble in water
Arsenic trichloride	Water, SNL Foam	Slightly soluble in cold water
Arsine	—	Slightly soluble in water
Chlorine	—	
Chloropicrin	SNL Foam	
Chlorosarin	Bleaching powder, supertropical bleach, SNL Foam	
Chlorosoman	Bleaching powder, supertropical bleach, SNL Foam	
Cyanogen chloride	SNL Foam	Soluble in water, alcohol, and ether
DA	—	Soluble in carbon tetrachloride
Diethyl phosphite	—	Soluble in water and most organic solvents
Dimethyl phosphite	—	Soluble in water, many organic solvents
Diphosgene	Decomposed by heat, alkalies, hot water	Alkalies, hot water
Hydrogen chloride	—	Very soluble in water, soluble in ether and alcohol
Hydrogen cyanide	—	Soluble in ether, miscible in water, alcohol
Lewisite	Neutralized by British antilewisite	
Methyldiethanolamine	—	Miscible with benzene and water
Methylphosphonyl dichloride	SNL Foam	
Mustard gas	Chlorines or bleach	
Mustard T-mixture	Bleaching powder, supertropical bleach	
Nitrogen mustard	Bleaching powder, supertropical bleach, SNL Foam	
O-mustard	Bleaching powder, supertropical bleach, SNL Foam	
Phosgene	SNL Foam	Soluble in benzene and toluene
Phosgene oxime	—	Soluble in water, alcohol, ether, and benzene

Compound	Decontamination	Solubility
Phosphorus oxychloride	SNL Foam, decomposed by water and alcohol in	Soluble in hot water and alcohol
Phosphorus pentachloride	SNL Foam	Soluble in carbon disulfide and CCl5
Phosphorus trichloride	SNL Foam, decomposes in moist air	Soluble in ether, benzene, carbon disulfide, and CCl$_5$
Sarin	NaOH bleaches, DS2, phenol, ethanol, SNL Foam, bleaching powder, supertropical bleach	
Sesqui mustard	Bleaching powder, supertropical bleach, SNL Foam	
Soman	Bleaches [NaOCl or Ca(OCl)$_2$], DS2, SNL Foam, bleaching powder, supertropical bleach	
Sulfur dioxide	—	Soluble in water, alcohol, ether
Sulfur monochloride	SNL Foam, decomposes on contact with water	Soluble in water
Sulfur trioxide	Combines with water to form sulfuric acid	Soluble in water
Sulfur mustard (distilled)	Common bleach and chloramine, SNL Foam, bleaching powder, supertropical bleach	
Tabun	Bleaches, DS2, phenol, aqueous NaOH, SNL Foam, bleaching powder, supertropical bleach	
Thiodiglycol	SNL Foam	SNL Foam, soluble in acetone, alcohol, chloroform, water
Thionyl chloride	SNL Foam, decomposes in water	Soluble in benzene, CCl$_5$
Titanium tetrachloride	—	Soluble in water with heat
Triethanolamine	—	Soluble in chloroform, miscible in water, alcohol
Triethylamine	—	Soluble in water and alcohol
Triethyl phosphite	—	Soluble in alcohol and ether
Trimethyl phosphite	—	Soluble in hexane, benzene, acetone, alcohol, ether
VX	Bleaching powder, supertropical bleach, SNL Foam	

Figure 13.2 Ozone generator designed for remediation applications. (*Image provided courtesy of Clarence Marsden, Trio3 Industries, Inc.*)

5. Engage the ozone generator and monitor building levels.

6. Operate the system for a specified time period (i.e., 1 to 2 days).

7. Upon completion, vent the building with outdoor air for an additional day.

8. Verify that no residual ozone is left in building before entering.

Figure 13.3 illustrates the setup of an ozone remediation system. The outside air (OA) intakes must be closed, as well as the doors and windows. The ventilation system is then run in full recirculation mode. The ozone generator can inject ozone directly into the duct or air-handling unit. It can also deliver ozone directly into the general areas if the duct is inaccessible. Ozone sensors should be distributed around the building to ensure that sufficient levels of ozone are reached everywhere.

Prior to performing remediation of a building with ozone, some estimation should be made of the ozone level needed to achieve sterilization over the specified time period. Sufficient information is available in the literature to approximate the disinfection rate constants for spores like *Aspergillus niger* and *Bacillus cereus,* and these can be used as models (Dyas et al. 1983, Ishizaki et al. 1986).

Monitoring of the ozone levels does not require expensive high-end sensors. Several low-cost ozone sensors are available with sufficient accuracy and sensitivity in the range of 1 to 10 ppm, such as the one shown in Fig.

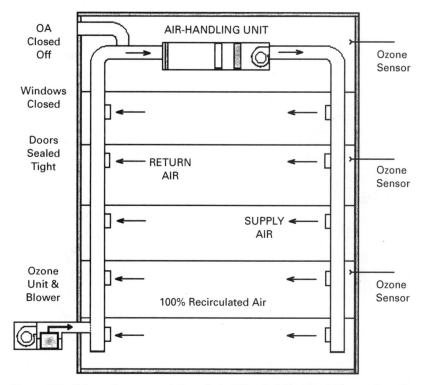

Figure 13.3 Schematic representation of a building in full recirculation mode, with ozone being injected into the supply air ductwork using an external blower.

13.4. Other low-cost sensors are available that can detect ozone at small fractions of 1 ppm (i.e., 0.02 ppm), and these sensors will be adequate to verify the absence of ozone inside a building once remediation is complete.

Figure 13.5 shows a graph of the decay rates of *Bacillus cereus* spores under exposure to ozone at various relative humidities (RHs). *Bacillus cereus* is morphologically identical to *Bacillus anthracis,* and genetically identical also except for two plasmids, making this spore a good model for anthrax spores. The ozone dose given in Fig. 13.5 is specified in units of ppm·h, which is the concentration in ppm multiplied by the hours of exposure. The original test was performed at 3 ppm (Ishizaki et al. 1986), so concentrations of ozone in this range or at least higher could have their survival rates predicted reasonably well by this chart.

The ozone dose is computed as follows:

$$\text{Dose} = C_a E_t \tag{13.1}$$

In Eq. (13.1) C_a = concentration of ozone in air and E_t = exposure time. This relationship is analogous to those used for ultraviolet light

Figure 13.4 Examples of two low-cost ozone sensors suitable for monitoring building remediation processes. The unit on the left measures ozone levels from 0 to 10 ppm, while the unit on the right can measure ozone levels up to 20 ppm and has a remote sensor. (*Images printed with permission of Larry Kilham of Ecosensors, Santa Fe, NM.*)

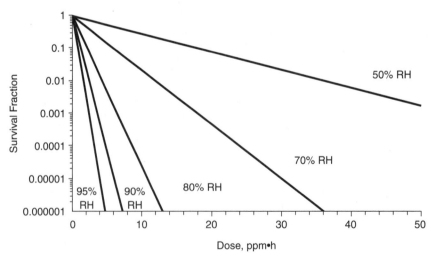

Figure 13.5 Survival of *Bacillus cereus* spores under ozone exposure at different relative humidities. Based on data from Ishizaki et al. (1986).

in Chap. 8, and for chemical exposure in Chap. 4. If the airborne concentration in ppm is known, then the time to achieve sterilization of anthrax spores can be computed using Fig. 13.5 as a guide. Sterilization is defined as a six-log (base 10) reduction, which is the abscissa, or lower axis, in Fig. 13.5.

For example, if the ozone concentration is 10 ppm, and the relative humidity is 70 percent, then the sterilization dose needed is approximately 36 ppm·h from Fig. 13.5. The exposure time can now be computed based on Eq. (13.1) as follows:

$$E_t = \frac{\text{dose}}{C_a} = \frac{36}{10} = 3.6 \text{ h} \tag{13.2}$$

It should be noted that the decay rate of *Bacillus cereus* spores under ozone increases with increasing relative humidity. The relative humidity can therefore be used to manipulate the decay rate. That is, if the building ventilation system has humidity control, then it would be desirable to maximize RH in conjunction with ozonation. This would minimize the damage to building materials as a result of high ozone levels.

Although the data on which Fig. 13.5 is based (Ishizaki et al. 1986) suggest only a single stage of decay, this may not reflect the complete decay curve. Many spores show evidence of two stages under ozone exposure (Dyas et al. 1983), just as they do under UVGI. Therefore, the following equation is proposed for use in estimating the decay of *Bacillus cereus*, which is essentially identical to *Bacillus anthracis*, under ozone exposure:

$$S(t) = 0.9991 \left[1 - (1 - e^{-1.153t})^{65} \right] + 0.0009 e^{-0.153t} \tag{13.3}$$

Equation (13.3) has an assumed second stage based on 13 percent of the rate constant of the first stage and a resistant fraction of 0.01 percent of the total population. These are reasonable estimates based on data for other spores and bacteria (Kowalski et al. 2002a). This equation is based on 3 ppm and 70 percent relative humidity, and produces the curve shown in Fig. 13.6. It can be scaled to other levels of ozone by adjusting the rate constant (1.153 in the exponent) as per the method previously described. In the event there is no second stage for *Bacillus cereus* or *Bacillus anthracis*, then this equation is conservative.

As stated previously, ozonation as a remediation technology is in a developmental stage and no actual field data are presently available. The degree to which ozone may damage building materials is unknown. Organic rubber is particularly susceptible to high levels of ozone exposure, but data on possible damage to other organic materials, or electrical and computer components, are not available at present.

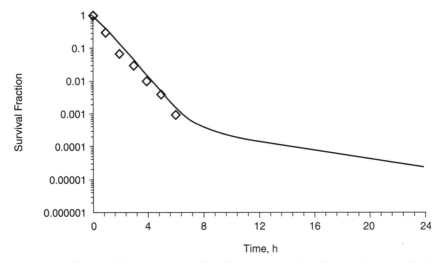

Figure 13.6 Proposed decay curve for *Bacillus cereus* and *Bacillus anthracis* under 3 ppm ozone exposure and 70 percent relative humidity. Data shown are from Ishizaki et al. (1986).

13.5   Chlorine Dioxide

Chlorine dioxide is a disinfectant that has been used to disinfect the Hart Senate Office Building in Washington, DC, of anthrax spores. This technology is relatively new in this application, however, and no data are available to use for prediction of disinfection rates. Chlorine dioxide is also an oxidant like ozone, but is used in both vapor and liquid form whereas ozone is always used in gaseous form.

Chlorine dioxide is used for odor and taste control in water supplies. It is also used as a bleaching agent. It has a boiling point of 11°C (51.8°F) and so can be stored as a liquid when kept cold or pressurized.

Chlorine dioxide's toxicity to humans has not been well studied, but the U.S. TLV (STEL) [threshold limit value (short-term exposure limit)] is 0.3 ppm, or similar to that of ozone. In one case, an exposure of a worker to 19 ppm caused fatality, but the duration of exposure was unspecified (Richardson and Gangolli 1992).

The application of chlorine dioxide for building remediation is as straightforward as that for ozone, except that chlorine dioxide would be provided in liquid form and sprayed into the air. The evaporation in air would produce gaseous chlorine dioxide, which would then be circulated throughout the building by the ventilation system. In the remediation of the Hart building, the apparatus was set up outside the building and a doorway was modified for use as a supply duct.

The actual effectiveness of the remediation of the Hart building with chlorine dioxide was less than expected, and the process had to be performed twice, with the exposure period being several days. Reports suggest that chlorine dioxide experiences the same susceptibility to relative humidity that ozone does, although data are not yet available to verify this. Anecdotal reports also suggest that residue deposited on some of the furnishings and caused some limited damage to them.

13.6 SNL Foam

An alternative to the use of gaseous ozone or chlorine dioxide is a foam developed at Sandia National Laboratories in Albuquerque, New Mexico, under the U.S. Department of Energy's DOE-NN-20 CBNP Program's Decontamination and Restoration Thrust Area (IITRI 1999; Modec 2001). The objective of this research was to develop foams, fogs, liquids, sprays, or aerosols that could be used to neutralize chemical agents and act as microbiocides against biological agents.

Figure 13.7 shows a mass of SNL Foam. SNL Foam consists of a combination of quaternary ammonium salts, cationic hydrotopes, and hydrogen peroxide. The hydrogen peroxide reacts with bicarbonates in the formulation to create effective oxidizing agents. Tests have suggested that the destruction of CW agent simulants occurred without the production of toxic by-products.

When generated as a foam, the foam remains stable and in contact for extended periods. The half-life of the foam is a measure of how much time it takes for the foam to lose half of its liquid mass to drainage or evaporation. The half-life of SNL Foam is several hours.

The pH of SNL Foam can be adjusted to suit the particular chemical or biological agent that is to be decontaminated. Powdered formulations are used to mix the foam base just prior to use, and these packets of powder can be selected to produce the desired pH. The optimum pH is 8.0 for sarin, soman, tabun, mustard gas, and anthrax spores, while the optimum pH is 10.5 for VX (Modec 2001).

Figure 13.8 shows the decay rate for anthrax spores (strain ANR-1) under exposure to SNL Foam. The actual survival at 1 h was below the detection limit, but the curve has been extended, based on the data at 30 min, to establish the time of complete sterilization, which is approximately 78 min.

Deployment methods for SNL Foam include delivery by equipment similar to that used with fire-fighting foam agents, spraying as a liquid, or aerosolization as a vapor. Generation of the foam can be accomplished through the use of compressed air injection systems. For use in remediating buildings, SNL Foam would likely be generated as a vapor or fog, and then routed through the ventilation system in the same way described previously for ozone or chlorine dioxide.

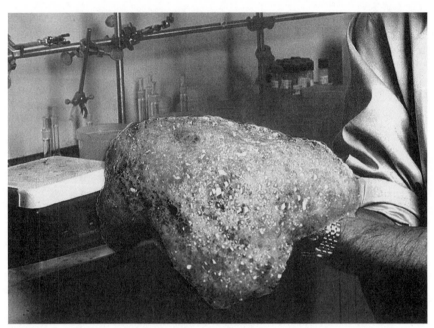

Figure 13.7   SNL Foam disinfectant. (*Image courtesy of John D. German, Sandia National Laboratories, Albuquerque, NM.*)

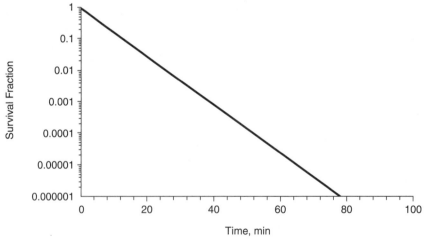

Figure 13.8   Survival of anthrax spores in contact with SNL Foam with a pH of 8.0. Based on data from IITRI (1999).

13.7 Summary

In summary, this chapter has presented several decontamination technologies that can be used for remediation in postattack scenarios. Information has been provided that will enable engineers to apply some of these technologies to the decontamination of entire buildings that have been contaminated with chemical or biological agents. Additional information is available from various references on emergency decontamination, and these include commercial sources of decontaminants and decontamination equipment (Dept. of the Army 1996; DOT 2000; Fatah et al. 2001).

14

Alternative Technologies

14.1 Introduction

The basic immune building technologies covered in previous chapters include dilution ventilation, filtration, ultraviolet germicidal irradiation (UVGI), and carbon adsorption. Other disinfection technologies exist that could be used in integrated immune building systems; however, most of these are either still undergoing research, expensive to apply for air disinfection, or limited to highly specialized applications. For completeness, and to assist the search for new solutions, these emerging technologies are described in this chapter. Readers should note that information on designing systems using the types of equipment presented in this chapter is, at present, limited or unpublished in most cases. A few of these technologies are, in fact, well understood but not widely used. The references cited in each section can be consulted for additional information.

The technologies presented here should not be considered all-inclusive, as research that is still unpublished continues in many areas and new technologies will surely arise in the future. Chemical disinfectants have been specifically excluded, since they were treated as decontaminants in Chap. 13. These technologies are roughly arranged from the simplest to the most complex. Table 14.1 summarizes the advantages and disadvantages inherent in each. Not all of these technologies necessarily have applications in CBW defense, but they may form components of an overall immune building system in which more mundane microbial hazards are controlled.

TABLE 14.1 Alternative Immune Building Technologies

Technology	Applications	Advantages	Disadvantages
Thermal disinfection	Sewage air treatment	Simple and effective	Costly
Cryogenic freezing	Food disinfection	Simple	Costly, preserves some microbes
Desiccation	Air cleaning	Cost-free by-product of desiccant or dehumidification systems	Not cost-effective as a stand-alone air disinfection technology
Passive solar exposure	Air and surface disinfection	Free	Slow, not always practical
Vegetation air cleaning	Air quality	Simple	Only partly effective, soil can breed fungi
Antimicrobial coatings	May enhance filters, protects surfaces	Prevents microbial growth, may facilitate remediation	Removal performance unknown, concerns about dislodging of coatings
Electrostatic filters	Air cleaning	Enhances filtration, removes dust	Increased energy consumption, susceptible to humidity
Negative ionization	Dust removal	Simple, not costly	Only partly effective
Ultrasonication	Water treatment	May enhance other technologies	Expensive, may not be practical in air
Photocatalytic oxidation (PCO)	Air cleaning	Removes biological and chemical agents	Performance largely unknown, may be comparatively expensive
Ozone air disinfection	Air cleaning, disinfection	Simple, effective, low cost	Possible leakage of ozone
Microwave irradiation	Air cleaning, disinfection	Simple, possibly cost-effective	Slow, performance unknown
Pulsed white light	Air cleaning	Rapid disinfection, well understood	Costly for air disinfection
Pulsed filtered light	Air cleaning, operating rooms, skin decontamination	Rapid disinfection, not harmful with UV removed	Costly for air disinfection
Pulsed electric fields	Air cleaning, food disinfection	Rapid disinfection, well understood	Costly for air disinfection
Gamma irradiation	Air cleaning, disinfection	Effective against BW agents	Hazardous, expensive, subject to regulatory control
Electron beams	Air cleaning, disinfection	Effective against BW agents	Costly

14.2   Thermal Disinfection

Heating can inactivate microorganisms, but high temperatures and long durations are required. The food industry has studied this matter extensively and guidelines for the sterilization of processed food are in common use. A variety of heating systems are available for sterilizing medical and laboratory equipment and for destroying biocontaminants. Steam heating, microwave heating, and dry-heating ovens are commonly used for sterilizing equipment, but they typically require 20 to 30 min at high temperature for complete inactivation of microbes. This precludes their use for sterilizing airstreams.

Sewage and waste processing facilities sometimes incinerate the exhaust air with direct gas flames to destroy bacteria and control odors (Waid 1969). Such systems would not be practical for use in buildings because of the high energy costs and the fact that after heating the air would have to be cooled down again, which incurs even greater costs.

14.3   Cryogenic Freezing

Cryogenic freezing temperatures to inactivate pathogens on surfaces have been used with success in certain areas of the food industry (Andrews et al. 2000). The destruction of microbes by freezing is species-dependent, and some species will simply become dormant (Mitscherlich and Marth 1984). The cost of cryogenic freezing, like the cost of heat destruction, is high and impractical for the sterilization of airstreams. However, as a means of remediating buildings, the use of freezing temperatures has the advantage of being simple to accomplish in cold climates without necessarily damaging the furnishings. That is, a building contaminated with certain pathogens susceptible to cold temperatures could be opened to the elements in winter. Freezing chemical agents would reduce the agents to liquid or solid form, precipitating them out of the air and facilitating remediation, but this is unlikely to be a practical or cost-effective approach.

14.4   Desiccation

Dehydration can be effective in destroying microorganisms, but the effect is very species-dependent, and may require extended durations. Spores have only a limited susceptibility to dehydration, and some bacterial species may be induced into dormancy by dehydration (Austin 1991; Russell 1982). The means of attaining dehydration is often simply heating, which results in the same basic high costs of energy discussed in Sec. 14.2 on heating.

Desiccant dehumidification usually involves the use of a desiccant wheel that adsorbs moisture and then releases it to a secondary airstream that then exhausts to the outside. Desiccant wheels can be used as part of evaporative cooling systems. Desiccant cooling has been shown to reduce levels of indoor species of microorganisms, and since this occurs as a by-product of the cooling process, the disinfection effect can be had for no additional cost. Desiccant cooling may be able to reduce bacterial levels in the entering airstream by 10 to 80 percent or more (Kovak et al. 1997). In the same tests, airborne fungal levels were reduced 30 to 65 percent or more.

Desiccant systems have also been shown to have the capacity to remove volatile organic compounds and other pollutants from indoor air (Hines et al. 1992). Although desiccant dehumidifiers are not intended to be air-cleaning systems, their presence in a ventilation system clearly offers some degree of potential protection against airborne contaminants.

14.5 Passive Solar Exposure

Passive exposure to solar irradiation as a means of destroying airborne pathogens (Fig. 14.1) is based on the fact that sunlight contains some ultraviolet radiation and is lethal to airborne human pathogens (Beebe 1958; El-Adhami et al. 1994; Fernandez 1996). In the ultraviolet light B band (UVB) spectrum at about 300 nm, for example, sunlight produces approximately 1 μW/cm^2 (Webb 1991). Even visible light has some biocidal effects (Futter and Richardson 1967; Griego and Spence 1978). Sunlight is one of the primary reasons that most human pathogenic microorganisms die rapidly in the outdoor air (Harper 1961; Mitscherlich and Marth 1984).

In a hypothetical design for a building using passive solar exposure to control airborne microbes, the windows form a plenum for return air, with the outside panes being UV transmitting glass (Ehrt et al. 1994). This design requires no additional space, but may increase the cooling load. Another option is the use of light pipes to bring solar radiation into areas that are normally dark, like cooling coils, and allow long-term exposure to control microbial growth.

One potential enhancement to passive solar exposure is the incorporation of titanium dioxide PCO (photocatalytic oxidation) units into the plenum. The ultraviolet light from the sun would activate the titanium dioxide and oxidize any microorganisms on the plenum surfaces, achieving an effect similar to antimicrobial coatings.

Passive solar exposure may have a limited effect on CBW agent releases but may allow for remediation of the aftereffects.

Figure 14.1   Both solar exposure and vegetation contribute to reducing the bacterial load of indoor air. (*Photo courtesy of Art Anderson, Penn State.*)

14.6   Vegetation Air Cleaning

Large amounts of living vegetation can act as a natural biofilter (Fig. 14.1), removing or reducing levels of airborne microorganisms for a variety of reasons (Darlington et al. 1998). The surface area of large amounts of vegetation may absorb or adsorb microbes or dust. The oxygen generation of the plants may have an oxidative effect on microbes. The increased humidity may have an effect on reducing some microbial species although it may favor others. The presence of symbiotic microbes such as *Streptomyces* may cause some disinfection of the air. Natural plant defenses against bacteria may operate against mammalian pathogens. In such systems, air is drawn through greenhouse-like areas before entering the ventilation system. One downside to keeping large amounts of vegetation indoors is that the potting soil may include potentially allergenic fungi.

Vegetation has also been shown to have the capability of removing indoor pollutants and chemicals (Darlington et al. 2001). The degree to which vegetation may protect against chemical weapons agents is probably limited, but the enhancement of indoor air quality by vegetation is a field worth further study as part of any immune building.

14.7   Antimicrobial Coatings

Antimicrobial coatings are available for a host of products, including filters. Antimicrobial filters are currently available, but their actual effectiveness is still under study. The primary purpose of most antimicrobial filters is to prevent microbial growth on the filter media if it should become moist. To this end they are effective, and public concerns about antimicrobial particulates becoming reentrained have not been borne out by industry experience (Foarde et al. 2000). Some sources debate the efficacy of such treatments, and comparisons of different filters in terms of their effectiveness in inhibiting microbial growth have given mixed results (Fellman 1999).

The primary concern with microbial growth on filters is that such growth, especially fungal growth, may penetrate the filter and release spores or other biological components downstream. Another concern is that contaminated filters may be a hazard to maintenance crews. To these ends, antimicrobial filters may be effective, but further study is warranted.

Antimicrobial coatings have applications where moisture or biofouling is a concern, such as in ductwork. Materials like silver and copper have long been known to inhibit the growth of microorganisms and even to destroy them in time (Thurman and Gerba 1989). Coatings made from silver ions can be bonded to metallic or other surfaces and they may have extremely long life spans, depending on wear.

Structural materials are often painted or coated with organically based compounds like latex that may be susceptible to microbial growth (Klens and Yoho 1984). The replacement of these materials or the use of antimicrobials can counteract this effect (Kennedy 2002).

Other antimicrobial materials that are currently available or under study include antimicrobial fabrics, antimicrobial coated steel ductwork, and antimicrobial duct fabric. This is a developmental field and research is ongoing. For CBW defense, one of the potential applications of antimicrobial ductwork is to facilitate remediation.

14.8 Electrostatic Filters

Electrostatic filters use an electrical charge maintained by an external source to attract particles, or else to precipitate them toward an oppositely charged surface (McLean 1988). Generally, such filters are used in industrial applications for controlling dust and other relatively large particles (White 1977) although they have become increasingly popular in residential markets. The efficiencies of such systems are not high for smaller particles, although some smaller commercial units are designed to attract smoke particles.

Some types of electrostatic filters, like electret filters, can maintain a charge from the passage of airflow and need no power supply (Tennal et al. 1991). The performance of these filters has been studied to a limited extent, and the results do not always suggest any improvement over standard filters. Furthermore, they are subject to losing their charge under high relative humidity. Little is known about their ability to remove airborne pathogens although studies abound on particulate removal (Cheng et al. 1981; Huang and Chen 2001). These units are reported to generate some ozone although the levels may not be hazardous unless they accumulate (Boelter and Davidson 1997; Hautenen et al. 1986). The performance of electrostatic filters can rival that of high-efficiency particulate air (HEPA) filters (Jaisinghani and Bugli 1991). Figure 14.2 shows one example of an electrostatic air filter for duct applications.

Some types of electrostatic filters use variations like the ionization of the air entering an electrostatic filter. In one such system, called an "electrically enhanced filter" (EEF), a series of ionizing wires is used to generate a high-intensity ionizing field. Airborne bacteria that survive the ionizing field are trapped by a grounded filter. Any bacteria that become attached to the filter are then subject to a continuous flow of ions and are ultimately destroyed (Jaisinghani 1999). Figure 14.3 shows an example of an EEF unit.

14.9 Negative Ionization

Ions are produced when electrons are stripped from or added to atoms, leaving a temporary charge imbalance. These charged atoms can cause particles like dust to clump together. Negative air ionization has an effect on reducing the incidence of respiratory infection transmission, but it is somewhat species-dependent and can be affected by relative humidity (Estola et al. 1979; Happ et al. 1966). The research has been

Figure 14.2 Electrically powered electrostatic filter for in-duct applications. (*Image courtesy of LakeAir International, Inc., Racine, WI.*)

limited, but several studies have determined that ionization has the potential to reduce the concentration of bioaerosols (Makela et al. 1979; Phillips et al. 1964).

In some situations where dust may carry microorganisms, negative air ionization can be used to reduce infections by precipitating the dust out of the air, such as in poultry houses. Some success has also been reported in burn wards, classrooms, and dental offices (Gabbay 1990; Makela et al. 1979).

Ionization units, like the one shown in Fig. 14.4, are available from various manufacturers. Ionization can also be used to enhance the efficiency of air filters in dry air by causing the agglomeration of airborne particulate matter and imparting a static charge to the particles. Such techniques are similar to those used with electrostatic filters (Chen and Huang 1998; Lee at al 2001).

14.10   Ultrasonication

Ultrasonic waves can exhibit germicidal effects (Hamre 1949; Scherba et al. 1991). When sufficient power is applied, viruses and bacteria can be atomized like liquids. There are two methods by which this may be accomplished: supersonic nozzles and sonic generators.

Figure 14.3 Bactericidal electrically enhanced filter. (*Photo courtesy of Technovation Systems, Inc., Midlothian, VA.*)

If the airstream is forced through a supersonic nozzle, a standing shock wave develops at the nozzle outlet. This shock wave dissipates energy by imparting it to the airstream, causing it to expand suddenly and rapidly. This results in the atomization, or reduction to gas, of all bioaerosols in the airstream. This technique is used effectively in ultrasonic humidifiers, but forcing high volumes of airflow through an ultrasonic nozzle would be prohibitively expensive.

A sonic generator that was tuned to resonate within a cavity would create a standing shock wave through which an airstream could pass, and in which bioaerosols could be atomized, but the power levels needed in air might be prohibitive, not to mention noisy. In liquids, a mechanism that might have biocidal effects is the generation of microscopic bubbles, which, on collapse, can damage bacterial cells (Vollmer et al. 1998).

The true value of ultrasonication may lie in its combined use with other agents such as ozone or UVGI, in which the enhancement of chemical reactions may occur. In particular, ultrasonication has been shown to reduce clumping, limiting this natural protective effect (Harakeh and Butler 1985). Mycobacteria tend to clump in moist environments (Ryan

Figure 14.4 Negative ion generator for indoor use. (*Unit provided courtesy of Electrocorp, Cotati, CA.*)

1994), suggesting the possibility that ultrasonics may be used to enhance the efficacy of other technologies. Ultrasonics may also decrease the heat resistance of spores (Sanz et al. 1985) and may also have the potential to break down some chemical compounds (Hirai et al. 1996; Hoffman et al. 1996).

14.11   Photocatalytic Oxidation

Photocatalytic oxidation (PCO) is a developmental technology that uses titanium dioxide (TiO_2), a semiconductor photocatalyst. When this material is irradiated with visible light or ultraviolet light, electrons are released for a brief period that may participate in chemical reactions that produce hydroxyl radicals and other ions (Goswami et al. 1997; Jacoby et al. 1998). Hydroxyl radicals are highly reactive species that will oxidize any volatile organic compounds (VOCs) that may contact the catalyst surface (Jacoby et al. 1996).

Hydroxyl radicals can decompose organic compounds like those on the surfaces of microorganisms, and they may also decompose some CBW agents. Pollutants, particularly VOCs, are preferentially adsorbed on the catalytic surface and oxidized to mostly harmless compounds like carbon dioxide. Since the catalytic material is not consumed in the process and oxidizes compounds, PCO is essentially a self-cleaning process.

PCO systems are currently sold as self-contained recirculation units in Japan and are made from filter materials coated with TiO_2 that is exposed to a UV lamp. A typical system is shown in Fig. 14.5.

PCO systems promise low power consumption, potentially long service life, and low maintenance requirements (Block and Goswami 1995). One study, however, suggests these systems are less cost-effective for gas phase filtration than carbon adsorption (Henschel 1998). Insufficient data are available at present to determine the effectiveness of PCO systems against biological agents, but because of the presence of filters and a UV lamp, it could be expected that this technology would operate against both microbial pathogens and chemical agents.

14.12 Ozone Air Disinfection

Ozone has been investigated previously for air disinfection, but with mixed and inconclusive results (Elford and van den Eude 1942; Hartman 1925). The disinfection of entire rooms with ozone shows some promise (Masaoka et al. 1982), but data on ozone disinfection of air remain sparse. The toxicity of ozone to humans presents an obvious obstacle to such applications, but the development of efficient ozone filters (Reiger et al. 1995) and catalytic removal of ozone (Oeurderni et al. 1996) have paved the way for the development of ozone systems that could produce sterilized, breathable air.

Low doses of ozone have proven highly effective in water (Beltran 1995; Chang et al. 1996; Hart et al. 1995; Katzenelson and Shuval 1973). Theoretical and empirical evidence suggests that most of the sterilization effect results from the radicals produced by ozone, and not necessarily the ozone itself (Beltran 1995; Glaze et al. 1980; Rice 1997). The decomposition reaction can be enhanced in air by the use of ultra-

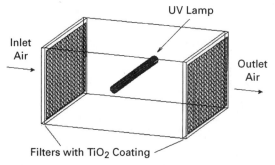

Figure 14.5 Typical arrangement of a PCO unit showing TiO_2-coated filters on the inlet and outlet.

violet irradiation and through controlled humidity (NIST 1992). Studies have shown that the effects of ozone in air parallel the effects of ozone in water, and the effectiveness of ozone for eliminating airborne pathogens in either medium should be comparable (Kowalski et al. 1998).

The threshold concentrations at which ozone inactivates viruses and bacteria in water appear to be relatively low. For example, the threshold for *E. coli* lies between 0.1 and 0.4 ppm (Fetner and Ingols 1956; Katzenelson and Shuval 1973). The Occupational Safety and Health Administration (OSHA) limit for human exposure is set at 0.1 ppm.

The decay of microbes under ozone exposure is mathematically analogous to that under UVGI exposure. Figure 14.6 shows the rate constants from experiments on *Escherichia coli* and *Staphylococcus aureus* compared with several water-based experiments performed with bacteria (Kowalski et al. 1998). The rate constants shown in Fig. 14.6 resulted from the first 15 to 20 s of exposure.

An ozone disinfection system has been developed in which the ozone is removed from the exhaust airstream with a catalytic material called Carulite™ (Kowalski and Bahnfleth 2003). Carulite is a carbon-like material impregnated with metallic salts and is one of several types of catalysts that are collectively known as "metallic salt impregnated" (MSI) carbon adsorbers. Although this system has only been used as a

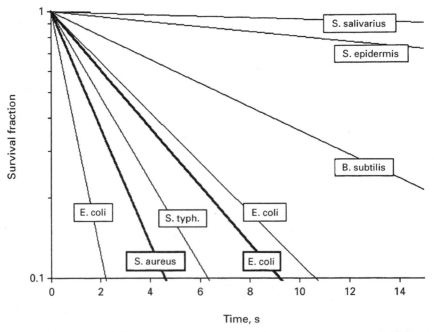

Figure 14.6 Ozone disinfection rates in air for several bacteria (Kowalski et al. 1998).

test chamber for measuring surface disinfection rates, it removes ozone completely from the exhaust air and can therefore be used as part of an air disinfection system. The economics appear to be viable in comparison with UVGI and filtration, but no full-scale model has been developed yet. A prototype ozone air disinfection system is shown in Fig. 14.7.

Ozone has other applications in which it has already proven highly successful, including water disinfection (AWWA 1971), clothes laundering, and cooling tower water disinfection. If a central ozone generator were used to provide all of these functions for a single commercial building, the added cost of an air disinfection system might prove to have viable economics, especially since no more than about 90 percent of ozone can be transferred to water (Mayes and Ruisinger 1998) and the excess could be used for air disinfection.

The expected life of the catalytic ozone removal unit also factors into the cost of operation. Experimental results suggest that about 0.11 to 0.40 kg of ozone can be removed for every 1 kg of MSI carbon prior to any efficiency decrease. At this rate, 5 kg of MSI carbon would

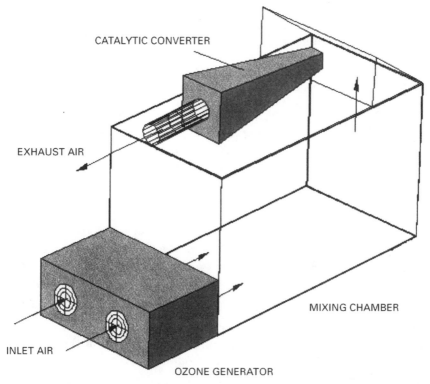

CATALYTIC CONVERTER

EXHAUST AIR

MIXING CHAMBER

INLET AIR

OZONE GENERATOR

Figure 14.7 Prototype ozone air disinfection system using a catalytic converter.

require a change-out every 6 months in a 25,000-cfm airstream with 0.1 ppm of ozone.

14.13 Microwave Irradiation

The effectiveness of microwaves for sterilization has been well established by numerous studies over the previous decades (Goldblith and Wang 1967; Latimer and Matson 1977). Microwaves heat liquids and the heating effect has often been considered to be the sole reason for the resulting disinfection. However, some studies have long suggested the existence of a "microwave effect," and these claims have caused controversy that has persisted to this day.

Microwaves consist of mutually perpendicular electric and electromagnetic waves that combine in their impact on biological systems. The primary effect of these fields is to induce the rotation of molecules that possess a dipole moment, such as water molecules (Pethig 1979). The dielectric effect on polar molecules has been known since 1912. Polar molecules are those that possess an uneven charge distribution and respond to an electromagnetic field by rotating, as illustrated in Fig. 14.8. The resulting angular momentum developed by these molecules results in friction with neighboring molecules, which then converts to linear momentum, which is the thermodynamic definition of heat in liquids and gases.

Microwave heating is not quite identical to ordinary thermal heating, or heating with external energy input. Microwaves force the molecules to rotate first, and there is a slight delay between the absorption of microwave energy and the development of random linear momentum. Studies by Kakita et al. (1995) and Rosaspina et al. (1993) have indicated effects that cannot be explained by thermal heating, such as DNA fragmentation. Further research is needed in this area, especially in light of the many claims that microwave radiation (i.e., from electric power lines and cell phones) may be a potential health hazard.

Regardless of the biocidal mechanism, microwaves have one advantage over UVGI and ozone when it comes to the problems of disinfecting mail or packages—microwaves tend to penetrate paper and would provide internal disinfection. Based on studies with spores (Cavalcante and Muchovej 1993), it is estimated that mail could be sterilized in as little as 30 min. Whether or not this is economical in comparison with other methods such as gamma irradiation is a matter that remains to be studied further. Also, some hazard exists because of the potential that metallic materials, such as CDs, charge cards, metal tape, etc., may be contained in letters.

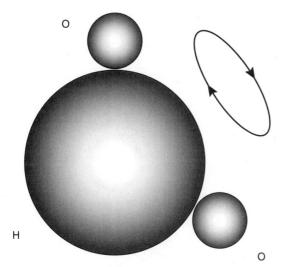

Figure 14.8 In a water molecule, the oxygen atoms (O) are located at 104.5° from each other relative to an axis through the center of the hydrogen (H) atom. The resulting dipole moment causes the water molecule to spin under the electromagnetic field created by microwaves.

Another development in microwave engineering is the use of pulsed microwaves. Pulsed microwaves can generate acoustic waves in fluids and may have biocidal effects (Kiel et al. 1999). In this capacity, they may be useful for sterilization of food or water.

14.14   Pulsed White Light

Pulsed white light (PWL), also called pulsed light or pulsed UV (PUV) light, involves the pulsing of a high-power xenon lamp for about 0.1 to 3 ms for some types of systems (Johnson 1982), or about 100 µs to 10 ms for other types (Wekhof 2000). The spectrum of light produced resembles the spectrum of sunlight but is momentarily 20,000 times as intense (Bushnell et al. 1997). The spectrum of PWL includes a large component of ultraviolet light, which may be responsible for most of the biocidal effects. Figure 14.9 shows one example of a pulsed-light unit. Figure 14.10 shows an example of a pulsed-light unit that is water-cooled.

This technology is currently being applied in the pharmaceutical packaging industry where translucent aseptically manufactured bottles and containers are sterilized in a once-through light treatment chamber (Bushnell et al. 1998). The chamber generates about 1.7 J/cm^2, or 1.7 $W \cdot s/cm^2$, at the surface of the exposed containers.

Figure 14.9 Pulsed-light disinfection system. (*Image courtesy of Alex Wekhof, Wektec Inc., Germany.*)

Only two or three pulses are needed to completely eradicate bacteria and fungal spores. Two pulses at 0.75 J/cm^2 each (1 J $= 1$ W·s) were sufficient to sterilize plate cultures of *Staphylococcus aureus* from more than 7 logs of colony-forming units (cfu's) (Dunn et al. 1997). Spores of *Bacillus subtilis, Bacillus pumilus, Bacillus stearothermophilus,* and *Aspergillus niger* were inactivated completely from 6 to 8 logs of cfu's with one to three pulses (Bushnell et al. 1998). One of the surprising aspects of PWL-exposed cultures is that they exhibit no tailing to their survival curves, at least in all the data available so far (Dunn et al. 1997).

Figure 14.11 shows the effects of pulsed light on *Bacillus subtilis* spores after exposure to a few pulses of high-intensity white light. Note the cratering effect.

Figure 14.12 shows a comparison of pulsed-light versus UVGI exposure on *Aspergillus niger* spores, from Dunn et al. (1997). According to this study, a few flashes of pulsed light were sufficient to greatly exceed the inactivation caused by UVGI. Pulsed light produces a much higher kill in less time than UVGI with a single stage of decay. UVGI exposure produces a classic two-stage curve, with a pronounced second stage.

Some reports have suggested that PWL is more efficient than UVGI, but this is not borne out by a comparison of test results. If the absorbed doses are computed for PWL exposures of microbes, they appear to be larger than the doses required for the same disinfection level by UVGI exposure. Table 14.2 summarizes the PWL rate constants from the indi-

Figure 14.10 Pulsed-light sterilization system showing hoses for cooling water. (*Image courtesy of PurePulse of California.*)

cated references and compares them with UVGI rate constants where these are known. The computed PWL rate constants are definitively lower than the computed UVGI rate constants. Although PWL may operate faster, it requires more total energy to inactivate microorganisms, and is, therefore, not more efficient than UVGI.

14.15   Pulsed Filtered Light

A secondary effect, called "pulsed-light disintegration," can occur from pulsed light when power levels are increased above those normally used. This effect is not dependent on the UV component of white light and can be achieved even after filtering out the UV spectrum. Filtering out the UV component of PWL and boosting total energy produces pulsed light that is not necessarily hazardous to humans and yet maintains biocidal properties (Wekhof et al. 2001). Bacterial cell wall rupture can result from sudden overheating with

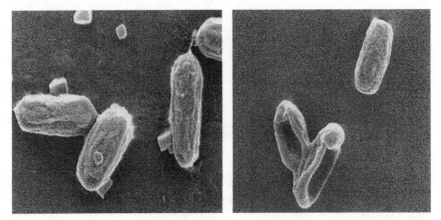

Figure 14.11 Effects of pulsed light on *Bacillus subtilis* spores. The left image shows spores before, and the right image shows spores after exposure. (*Images courtesy of Alex Wekhof, Wektec Inc., Germany.*)

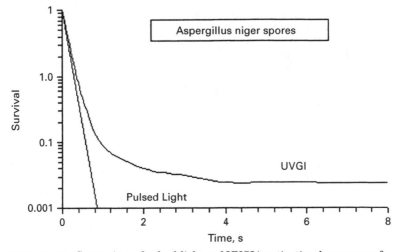

Figure 14.12 Comparison of pulsed-light and UVGI inactivation decay curves for *Aspergillus niger*.

or without the presence of the UVC band. The overheating results primarily from the UVA content, and under high doses the disintegration effect can be dominant.

Figure 14.13 shows *Aspergillus niger* spores before and after exposure to two different levels of pulsed filtered light. In the middle image, the spore has been collapsed by five pulses of 5 kW/cm^2 each. The spore on the right, which is clearly ruptured, was subjected to two pulses of

TABLE 14.2   PWL Rate Constants versus UVGI Rate Constants

Researcher	Microbe	PWL k,* m²/J	UVGI k, m²/J
Dunn 1997	*Salmonella enteritidis*	0.000351	0.22300
	Staphylococcus aureus	0.001075	0.08860
	Bacillus subtilis spores	0.000403	0.03240
Dunn 2000	*Aspergillus niger* spores	0.000261	0.00170
	Staphylococcus aureus	0.005263	0.08860
Bushnell et al. 1998	*Aspergillus niger* spores	0.000307	0.00170
	Bacillus subtilis spores	0.000307	0.03240
Wekhof et al. 2001	*Aspergillus niger* spores	0.000237	0.00170
	Bacillus subtilis spores	0.000184	0.03240
Wekhof 1991	*Escherichia coli*	0.017270	0.07675
	Bacillus subtilis spores	0.001473	0.03240

*k = rate constant.

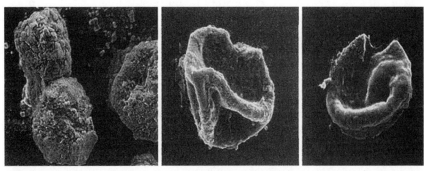

Figure 14.13   Pulsed-light effects on *Aspergillus niger* spores. The leftmost image shows spores before pulsing, the middle image shows normal levels of pulsed light, while the image on the right shows the disintegrating effects of increased pulsed-light intensity. (*Photos courtesy of Alex Wekhof, Wektec Inc., Germany.*)

33 kW/cm² each. It is not yet known at what intensity levels the disintegration effect becomes dominant.

Pulsed-light disintegration may be able to destroy bacterial cells on skin surfaces without necessarily harming skin cells due to the fact that skin cells are packed in a matrix, giving them protection against sudden overpressure. Even if skin cells were damaged, they might eventually recover from natural regeneration processes.

Applications of pulsed filtered light exist in the medical industry where the technology could be used in operating rooms during operations to control nosocomial infections. This technology has potential applications in personnel decontamination since contaminated skin surfaces can be exposed without posing any hazards.

14.16   Pulsed Electric Fields

Pulsed electric fields (PEFs) can be used to disinfect fluids, and are currently in use in the food industry for such applications. Pulsed light is actually a variation of pulsed electric field technology. Electric fields and light are both electromagnetic radiation; however, the mechanism of inactivation due to electric fields is distinctly different. In addition, spores seem to be resistant to pulsed electric fields. PEF sterilization requires an electric field of no less than 8 kV/cm. PEF exposure exhibits the characteristic survival tail and conforms to the standard logarithmic decay rate of microbes subjected to other lethal mechanisms such as radiation, biocides, and heating.

This technology has been applied to water systems, such as for the eradication of *Cryptosporidium,* and PEF systems are currently available for such applications (Clark et al. 1997). Water may attenuate the effects to some degree, but PEF may be more suitable for this application as it suffers less attenuation than PWL.

PEF involves the pulsing of an electric field of about 4 to 14 kV/cm through a liquid medium. The result of this momentary field is a membrane potential across the bacterial cell wall of more than 1.0 V, which is sufficient to lyse or damage the cell irreparably. The inactivation of various microbes, including *Escherichia coli, Lactobacillus brevis, Pseudomonas fluorescens, Bacillus cereus* spores, and *S. cerevisiae* has been found to be dependent on field strength and treatment times that are unique to each species. Since this method has little effect on proteins, enzymes, or vitamins, it is perfectly suited for food processing where the liquid medium may be anything from bullion soup to milk.

In CBW applications, PEF may have potential use for sterilizing water, but it is unlikely to be useful for disinfecting air or surfaces due to the inherent difficulty of applying it in such situations.

14.17   Gamma Irradiation

Gamma rays are a form of ionizing radiation and can be used to destroy microorganisms and to sterilize equipment (Smith et al. 2001). Gamma rays are particularly hazardous to all forms of life and can cause cellular and genetic damage. Gamma rays are emitted by radioactive materials and have the ability to penetrate solid and liquid matter. This gives them a unique advantage in the sterilization of packages and envelopes. However, materials can be damaged by gamma rays and because of the health hazards they pose they must be used with caution.

The biocidal effects of gamma rays reduce microbial populations according to relations that are similar to the effects of UVGI. The decay rate is exponential, and the response of any species to gamma rays can be defined with a rate constant, just as with UVGI. Gamma rays are used to sterilize some packaged foods (Alimov et al. 2000) and have been considered as one method for sterilizing mail, but the expense and hazards may limit their use as part of any immune building system. Figure 14.14 shows a gamma irradiation unit with heavy shielding.

14.18   Electron Beams

The use of electron beams as a sterilization technology has become more common as the technology has advanced (VanLancker and Bastiaansen 2000). Electron beams are related to gamma ray technology and to X-ray technology, but they don't depend on radioactive sources (Alimov et al. 2000). They have the same basic penetrating ability as gamma rays and have applications in equipment and food sterilization. They have also been tested in applications involving the sterilization of contaminated mail.

Figure 14.14 Gamma irradiation system from J. L. Shepherd. (*Photo courtesy of Defense Microelectronics Activity, U.S. Secretary of Defense.*)

14.19 Summary

This chapter has presented an overview of the variety of alternative technologies that are available or under development and that may be used for protecting building occupants against CBW agents. Except for some of these technologies like pulsed light, little information is available to establish their performance against chemical or biological agents, but data may eventually be available as research and development continues.

Economics and Optimization

15.1 Introduction

Any immune building system can be optimized to improve performance, economics, or both. Given some defining criteria for performance, and a finite budget, the best system for the application can be selected or designed. The basic principles of engineering economics can be applied to any system to obtain the highest removal rates of CBW agents, or lowest fatality estimates, for the money. The removal rates form the criteria for assessing performance, and life cycle costs form the basis for assessing economics.

15.2 Selecting Performance Criteria

The analysis in Chap. 9 suggests that moderately sized filtration and ultraviolet germicidal irradiation (UVGI) systems provide a high degree of protection for building occupants. Results also suggest that increasing filter size above a nominal filter efficiency of 80 percent offers no significant further reduction in fatalities. Figure 15.1 exemplifies this effect. In this figure, each of the six building types under study shows a significant decrease in anthrax fatalities with a 60 or 80 percent filter, but no additional benefits from a 90 percent filter. The same diminishing returns effect applies to UVGI systems as well, and applies to all pathogens.

The conclusion to be drawn from the attack simulations in Chap. 9 is that investing in excessive filtration and higher UVGI system power than is needed will not provide cost-effective reductions in predicted fatalities. The reasons for this are largely due to the fact that the ven-

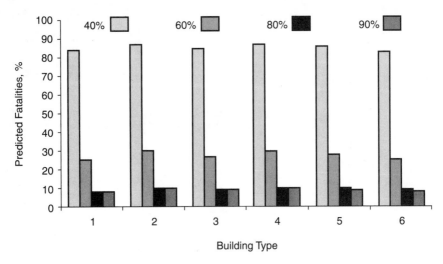

Figure 15.1   Effect of increasing nominal filter size (DSP) on reduction of fatalities due to anthrax spore release. Building types correspond to those in Tables 9.14 and 9.15.

tilation system flowrate limits the effectiveness of any installed air-cleaning system. Past some point, only increasing the total system airflow or the total outside airflow will improve performance. Since this is likely to be uneconomical in most cases, the most cost-effective system that can be installed is defined by the maximum reduction in predicted fatalities for any particular building. Of course, this greatly simplifies the search for a balance between risk and system cost, since the most effective system possible is not likely to be costly.

The most practical approach for protecting a building against CBW agents is to design an affordable system that will offer significant protection to most building occupants. Although a specific cutoff or minimum target for predicted fatalities cannot be assigned in advance for all buildings, Table 15.1 provides some general guidelines for sizing CBW defense systems based on the previous analyses. These are not absolute limits since situations can vary with each building and each ventilation system, but these values should provide a starting point for determining what components may be effective for each building and what the basic life cycle costs may be.

15.3   Economics of Filtration

Filtration economics are fairly straightforward to evaluate. Life cycle costs depend on the first costs of the filter and framing, the replacement costs of the filters, the annual maintenance costs, and the energy costs due to possible increased fan motor power.

TABLE 15.1   Suggested Criteria for CBW Defense Systems

Filter size (minimum)	60%	MERV 11
Filter size (maximum)	80%	MERV 13
UVGI average intensity (minimum)	1000	URV 11; μW/cm^2
UVGI average intensity (maximum)	2000	URV 13; μW/cm^2

MERV = minimum efficiency reporting value, URV = UVGI rating value. Based on ASHRAE Standard 52.2—1999 (ASHRAE 1999a).

Typically, ventilation systems already include dust filters and/or pre-filters, so there are no significant first costs associated with upgrading to higher-efficiency filters, only energy and maintenance costs. Therefore, in the following examples, only annual energy and maintenance costs are included. However, for completeness, following is the general equation for the total first cost, FC\$, of a filtration system:

$$FC\$ = Frame\$ + Labor\$ + FilterFC\$ \qquad (15.1)$$

In Eq. (15.1), Frame\$ refers to the cost of a filter frame for holding the filter cartridges, Labor\$ is the cost of installation, and FilterFC\$ is the first cost of the filters.

The annual cost of owning and operating a filter can be written in terms of the annual energy and maintenance costs, ME\$, as follows:

$$ME\$ = FP\$ + FR\$ + MAINT\$ \qquad (15.2)$$

In Eq. (15.2), FP\$ refers to the fan power consumed by the additional system pressure loss due to the filter. Ideally this will be determined from the impact on the fan, based on the manufacturer's fan performance curve. In the following example, however, it is estimated by assuming that the air horsepower represents the increased power consumption of the fan. Although this will not be correct for all types of fans, it is a conservative estimate for most commonly used ventilation fans. The value FR\$ represents the filter replacement cost and the term MAINT\$ represents the maintenance costs.

Equations (15.1) and (15.2) can be used as a starting point for evaluating the economics of filtration, but in lieu of writing equations, the following calculations provide good examples of the actual computations.

Table 15.2 summarizes the life cycle costs of four filters—nominal 25, 80, and 90 percent filters, and a high-efficiency particulate air (HEPA) filter. In this comparison, the outside air is assumed to have a constant concentration of 100 spores per cubic meter, which are added to the building via the 25 percent outside air. It is also assumed that 2 percent of the 6000 building occupants are producing airborne bacteria

TABLE 15.2 Life Cycle Costs of Filters

Design airflow	100,000	100,000	100,000	100,000	100,000	100,000	cfm
Model	20–25%	40–45%	60–65%	80–85%	90–95%	HEPA	
Height	24	24	24	24	24	24	in
Width	24	24	24	24	24	24	in
Depth	1	2	4	6	12	11.5	in
Face area	4	4	4	4	4	4	ft^2
Air velocity	350	500	500	500	500	250	fpm
Media velocity	1.023	1.100	0.750	0.350	0.097	0.025	m/s
Pressure losses at design conditions							
dP average	0.56	0.62	0.78	0.955	0.975	1.25	in.w.g.
Air HP	8.792	9.734	12.246	14.994	15.308	19.625	hp
Fan efficiency	84.0	84.0	84.0	84.0	84.0	84.0	%
Motor efficiency	95.0	95.0	95.0	95.0	95.0	95.0	%
Combined efficiency	79.8	79.8	79.8	79.8	79.8	79.8	%
Additional fan motor HP	11.02	12.20	15.35	18.79	19.18	24.59	hp
Total energy cost	14.77	16.36	20.58	25.20	25.72	32.98	kW
Operating period	3,744	3,744	3,744	3,744	3,744	3,744	h
Total fan energy	55,316	61,243	77,047	94,333	96,309	123,473	kW·h
Rate	0.08	0.08	0.08	0.08	0.08	0.08	$/kW·h
Annual energy cost	4,425	4,899	6,164	7,547	7,705	9,878	$
Replacement costs							
Average filter life	8,760	8,760	8,760	8,760	8,760	8,760	h
Number of filters	72	50	50	50	50	100	
Filter hours/year	269,568	187,200	187,200	187,200	187,200	374,400	h
Replacements/year	31	21	21	21	21	43	
Cost/filter	13	23	38	49	62	75	
Annual energy cost	400	492	812	1,047	1,325	3,205	$
Prefilter energy cost						800	$
Maintenance cost	200	200	200	200	200	300	
Subtotal energy and maintenance	600	692	1,012	1,247	1,925	4,306	
Total annual cost	5,025	5,591	7,176	8,794	9,630	14,183	

and viruses at a rate of 100 microbes per hour. The microbial challenge to this building approximates normal winter conditions and provides an aerobiologically balanced array of test pathogens. Such simulations can be adjusted to target any scenario, whether it be protecting a residence from allergens, a hospital from nosocomial infections, or a commercial building from a bioterrorist attack.

Note that although the energy costs increase roughly linearly from the 25 percent to the HEPA filter, the annual costs increase almost exponentially. This is partially due to the fact the 90 percent and HEPA filters each include a 25 percent prefilter, but also because the HEPA operates at 250 fpm compared to 500 fpm for the other filters. This velocity difference necessitates twice as many HEPA filters for the same total airflow.

The model building for this economic study assumes a 100,000-cfm system with a fan efficiency of 84 percent and a motor efficiency of 95 percent. Filter pressure drops varying from clean and dirty for the life of each filter were accounted for, with fans operating continuously. Representative filter costs from a leading manufacturer were used.

The average pressure loss over the life of the filter is normally provided in the manufacturer's catalogs. The fan curve for each system should be used to determine the increase, if any, in fan power consumption and the impact on total system airflow. As an estimate of the change in fan energy consumption, the air horsepower (hp) required to move the design airflow through the new filter can be computed as follows:

$$\text{hp} = \frac{hQ}{6350} \tag{15.3}$$

In Eq. (15.3), h is the head loss through the filter in units of inches of water gauge (in.w.g.) and Q is the total airflow in cfm. This estimate of the energy consumption due to the filter assumes that total airflow remains the same, which may not always be the case.

Given the air horsepower from Eq. (15.3), the fan motor horsepower, P, can be computed by factoring in the fan efficiency and the fan motor efficiency as follows:

$$P = \frac{\text{hp}}{\eta_{\text{fan}}\eta_{\text{motor}}} \tag{15.4}$$

In Eq. (15.4), η_{fan} is the fan efficiency and η_{motor} is the motor efficiency. The total power, P, can be converted to kW by multiplying by the conversion factor 1.341, after which the cost can be computed if the energy cost, in \$/kW, is known. In the example in Table 15.1, the energy cost was assumed to be 0.08 \$/kW, but this may not be true in every area of the country. California, for example, has recently been subject to much higher energy costs.

To round out the economic evaluation of filters, the bottom section of Table 15.1 shows the filter first costs, the replacement costs, and the maintenance costs. Summed together with the annual energy and maintenance costs previously computed, the last line in Table 15.1 shows the total annual operating cost, or annualized life cycle cost, of each filter.

The life cycle costs for the six filters in Table 15.1 are graphically summarized in Fig. 15.2. As was previously demonstrated, there was no significant increase in performance, in terms of reducing predicted fatalities, from the use of a 90 percent filter in place of an 80 percent filter. Since the 90 percent filter has a measurable cost increase, it can be concluded that the 90 percent filter will prove to be less economical for normal applications. This would be even truer for the HEPA filter, and the economics of HEPA filtration have been shown elsewhere to have limited justification (Kowalski and Bahnfleth 2002a). HEPA filters can be justified where the performance criteria may require them, but they are difficult to justify in terms of economics for applications in normal (i.e., nonhealthcare and nonmilitary) applications.

15.4   Economics of UVGI

There are basically two types of UVGI systems—in-duct systems and stand-alone recirculation units. The type of UVGI system used for

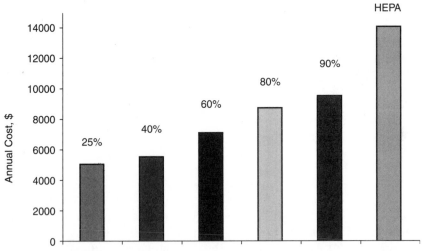

Figure 15.2   Comparison of life cycle costs for four filters. The 90 percent and HEPA filters include a prefilter. Based on the results of Table 15.1.

microbial growth control (MGC) is virtually identical to in-duct systems, the only difference being in the lamp power and possibly the arrangement of lamps. If the disinfection rate, or the kill rate, is specified as the design criterion, the average intensity of any UVGI system can be determined. If the average intensity is known, or is specified as the design criterion, then the first costs and operating costs can be determined in a straightforward manner.

The first costs of a UVGI system include the materials and components—the UV lamp assemblies, ductwork, reflective materials, prefilters, and installation costs. For retrofitted systems, there is probably no ductwork to be installed.

The operating costs of a UVGI system include UV lamp energy consumption, fan energy consumption, cooling energy costs, maintenance costs, lamp replacement costs, and prefilter replacement costs. The fan energy consumption is usually negligible, and the prefilters may not be needed for in-duct systems since these often already contain filters.

In order to facilitate the economic evaluation of UVGI systems, the various cost relationships must be quantified. General relationships can be established between lamp assembly first cost and lamp UV power, the cost of reflective materials as a function of reflectivity, the cost of sheet metal ductwork, the cost of a filter assembly, the cost of a fan and motor, and the cost of assembly or construction of a UVGI system. These particular cost estimates are based on quotes from various sources, including manufacturers' catalogs, Internet sources, and quotes from suppliers.

The cost estimates used in these relationships are representative costs, and do not necessarily reflect the lowest-cost systems since only a limited number of manufacturers were able to provide complete information. These are provided for estimation purposes and to illustrate the methods of UVGI cost analysis. If a system is being designed for a specific application, quotes should be obtained for the components in order to determine the life cycle costs with as much precision as possible.

Figure 15.3 illustrates the basic configuration of a typical UVGI system. The dimensions are defined as W = width, H = height, and L = length. This configuration, known as crossflow, in which the airflow direction is perpendicular to the lamp axis, is the most common configuration.

The total system UV power depends on the model of lamps used and the number of lamps. A survey was performed of 33 commercially available systems, and Fig. 15.4 shows a summary plot of their associated first costs. These costs include the lamp, ballast, and assemblies, but no filters or ductwork since they are in-duct systems designed for installation in existing ventilation systems. All of these in-duct systems

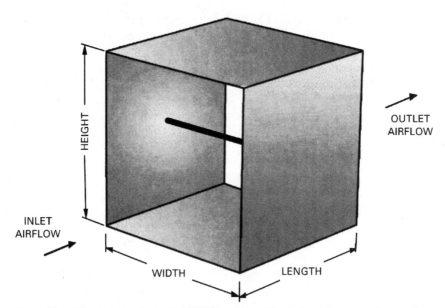

Figure 15.3 Dimensions of a typical UVGI system with one lamp in a crossflow configuration.

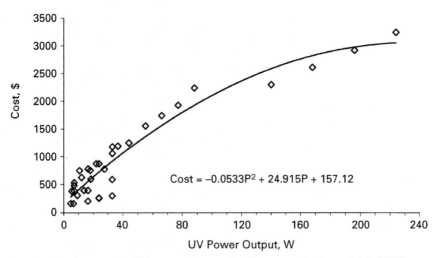

$$Cost = -0.0533P^2 + 24.915P + 157.12$$

Figure 15.4 Cost versus UV power in watts for 33 commercially available UVGI systems. System airflow is based on a design velocity of 500 fpm.

are, in fact, merely lamp assemblies that require only attachment and a power source for operation.

Figure 15.4 includes a curve for a second-order polynomial least-squares curve fit to the data. Defining UVP$ as the cost of UV lamp power, the equation defining the cost versus power relationship can be written as follows:

$$\text{UVP\$} = -0.0533P^2 + 24.915P + 157.12 \qquad (15.5)$$

Equation (15.5) is obviously an approximation and may diverge by an order of magnitude at low UV power levels. This is because of the wide variation in costs of low-power UVGI systems. Some of these UV lamps come in specialized types like non-ozone-producing lamps, biaxial lamps with single-ended connections, and lamps that are suited for low- or high-temperature operation. These factors are not accounted for in Eq. (15.5) or in the following economic analysis and must be dealt with on a case-by-case basis.

The reflective material cost depends on the total surface area and type of reflective material. Figure 15.5 shows a sampling of reflective material costs based on quotes from a handful of sources. The relation for the reflective material is based on the actual costs for three materials—ordinary galvanized duct, sheet aluminum, and polished sheet aluminum—and the estimated costs for coated sheet aluminum. The UV reflectivities for these materials are approximately 54, 65, 75, and 85 percent, respectively. The galvanized duct has a cost of zero since it is the duct surface already present in the system. Because these reflective materials are

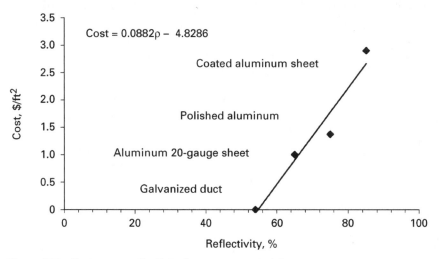

Figure 15.5 Cost versus reflectivity for common materials.

likely to be purchased in small, not bulk, quantities, large variations in price are possible and specific quotes should be obtained for any actual installation.

A representative cost for the duct was found from a survey of duct and sheet metal manufacturers. Typically, 16- to 20-gauge galvanized sheet metal at about \$1.70 per square foot is used for such duct, although it is possible to use other sheet metal thicknesses, depending on the application. Since the area of the duct will be the circumference multiplied by the length, the cost of the duct, designated Duct\$, can be written as follows:

$$\text{Duct\$} = 1.70 \, (2W + 2H) \, L = 3.40L \, (W + H) \qquad (15.6)$$

The linear equation in Fig. 15.5 is a least-squares curve fit of the data, and can be written with RF\$ defined as the first cost of the reflective material per square foot:

$$\text{RF\$/ft}^2 = 0.0882\rho - 4.8286 \qquad (15.7)$$

The area of the reflective surface is the same as the area of the duct, and so we can write the first cost of reflectivity in terms of the dimensions W, H, and L as follows:

$$\text{RF\$} = (0.1764\rho - 3.40) \, L \, (W + H) \qquad (15.8)$$

A representative cost of labor was obtained by surveying duct fabricators. The baseline cost for UVGI systems was estimated as \$460, and the labor cost is a function of any increase in size over the cost of this baseline system. The baseline system cost was estimated to be 0.5 \$/ft^2. The labor cost, called Labor\$, includes the cost of installing the reflective materials, and is defined by the following relationship:

$$\text{Labor\$} = 460 + 0.5 \, [2L \, (W + H) \,] = 460 + L \, (W + H) \quad (15.9)$$

The fan first cost is designated FanFC\$ and depends mainly on the fan power, type of fan, and volume of air moved. From a survey of available fan models from several manufacturers, the cost per cfm was estimated to be 0.01452 \$/cfm. The fan first cost for a UVGI system is

$$\text{FanFC\$} = 0.01452Q \qquad (15.10)$$

In Eq. (15.10), the airflow, Q, is given in cfm. If the airflow is specified in m^3/min, the equation can be written in metric form as follows:

$$\text{FanFC\$} = 0.01452 \, (35.315) \, Q = 0.5128Q \qquad (15.11)$$

The filter first cost, designated FilterFC$, was determined by surveying commercially available filter models and was found to be approximately proportional to the filter face area. The cost was estimated as $10/ft², and so the filter first cost can be defined as follows:

$$\text{FilterFC\$} = 10WH \tag{15.12}$$

The total first cost, FC$, of any UVGI system is the sum of all the component first costs:

$$\text{FC\$} = \text{UVP\$} + \text{Duct\$} + \text{RF\$} + \text{Labor\$} + \text{FanFC\$} + \text{FilterFC\$} \tag{15.13}$$

Substituting Eqs. (15.5), (15.6), (15.8), (15.9), (15.11), and (15.12):

$$\text{FC\$} = -0.0533P^2 + 24.915P + 157.12 + 3.4L\,(W + H)$$
$$+ (0.1764\rho - 9.66)\,L\,(W + H) + 460 + L\,(W + H)$$
$$+ 0.5128Q + 10WH \tag{15.14}$$

Simplifying equation (15.14) and converting L, W, and H from feet to centimeters produces the following expression for the first cost of a UVGI system:

$$\text{FC\$} = -0.0533P^2 + 24.915P + 617.12 + (0.0001899\rho$$
$$- 0.001076)\,L\,(W + H) + 0.5128Q + 0.1076WH \tag{15.15}$$

The first cost can be annualized (AFC), at 8 percent interest, with the capital recovery factor as follows:

$$\text{AFC\$} = \frac{0.08\,(1.08)^{20}}{(1.08)^{20} - 1}\ \text{FC\$} = 0.10185\text{FC\$} \tag{15.16}$$

Combining Eqs. (15.15) and (15.16), the annualize first cost becomes

$$\text{AFC\$} = -0.00543P^2 + 2.538P + 62.85 + (0.0000193\rho$$
$$- 0.00011)\,L\,(W + H) + 0.0522Q + 0.01096WH \tag{15.17}$$

The energy and maintenance costs must also be evaluated. The energy cost due to the total power consumption of the UVGI lamp is PL$. There is also an energy cost due to the excess heat the lamp puts into the ventilation system, termed LH$. Finally, there is a pressure loss associated with both the lamp assembly and the duct, although this will generally be negligible. It is assumed, furthermore, that the UVGI system comes complete with a prefilter, or that it already exists in the system, to maintain cleanliness.

The energy cost of the lamp, LP\$, is computed based on the total power consumption (at an assumed efficiency of 76 percent), and at a cost of 0.08 \$/kWh for full-time operation. The lamp energy cost is then defined as

$$LP\$ = \frac{(0.08)\,(8760)\,P}{0.76} = 2.06118P \qquad (15.18)$$

The cooling cost of the lamp, CP\$, is based on a location of Philadelphia, Pennsylvania, where the cooling load will be present for approximately 70 percent of the year, based on analysis by the BLAST program (DOE 1994; Kowalski 1998), and in which the lamp energy is assumed to be ultimately converted entirely to heat:

$$CP\$ = \frac{0.08\,(0.70)\,8760P}{0.34\,(1000)} = 1.443P \qquad (15.19)$$

Pressure losses are based on combined entry and exit losses that have a combined loss coefficient of $C_0 = 1.5$. The lamp losses depend on the cross-sectional area of the lamp. Assuming the lamp spans the entire duct, which is a conservative assumption, the loss coefficient for the entry, exit, and lamp can be written as a function of the radius. The loss coefficient for lamps in the size range under consideration, and in the air velocity range of interest, is approximately 0.11, conservatively, for all cases (ASHRAE 1985). Incorporating the definition of the velocity pressure, the lamp pressure loss, dP_L, can be written as

$$dP_L = 1.61\,(0.075)\,Pv = 0.12075\,Pv \qquad (15.20)$$

The pressure loss due to filters associated with the UVGI systems accounts for the energy costs. The design pressure loss of a nominal 60 percent DSP (dust spot efficiency) filter over its life is estimated as approximately 0.56 in.w.g. at a design velocity of 500 fpm. This yields a pressure loss coefficient of $C_0 = 35.94$, which is used to account for variations in air velocity. The filter pressure loss, dP_F, is written as follows:

$$dP_F = 35.94Pv \qquad (15.21)$$

The velocity pressure can then be written as

$$Pv = 0.075 \left(\frac{Q35.315/\,(10.764WH/10,000)}{1097} \right)^2 = 67.0822 \left(\frac{Q}{WH} \right)^2 \quad (15.22)$$

The total pressure loss for the lamp and the filter can now be written as

$$dP = (0.12075 + 35.94)\,Pv = 2419\left(\frac{Q}{WH}\right)^2 \qquad (15.23)$$

The total energy loss due to the pressure drop of the duct, lamp, and filters together, termed P_F, is then calculated based on the air horsepower with an assumed fan efficiency of 0.75 and a fan motor efficiency of 0.75. Including all the appropriate metric unit conversions, the equation for the total power consumption is as follows:

$$P_F = (2419)\left(\frac{Q}{WH}\right)^2 \frac{745.7\,(0.0001575)\,35.31Q}{(0.75)\,(0.75)\,(1000)} \qquad (15.24)$$

Dimensions W, H, and L are specified in cm, and Q is specified in m³/min. Simplifying Eq. (15.24) results in the following equation for the power consumption of the fan due to the pressure losses:

$$P_F = 17.837\,\frac{Q^3}{(WH)^2} \qquad (15.25)$$

The annual fan power cost, designated FP\$, becomes

$$FP\$ = (0.08)\,(8760)\,17.837\,\frac{Q^3}{(WH)^2} = 12,500\,\frac{Q^3}{(WH)^2} \qquad (15.26)$$

Maintenance is assumed at a flat rate of \$50/year, regardless of the system size. Lamp replacements are assumed to occur at a rate of 1.0274/year, based on the normal life of 9000 h for such lamps, with all UV lamps operating for 3744 h/year (assuming 12 h/day and 6 days/week of operation). It is further assumed that the basic replacement lamp is 5.33 UV watts and costs \$28.23. This will, of course, vary from lamp to lamp, and actual values for UV lamps should be used where they are known. Assuming further that the relationship between replacement lamp costs and lamp size is continuous between the minimum and maximum power levels, we can write the lamp replacement cost, LR\$, as follows:

$$LR\$ = \frac{1.0274\,(28.23)\,P}{5.33} = 5.442P \qquad (15.27)$$

Filter replacement costs, FR\$, are based on an estimated cost of \$4.09/ft² of filter surface area (for a 60 percent filter), or \$44.025/m². Assuming again that they are continuously distributed over the size range and that they are replaced annually, the following relation can be written:

$$FR\$ = \frac{44.025\,(WH)}{10{,}000} = \frac{WH}{227.14} \qquad (15.28)$$

The total annual maintenance and energy cost is

$$ME\$ = LP\$ + CP\$ + FP\$ + LR\$ + FR\$ + 50 \qquad (15.29)$$

Inserting Eqs. (15.18), (15.19), (15.26), (15.27), and (15.28) produces the following equation:

$$ME\$ = 2.06118P + 1.443P + 12{,}500\,\frac{Q^3}{(WH)^2} + 5.442P + \frac{WH}{227.14} + 50$$
$$(15.30)$$

Simplifying Eq. (15.30), we get

$$ME\$ = 8.946P + 12{,}500\,\frac{Q^3}{(WH)^2} + \frac{WH}{227.14} + 50 \qquad (15.31)$$

The annual cost, AC\$, of a UVGI system can be written by adding Eqs. (15.17) and (15.31):

$$AC\$ = -0.00543P^2 + 2.538P + 62.85 + 0.0522Q + 0.01096WH$$

$$+\,(0.0000193\rho - 0.00011)\,L\,(W + H) + 8.946P$$

$$+\,12{,}500\,\frac{Q^3}{(WH)^2} + \frac{WH}{227.14} + 50 \qquad (15.32)$$

Simplifying Eq. (15.32) produces the following:

$$AC\$ = 11.48P - 0.00543P^2 + \frac{12{,}500Q^3}{(WH)^2} + 0.0522Q + 112.5$$

$$+\,(0.0000193\rho - 0.00011)\,L\,(W + H) + 0.01536WH \qquad (15.33)$$

In Eq. (15.33), the units are metric. Dimensions W, H, and L are specified in cm, and Q is specified in m³/min. Power, P, is specified in W. The unit convention for SI could be used, in which power is specified in J or J/s, but UV lamps are rated in μW/cm², which makes metric units more convenient to use.

In summary, the above economic evaluation provides an example based on estimated costs for all aspects of UVGI systems. These costs will vary from project to project, but this method can be used to establish the life cycle costs for any specific application. Together with the method for evaluating the economics of filters presented in Sec. 15.3, the life cycle cost of a combined system of filtration and UVGI can be determined.

15.5 Economics of Carbon Adsorbers

The sizing of carbon adsorbers is limited by the availability of information on the performance of carbon adsorbers against the array of possible CBW agents. Because so few CBW agents have been tested against carbon adsorbers, and because there are several types of carbon adsorbers that could be selected depending on the agent to be targeted, it is impossible at present to define the economics of carbon adsorption in relation to CBW agent removal. Instead, a general treatment of carbon adsorbers is presented here to assist designers in lieu of future developments in this rapidly changing area of research.

The first costs of a carbon adsorber vary considerably depending on the system size and type. There are several types of carbon adsorbers available, but the choice of which to use depends on the application. Standard granulated activated carbon (GAC) systems are available with refillable frames or as solid media cartridges that resemble filters. The cost of refilling a metal frame with GAC is relatively low, but may have high maintenance costs. The replacement of cartridge-type filters has low maintenance but these are more expensive. It is difficult to say in advance which alternative is more economical because of the wide variety of systems and costs.

In addition to standard GAC units, carbon adsorbers impregnated with various compounds can be obtained to target specific chemicals. The choice of such added protection entails costs that can only be assessed after obtaining quotes from manufacturers. Determining cost-effectiveness based on performance is, however, almost impossible due to the lack of information on how well these systems remove the broad range of chemical agents.

No specific examples of life cycle cost calculations can be provided because of the complexity of gas phase filtration equipment and performance, but the general economics of GAC can be developed in a similar manner to filtration. The total first cost, FC\$, of any GAC system is the sum of all the component first costs:

$$FC\$ = Frame\$ + Labor\$ + FilterFC\$ \qquad (15.34)$$

In Eq. (15.34), the first cost of the frame is either the cost for a refillable perforated metal housing or a frame for holding GAC cartridges. It is assumed the GAC unit is retrofitted to an existing ventilation system and so there is no fan first cost. The filter first cost, FilterFC\$, is due to the fact that some type of prefilter and/or downstream dust filter is often included in such systems. For cartridge-type GAC units, the filters are often integral and so only one cost is needed.

The annual maintenance and energy costs, ME\$, for a GAC unit can be written as follows:

$$ME\$ = FP\$ + CR\$ + FR\$ + MAINT\$ \qquad (15.35)$$

In Eq. (15.35), FP\$ refers to the fan power cost that results from the increased pressure loss through the system due to the GAC unit. The carbon replacement cost or cartridge replacement cost is designated CR\$. The filter replacement cost, if any, is designated FR\$, and the maintenance costs are designated MAINT\$. All of these values must be obtained from the manufacturer or the supplier, or must otherwise be estimated.

The fan pressure loss can be estimated in the same manner as for filters and UVGI systems. Carbon adsorbers have a pressure loss associated with them that depends on the type and depth of the media. The pressure loss data are provided by the manufacturer of the specific media and cannot be generalized, although pressure losses will generally be in the range of 0.01 to 1 in.w.g. or sometimes higher. The pressure losses are a function of air velocity, as is the residence time. Manufacturers normally supply a graph of pressure loss versus air velocity, and this can be used to determine the fan power consumption with Eqs. (15.24) and (15.25).

Once the first costs are established, the annualized first cost can be computed using Eq. (15.16). The life cycle cost can then be found by adding the annualized first cost, AFC\$, to the annual energy and maintenance costs, as in the preceding section for filters.

Optimization of GAC systems is possible. Performance can be improved, and fan energy saved, with lower air velocities. This can be accomplished by increasing the face area. Larger first costs may result, so any attempt to optimize the economics of a carbon adsorber should account for these factors.

In the event a carbon unit is to be sized or selected, it is recommended that designers consult with manufacturers to establish the size, performance characteristics, and economics of operation of the units they are considering. It should be noted here that since GAC systems perform a different function, removing chemical agents, than filters and UVGI systems that remove biological agents, combined optimization is not necessary. However, filters may remove some vapors, and so there is some theoretical room for the three-way optimization of a combined system of filters, UVGI, and GAC, although insufficient data are available at present to demonstrate any such examples.

15.6 Energy Analysis

The addition of any equipment such as filters, UVGI systems, or carbon adsorbers to a building heating, ventilating, and air conditioning (HVAC) system will have some impact on overall building energy consumption. Filters and carbon adsorbers will add a pressure loss to the system that will convert to either increased fan power consumption or to decreased airflow, as discussed previously. Increased fan power consumption may increase the amount of heat released into the supply air if the fan motor is cooled by the airstream. If the fan motor is cooled by outside air or is external to the airstream in a separately ventilated equipment room, there may be no additional cooling load to be considered. In either case the system should be evaluated to determine whether the building cooling load is affected.

If the fan motor is cooled by the same ventilation airstream it is delivering, then the amount of heat added can be computed from the fan efficiency and the motor efficiency. If the fan curves are available, the increased pressure loss due to the addition of the filters and/or carbon adsorbers can be read directly from the curve. Otherwise, the increased fan power can be estimated by air horsepower as defined by Eq. (15.3). Based on the latter, the following equation can be written to define the total additional electrical energy, in kW, required by the fan motor:

$$P = 1.341 \, \frac{hp}{\eta_{fan} \eta_{motor}} \tag{15.36}$$

In Eq. (15.36), the amount of this energy which converts to heat is simply the inefficiency, or the remainder. The heat added to the system, Δh, can be written as follows:

$$\Delta h = [1 - \eta_{fan} \eta_{motor}] \, P \tag{15.37}$$

The efficiency of the fan and the motor can be obtained from the manufacturers. These efficiencies will vary with the size and type of the motor and usually increase with increasing motor size. Efficiencies in the range of 75 to 90 percent would be typical for normal building fans and motors.

Although UVGI systems usually have negligible pressure losses, they will add some heat to the system because of the fact that essentially all the electrical energy converts to heat. UV lamps typically convert about 25 to 33 percent of the total electrical input into UV light, but as this UV light is absorbed or reflected it converts to wavelengths in the infrared. It is reasonable and conservative to assume that 100 percent of the total lamp power (not just the UV power) converts to heat. Therefore, the heat added to a ventilation system by a UV lamp can be written as

$$\Delta h = P_{\text{lamp}} \qquad (15.38)$$

In Eq. (15.38), both the heat added Δh and total lamp wattage P_{lamp} are specified in kW.

The heat added by these systems becomes part of the cooling load of the building. During the cooling season for the particular climatic location of the building, this will have some impact on the cooling capacity that must be evaluated. During the heating season, this heat may reduce the heating load of the building and credit may be taken for it, depending on various factors like the location of the added heat and whether humidity is being controlled.

The impact on cooling load, heating load, and building relative humidity must be evaluated manually to determine if it is significant. If the impact is significant, the best approach is to perform a building energy analysis using one of the software packages that are currently in use. Some of the programs that are used today include BLAST (DOE 1994) and HAP (Carrier 1993 and 1994). Consult the references for additional information on obtaining or using these programs.

15.7 Summary

This chapter has presented the details of economic analysis for filtration and UVGI in a way that both defines the range of typical costs and summarizes the methods for evaluating life cycle costs. In addition, methods have been discussed to enable economic optimization of combined systems, and for evaluating the impact of air-cleaning technologies on building energy analysis. The information on carbon adsorption performance is, again, too limited to establish an appropriate or effective size for such systems, and so the economics have been treated only generally.

Chapter

16

Mailrooms and CBW Agents

16.1 Introduction

The subject of mailrooms and CBW agents requires special treatment because of the unique circumstances and hazards involved, and the difficulty of disinfecting mail (Fig. 16.1). The subject of remediating, or decontaminating, mailrooms has been addressed generally in Chap. 13 and will not be revisited here. This chapter focuses on preventive technologies and alternatives for Post Offices and commercial mailrooms.

16.2 Mailroom Contamination

In the events following the September 11 attacks, a number of letters containing anthrax spores were mailed to government officials and others. Although the intended victims were high-profile individuals, the casualties were predominantly Post Office employees or innocent recipients of mail that had been cross-contaminated in Post Office mail-sorting equipment. The terrorist(s) who perpetrated these acts took advantage of the existing infrastructure and functions of the Post Office, literally using them as a delivery weapon. Because of the fact that anthrax spores can penetrate paper envelopes, everyone who handles mail becomes at risk. Furthermore, since the anthrax spores easily become aerosolized once out of the envelope, everyone in the building may be at risk when contaminated mail passes through a facility.

In addition to contaminated mail and airborne spores, the equipment that processes the mail may become contaminated, creating a contact hazard for postal employees. The problems of preventing mailroom contamination and protecting postal employees are difficult and com-

Figure 16.1 Post Offices and postal employees are subject to unique hazards when their facilities are used as delivery vehicles for CBW agents.

pounded by the multifaceted aspects of the mail processing system. The following major aspects of these problems can be identified as points where preventive technologies may be needed:

- Building ventilation systems
- Building general areas
- Mail handling by employees
- Contaminated letters and packages
- Mail processing equipment
- Delivery vehicles

Each of the above subjects will be dealt with separately in the following sections. In addition, basic protocol for mailrooms will be reviewed in the final section.

16.3 Building Ventilation Systems

CBW agents that are delivered by mail impose the distinct risk of a release during handling. Such releases invariably involve aerosolization. Ventilation systems in Post Offices or commercial mailrooms are no different from those used in other similarly sized facilities. The options for controlling any indoor airborne contamination are those that have been discussed previously in Chap. 8, as well as the alternative technologies discussed in Chap. 14. The primary options for air cleaning include ventilation air purging, filtration, ultraviolet germicidal irradiation (UVGI), and, for chemical concerns, carbon adsorption.

The normally operating ventilation system of a Post Office or commercial mailroom offers a limited amount of protection against airborne contaminants. Protection can be improved by increasing the rate of outside airflow, but filtration and UVGI are generally more economical to use for processing the recirculated air of postal facilities. The effectiveness of filters against the various potential pathogens has been described in detail in Chap. 8. Those pathogens that cannot be easily or economically filtered out can usually be removed by UVGI. It should be emphasized here that the often noted resistance of anthrax spores to UVGI should not be used to overshadow the effectiveness of UVGI against other BW agents like smallpox. Many other agents could conceivably be sent through the mail and become airborne in postal rooms, and a complete solution to the problem should include a combination of filtration and UVGI.

Neither filters nor UVGI have any significant effect against chemical agents. Only gas phase filtration equipment like carbon adsorbers can be used for this application. However, the likelihood of chemical agents being shipped through the mail is less than that of biological agents. For one thing, chemical agents are either liquids or gases, and require containers. These containers would offer some protection to mailroom employees if they were used. As a caveat, chemical-containing devices that were mailed might contain release mechanisms or even explosives, which would put employees at some risk. In any event, these would be larger, heavier packages and thus would be limited to certain areas and to specific types of handling.

16.4 Building General Areas

In addition to airborne contamination, mail containing BW agents can contaminate surfaces like tabletops, desks, mail-handling equipment, mail slots, etc. Only a few technologies exist that provide options for disinfecting or decontaminating surfaces that are in constant use, but at least one of them has been put into practical application—upper-air UVGI systems.

Upper-air UVGI systems provide a zone of UV irradiation near the ceiling which will tend to destroy microbes that drift into the irradiation zone due to the natural circulation of air (Dumyahn and First 1999; First et al. 1999; Nicas and Miller 1999). Figure 16.2 shows a

Figure 16.2 A commercial office mailroom with an upper-air UVGI system (upper right of photo) installed by Lumalier/Commercial Lighting Design, Inc., of Memphis, TN.

commercial office mailroom with an upper-air UVGI system on the back wall, and Fig. 16.3 shows an individual unit. This system is designed to create a microbial "kill zone" above the heads of the employees. Since the UV irradiation levels below this height are within an acceptable and safe limit, the system can be operated continuously. Table 16.1 summarizes the allowable ACGIH limits for daily human exposure to UV (ACGIH 1973).

Not only does the air circulating through the upper room receive constant exposure, but the various surfaces in the room receive continuous exposure at some level. This constant dosing can have a definitive biocidal effect in the long term even though it is not hazardous to personnel.

Upper-air UVGI systems can be adapted to various environments. In office areas like that shown in Fig. 16.4, the levels should not be allowed to exceed the ACGIH limits for UV exposure. However, during the hours when the facilities are not occupied, UV levels can be raised by the addition of UV lamps mounted around the perimeter in the same fashion as upper-air systems. These types of systems are called "after-hours" UVGI systems and can be engaged by timers. High levels of UV can be used in an after-hours system to guarantee that even anthrax spores would not be likely to survive overnight. After-hours UVGI systems can be shut down by motion detectors upon anyone's entry into the area, or can be manually turned off.

Figure 16.3 Upper-air UVGI unit. (*Image provided courtesy of Lumalier/Commercial Lighting Design, Inc., Memphis, TN.*)

TABLE 16.1 Permissible Human UV Exposures

UV irradiance, μW/cm^2	Duration of exposure per day
0.2	8 h
0.4	4 h
0.8	2 h
1.6	1 h
3.4	30 min
6.6	15 min
10	10 min
20	5 min
100	1 min

Figure 16.4 Mailroom office area with two upper-air UVGI systems positioned near the ceiling in the back.

Few other options exist for the preventive control of surface contamination in mailrooms. The regular cleaning of surfaces with bleach or disinfectants is one alternative. Pulsed-light systems using non-UV irradiation are another potential alternative, but this is a developmental technology (see Chap. 14). Ozone could be recirculated through the building by the ventilation system and would disinfect all surfaces, including mail, but it could only be used when the facilities were unoccupied (i.e., at night) and would have to be purged prior to the entry of employees in the morning.

16.5 Mail Handling by Employees

Perhaps those in the greatest danger from contaminated mail are the postal employees who handle the mail directly since many BW agents, including anthrax spores, may leak through the paper of normal envelopes. Even if mail decontamination is done on a large scale, employees may still have to handle mail prior to such a process, and the decontamination process may not be perfect anyway. Still, there are a few options available to postal facilities that will provide considerable, if not complete, protection to workers.

The possibility of vaccinating postal employees against anthrax has been reviewed by officials of the U.S. Post Office, who are providing such medical treatment to employees who have been put at risk (USPS 2002). Vaccinations may also be provided in the future to employees requesting them, but the number of BW agents against which vaccinations are available is limited. Vaccinations for anthrax and smallpox are no help against Ebola and tularemia, and are not necessarily a cost-effective alternative in relation to the actual risk.

Perhaps the most obvious option for protecting mail workers is the use of surgical face masks (also called filtering face pieces or FFPs) and latex gloves, as well as clothing with full-body coverage. Basic face masks such as those used in operating rooms offer a degree of protection against aerosols in approximate proportion to the risk without undue expense or discomfort (Fig. 16.5). Latex gloves and full-body coverage of clothing will protect against contact with agents without overburdening employees or increasing operating costs significantly. Some of these items are currently provided to U.S. Post Office employees as optional equipment but are usually not provided in commercial mailrooms.

Another option for mail handlers is the use of laboratory-style exhaust hoods for sorting mail. These hoods draw air through the front opening and create a curtain of air that prevents the escape of any aerosols. Employees are protected since any loose particles will be drawn away from them and into the unit. The air can be exhausted to

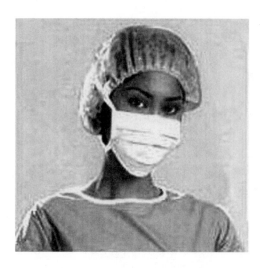

Figure 16.5 Surgical masks offer a certain degree of protection against airborne pathogens. (*Image courtesy of Med-Con, Ltd., Thailand.*)

the outside or else it is processed by air cleaners and then recirculated. Figure 16.6 shows a mailroom hood designed to filter and process the air. A cost-effective option for large facilities is to use an industrial-scale hooded exhaust system large enough for dozens or hundreds of employees. Such systems could be arranged in lines with ductwork and be serviced by a single high-volume fan. In order to save energy costs, they could be designed as recirculation systems with integral filters and UVGI lamps.

Other technologies could be adapted, such as photocatalytic oxidation (PCO) or pulsed-light systems, but research would be required to adapt such experimental technologies. Ultimately, the Post Office may switch to leakproof envelopes with airtight seals as a means of eliminating the problem, but such a changeover could take years.

16.6 Contaminated Letters and Packages

The U.S. Post Office has considered a variety of technologies for disinfecting or decontaminating mail in bulk quantities, including pulsed light, pulsed electric fields, gamma rays, UVGI systems, ozone, and microwaves, among others. All of these options have advantages and disadvantages but so far none of them have stood out as both efficacious and economic in this application.

UVGI systems can be used to disinfect the exterior surfaces of mail and packages. This may be sufficient to protect the employees handling the mail, but mail may continue to leak BW agents after irradiation and so this approach does not guarantee protection. However, it can be a cost-effective approach that provides a means of reducing the risk. Mail decontamination units have already been designed and marketed

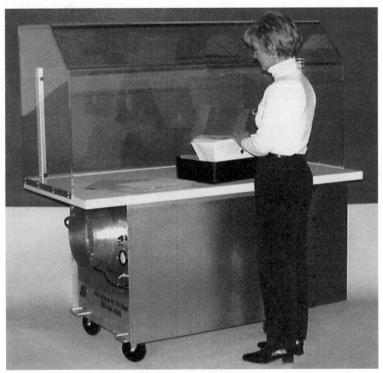

Figure 16.6 Mailroom biosecurity system with filtration and carbon adsorption for removing airborne biological and toxic hazards from contaminated mail. (*Photo courtesy of Jerry Straily, International Air Filtration, Illinois.*)

by several sources. Figure 16.7 shows one example of such a UVGI mail decontamination unit.

One of the often asked questions in the postal industry is whether UVGI is effective against anthrax spores. As explained in previous chapters, UVGI is effective against a wide range of pathogens, but its effectiveness depends on the irradiation dose or lamp UV power and exposure time. Figure 16.8 shows the results of tests performed on anthrax spores. These results indicate a 3 log reduction is possible within about 50 min at a test intensity of 90 μW/cm^2. Mail decontamination units operate at much higher intensities than this and can offer surface sterilization times of 5 to 10 min. Since spores are the most UV-resistant of microbes, higher rates of disinfection could be expected against other pathogens.

The rate constant for anthrax spores on plates is 0.000031 cm^2/μW·s based on a single-stage rate-constant curve fit of data from Knudson (1986). Fitting a two-stage curve with a shoulder to the same data provides a slightly higher value for the first-stage rate constant but pre-

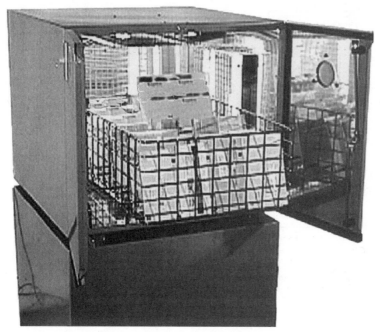

Figure 16.7   Mail decontamination unit using high-intensity UVGI. (*Image provided courtesy of Charley Dunn of Lumalier/Commercial Lighting Design, Inc.*)

dicts longer times for sterilization. The two-stage model should be used when kill rates are expected to be in the second-stage region (see Chap. 2). A multihit two-stage model was fitted to the Knudson data by minimizing the sum of squares error. The complete two-stage surface disinfection equation for anthrax spores is as follows:

$$S(t) = 0.9984 \left[1 - (1 - e^{-0.0000424It})^{2.6}\right]$$
$$+ 0.0016 \left[1 - (1 - e^{-0.000006It})^{2.6}\right] \quad (16.1)$$

The constants in Eq. (16.1) are as follows:

- First-stage rate constant: 0.0000424 cm^2/μW·s
- Second-stage rate constant: 0.000006 cm^2/μW·s
- Resistant fraction: 0.0016 (0.16 percent)
- Multihit exponent n: 2.6

Figure 16.9 provides some additional data on UVGI irradiation of anthrax spores on plates. This chart is graphed in terms of dose instead

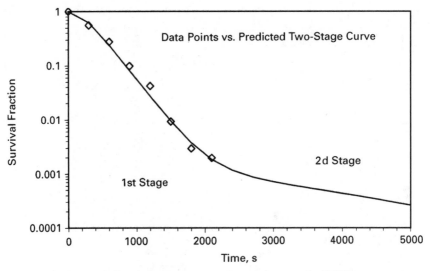

Figure 16.8 Survival of anthrax spores on surface of plates under UVGI exposure at 90 μW/cm^2. Based on data from Knudson (1986).

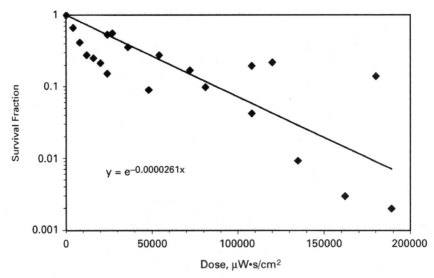

Figure 16.9 Survival versus dose for anthrax spores under UVGI irradiation. Based on combined data for exposure on three types of surfaces from Dietz et al. (1980).

of time, but this does not affect the computed rate constant. The data come from Dietz et al. (1980) and represent the combined results of three tests in which different surfaces were used. The data are scattered, especially beyond that shown on the chart, but the surfaces may have impacted the survival in various ways. One of the surfaces used was aluminum, which has high UV reflectivity. The average rate constant for the data shown is 0.0000261 $cm^2/\mu W \cdot s$, which is in reasonable agreement with the rate constant for the Knudson (1986) data.

Gamma ray devices use ionizing radiation to disinfect and destroy microorganisms. The effect of gamma rays on microorganisms is similar to that of UV irradiation except that they can penetrate envelopes and packages. Gamma rays are also hazardous to humans and can cause cellular and genetic damage (Alpen 1990; Alper 1979; Casarett 1968; Coggle 1971). Gamma ray equipment is also relatively expensive and may not be an affordable option for most postal facilities. Only a few have been procured or leased for use in Post Office facilities, and they are only used on targeted mail (USPS 2002). Early reports on the use of gamma ray exposure of mail have suggested that the exposed envelopes and packaging materials become damaged and may lose some structural integrity.

Pulsed light operates in a similar fashion to UVGI, but the disinfection rate occurs much more rapidly. Such equipment is suited to assembly line operation where only seconds, or fractions of a second, are available for decontaminating materials (Bushnell et al. 1998; Dunn 1997; Wekhof 1991). The drawbacks of pulsed light are that it can be expensive and probably only disinfects the surface of mail. It is, however, a developing technology and may prove viable in such applications in the future.

Pulsed electric fields are similar to pulsed-light technology except that they generate an electric field or current that is capable of disinfecting materials internally and externally. It is used in the food industry for eliminating potential foodborne pathogens. It has not been applied to the disinfection of mail, so little is known about its ability to penetrate mail or disinfect paper surfaces. Like pulsed light, it can be an expensive alternative but is still in a developmental stage.

Ozone is a potentially economical alternative to the previously mentioned technologies. It has the ability to disinfect spores, and the disinfection rate is roughly proportional to the concentration (Ishizaki et al. 1986; Kowalski and Bahnfleth 1998). Its ability to penetrate paper envelopes is probably limited, and therefore it would act only as a surface disinfectant, like UVGI. Ozone could be used inside large rooms or chambers (i.e., trucks) into which bulk quantities of mail were placed. See the ozonation data on *Bacillus cereus* presented in Chap. 13 for details on disinfection rates.

Microwave irradiation of mail provides an option that will disinfect both the inside and outside of mail, and it is comparatively inexpensive. Research has shown that the destruction of spores by microwaves can be accomplished with ordinary microwave ovens in about 30 to 50 min (Cavalcante and Muchovej 1993). Figure 16.10 shows the results of one test on *Aspergillus niger* spores using a microwave oven on full power. *Aspergillus niger* spores are somewhat similar to *Bacillus anthracis* spores in terms of their resistance to biocidal factors. No data are currently available on microwave-irradiated *Bacillus anthracis* spores. One advantage of this technology is that microwave ovens are common, but one disadvantage is that they may ignite fires if the mail being disinfected—such as CDs, credit cards, or metal tape—contains metal.

Other technologies have the potential to be used for the bulk disinfection of mail, including some of those discussed in Chap. 14, but no data are available for most of these.

16.7 Mail Processing Equipment

One of the problems with processing contaminated mail is that the equipment used for automatic sorting and routing may result in the cross-contamination of other mail. Disinfecting the mail prior to sorting is one option, but this has the limitations discussed previously.

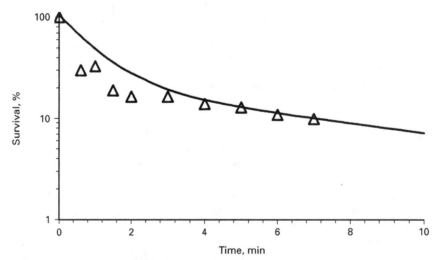

Figure 16.10 Reduction of spores of *Aspergillus niger* under microwave irradiation. Data points are shown versus fit of a two-stage exponential decay curve. Based on data from Cavalcante and Muchovej (1993).

Disinfecting the mail during sorting and processing would require the integration of disinfection technologies with the equipment used for sorting and processing. Some of the technologies discussed earlier, such as pulsed light, have the potential to be adapted to such an application. Another option that would minimize the danger of cross-contamination might be the use of downdrafts or "plug-flow" to draw a constant airstream across the mail being processed. Operating like an upside-down lab hood, the airstream would entrain any airborne particles that came off the surface of envelopes and draw them into an exhaust airstream leading to a filter. How effective this might be is unknown at present.

The use of plastic or airtight envelopes would presumably eliminate this problem, but such a change is not likely to happen in the near future. Clear envelopes have also been suggested, but this compromises privacy. The use of biosensors to detect the presence of BW agents on the surface of envelopes is another option that must await technological advances in the detection speed and the number of agents that can be detected.

16.8   Delivery Vehicles

Mail delivery vehicles can also become contaminated if they transport mail containing any BW agents (Fig. 16.11). The problem with protecting the drivers of such vehicles, or the workers who unload them, is

Figure 16.11   Mail delivery vehicles may be subject to contamination from hazardous mail, but also offer opportunities for decontamination in transit.

much the same as the problem of protecting employees against the indoor contamination of postal facilities, except that trucks usually have no ventilation systems other than natural ventilation.

The use of ozone in mail trucks during transport has been suggested (Marsden 2001). This approach has certain advantages, such as the fact that the transport time would be efficiently used as disinfection time and the truck volume itself would serve as the exposure chamber. Such an approach might be limited to interstate transport vehicles since local vehicles have irregular schedules and driver exposure is not desirable. Venting of the ozone to the ambient air might be a problem, but ozone removal technologies are currently available (Kowalski et al. 2002a).

Like mail sorting and handling equipment, trucks that have been exposed to BW agents can be cleaned with bleaches and disinfectants.

16.9   Mailroom Protocol

Based on the previous review of problems and potential applications of disinfection technologies, some general guidelines can be suggested for mailrooms. Some of these guidelines have been adapted from other sources (Kroll 2001). These are summarized below:

Personal protection (suggested or optional):

- Workers handling mail are advised to wear latex or other nonpermeable gloves.

- Surgical face masks, or other suitable protection against aerosols, should be available for use by those workers handling mail directly.

- Workers handling mail are advised to wear long-sleeve shirts and full-length pants so as not to expose skin to possible contact with pathogens.

- Optional jumpsuits made from nonpermeable materials or antimicrobial fabrics could be made available for worker use or in the event of emergencies.

- Optional respirators could be provided for emergency use.

Work environment:

- Adequate ventilation, in which the amount of ventilation air per person meets or exceeds the applicable standards, should be provided in all mail-handling areas.

- Consideration should be given to the use of filters and in-duct UVGI in the ventilation systems of all areas in which mail is handled or processed.

- Consideration should be given to the use of upper-air UVGI systems in all areas where mail is handled or stored.

- Consideration should be given to the use of after-hours disinfection systems (if upper-air UVGI is not installed) in all areas where mail is handled or stored.

- Disinfectants should be provided in washing and cleaning facilities for emergency use. Disinfectant sprays should be provided for use by employees in the event of an emergency.

Monitoring and emergencies:

- Employees should keep vigilant for suspicious mail, including letters or packages leaking powder or fluids, packages spraying liquids, packages containing aerosolizer-type equipment, or packages producing irritating odors or gases.

- Suspicious mail should be abandoned in place, equipment processing stopped, and the area or building evacuated.

- Suspect mail should be handled by the appropriately trained responsible individuals or emergency response personnel. Mail should be dropped in an appropriate container, covered, wrapped, or otherwise sealed.

- Employees encountering suspect mail should immediately cleanse themselves with disinfectants and then wash themselves and change clothes.

- In smaller mailroom facilities, mail can be sorted into two categories: (1) expected, recognizable, or typical business correspondence and (2) unexpected, unfamiliar-looking, or suspicious mail. The latter category may include mail to addressees no longer with the company.

- Mail falling into the second category, if addressed to specific individuals, should be checked with the recipients to verify that they are expecting such deliveries from the indicated senders prior to delivering the mail.

- Suspicious mail can be x-rayed or run through a metal detector.

- The opening of suspicious mail should be performed in an isolated area by individuals wearing respirators and full-body protection.

Training and responsibility:

- An incident response manager should be appointed and trained to manage hazardous chemical or biological situations.

- All employees should be trained to understand and recognize the hazards associated with chemical and biological weapon agents.

- Employees should report any illness at the moment they become ill and not the next business day. A specific security or managerial officer should be appointed to receive phone calls or communication (i.e., email) at any time of the day or night and on weekends. It will be this person's responsibility to recognize unusual trends or commonalties in illness symptoms that may be due to BW agents. This person should be equipped with proper information for making such epidemiological assessments and trained accordingly.

Many additional procedures and protocols could be developed for handling emergencies or identifying hazards. The above are merely some basic recommendations. Each facility, whether it is a postal office or a commercial mailroom, should review its emergency procedures and vulnerabilities to determine the best plan of action that would be appropriate for the expected risk.

16.10   Summary

This chapter has addressed the problematic issue of mailroom contamination in the hopes of simplifying the decisions regarding appropriate options for protecting mailroom workers and decontaminating mail. Recommendations have been provided and equipment for such applications has been discussed. This is an area that is under study and so developments and new solutions may yet arise in the future.

Chapter

17

Epilogue

17.1 Introduction

This book has provided extensive information on how to protect buildings against CBW agents and turn them into immune buildings through the application of air-cleaning technologies. The databases in the Appendices provide a concentration of useful information to assist engineers and other professionals in designing systems for immune buildings and assessing both the threats and the degree of protection offered by these systems under various CBW attack scenarios.

Chapters 2 and 3 introduced a broad range of potential biological and chemical weapons and their properties as they relate to filtration and ultraviolet germicidal irradiation (UVGI). Chapter 4 presented a simplified mathematical method of approximating the lethal dose curves of CBW agents, and disease curve progressions for BW agents.

Chapter 5 presented a variety of hypothetical delivery means for CBW agents and described ways in which terrorists might use them against buildings. Building and ventilation system vulnerabilities were addressed in Chaps. 6 and 7, respectively, and in the former chapter a mathematical method of approximating the risk of individual buildings was proposed.

Chapter 8 presented detailed methods for sizing filtration and UVGI systems, and summarized the current state of gas phase filtration in regard to CW agents. The design information provided on gas phase filtration in Chap. 8 is, unfortunately, not as detailed or specific as that for ventilation, filtration, and UVGI due the paucity of such information that relates to CW agent removal. As a result, no simulations of CW agent removal rates are possible at this time. The exact level of protection offered by carbon adsorbers cannot be known at present and must await future research. However, the removal rates of carbon adsorbers that are provided in traditional sizes by manufacturers for the purpose of controlling air quality, coupled with the potential removal of liquid aerosols by filters, may be sufficient to provide a reasonable degree of protection against most CW agents, and therefore their installation can be pursued as a protective measure. It is only their performance that cannot be verified at present.

The results of the CBW attack simulations performed in Chap. 9 and the economic evaluations in Chap. 15 indicate that such protection can be effective without being prohibitively expensive. Because of the fact that ventilation system characteristics limit the total reduction of casualties or fatalities that can be achieved by increasing the level of filtration or UVGI system power, or what is herein termed the "diminishing returns effect," increasing system disinfection rates only work up to a point unless the ventilation system itself is redesigned. Nevertheless, the reduction of potential fatalities possible through the use of moderate levels of filtration and UVGI power are impressive enough, being able to guarantee protection of approximately 90 percent of building occupants, that a happy medium may be found between system performance, cost, and the relative risk for any facility. In short, these results strongly suggest that filtration levels between 60 to 80 percent DSP (dust spot efficiency) [MERV (or minimum efficiency reporting value) 10 to 15] and UVGI system average intensities between 500 to 1000 μW/cm^2 will provide adequate protection to most building occupants without excessive cost. The question of how much filtration or UV power is needed is seen to be a function limited and defined by the existing building and ventilation system characteristics. Of course, every building is unique and there is no substitute for a well-engineered immune building system. To this end, this textbook should provide the possibility of achieving the state-of-the-art in immune building design for any given application.

In Chaps. 10 and 11, information has been provided to facilitate the design and installation of detection and control systems. This information contains a sufficient amount of detail and number of examples that the engineer can pursue the development of a CBW detection system if it is desired to do so. Current detection technologies are limited by the range of agents that can be detected, but the design of such systems is still possible if budgets permit. Specifically, a combination of chemical sensors and a particle detectors can be used to provide isolation and alarm capabilities. Alternatively, or in combination, an inexpensive air-sampling procedure can be implemented to return regular reports on the microbial content of indoor air, a measure that may detect BW agents in sufficient time to obtain treatment and prevent fatalities.

Chapter 12 addressed the issue of building security as an integral part of an immune building system designed for protection against terrorist releases of CBW agents. A variety of information was provided for security personnel to assist them in protecting buildings, identifying threats, or dealing with actual incidents. In addition, design information was provided in this same chapter on the option of sheltering-in-place. Chapter 13 provided decontamination and remediation information for postattack scenarios.

Chapter 14 introduced alternative technologies for possible applications in immune buildings, and Chap. 15 provided details and examples of economic analysis and information on optimization of immune building technologies. Chapter 16 covered the special topic of mailrooms and provided some specific information and suggestions on what designers or mailroom managers can do to disinfect mail and protect personnel.

Although some areas of this book could be treated in more detail, the information provided here should be sufficient to assist any engineer or building manager in the design of functional immune building systems that will offer a considerable degree of protection to occupants without necessarily incurring high costs. In fact, some of the collateral benefits of immune building systems may even offer a payback, as will be discussed shortly. But first, some digression on the nature of future CBW terrorist attacks, although somewhat outside the realm of engineering, may be of use to certain building owners and government planners.

17.2   The Future of Bioterrorism

The consensus of the authorities and experts today is that the recent biological and chemical attacks that have occurred in recent years are merely the beginning of a new phase of catastrophic terrorism (Drell et al. 1999). The use of weapons of mass destruction that began in the 18th century with smallpox attacks on Native Americans may culminate in mass disasters visited upon innocent civilians in the future. The nature of the agents used could range from the worst known human pathogens to unimaginable nightmares created by genetic engineering. Indeed, there is nothing in nature or science that precludes the possibility of anyone creating an incurable disease capable of extinguishing the entire human race.

The future of bioterrorism is likely to be primarily a home-grown phenomenon, as most domestic terrorism has been in the past, and it can be expected that the most accessible CBW agents will be developed into weapons by those organizations that have the motivation to use them (CSIS 1995; Stern 1999; Tucker 1999; Zilinskas 1999b). As the history of biological warfare has shown (see Chap. 1), offensive use of biological weapons has been almost the exclusive domain of those who dehumanize their intended victims to the greatest extent (Ellis 1999).

An inordinate number of organizations exist today that promote the dehumanization of others, and they are not limited to overseas fanatical Islamic extremists. The United States has allowed hundreds of hate groups to proliferate in its own backyard. American Nazis, neo-Nazi skinheads, Aryan supremacists, Christian Identity, the Ku Klux Klan, neo-fascists, and countless other groups are permitted to encourage oppression of other Americans. Such groups are often cited as actual or potential sources of domestic bioterrorism (Cole 1996; CSIS 1995; Drell et al. 1999; Siegrist 1999; Stern 1999; Tucker 1999; Zilinskas 1999b, 1999c; Zink 1999). Members of one group, the so-called "White Aryan Resistance," have even suggestively condoned the World Trade Center bombing, circulating the idea that it was Jewish property (Cooke 2001). Although America has unilaterally declared war on terrorists abroad, we simultaneously tolerate threats of terrorism against minorities, foreigners, and religious groups in our own homeland.

Hate groups often claim freedom of speech in defense of their overt desire to deny other Americans life, liberty, and the pursuit of happiness. Such specious arguments should not obscure the reality of their purposes. We can take a cue by analogy from the world of microorganisms, where a sharp line can be drawn between commensal bacteria that live in harmony with our bodies and contribute to our well-being, and those dangerous parasites that have the sole purpose of doing us harm. The human immune system will use every weapon at its disposal to rid itself of pathogenic threats, and society should do no less.

Such groups should be carefully scrutinized and denied access to materials and information that would allow them to develop weapons of mass destruction. Indeed, the access to educational and research facilities currently enjoyed by those who have agendas of racism, hatred, or fanaticism should be curtailed. Our institutions of higher learning should not be used as weapons against society like our airliners were on September 11. Furthermore, information and materials that would facilitate the development of CBW agents should be denied to such groups and removed from their hands wherever it is found. This includes restricting access to university libraries and the prevention of Internet publication of sensitive information by irresponsible persons. Certainly the legislative means exist to effect such controls, and it would be most prudent to do so sooner rather than too late.

In the long term, there may be little that can be done to prevent determined individuals from launching bioterrorist attacks, but through immune building technology and constraints on potential terrorists, it may be possible to offer the greatest protection to the largest number of our citizens.

17.3 Collateral Indoor Air Quality Benefits

Implicit throughout the previous chapters is the potential benefits of immune building technologies in terms of improving indoor air quality (IAQ) and reduction in the incidence of natural diseases. Although it is difficult to quantify the economic benefits of reduced disease incidence, some general information and statistics can be introduced to highlight the available room for improvement.

Microorganisms in the home and work environments consist of allergenic and toxigenic fungi, bacteria, and pathogenic viruses (Flannigan et al. 2001; Yang et al. 1993). Some fungi and bacteria produce volatile organic compounds (VOCs) that are a contributing factor in poor IAQ and are often associated with building-related illness (BRI) or sick building syndrome (SBS). Even at low levels, airborne microbes and the growth of microbes on surfaces contribute to a variety of human health problems (Haas 1983; Spendlove and Fannin 1983). Buildings sustain airborne microbes because of the controlled temperature and humidity inside the structures and allow the microbes to transmit to new hosts, even transporting microorganisms through their ventilation systems to distant new hosts (Zeterberg 1973). Some buildings, especially those with moisture problems, can acts as amplifiers for fungal spores (Woods et al. 1997). The use of filtration, UVGI, and carbon adsorption will reduce indoor levels of airborne microbes and VOCs and produce a perceivable improvement in IAQ, which will, in turn, contribute to improved health and productivity (Dorgan et al. 1998).

Approximately one-fourth of medical costs and nearly one-half of lost earnings are due to respiratory illness (Tolley et al. 1994). Healthcare costs for acute respiratory infections total $30 billion annually in the United States, and the lost work time and reduced productivity due to on-the-job morbidity are estimated to be even greater, at $35 billion annually (Fisk and Rosenfeld 1997). Clearly, there is much to be gained from immunizing our buildings against airborne diseases.

Since few large commercial buildings have installed air-cleaning systems for the purpose of disinfecting indoor air, no epidemiological data are currently available that would allow an estimation of the actual amount of reduced illness time that could be expected from retrofitting such equipment. Although air disinfection in a commercial building would not reduce illnesses to zero, since some infections would still occur outside the building, measurable reductions in illness could still be expected. Once data become available, it may be possible to compute a payback period for air disinfection systems in terms of the savings in lost work time, health insurance, and improved productivity, and this would give managers a handle on how much to spend on immunizing a particular building. Until then, the system sizes provided in the Chap. 9 simulations could be used as a starting point for costing. In any event, it should be noted that many previous studies on the psychology of the workplace indicate that the mere attempt to improve the work environment is often sufficient to boost employee morale and productivity, and the installation of a well-designed air disinfection system should certainly have more than just a placebo effect.

17.4 The Engineer and the Future of Disease Control

In the war against airborne disease, engineers may understand little about molecular biology, medical pathology, or epidemiology, but they are in the unique position that their engineered solutions can be effective against a wide range of pathogens. Air disinfection technologies do not discriminate between the microorganisms that are destroyed or removed. Some degree of discrimination may occur based on size or susceptibility, but generally speaking these technologies are designed to be effective against broad ranges of airborne pathogens.

A well-designed and integrated immune building system will perform the function of suppressing airborne disease transmission regardless of the specific pathogens that may appear in the building. Buildings act like vectors, or giant incubators, for airborne diseases and toxigenic fungi (Fig. 17.1), but the addition of air disinfection technologies reverses this propensity. Any system designed for biodefense should also tend to be effective in reducing the incidence of respiratory disease transmission.

In theory, the installation of air disinfection technologies in a commercial office building should result in fewer cold virus infections, fewer cases of the flu, and even protection against TB where it is a problem. In actuality, only a fraction, perhaps about one-third, of these infections occur inside the office building. The rest of the infections, perhaps the majority, occur at home, in bars, in restaurants, in schools, or at other, unidentifiable locations. In addition, some respiratory infections transmit by direct contact, a mode that is determined by the individual's behavior and hygiene, and that is not very controllable by any technology.

The real solution to the problem of airborne disease transmission is the widespread application of air disinfection technologies. When a sufficiently large number of buildings have been immunized against airborne disease transmission, a synergy comes into play that is known in epidemiology as "herd immunity" (Fig. 17.2). This term refers to the fact that when a sufficient percentage of a herd or population has been immunized against some specific contagious disease, the disease can no longer propagate and any would-be epidemic simply stops dead in its

Figure 17.1   Modern buildings, with their controlled indoor climates, act like giant, short-term incubators for airborne pathogens. (*Photo courtesy of Art Anderson, Penn State.*)

tracks. The actual percentage of the population that must be immunized is a function of the disease transmission rate, the infection rate, the population susceptibility, and a few other minor factors. It is only necessary to immunize a finite percentage of individuals to stop a disease from spreading. In other words, it is not necessary to immunize everyone to achieve immunity for the population.

Figure 17.2 Based on the herd immunity concept, immunizing a certain percentage of buildings may drive many respiratory diseases to extinction.

In analogous fashion, it is not necessary for every building to be immunized in order to prevent the transmission of contagious airborne diseases. It is only necessary to immunize a certain percentage. The computation of the critical percentage of buildings that might have to be immunized is a challenging mathematical and epidemiological problem that depends on properties of buildings as well as properties of each of the subject diseases. Although the percentage of buildings that would have to be immunized is not known, some value must exist that defines this critical limit.

The implications of the herd immunity hypothesis as applied to buildings lead directly to the notable conclusion that many airborne diseases we have accepted as inevitable facts of life could actually be driven to extinction by the proliferation of immune building technology. At some point, when enough buildings are immunized, these diseases will begin to disappear. No vaccines, no remedies, no microbiological research will be necessary to reach this goal—it may be achieved by engineering alone. The engineer's place in the future of disease control is not merely the protection of buildings against CBW threats, but the complete eradication of airborne disease, perhaps one of the most important goals in the world today.

A

Database of Biological Weapon Agents

Pathogen	Group	Disease group
Bacillus anthracis	Bacterial spore	Noncommunicable
Blastomyces dermatitidis	Fungal spore	Noncommunicable
Bordetella pertussis	Bacteria	Communicable
Brucella	Bacteria	Noncommunicable
Burkholderia mallei	Bacteria	Noncommunicable
Burkholderia pseudomallei	Bacteria	Noncommunicable
Chikungunya virus	Virus	Vector borne
Chlamydia psittaci	Bacteria	Noncommunicable
Clostridium botulinum	Bacteria	Noncommunicable
Clostridium perfringens	Bacteria	Noncommunicable
Coccidioides immitis	Fungal spore	Noncommunicable
Corynebacterium diphtheriae	Bacteria	Communicable
Coxiella burnetii	Bacteria/Rickettsiae	Noncommunicable
Crimean-Congo hemorrhagic fever	Virus	Communicable
Dengue fever virus	Virus	Vector borne
Ebola	Virus	Communicable
Francisella tularensis	Bacteria	Noncommunicable
Hantaan virus	Virus	Noncommunicable
Hepatitis A	Virus	Communicable
Histoplasma capsulatum	Fungal spore	Noncommunicable
Influenza A virus	Virus	Communicable
Japanese encephalitis	Virus	Vector borne
Junin virus	Virus	Noncommunicable
Lassa fever virus	Virus	Communicable
Legionella pneumophila	Bacteria	Noncommunicable
Lymphocytic choriomeningitis	Virus	Noncommunicable
Machupo	Virus	Noncommunicable
Marburg virus	Virus	Communicable

Pathogen	Group	Disease group
Mycobacterium tuberculosis	Bacteria	Communicable
Mycoplasma pneumoniae	Bacteria	Endogenous
Nocardia asteroides	Bacterial spore	Noncommunicable
Paracoccidioides	Fungal spore	Noncommunicable
Rickettsia prowazeki	Bacteria/Rickettsiae	Vector borne
Rickettsia rickettsii	Virus	Vector borne
Rift Valley fever	Virus	Vector borne
Salmonella typhi	Bacteria	Food borne
Shigella	Bacteria	Food borne
Stachybotrys chartarum	Fungal spore	Noncommunicable
Streptococcus pneumoniae	Bacteria	Communicable
Variola (smallpox)	Virus	Communicable
VEE	Virus	Vector borne
Vibrio cholerae	Bacteria	Food borne
WEE, EEE	Virus	Vector borne
Yellow fever virus	Virus	Vector borne
Yersinia pestis	Bacteria	Communicable

Bacillus anthracis

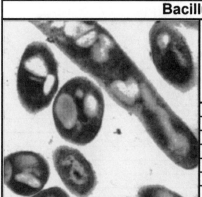

GROUP	Bacterial Spore
TYPE	Gram+
GENUS	Bacillus
FAMILY	Bacillaceae
DISEASE GROUP	Non-communicable
BIOSAFETY LEVEL	Risk Group 2-3
Infectious Dose (ID$_{50}$)	10000
Lethal Dose (LD$_{50}$)	28000
Infection Rate	none
Incubation Period	2-3 days
Peak Infection	na
Annual Cases	rare
Annual Fatalities	rare

Causes anthrax, which is primarily a disease of lower animals that can be transmitted to man by contact and by inhalation of spores. Pulmonary anthrax was commonly known as "woolsorter's disease". Anthrax rapidly spreads from the lungs to other organs, and is deadly if untreated. The US, Japan, Iraq, and the Soviet Union have researched the use of anthrax as a biological weapon.The use of anthrax as a terrorist weapon remains to this day a potentially serious terrorist threat.

Disease or Infection	anthrax, woolsorter's disease			
Natural Source	Cattle, sheep, other animals, soil.			
Toxins	none			
Point of Infection	Upper Respiratory Tract			
Symptoms	Inhalation anthrax: respiratory distress, fever and shock with death shortly thereafter.			
Treatment	Ciprofloxacin, doxycycline, tetracyclines, erythromycin, chloramphenicol			
Untreated Fatality Rate	5-20%	**Prophylaxis:** Antibiotics	**Vaccine:**	Available
Shape	spherical spore			
Mean Diameter, μm	1.118	**Size Range:**	1-1.25 microns	
Growth Temperature	37 C	**Survival Outside Host:**	years	
Inactivation	Moist heat: 121 C for 30 minutes.			
Disinfectants	5% formalin, 2% glutaraldehyde formaldehyde			
Nominal Filter Rating	**MERV 6**	**MERV 8**	**MERV 11**	**MERV 13** / **MERV 15**
Estimated Removal, %	18.8	40.4	57.1	92.5 / 100.0
UVGI Rate Constant	0.000031	cm^2/μW-s	**Media**	Plates
Dose for 90% Kill	74277	μW-s/cm^2	**Ref.**	Knudson 1986
Suggested Indoor Limit	0 CFU/cu.m.			
Genome Base Pairs				
Related Species	B. cereus, Type Species: B. subtilis (B. globigii)			
Notes	CDC Reportable.			
Photo Credit	Centers for Disease Control, PHIL# 1823, Dr. Sharif Zaki, Elizabeth White.			
References	Brachman 1966, Braude 1981, Freeman 1985, Inglesby 1999, Mitscherlich 1984, Murray 1999, Prescott 1996, Ryan 1994, Canada 2001			

Blastomyces dermatitidis

GROUP	Fungal Spore
TYPE	Hyphomycetes
GENUS	Moniliales
FAMILY	
DISEASE GROUP	Non-communicable
BIOSAFETY LEVEL	Risk Group 2-3

Infectious Dose (ID$_{50}$)	11000
Lethal Dose (LD$_{50}$)	unknown
Infection Rate	none
Incubation Period	weeks
Peak Infection	na
Annual Cases	rare
Annual Fatalities	-

Can resemble TB and spread beyond the lungs. Found mainly in north central and eastern US. Causes blastomycosis. This malady is also known as Gilchrist's disease and Chicago disease. Entry is through the upper respiratory tract, but can spread to other locations. Males are more susceptible to this progressive disease than females. Like other fungi pathogenic for man, this fungi exhibits dimorphism, existing in one form in nature and another when causing infection.

Disease or Infection	blastomycosis, Gilchrist's Disease, Chicago disease.				
Natural Source	Environmental, nosocomial.				
Toxins	none				
Point of Infection	Upper Respiratory Tract, skin				
Symptoms	Acute or chronic granulomatous mycosis of the lungs or skin. Indolent onset becoming chronic pulmonary infection.				
Treatment	Amphotericin B, itraconazole, ketoconazole, hydroxystilbamidine				
Untreated Fatality Rate	-	**Prophylaxis:** none		**Vaccine:** none	
Shape	spherical spore				
Mean Diameter, μm	12.649	**Size Range:** 8-20 microns			
Growth Temperature	37 C	**Survival Outside Host:** years			
Inactivation	Moist heat: 121 C for 15 minutes.				
Disinfectants	1% sodium hypochlorite, phenolics, formaldehyde, 10% formalin.				
Nominal Filter Rating	MERV 6	MERV 8	MERV 11	MERV 13	MERV 15
Estimated Removal, %	50.0	77.3	94.0	96.0	100.0
UVGI Rate Constant	0.000247	cm^2/μW-s	**Media**	Plates (estimated)	
Dose for 90% Kill	9322	μW-s/cm^2	**Ref.**	Chick 1963	
Suggested Indoor Limit	150-500 CFU/cu.m.				
Genome Base Pairs					
Related Species					
Notes	CDC Reportable.				
Photo Credit	Centers for Disease Control, PHIL# 493.				
References	Collins 1993, Freeman 1985, Howard 1983, Lacey 1988, Murray 1999, Ryan 1994, Smith 1989, Miyaji 1987, Sorensen 1999, DiSalvo 1983, Canada 2001				

Bordetella pertussis

GROUP	Bacteria
TYPE	Gram-
GENUS	Bordetella
FAMILY	
DISEASE GROUP	Communicable
BIOSAFETY LEVEL	Risk Group 2

Infectious Dose (ID$_{50}$)	(4)
Lethal Dose (LD$_{50}$)	(1314)
Infection Rate	high
Incubation Period	7-10 days
Peak Infection	7-14 days
Annual Cases	6,564
Annual Fatalities	15

Bordetella pertussis is the cause of whooping cough. It produces microbial toxins, which are primarily responsible for the disease symptoms. Occurring worldwide, this infection almost exclusively affects children. Almost two-thirds of cases are under 1 year of age. Asymptomatic cases, however, are more frequent. It is highly contagious and transmits via airborne bioaerosols, and fomites (direct contact).

Disease or Infection	whooping cough, toxic reactions				
Natural Source	Humans, nosocomial				
Toxins	none				
Point of Infection	Upper Respiratory Tract, trachea				
Symptoms	Three stages occur: coughing (1-2 weeks), violent whooping cough (2-6 weeks), recovery for several weeks.				
Treatment	14 days treatment with erythromycin or TMP-SMX, oxygenation, hydration & electrolyte balance				
Untreated Fatality Rate	-	**Prophylaxis:** Antibiotics		**Vaccine:**	Available
Shape	coccobacilli				
Mean Diameter, μm	0.245	**Size Range:**	0.2-0.3 x 0.5-1 microns		
Growth Temperature	35-37 C	**Survival Outside Host:**	1 hrs - 7 days		
Inactivation	Moist heat: 121 C for 15 minutes. Dry heat: 170 C for 1 hour.				
Disinfectants	1% sodium hypochlorite, 70% ethanol, iodines, phenolics, glutaraldehyde, formaldehyde.				
Nominal Filter Rating	**MERV 6**	**MERV 8**	**MERV 11**	**MERV 13**	**MERV 15**
Estimated Removal, %	4.3	8.7	14.5	39.4	70.4
UVGI Rate Constant	(unknown)	cm^2/μW-s	**Media**		
Dose for 90% Kill	-	μW-s/cm^2	**Ref.**		
Suggested Indoor Limit	0 CFU/cu.m.				
Genome Base Pairs					
Related Species					
Notes	CDC Reportable.				
Photo Credit	Centers for Disease Control, PHIL# 254, Janice Carr.				
References	Braude 1981, Freeman 1985, Mitscherlich 1984, Murray 1999, Prescott 1996, Ryan 1994, Castle 1987, Weinstein 1991, Canada 2001				

Brucella

GROUP	Bacteria
TYPE	Gram-
GENUS	Brucella
FAMILY	Brucellaceae
DISEASE GROUP	Non-communicable
BIOSAFETY LEVEL	Risk Group 2-3
Infectious Dose (ID$_{50}$)	1300
Lethal Dose (LD$_{50}$)	-
Infection Rate	-
Incubation Period	5-60 days
Peak Infection	-
Annual Cases	98
Annual Fatalities	rare

Contact with animals is usually required for infection but airborne infection has occurred. May become chronic. Predominantly an occupational disease of those who work with infected animals. Route of infection may be via ingestion, abraded skin, or through the mucous membranes.

Disease or Infection	Brucellosis, undulant fever
Natural Source	Goats, cattle, swine, dogs, sheep, caribou, elk, coyotes, camels.
Toxins	none
Point of Infection	Upper Respiratory Tract, skin
Symptoms	Acute onset with intermittent fever, fatigue, headache, profuse sweating, chills, arthralgia, local infections, long recovery period.
Treatment	Tetracyclines, streptomycin usually combined with doxycycline, TMP-SMX, combined with aminoglycoside
Untreated Fatality Rate	<2% **Prophylaxis:** none **Vaccine:** none
Shape	short rods
Mean Diameter, μm	0.566 **Size Range:** 0.4-0.8 x 0.4-1.5 microns
Growth Temperature	10-40 C **Survival Outside Host:** 32-135 days
Inactivation	Moist heat: 121 C for 15 minutes. Dry heat: 170 C for 1 hour.
Disinfectants	1% sodium hypochlorite, 70% ethanol, iodines, glutaraldehyde, formaldehyde.

Nominal Filter Rating	MERV 6	MERV 8	MERV 11	MERV 13	MERV 15
Estimated Removal, %	8.0	17.8	29.2	70.2	96.4
UVGI Rate Constant	(unknown)	cm^2/μW-s	Media		
Dose for 90% Kill	-	μW-s/cm^2	Ref.		
Suggested Indoor Limit					
Genome Base Pairs					
Related Species	B. melitensis, B.suis, B.abortus, B.canis.				
Notes	CDC Reportable.				
Photo Credit	Centers for Disease Control, PHIL# 734, Dr. Marshall Fox.				
References	Braude 1981, Murray 1999, Mitscherlich 1984, Prescott 1996, Ryan 1994, Canada 2001, Franz 1997, NATO 1996, Mandell 2000, Ellison 2000.				

Burkholderia mallei

GROUP	Bacteria
TYPE	Gram-
GENUS	Burkholderia
FAMILY	
DISEASE GROUP	Non-communicable
BIOSAFETY LEVEL	Risk Group 3
Infectious Dose (ID$_{50}$)	3200
Lethal Dose (LD$_{50}$)	-
Infection Rate	none
Incubation Period	1-14 days
Peak Infection	na
Annual Cases	-
Annual Fatalities	none

Formerly Pseudomonas mallei, B. mallei causes a disease called Glanders, which is a disease of horses and mules. It has, on rare occasions been transmitted to humans, but has been eradicated in the West. The infection route is most probably through skin abrasions.

Disease or Infection	Glanders, fever, opportunistic infections
Natural Source	Environmental, horses, mules, nosocomial.
Toxins	none
Point of Infection	skin
Symptoms	Chronic form with cough and mucous discharge. Septicemic form with fever, chills, death within 7-10 days.
Treatment	Ceftazidime, imipenem, doxycycline, minocycline, ciprofloxacin, gentamicin

Untreated Fatality Rate	-	Prophylaxis:	none	Vaccine:	none
Shape	rods				
Mean Diameter, μm	0.674	Size Range:	0.3-0.8 x 1.4-4 microns		
Growth Temperature	20-42 C	Survival Outside Host:	30 days in water		
Inactivation	Moist heat: 55 C for 10 minutes.				
Disinfectants	1% sodium hypochlorite, 70% ethanol, 2% glutaraldehyde.				

Nominal Filter Rating	MERV 6	MERV 8	MERV 11	MERV 13	MERV 15
Estimated Removal, %	9.9	22.2	35.2	77.8	98.5
UVGI Rate Constant	(unknown)	cm^2/μW-s	Media		
Dose for 90% Kill	-	μW-s/cm^2	Ref.		
Suggested Indoor Limit	50 CFU/cu.m.				
Genome Base Pairs					
Related Species	(previously Pseudomonas mallei)				
Notes					
Photo Credit	Image shows B. cepacia. Centers for Disease Control, PHIL# 255, Janice Carr.				
References	Braude 1981, Freeman 1985, Mitscherlich 1984, Murray 1999, Prescott 1996, Ryan 1994, Castle 1987, Weinstein 1991, Canada 2001				

Burkholderia pseudomallei

GROUP	Bacteria
TYPE	Gram-
GENUS	Burkholderia
FAMILY	
DISEASE GROUP	Non-communicable
BIOSAFETY LEVEL	Risk Group 2-3

Infectious Dose (ID_{50})	-
Lethal Dose (LD_{50})	-
Infection Rate	none
Incubation Period	2 days min.
Peak Infection	na
Annual Cases	rare
Annual Fatalities	rare

Formerly Pseudomonas pseudomallei, B. pseudomallei occurs primarily in rodents and other animals in the South Pacific. In these areas, it is endemic and can be acquired by humans, but is very rare in the West. Being an animal respiratory disease, like brucellosis, it can be transmitted to humans outdoors via direct contact or inhalation. It can also be found in moist soil or warm water, from which it may contaminate open wounds.

Disease or Infection	meliodosis, opportunistic infections
Natural Source	Environmental, rodents, soil, water, nosocomial.
Toxins	none
Point of Infection	Upper Respiratory Tract
Symptoms	May resemble typhoid fever or tuberculosis. Can vary from chronic infection to rapidly fatal septicemia.
Treatment	TMP-SMX, ceftazidime, imipenem, doxycycline, ciprofoxacinsulphas, tetracycline, chloramphenicol

Untreated Fatality Rate	-	Prophylaxis:	none	Vaccine:	none
Shape	rods				
Mean Diameter, μm	0.494	Size Range:	0.3-0.8 x 1-3 microns		
Growth Temperature	5-42 C	Survival Outside Host:	years in soil & water		
Inactivation	Moist heat: 121 C for 15 min. Dry heat: 170 C for 1 hr.				
Disinfectants	1% sodium hypochlorite, 70% ethanol, glutaraldehyde.				

Nominal Filter Rating	MERV 6	MERV 8	MERV 11	MERV 13	MERV 15
Estimated Removal, %	6.8	15.1	25.2	63.9	93.6

UVGI Rate Constant	(unknown)	$cm^2/\mu W$-s	Media	
Dose for 90% Kill	-	μW-s/cm^2	Ref.	

Suggested Indoor Limit	50 CFU/cu.m.
Genome Base Pairs	
Related Species	(previously Pseudomonas pseudomallei)
Notes	
Photo Credit	Donald Woods of the Canadian Bacterial Diseases Network, University of Calgary, Alberta.
References	Braude 1981, Freeman 1985, Mitscherlich 1984, Murray 1999, Prescott 1996, Ryan 1994, Canada 2001

Chikungunya virus

GROUP	Virus
TYPE	RNA
GENUS	Alphavirus
FAMILY	Togaviridae
DISEASE GROUP	Vector-borne
BIOSAFETY LEVEL	Risk Group 2

Infectious Dose (ID$_{50}$)	unknown
Lethal Dose (LD$_{50}$)	unknown
Infection Rate	none
Incubation Period	1-12 days
Peak Infection	na
Annual Cases	rare
Annual Fatalities	rare

Non-respiratory and non-airborne. Transmitted by the bite of infected mosquitoes. No evidence of person-to-person transmission. May originate from primates as the ultimate reservoir. May present hemorrhagic symptoms. A rash may develop in 1-10 days. Recovery may be prolonged. Endemic to Africa and Asia. Closely related to VEE, WEE, and EEE.

Disease or Infection	Chikungunya fever, Epidemic polyarthritis and rash, CHIK			
Natural Source	Primates, humans, birds			
Toxins	none			
Point of Infection	Insect bite.			
Symptoms	High fever, with arthralgia or arthritis in small joints of the extremities, maculopapular rash, possible buccal and palatal enanthema, nausea and vomiting.			
Treatment	No antiviruses, supportive treatment only.			
Untreated Fatality Rate	Prophylaxis: none Vaccine: none			
Shape	spherical, enveloped virion			
Mean Diameter, μm	0.06 Size Range: 0.05-0.07 microns			
Growth Temperature	na Survival Outside Host: less than 1 day at 37 C			
Inactivation	Moist or dry heat: 60 C for prolonged periods.			
Disinfectants	70% ethanol, 1% sodium hypochlorite, 2% glutaraldehyde.			
Nominal Filter Rating	MERV 6	MERV 8	MERV 11	MERV 13	MERV 15
Estimated Removal, %	10.9	19.5	28.7	58.9	84.0
UVGI Rate Constant	(unknown)	cm^2/μW-s	Media		
Dose for 90% Kill		μW-s/cm^2	Ref.		
Suggested Indoor Limit	na			
Genome Base Pairs	9700-11800 nt, 11700 bp			
Related Species				
Notes	CDC Reportable.			
Photo Credit	Photo shows the Kunjin alphavirus. Image provided courtesy of Ed Westaway of SASVRC, The Royal Children's Hospital, Australia.			
References	Fraenkel-Conrat 1982, Braude 1981, Freeman 1985, Canada 2001, Ellison 2000			

Chlamydia psittaci

GROUP	Bacteria
TYPE	Gram-
GENUS	Chlamydia
FAMILY	Chlamydiaceae
DISEASE GROUP	Non-communicable
BIOSAFETY LEVEL	Risk Group 2-3

Infectious Dose (ID$_{50}$)	-
Lethal Dose (LD$_{50}$)	–
Infection Rate	none
Incubation Period	5-15 days
Peak Infection	na
Annual Cases	33
Annual Fatalities	rare

Generally parasitizes animals and birds in the wild. Man is but an incidental host who becomes infected as the result of contact with domesticated birds and fowl. Human infections are rare, and often mild. These bacteria are obligate intracellular parasites that use ATP produced by the host cell, hence, they are termed "energy parasites."

Disease or Infection	psittacosis/ornithosis (parrot fever), pneumonitis			
Natural Source	Birds, fowl			
Toxins	none			
Point of Infection	Upper Respiratory Tract			
Symptoms	Fever, myalgia, headache, chills, respiratory distress, pneumonia, fatigue, anorexia, encephalitis.			
Treatment	Penicillin, tetracyclines, erythromycin			
Untreated Fatality Rate	<6% **Prophylaxis:** Anitbiotics **Vaccine:** none			
Shape	spherical			
Mean Diameter, μm	0.28 · **Size Range:** 0.2-0.4 x 0.4-1 microns			
Growth Temperature	33-41 C **Survival Outside Host:** 2-20 days			
Inactivation	Moist heat: 121 C for 15 min. Dry heat: 170 C for 1 hr.			
Disinfectants	1% sodium hypochlorite, 70% ethanol, glutaraldehyde.			

Nominal Filter Rating	MERV 6	MERV 8	MERV 11	MERV 13	MERV 15
Estimated Removal, %	4.4	9.2	15.5	42.6	74.9

UVGI Rate Constant	(unknown)	$cm^2/\mu W\text{-}s$	**Media**
Dose for 90% Kill	-	$\mu W\text{-}s/cm^2$	**Ref.**
Suggested Indoor Limit	0 CFU/cu.m.		
Genome Base Pairs			
Related Species	Type Species: C. trachomatis		
Notes	CDC Reportable.		
Photo Credit	Centers for Disease Control, PHIL# 1351, Dr. Martin Hicklin.		
References	Braude 1981, Freeman 1985, Mitscherlich 1984, Murray 1999, Prescott 1996, Ryan 1994, Storz 1971, Zhang 1993, Canada 2001, Mandell 2000		

Clostridium botulinum

GROUP	Bacteria
TYPE	Gram+
GENUS	Clostridium
FAMILY	
DISEASE GROUP	Non-communicable
BIOSAFETY LEVEL	Risk Group 2-4

Infectious Dose (ID$_{50}$)	-
Lethal Dose (LD$_{50}$)	-
Infection Rate	-
Incubation Period	12-36 hours
Peak Infection	-
Annual Cases	-
Annual Fatalities	-

A bacterial spore that can cause food poisoning when conditions are right for toxin production. May settle on food. May also settle on open wounds causing wound botulism. Infant botulism occurs when infants under one year of age ingest spores. Produces a neurotoxin under anaerobic conditions and in low acid foods.

Disease or Infection	Botulism, toxic poisoning.
Natural Source	Environmental
Toxins	Botulinum
Point of Infection	Ingested.
Symptoms	Acute flaccid paralysis involving the muscles of the face, head, and pharynx, down to the thorax and extremities, respiratory failure.
Treatment	Antibiotic treatment generally not effective, trivalent (types A, B, & E) equine antitoxin, Penicillin G sometimes used

Untreated Fatality Rate	-	**Prophylaxis:**	Antitoxin	**Vaccine:**	B. Toxoid
Shape	Rods				
Mean Diameter, μm	1.97	**Size Range:**	1.3-3 microns		
Growth Temperature	100	**Survival Outside Host:**		indefinitely in soil, water	
Inactivation	Boiling 10 mins. Moist heat: 120 C for 15 min.				
Disinfectants	1% sodium hypochlorite, 70% ethanol.				

Nominal Filter Rating	MERV 6	MERV 8	MERV 11	MERV 13	MERV 15
Estimated Removal, %	33.2	63.9	80.8	95.9	100.0
UVGI Rate Constant	(unknown)	cm^2/μW-s	Media		
Dose for 90% Kill	-	μW-s/cm^2	Ref.		
Suggested Indoor Limit	-				
Genome Base Pairs					
Related Species					
Notes					

Photo Credit	Photo courtesy of Upjohn Company, Scope monograph.
References	Braude 1981, Freeman 1985, Mitscherlich 1984, Murray 1999, Prescott 1996, Ryan 1994, Canada 2001, Mandell 2000

Clostridium perfringens

GROUP	Bacteria
TYPE	Gram+
GENUS	Clostridium
FAMILY	
DISEASE GROUP	Non-communicable
BIOSAFETY LEVEL	Risk Group 2

Infectious Dose (ID$_{50}$)	10 per g of food
Lethal Dose (LD$_{50}$)	-
Infection Rate	none
Incubation Period	6-24 hrs.
Peak Infection	na
Annual Cases	10,000
Annual Fatalities	10

Non-respiratory but may settle on exposed foods. A food-borne pathogen that produces enterotoxins. May grow on foods like meat. Often found in the intestines and in feces. Can cause wound contamination.

Disease or Infection	sepsis, toxic reactions, food poisoning.
Natural Source	Environmental, Humans, animals, soil. Spp: C. novyi, septicum.
Toxins	none
Point of Infection	intestines
Symptoms	Sudden onset of colic, diarrhea, nausea, short duration, rarely fatal.
Treatment	Penicillin

Untreated Fatality Rate	-	Prophylaxis:	none	Vaccine:	none
Shape	spherical spore, "box-car" vegetative cell				
Mean Diameter, μm	5	Size Range:	2.5-10 microns		
Growth Temperature	6-47 C	Survival Outside Host:	years		
Inactivation	Moist heat: 121 C for >15 min.				
Disinfectants	1% sodium hypochlorite, prolonged contact with glutaraldehyde.				

Nominal Filter Rating	MERV 6	MERV 8	MERV 11	MERV 13	MERV 15
Estimated Removal, %	48.8	77.3	93.8	96.0	100.0

UVGI Rate Constant	0.00017	cm^2/μW-s	Media	Plates
Dose for 90% Kill	13545	μW-s/cm^2	Ref.	
Suggested Indoor Limit	0 CFU/cu.m.			
Genome Base Pairs				
Related Species	Type Species: C. butyricum			
Notes				
Photo Credit	Centers for Disease Control (Clostridium novyi shown), PHIL #1033, Dr. William A. Clark.			
References	Braude 1981, Freeman 1985, Mitscherlich 1984, Murray 1999, Prescott 1996, Ryan 1994, Canada 2001, Mandell 2000			

Coccidioides immitis

GROUP	Fungal Spore
TYPE	Hyphomycetes
GENUS	Hyphomycetales
FAMILY	
DISEASE GROUP	Non-communicable
BIOSAFETY LEVEL	Risk Group 3
Infectious Dose (ID$_{50}$)	1350
Lethal Dose (LD$_{50}$)	-
Infection Rate	none
Incubation Period	1-4 weeks
Peak Infection	na
Annual Cases	uncommon
Annual Fatalities	-

The most dangerous fungal infection. Causes the disease coccidioidomycosis. Estimates are that 20-40 million people in the Southwest have had coccidioidomycosis. Only about 40% of infections are symptomatic, and only 5% are clinically diagnosed. A self-limiting, non-progressive form of the infection is commonly known as valley fever or desert rheumatism. Transmission occurs by inhalation and no direct transmission from animals occurs. The natural reservoir is the soil.

Disease or Infection	coccidioidomycosis, valley fever, desert rheumatism
Natural Source	Environmental, soil, nosocomial.
Toxins	none
Point of Infection	Upper Respiratory Tract
Symptoms	Respiratory influenza-like infection, erythema nodosum in 20% cases, may progress to lung lesions, abcesses. Some 60% of infections are normally asymptomatic.
Treatment	Amphotericin B, ketoconazole, itraconazole, fluoconazole for meningeal infections

Untreated Fatality Rate	0.9	**Prophylaxis:** none	**Vaccine:**	none
Shape	barrel-shaped			
Mean Diameter, μm	3.46	**Size Range:** 2-6 microns		
Growth Temperature		**Survival Outside Host:** years		
Inactivation	Moist heat: 121 C for >15 min.			
Disinfectants	1% sodium hypochlorite, phenolics, glutaraldehyde, formaldehyde.			

Nominal Filter Rating	MERV 6	MERV 8	MERV 11	MERV 13	MERV 15
Estimated Removal, %	45.2	75.8	92.1	96.0	100.0

UVGI Rate Constant	(unknown)	cm^2/μW-s	**Media**
Dose for 90% Kill	-	μW-s/cm^2	**Ref.**
Suggested Indoor Limit	150-500 CFU/cu.m.		
Genome Base Pairs			
Notes	CDC Reportable.		
Photo Credit	Centers for Disease Control, PHIL #476, Dr. Hardin.		
References	Freeman 1985, Howard 1983, Lacey 1988, Murray 1999, Ryan 1994, Smith 1989, Sorensen 1999, Miyaji 1987, Canada 2001		

Corynebacterium diphtheriae

GROUP	Bacteria
TYPE	Gram+
GENUS	Corynebacterium
FAMILY	
DISEASE GROUP	Communicable
BIOSAFETY LEVEL	Risk Group 2

Infectious Dose (ID_{50})	-
Lethal Dose (LD_{50})	-
Infection Rate	varies
Incubation Period	2-5 days
Peak Infection	-
Annual Cases	10
Annual Fatalities	-

Primarily affects children but less prevalent today. Harbored in the upper respiratory tracts of healthy carriers. Corynebacterium diptheria is the causative agent of diptheria, which was historically a disease of children. In modern times this disease is less prevalent, but increasingly afflicts those in older age groups. Healthy carriers may harbor the bacteria in their throats asymptomatically.

Disease or Infection	diphtheria, toxin produced, opportunistic.
Natural Source	Humans (only known reservoir), nosocomial
Toxins	diphtheria toxin
Point of Infection	Upper Respiratory Tract
Symptoms	Pharyngitis, fever, malaise, swelling of the neck, headache, hypoxia, toxic effects on nervous system.
Treatment	Administer antitoxin in conjunction with erythromycin, penicillin
Untreated Fatality Rate	5-10% Prophylaxis: DTP Vaccine: Available
Shape	rods
Mean Diameter, µm	0.70 Size Range: 0.3-0.8 x 1-6 microns
Growth Temperature	34-36 C Survival Outside Host: 2.5 hrs in air, < 1 year soil
Inactivation	Moist heat: 121 C for 15 minutes. Dry heat: 170 C for 1 hour.
Disinfectants	1% sodium hypochlorite, phenolics, glutaraldehyde, formaldehyde, iodines.

Nominal Filter Rating	MERV 6	MERV 8	MERV 11	MERV 13	MERV 15
Estimated Removal, %	10.4	23.2	36.6	79.3	98.8

UVGI Rate Constant	0.000701	$cm^2/\mu W\text{-}s$	Media	Water
Dose for 90% Kill	3285	$\mu W\text{-}s/cm^2$	Ref.	Westinghouse
Suggested Indoor Limit	0 CFU/cu.m.			
Genome Base Pairs				
Related Species				
Notes	CDC Reportable.			
Photo Credit	Centers for Disease Control (Blood agar culture), PHIL 1566, Dr. W.A. Clark.			
References	Braude 1981, Freeman 1985, Mitscherlich 1984, Murray 1999, Prescott 1996, Ryan 1994, Canada 2001, Mandell 2000			

Coxiella burnetii

GROUP	Bacteria / Rickettsiae
TYPE	Gram-
GENUS	Coxiella
FAMILY	Rickettsiae
DISEASE GROUP	Non-communicable
BIOSAFETY LEVEL	Risk Group 2-3

Infectious Dose (ID$_{50}$)	10
Lethal Dose (LD$_{50}$)	-
Infection Rate	none
Incubation Period	9-18 days
Peak Infection	na
Annual Cases	rare
Annual Fatalities	-

Rickettsiae -- a bacteria-like microorganism. Transmitted from animals to humans by inhalation. Can survive in fomites for years. A high risk for slaughterhouses and animal research. Endemic in many areas. Resistant to heat and dessication. Potential biological weapon.

Disease or Infection	Q fever				
Natural Source	Cattle, sheep, goats.				
Toxins	none				
Point of Infection	Upper Respiratory Tract				
Symptoms	Sudden onset of acute febrile disease, chills, headache, fatigue, sweats, pneumonitis, pericarditis, hepatitis.				
Treatment	Tetracycline, chloramphenicol, rifampin				
Untreated Fatality Rate	<1% **Prophylaxis:** ineffective **Vaccine:** Available				
Shape	short rods, highly pleomorphic				
Mean Diameter, μm	0.283 **Size Range:** 0.2-0.4 x 0.4-1 microns				
Growth Temperature	32-35 C **Survival Outside Host:** years				
Inactivation	Moist heat: 130 C for 1 hour.				
Disinfectants	Ethanol, glutaraldehyde, gaseous formaldehyde.				
Nominal Filter Rating	MERV 6	MERV 8	MERV 11	MERV 13	MERV 15
Estimated Removal, %	4.4	9.2	15.5	42.6	74.9
UVGI Rate Constant	0.001535	cm^2/μW-s	**Media**	Water	
Dose for 90% Kill	1500	μW-s/cm^2	**Ref.**	Little 1980	
Suggested Indoor Limit	0 CFU/cu.m.				
Genome Base Pairs					
Related Species					
Notes	CDC Reportable.				
Photo Credit	Image shows rickettsia in an endothelial cell. Centers for Disease Control, PHIL# 1349, Martin Hicklin.				
References	Braude 1981, Freeman 1985, McCaul 1981, Murray 1999, Prescott 1996, Ryan 1994, Walker 1988, Canada 2001, NATO 1996, Mandell 2000				

Crimean-Congo hemorrhagic fever

GROUP	Virus
TYPE	RNA
GENUS	Nairovirus
FAMILY	Bunyaviridae
DISEASE GROUP	Communicable
BIOSAFETY LEVEL	Risk Group 4

Infectious Dose (ID$_{50}$)	unknown
Lethal Dose (LD$_{50}$)	unknown
Infection Rate	unknown
Incubation Period	1-3 days
Peak Infection	-
Annual Cases	rare
Annual Fatalities	-

Normally a vector-borne virus endemic to western Crimea, the Congo, western China and other areas. Transmission by ticks, but nosocomial outbreaks have occurred and can be transmitted by exposure to blood and secretions of infected animals. Has been studied as a potential biological weapon.

Disease or Infection	CCHF, hemorrhagic fever, Central Asian hemorrhagic fever
Natural Source	Birds, ticks, domestic animals, rodents, mosquitoes.
Toxins	none
Point of Infection	Upper Respiratory Tract
Symptoms	Sudden onset of fever and fatigue, irritability, headache, severe pain in loins, anorexia, vomiting, diarrhea, bleeding from nose, lungs, gumd, uterus.
Treatment	Ribavarin
Untreated Fatality Rate	2-50% **Prophylaxis:** possible **Vaccine:** none
Shape	spherical enveloped virion
Mean Diameter, μm	0.09 **Size Range:** 0.085-0.1 microns
Growth Temperature	na **Survival Outside Host:** <10 days in blood
Inactivation	Moist heat: 56 C for 30 minutes.
Disinfectants	1% sodium hypochlorite, 2% glutaraldehyde.

Nominal Filter Rating	MERV 6	MERV 8	MERV 11	MERV 13	MERV 15
Estimated Removal, %	7.5	13.6	20.4	45.9	72.2

UVGI Rate Constant	unknown	cm^2/μW-s	**Media**
Dose for 90% Kill	-	μW-s/cm^2	**Ref.**
Suggested Indoor Limit	0 CFU/cu.m.		
Genome Base Pairs	10,500-22,700		
Related Species			
Notes	CDC Reportable.		
Photo Credit	Photo shows the bunyavirus Tospovirus. Image provided courtesy of Cornelia Buchen-Osmond of Columbia University, Roy Woods, Phil Jones, Sheila Roberts, & Adrian Bell, of the International Committee on Taxonomy of Viruses Universal Virus Database & John Antoniw of The Institute Of Arable Crops Research in Rothamsted.		
References	Dalton 1973, Fraenkel-Conrat 1985, Freeman 1985, Mahy 1975, Murray 1999, Ryan 1994, Canada 2001		

Dengue fever virus

GROUP	Virus
TYPE	RNA
GENUS	Flavivirus
FAMILY	Flaviviridae
DISEASE GROUP	Vector-borne
BIOSAFETY LEVEL	Risk Group 2

Infectious Dose (ID_{50})	unknown
Lethal Dose (LD_{50})	unknown
Infection Rate	na
Incubation Period	3-14 days
Peak Infection	na
Annual Cases	rare
Annual Fatalities	rare

Non-respiratory and non-airborne. Normally transmitted by the bite of infected mosquitoes. Host becomes infectious for other mosquitoes for 3-5 days. Some secondary transmission has been suggested by close contact. Endemic to regions of the tropics in Asia, Africa, and America.

Disease or Infection	Dengue fever, breakbone fever, Dengue hemorrhagic fever
Natural Source	Primates, humans, mosquitoes.
Toxins	none
Point of Infection	Insect bite.
Symptoms	Acute febrile disease with sudden onset of fever, intense headache, myalgia, arthralgia, anorexia, rash, and the feeling the bones are breaking.
Treatment	No antiviral agents, maintain hydration and avoid salicylates.

Untreated Fatality Rate	40-50%	Prophylaxis:	none		Vaccine:	none
Shape	spherical enveloped virion					
Mean Diameter, µm	0.04472136	Size Range:	0.04-0.05 microns			
Growth Temperature	na	Survival Outside Host:		several days		
Inactivation	-					
Disinfectants	1% sodium hypochlorite, 2% glutaraldehyde, 70% ethanol					

Nominal Filter Rating	MERV 6	MERV 8	MERV 11	MERV 13	MERV 15
Estimated Removal, %	13.5	24.0	35.0	67.5	90.3
UVGI Rate Constant	unknown	$cm^2/\mu W\text{-}s$	Media		
Dose for 90% Kill		$\mu W\text{-}s/cm^2$	Ref.		

Suggested Indoor Limit	na
Genome Base Pairs	9500-12500 nt, 10500 bp
Related Species	
Notes	CDC Reportable.
Photo Credit	Centers for Disease Control, Atlanta.
References	Dalton 1973, Fraenkel-Conrat 1985, Freeman 1985, Mahy 1975, Ryan 1995, Canada 2001

Ebola

GROUP	Virus
TYPE	ssRNA
GENUS	
FAMILY	Filoviridae
DISEASE GROUP	Communicable
BIOSAFETY LEVEL	Risk Group 4

Infectious Dose (ID$_{50}$)	10
Lethal Dose (LD$_{50}$)	-
Infection Rate	unknown
Incubation Period	2-21 days
Peak Infection	-
Annual Cases	rare
Annual Fatalities	-

A blood-borne virus suspected of potential for airborne transmission. Periodically causes epidemics in Central Africa. Several sub-types exist, including Reston which does not infect man. No evidence for airborne transmission but reportedly genetically engineered for airborne transmission and toxicity by Soviet researchers.

Disease or Infection	African hemorrhagic fever, EHF, Ebola disease, EBO, EBOV
Natural Source	Monkeys, chimpanzees, ultimate source unknown.
Toxins	none
Point of Infection	Upper Respiratory Tract, mouth.
Symptoms	Sudden onset of high fever, fatigue, abdominal pain, myalgias, diarrhea and vomiting, rash, hemorrhagic diathesis of internal organs.
Treatment	No effective treatment Maintain renal function and electrolyte balance
Untreated Fatality Rate	50-90% **Prophylaxis:** none **Vaccine:** none
Shape	pleomorphic
Mean Diameter, μm	0.09 **Size Range:** 0.080-0.097 microns
Growth Temperature	na **Survival Outside Host:** weeks in blood samples
Inactivation	Moist heat: 60 C for 1 hr.
Disinfectants	2% sodium hypochlorite, 2% glutaraldehyde, 5% peracetic acid, 1% formalin.

Nominal Filter Rating	MERV 6	MERV 8	MERV 11	MERV 13	MERV 15
Estimated Removal, %	7.8	14.1	21.1	47.1	73.4
UVGI Rate Constant	unknown	cm^2/μW-s	**Media**		
Dose for 90% Kill	-	μW-s/cm^2	**Ref.**		
Suggested Indoor Limit	0 CFU/cu.m.				
Genome Base Pairs					
Related Species	Marburg, Reston				
Notes	CDC Reportable.				
Photo Credit	Centers for Disease Control, PHIL# 1181, C. Goldsmith.				
References	Dalton 1973, Fraenkel-Conrat 1985, Freeman 1985, Mahy 1975, Murray 1999, Ryan 1994, Salvato 1993, Malherbe 1980, Canada 2001				

Francisella tularensis

GROUP	Bacteria
TYPE	Gram-
GENUS	Francisella
FAMILY	
DISEASE GROUP	Non-communicable
BIOSAFETY LEVEL	Risk Group 2-3

Infectious Dose (ID$_{50}$)	10
Lethal Dose (LD$_{50}$)	-
Infection Rate	none
Incubation Period	1-14 days
Peak Infection	na
Annual Cases	rare
Annual Fatalities	-

Primarily causes a blood infection (tularemia) contracted from rabbits or other wild animal, but sometimes from insect vectors. It can be transmitted via the airborne route, but cases are rare. Common throughout North America, Europe (except UK), and Asia throughout the year.

Disease or Infection	tularemia, pneumonia, fever				
Natural Source	wild animals, natural waters				
Toxins	none				
Point of Infection	Upper Respiratory Tract, skin				
Symptoms	Indolent ulcer at site of infection, with swelling of local lymph nodes, pain, fever, pneumonic disease may follow.				
Treatment	Gentamicin, aminoglycosides, streptomycin, tobramicin, kanamycin, tetracyclines, chloramphenicol				
Untreated Fatality Rate	5-15%	Prophylaxis:	Antibiotics	Vaccine:	Available
Shape	pleomorphic coccobacillus				
Mean Diameter, μm	0.2	Size Range:	0.2 x 0.2-0.7 microns		
Growth Temperature	37 C	Survival Outside Host:	31-133 days		
Inactivation	Moist heat: 121 C for 15 min. Dry heat: 170 C for 1 hr.				
Disinfectants	1% sodium hypochlorite, 70% ethanol, glutaraldehyde, formaldehyde.				
Nominal Filter Rating	MERV 6	MERV 8	MERV 11	MERV 13	MERV 15
Estimated Removal, %	4.4	8.7	14.0	36.9	65.8
UVGI Rate Constant	(unknown)	cm^2/μW-s	Media		
Dose for 90% Kill	-	μW-s/cm^2	Ref.		
Suggested Indoor Limit	0 CFU/cu.m.				
Genome Base Pairs					
Related Species					
Notes	CDC Reportable.				
Photo Credit	Photo shows tularemia lung infection. Reprinted from Baskerville and Hambleton (1976), Brit. J. Exp. Path. 57, p339 with permission from Blackwell Publishing.				
References	Braude 1981, Freeman 1985, Mitscherlich 1984, Murray 1999, Prescott 1996, Ryan 1994, Canada 2001, NATO 1996, Mandell 2000				

Hantaan virus

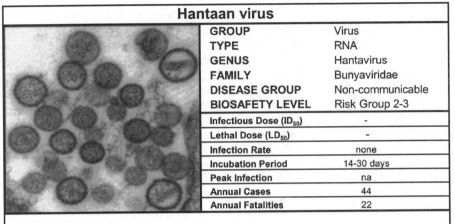

GROUP	Virus
TYPE	RNA
GENUS	Hantavirus
FAMILY	Bunyaviridae
DISEASE GROUP	Non-communicable
BIOSAFETY LEVEL	Risk Group 2-3

Infectious Dose (ID$_{50}$)	-
Lethal Dose (LD$_{50}$)	-
Infection Rate	none
Incubation Period	14-30 days
Peak Infection	na
Annual Cases	44
Annual Fatalities	22

Occurs from inhalation of infected rodent feces. Rapidly incapacitates and fatal without treatment. This unusually deadly pathogen emerged in the 1980s where it killed a number of people in the Southwest. It was subsequently identified as existing in rodent populations across the country. In dry climates, the feces of infected mice living indoors become airborne. Inhalation can then rapidly incapacitate and prove fatal without treatment. Has potential as a biological weapon.

Disease or Infection	Korean hemorrhagic fever, Hantavirus, HFRS, HPS				
Natural Source	Field rodents, Deer mouse, Rattus spp.				
Toxins	none				
Point of Infection	Upper Respiratory Tract				
Symptoms	Abrupt onset of fever lasting 3-8 days, conjunctival infection, prostration, lower back pain, headache, abdominal pain, anorexia, vomiting, respiratory distress.				
Treatment	Ribavarin (IV) during early phase if HFRS				
Untreated Fatality Rate	5-15% **Prophylaxis:** none **Vaccine:** none				
Shape	spherical				
Mean Diameter, μm	0.095 **Size Range:** 0.08-0.115 microns				
Growth Temperature	na **Survival Outside Host:** 2-8 years				
Inactivation	Moist heat: 60 C for 1 hr.				
Disinfectants	1% sodium hypochlorite, 70% ethanol, 2% glutaraldehyde.				
Nominal Filter Rating	MERV 6	MERV 8	MERV 11	MERV 13	MERV 15
Estimated Removal, %	7.3	13.2	19.9	45.1	71.4
UVGI Rate Constant	(unknown)	cm^2/μW-s	**Media**		
Dose for 90% Kill	-	μW-s/cm^2	**Ref.**		
Suggested Indoor Limit	0 CFU/cu.m.				
Genome Base Pairs	10,500-22,700				
Related Species	Dobrova-Belgrade virus, Sin Nombre virus, Seoul virus				
Notes	CDC Reportable.				
Photo Credit	Centers for Disease Control, PHIL 1137, Cynthia Goldsmith.				
References	Dalton 1973, Fraenkel-Conrat 1985, Freeman 1985, Mahy 1975,Murray 1999, Ryan 1994, Canada 2001				

Hepatitis A

GROUP	Virus
TYPE	ssRNA
GENUS	Hepatovirus
FAMILY	Picornaviridae
DISEASE GROUP	Communicable
BIOSAFETY LEVEL	Risk Group 2

Infectious Dose (ID$_{50}$)	10-100
Lethal Dose (LD$_{50}$)	-
Infection Rate	-
Incubation Period	10-50 days
Peak Infection	
Annual Cases	-
Annual Fatalities	-

Non-respiratory and non-airborne. HAV is spread is by the fecal-oral route, ingestion in food or water, and person-to-person contact, and so could be considered both a food-borne pathogen and a communicable disease, but not an airborne pathogen. Causes worldwide cycles of epidemics in institutions and where sanitation is poor. Often affects the young. HAV is excreted in concentrated form before the onset of symptoms. Relapses up to 1 year in 15% cases.

Disease or Infection	HAV, VHA, Infectious hepatitis, epidemic jaundice, Botkins disease, type A viral hepatitis, HA, MS-1				
Natural Source	Humans, sometimes primates.				
Toxins	none				
Point of Infection	oral				
Symptoms	Often asymptomatic, but onset can be sudden with fever, malaise, nausea, anorexia, and abdominal discomfort, followed by jaundice, and a prolonged convalescent period.				
Treatment	No antibiotics, rest only.				
Untreated Fatality Rate	-	Prophylaxis: Vaccine		Vaccine: Available	
Shape	non-enveloped spherical				
Mean Diameter, μm	0.03	Size Range: 0.027-0.030 microns			
Growth Temperature	na	Survival Outside Host: 7 days in fomites			
Inactivation	In saline solution at 85 C, or 4 mins. at 70 C.				
Disinfectants	1% sodium hypochlorite, formaldehyde, 2% glutaraldehyde.				
Nominal Filter Rating	MERV 6	MERV 8	MERV 11	MERV 13	MERV 15
Estimated Removal, %	18.3	32.0	45.9	79.2	96.4
UVGI Rate Constant	(unknown)	cm^2/μW-s	Media		
Dose for 90% Kill	-	μW-s/cm^2	Ref.		
Suggested Indoor Limit	-				
Genome Base Pairs					
Related Species	hepatitis B,C,D, and E				
Photo Credit	Centers for Disease Control, Hepatitis B virus shown, PHIL# 270, Dr. Erskine Palmer.				
References	Dalton 1973, Fraenkel-Conrat 1985, Freeman 1985, Mahy 1975,Murray 1999, Ryan 1994, Canada 2001				

Histoplasma capsulatum

GROUP	Fungal Spore
TYPE	Ascomycetes
GENUS	Moniliales
FAMILY	
DISEASE GROUP	Non-communicable
BIOSAFETY LEVEL	Risk Group 3

Infectious Dose (ID$_{50}$)	10
Lethal Dose (LD$_{50}$)	40000
Infection Rate	none
Incubation Period	4-22 days
Peak Infection	na
Annual Cases	common
Annual Fatalities	-

Histoplasma capsulatum causes histoplasmosis, an infection estimated to have afflicted 40 million Americans, mostly in the Southeast. It most often causes mild fever and malaise, but in 0.1-0.2 % of cases the disease becomes progressive. The infection is inevitably airborne and enters through the lungs, from where it may spread to other areas. In the environment, it is most often found in pigeon roosts, bat caves, or old buildings. This infection can become fatal in some cases.

Disease or Infection	histoplasmosis, fever, malaise				
Natural Source	Environmental, nosocomial, pigeon roosts, bat caves, old buildings.				
Toxins	none				
Point of Infection	Upper Respiratory Tract				
Symptoms	Respiratory infection, mild cold-like symptoms, mild fever or cough.Severe cases may have chills, chest pain, malaise.				
Treatment	Amphotericin B.				
Untreated Fatality Rate	-	**Prophylaxis:** none		**Vaccine:** none	
Shape	spherical spore				
Mean Diameter, μm	2.236	**Size Range:**	1-5 microns		
Growth Temperature	37 C	**Survival Outside Host:**		indefinite	
Inactivation	Moist heat: 121 C for 15 min.				
Disinfectants	1% sodium hypochlorite, phenolics, formaldehyde, glutaraldehyde.				
Nominal Filter Rating	**MERV 6**	**MERV 8**	**MERV 11**	**MERV 13**	**MERV 15**
Estimated Removal, %	36.4	67.9	84.5	96.0	100.0
UVGI Rate Constant	0.000247	cm^2/μW-s	**Media**	Plates (estimated)	
Dose for 90% Kill	9322	μW-s/cm^2	**Ref.**	Chick 1963	
Suggested Indoor Limit	150-500 CFU/cu.m.				
Genome Base Pairs					
Related Species					
Notes	CDC Reportable.				
Photo Credit	Centers for Disease Control, PHIL# 299.				
References	Freeman 1985, Howard 1983, Lacey 1988, Murray 1999, Ryan 1994, Smith 1989, Ashford 1999, Fuortes 1988, Sorensen 1999, Miyaji 1987, Canada 2001				

Influenza A virus

GROUP	Virus
TYPE	RNA
GENUS	Influenza virus A, B
FAMILY	Orthomyxoviridae
DISEASE GROUP	Communicable
BIOSAFETY LEVEL	Risk Group 2
Infectious Dose (ID$_{50}$)	20
Lethal Dose (LD$_{50}$)	-
Infection Rate	0.2-0.83
Incubation Period	2-3 days
Peak Infection	3-4 days
Annual Cases	2,000,000
Annual Fatalities	20,000

Causes periodic flu pandemics and can cause widespread fatalities, and sometimes many millions, dead. Constant antigenic variations among the main types of Influenza, Type A and Type B, ensure little chance of immunity developing. Pneumonia can result from secondary bacterial infections, usually staphylococcus or streptococcus. Current theory suggests that the virus passes to and from humans, pigs, and birds, in agricultural areas of Asia where their close association is common.

Disease or Infection	flu, secondary pneumonia				
Natural Source	Humans, birds, pigs, nosocomial				
Toxins	none				
Point of Infection	Upper Respiratory Tract				
Symptoms	Acute fever, chills, headache, myalgia, weakness, sore throat, cough, runny nose.				
Treatment	No antibiotic treatments, fluids and rest				
Untreated Fatality Rate	low	**Prophylaxis:** possible		**Vaccine:**	Available
Shape	enveloped helical				
Mean Diameter, μm	0.098	**Size Range:** 0.08-0.12 microns			
Growth Temperature	na	**Survival Outside Host:** Several hours in mucous			
Inactivation	Moist heat: 56 C for 30 min.				
Disinfectants	1% sodium hypochlorite, 70% ethanol, glutaraldehyde, formaldehyde.				
Nominal Filter Rating	MERV 6	MERV 8	MERV 11	MERV 13	MERV 15
Estimated Removal, %	7.1	12.9	19.5	44.4	70.7
UVGI Rate Constant	0.00119	cm^2/μW-s	**Media**	Air	
Dose for 90% Kill	1935	μW-s/cm^2	**Ref.**	Jensen 1964	
Suggested Indoor Limit	0 CFU/cu.m.				
Genome Base Pairs	13588				
Related Species	Influenza Type A, B.				
Notes	CDC Reportable.				
Photo Credit	Centers for Disease Control (PHIL 279).				
References	Dalton 1973, Fraenkel-Conrat 1985, Freeman 1985, Gorman 1990, Mahy 1975, Murray 1999, Ryan 1994, Malherbe 1980, Canada 2001				

Japanese encephalitis

GROUP	Virus
TYPE	RNA
GENUS	
FAMILY	Flaviviridae
DISEASE GROUP	Vector-borne
BIOSAFETY LEVEL	Risk Group 3

Infectious Dose (ID$_{50}$)	unknown
Lethal Dose (LD$_{50}$)	-
Infection Rate	-
Incubation Period	5-15 days
Peak Infection	na
Annual Cases	rare
Annual Fatalities	-

Non-respiratory and non-airborne. Transmitted by the bite of an infected mosquito. Endemic to many areas of Asia. Cases occur in summer and early fall, and are limited to periods of high temperatures and dense mosquito populations. Various domestic animals may act as amplifying reservoirs.

Disease or Infection	JE, JEV, JBE, Japanese B encephalitis, Mosquito-borne encephalitis
Natural Source	Humans, birds, pigs, horses, bats, reptiles.
Toxins	none
Point of Infection	Insect bite.
Symptoms	Acute onset of high fever, chills, headache, nausea and vomiting, photophobia, stupor, tremors, coma, convulsions possible.
Treatment	No treatment available.
Untreated Fatality Rate	5-40% **Prophylaxis:** Possible **Vaccine:** none
Shape	enveloped spherical
Mean Diameter, μm	0.04472136 **Size Range:** 0.04-0.05 microns
Growth Temperature	na **Survival Outside Host:** Winters in mosquito eggs.
Inactivation	Moist heat: 56 C for 30 min.
Disinfectants	1% sodium hypochlorite, 70% ethanol, 2% glutaraldehyde, iodine, phenol idophors, formaldehyde.

Nominal Filter Rating	MERV 6	MERV 8	MERV 11	MERV 13	MERV 15
Estimated Removal, %	13.5	24.0	35.0	67.5	90.3

UVGI Rate Constant	(unknown) cm^2/μW-s	**Media**
Dose for 90% Kill	μW-s/cm^2	**Ref.**
Suggested Indoor Limit	na	
Genome Base Pairs	10000-11000	
Related Species	West Nile virus	
Notes	CDC Reportable.	
Photo Credit	Photo shows the Kunjin alphavirus. Image provided courtesy of Ed Westaway of SASVRC, The Royal Children's Hospital, Australia.	
References	Dalton 1973, Fraenkel-Conrat 1985, Freeman 1985, Mahy 1975, Murray 1999, Ryan 1994, Malherbe 1980, Canada 2001	

Junin virus

GROUP	Virus
TYPE	RNA
GENUS	Arenavirus
FAMILY	Arenaviridae
DISEASE GROUP	Non-communicable
BIOSAFETY LEVEL	Risk Group 4

Infectious Dose (ID$_{50}$)	unknown
Lethal Dose (LD$_{50}$)	10-100000
Infection Rate	low
Incubation Period	2-14 days
Peak Infection	7 days
Annual Cases	2,000
Annual Fatalities	rare

Occurs primarily in South American farm workers, in late summer or fall. Inhaled from rodent feces. Fatality rate 3-15%. Like several of the other arenaviruses, it causes hemorrhagic fever. It is relatively uncommon outside of South America. It occurs during late summer and fall, in concert with an increase in the rodent population. Rodent excreta contain the virus and are inhaled or ingested.

Disease or Infection	Argentinean hemorrhagic fever				
Natural Source	Rodents				
Toxins	none				
Point of Infection	Upper Respiratory Tract				
Symptoms	Slow onset of fever, fatigue, headache, muscular pain, bleeding may occur from nose, gums, intestines.				
Treatment	Ribavarin, human plasma treatment				
Untreated Fatality Rate	10-50%	**Prophylaxis:** none		**Vaccine:**	Available
Shape	enveloped helical				
Mean Diameter, μm	0.12	**Size Range:**	0.05-0.3 microns		
Growth Temperature	na	**Survival Outside Host:**	-		
Inactivation	-				
Disinfectants	1% sodium hypochlorite, 2% glutaraldehyde				
Nominal Filter Rating	**MERV 6**	**MERV 8**	**MERV 11**	**MERV 13**	**MERV 15**
Estimated Removal, %	6.0	11.1	16.9	40.1	66.5
UVGI Rate Constant	(unknown)	cm^2/μW-s	**Media**		
Dose for 90% Kill	-	μW-s/cm^2	**Ref.**		
Suggested Indoor Limit	0 CFU/cu.m.				
Genome Base Pairs	5000-7400 nt				
Related Species	Machupo, Lassa				
Notes	CDC Reportable.				
Photo Credit	Photo shows a cell infected with Cupixi virus, an arenavirus. Cynthia Goldsmith & Michael Bowen, Centers for Disease Control, Atlanta.				
References	Dalton 1973, Fraenkel-Conrat 1985, Freeman 1985, Mahy 1975, Murray 1999, Ryan 1994, Salvato 1993, Kenyon 1988, Malherbe 1980, Canada 2001				

Lassa Fever virus

GROUP	Virus
TYPE	
GENUS	Arbovirus
FAMILY	Arenaviridae
DISEASE GROUP	Communicable
BIOSAFETY LEVEL	Risk Group 4

Infectious Dose (ID$_{50}$)	15
Lethal Dose (LD$_{50}$)	2-200000
Infection Rate	high
Incubation Period	7-14 days
Peak Infection	-
Annual Cases	-
Annual Fatalities	-

This virus causes a hemorrhagic fever endemic to West Africa and which is similar to Junin and Machupo. Person-to-person spread occurs by contact with body fluids, but evidence exists for airborne spread.

Disease or Infection	Lassa fever				
Natural Source	Rodents				
Toxins					
Point of Infection	Upper Respiratory Tract				
Symptoms	Fever accompanied by hemorrhagic manifestations, shock, neurologic disturbances, and bradycardia.				
Treatment	Intravenous ribavarin within 6 days can help				
Untreated Fatality Rate	10-50%	**Prophylaxis:** none		**Vaccine:**	none
Shape	enveloped helical				
Mean Diameter, μm	0.12	**Size Range:**	0.05-0.3 microns		
Growth Temperature	na	**Survival Outside Host:**		-	
Inactivation	-				
Disinfectants	-				
Nominal Filter Rating	**MERV 6**	**MERV 8**	**MERV 11**	**MERV 13**	**MERV 15**
Estimated Removal, %	6.0	11.1	16.9	40.1	66.5
UVGI Rate Constant	(unknown)	cm^2/μW-s	**Media**		
Dose for 90% Kill	-	μW-s/cm^2	**Ref.**		
Suggested Indoor Limit	0 CFU/cu.m.				
Genome Base Pairs	3400-4800 nt				
Related Species	Machupo, Marburg	**Notes**	CDC Reportable.		
Photo Credit	Lassa virus in Vero cells (800x) Reprinted w/ permission from Malherbe and Strickland 1980, Viral Cytopathology, Copyright CRC Press, Boca Raton, FL				
References	Dalton 1973, Fraenkel-Conrat 1985, Freeman 1985, Mahy 1975, McCormick 1987, Ryan 1994, Schaal 1979, Peters 1987, Malherbe 1980, Franz 1997				

Legionella pneumophila

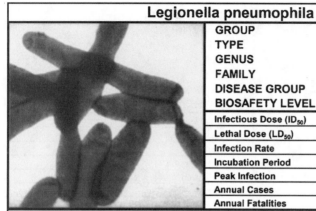

GROUP	Bacteria
TYPE	Gram-
GENUS	Legionella
FAMILY	Legionellaceae
DISEASE GROUP	Non-communicable
BIOSAFETY LEVEL	Risk Group 2
Infectious Dose (ID$_{50}$)	<129
Lethal Dose (LD$_{50}$)	140000
Infection Rate	<0.01
Incubation Period	2-10 days
Peak Infection	na
Annual Cases	1,163
Annual Fatalities	10

Exist naturally in warm outdoor ponds but can colonize warm water indoors. The well known cause of Legionnaire's Disease, Legionella pneumophila exists in warm outdoor ponds naturally. It only becomes a problem when amplified by air conditioning equipment and aerosolized in ventilation systems.

Disease or Infection	Legionnaire's Disease, Pontiac fever, opportunistic infections				
Natural Source	Enviro., Growth in cooling tower water, spas, potable water, nosocomial.				
Toxins	none				
Point of Infection	Upper Respiratory Tract				
Symptoms	Acute pneumonitis with malaise, myalgia, anorexia, headache, fever, chills, nonproductive cough, abdominal pain and diarrhea.				
Treatment	Erythromycin, rifampin, ciprofloxacin, oxygen and fluid replacement				
Untreated Fatality Rate	39-50% **Prophylaxis:** Antibiotics **Vaccine:** none				
Shape	rods				
Mean Diameter, µm	0.52 **Size Range:** 0.3-0.9 x 0.6-2 microns				
Growth Temperature	25-35 C **Survival Outside Host:** months in water				
Inactivation	Moist heat: 121 C for 15 min. Dry heat: 160 C for 1 hr.				
Disinfectants	1% sodium hypochlorite, 70% ethanol, glutaraldehyde, formaldehyde.				
Nominal Filter Rating	MERV 6	MERV 8	MERV 11	MERV 13	MERV 15
Estimated Removal, %	7.2	16.1	26.7	66.3	94.8
UVGI Rate Constant	0.00182	cm^2/µW-s	**Media**	Plates	
Dose for 90% Kill	1265	µW-s/cm^2	**Ref.**	Antopol 1979	
Suggested Indoor Limit	0 CFU/cu.m.				
Genome Base Pairs					
Related Species	L. parisiensis (suspected)				
Notes	CDC Reportable.				
Photo Credit	Centers for Disease Control, PHIL# 1187.				
References	Braude 1981, Freeman 1985, Gilpin 1984, Murray 1999, Prescott 1996, Ryan 1994, Thornsberry 1984, Berendt 1980, Canada 2001, Mandell 2000				

Lymphocytic choriomeningitis

GROUP	Virus
TYPE	ssRNA
GENUS	
FAMILY	Arenaviridae
DISEASE GROUP	Non-communicable
BIOSAFETY LEVEL	Risk Group 2-3

Infectious Dose (ID$_{50}$)	unknown
Lethal Dose (LD$_{50}$)	<1000
Infection Rate	na
Incubation Period	8-13 days
Peak Infection	3-7 days
Annual Cases	rare
Annual Fatalities	-

Can be inhaled or ingested from contaminated rodent feces. Prevalence in humans is 2-10%. Occasionally causes outbreaks in Europe, America, Australia, and Japan, sometimes by pet hamsters or laboratory animals. Rodents may harbor the virus for life and transmit to progeny.

Disease or Infection	LCM, lymphocytic meningitis				
Natural Source	House mouse, swine, dogs, hamsters, guinea pigs.				
Toxins	none				
Point of Infection	Inhaled or oral.				
Symptoms	Mild influenza-like symptoms, asymptomatic in 1/3 of cases, may progress to meningitis, with transverse myelitis, orchitis or protitis.				
Treatment	No treatment, ribavarin and anti-inflammatories helpful.				
Untreated Fatality Rate	<1% **Prophylaxis:** none **Vaccine:** none				
Shape	enveloped				
Mean Diameter, μm	0.08660254 **Size Range:** .05-15 microns				
Growth Temperature	na **Survival Outside Host:** in mouse droppings				
Inactivation	-				
Disinfectants	1% sodium hypochlorite, 70% ethanol, 2% glutaraldehyde, formaldehyde.				
Nominal Filter Rating	MERV 6	MERV 8	MERV 11	MERV 13	MERV 15
Estimated Removal, %	7.9	14.3	21.4	47.6	73.8
UVGI Rate Constant	(unknown)	cm^2/μW-s	Media		
Dose for 90% Kill		μW-s/cm^2	Ref.		
Suggested Indoor Limit	na				
Genome Base Pairs	3400-4800 nt				
Related Species					
Notes					
Photo Credit	Photo of LCV budding from infected cells. Reprinted with permission from M.S.Salvato, 1993, The Arenaviridae, Plenum Press, New York.				
References	Dalton 1973, Fraenkel-Conrat 1985, Freeman 1985, Mahy 1975, Murray 1999, Ryan 1994, Schaal 1979, Peters 1987, Malherbe 1980, Canada 2001				

Machupo

GROUP	Virus
TYPE	RNA
GENUS	Arenavirus
FAMILY	Arenaviridae
DISEASE GROUP	Non-communicable
BIOSAFETY LEVEL	Risk Group 4

Infectious Dose (ID_{50})	unknown
Lethal Dose (LD_{50})	<1000
Infection Rate	-
Incubation Period	7-16 days
Peak Infection	-
Annual Cases	-
Annual Fatalities	-

Similar to Junin virus. Inhaled from rodent feces. Fatality rate 3-15%. This virus causes hemmorhagic fever and is relatively uncommon outside of South America. It occurs during late summer and fall, in concert with an increase in the rodent population. Rodent excreta contain the virus and are inhaled or ingested.

Disease or Infection	Bolivian hemorrhagic fever			
Natural Source	Rodents			
Toxins	none			
Point of Infection	Upper Respiratory Tract			
Symptoms	Slow onset of fever, fatigue, headache, muscular pain, bleeding may occur from nose, gums, intestines.			
Treatment	Ribavarin, human plasma treatment			
Untreated Fatality Rate	5-30%	**Prophylaxis:** none	**Vaccine:**	none
Shape	pleomorphic			
Mean Diameter, μm	0.12	**Size Range:** 0.110-0.13 microns		
Growth Temperature	na	**Survival Outside Host:**	-	
Inactivation	-			
Disinfectants	1% sodium hypochlorite, 2% glutaraldehyde			

Nominal Filter Rating	**MERV 6**	**MERV 8**	**MERV 11**	**MERV 13**	**MERV 15**
Estimated Removal, %	6.0	11.1	16.9	40.2	66.5

UVGI Rate Constant	(unknown)	$cm^2/\mu W\text{-}s$	**Media**
Dose for 90% Kill	-	$\mu W\text{-}s/cm^2$	**Ref.**
Suggested Indoor Limit	0 CFU/cu.m.		
Genome Base Pairs	3400-4800 nt		
Related Species	Junin, Lassa		
Notes	CDC Reportable.		
Photo Credit	Photo shows the arenavirus Tacaribe. Reprinted with permission from M.S.Salvato, 1993, The Arenaviridae, Plenum Press, New York.		
References	Dalton 1973, Fraenkel-Conrat 1985, Freeman 1985, Mahy 1975, Murray 1999, Ryan 1994, Salvato 1993, Wagner 1977, Franz 1997		

Marburg virus

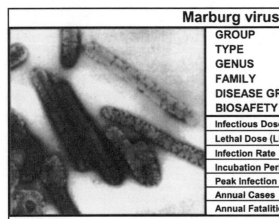

GROUP	Virus
TYPE	RNA
GENUS	Filovirus
FAMILY	Filoviridae
DISEASE GROUP	Communicable
BIOSAFETY LEVEL	Risk Group 4

Infectious Dose (ID$_{50}$)	-
Lethal Dose (LD$_{50}$)	-
Infection Rate	-
Incubation Period	7 days
Peak Infection	-
Annual Cases	rare
Annual Fatalities	rare

Resembles Lassa fever epidemiologically. Identical to Ebola morphologically. Primarilly transmitted by contact. Marburg virus was originally identified in outbreaks traced to contact with infected green monkeys from Uganda. It causes hemorrhagic fever and has a case fatality rate between 22% and 88%. Infections, though ultimately deriving from rodents or other animals, has mainly been transmitted between humans by direct contact, airborne inhalation, or contact with blood.

Disease or Infection	hemorrhagic fever			
Natural Source	Humans, monkeys			
Toxins	none			
Point of Infection	Upper Respiratory Tract			
Symptoms	Sudden onset of high fever, weakness, myalgia, vomiting, diarrhea, maculopapular rash, hemorrhagic diathesis, leukopenia.			
Treatment	No effective treatment Maintain renal function, electrolyte balance, transfusions			
Untreated Fatality Rate	25%	**Prophylaxis:** none	**Vaccine:**	none
Shape	enveloped helical			
Mean Diameter, μm	0.039	**Size Range:**	0.05-0.03 microns	
Growth Temperature	na	**Survival Outside Host:**	2 weeks in warm blood	
Inactivation	-			
Disinfectants	1% sodium hypochlorite, 2% glutaraldehyde, formaldehyde.			
Nominal Filter Rating	MERV 6	MERV 8	MERV 11	MERV 13	MERV 15
Estimated Removal, %	14.9	26.4	38.3	71.4	92.6
UVGI Rate Constant	(unknown)	cm^2/μW-s	**Media**	
Dose for 90% Kill	-	μW-s/cm^2	**Ref.**	
Suggested Indoor Limit	0 CFU/cu.m.			
Genome Base Pairs	19112			
Related Species	related to Ebola (non-airborne)			
Notes	CDC Reportable.			
Photo Credit	Centers for Disease Control, PHIL# 275, Dr. Erskine Palmer.			
References	Dalton 1973, Fraenkel-Conrat 1985, Freeman 1985, Johnson 1995, Mahy 1975, Murray 1999, Ryan 1994, Malherbe 1980, Canada 2001, Franz 1997, Martini 1971			

Mycobacterium tuberculosis

GROUP	Bacteria
TYPE	Gram+ (acid fast)
GENUS	Mycobacterium
FAMILY	Mycobacteriaceae
DISEASE GROUP	Communicable
BIOSAFETY LEVEL	Risk Group 2-3
Infectious Dose (ID_{50})	1-10
Lethal Dose (LD_{50})	-
Infection Rate	0.33
Incubation Period	4-12 weeks
Peak Infection	varies
Annual Cases	20,000
Annual Fatalities	-

Tuberculosis infects over 1/3 of the world's population. This bacteria causes TB, once called consumption because of the way it seemed to deplete a person till death, and was an ancient disease even to the Egyptians. Estimated to be at least 15,000 years old, this parasite poses one of the greatest modern health hazards due to the recent emergence of drug-resistant strains. It is highly contagious and a single bacilli is capable of causing an infection in lab animals.

Disease or Infection	tuberculosis, TB			
Natural Source	Humans, sewage (potential), nosocomial.			
Toxins	none			
Point of Infection	Upper Respiratory Tract			
Symptoms	Slow progress to pulmonary infection, fatigue, fever, cough, chest pain, hemoptysis fibrosis, cavitation.			
Treatment	Isoniazid, rifampin, streptomycin, ethambutol, pyrazinamide, if not resistant			
Untreated Fatality Rate	-	Prophylaxis: possible	Vaccine:	Available
Shape	rods	-		
Mean Diameter, μm	0.637	Size Range:	0.2-0.6 x 1-5 microns	
Growth Temperature	30-38 C	Survival Outside Host:	40-100 days	
Inactivation	Moist heat: 121 C for 15 min.			
Disinfectants	5% phenol, 1% sodium hypochlorite, iodine solutions, glutaraldehyde, formaldehyde.			

Nominal Filter Rating	MERV 6	MERV 8	MERV 11	MERV 13	MERV 15
Estimated Removal, %	9.3	20.7	33.2	75.5	98.0

UVGI Rate Constant	0.002132	$cm^2/\mu W\text{-}s$	Media	Plates
Dose for 90% Kill	1080	$\mu W\text{-}s/cm^2$	Ref.	David 1973
Suggested Indoor Limit	0 CFU/cu.m.			
Genome Base Pairs	4411529			
Related Species				
Notes	CDC Reportable.			
Photo Credit	Centers for Disease Control, PHIL# 837, Dr. Edwin P. Ewing, Jr.			
References	Braude 1981, David 1973, Freeman 1985, Kapur 1994, Higgins 1975, Murray 1999, Prescott 1996, Ryan 1994, Youmans 1979, Canada 2001, Mandell 2000			

Mycoplasma pneumoniae

GROUP	Bacteria
TYPE	no wall
GENUS	Mycoplasma
FAMILY	Mycoplasmataceae
DISEASE GROUP	Endogenous
BIOSAFETY LEVEL	Risk Group 2

Infectious Dose (ID_{50})	100
Lethal Dose (LD_{50})	-
Infection Rate	-
Incubation Period	6-23 days
Peak Infection	na
Annual Cases	uncommon
Annual Fatalities	rare

Weakly pathogenic for man and often found as commensals. Immunodeficiency predisposes one to infection. Mycoplasma pneumoniae is a member of a class called Mollicutes, which are considered to be different from bacteria since they contain no cell wall. Immune system disruption, usually by another disease, is required to produce an infection. Only about 3-10 % of infections result in apparent pneumonia. Possible candidate for genetic alteration as a bioweapon.

Disease or Infection	pneumonia, PPLO, walking pneumonia, (Gulf War Syndrome?)
Natural Source	Humans
Toxins	none
Point of Infection	Upper Respiratory Tract
Symptoms	Slow onset with malaise, headache, paraxysmal cough, substernal pain, leukocytosis possible, pneumonia.
Treatment	Tetracyclines, gentamicin, doxycycline, macrolides
Untreated Fatality Rate	- **Prophylaxis:** Antibiotics **Vaccine:** none
Shape	pleomorphic
Mean Diameter, μm	0.177 **Size Range:** 0.125-0.25 microns
Growth Temperature	36-37 C **Survival Outside Host:** 10-50 hours in air
Inactivation	Moist heat: 121 C for 15 min. Dry heat: 170 for 1 hr.
Disinfectants	1% sodium hypochlorite, 70% ethanol, glutaraldehyde, formaldehyde.

Nominal Filter Rating	MERV 6	MERV 8	MERV 11	MERV 13	MERV 15
Estimated Removal, %	4.6	8.9	14.2	36.5	64.3
UVGI Rate Constant	(unknown)	$cm^2/\mu W\text{-}s$	Media		
Dose for 90% Kill	Low	$\mu W\text{-}s/cm^2$	Ref.	Kundsin 1968	
Suggested Indoor Limit	0 CFU/cu.m.				
Genome Base Pairs	816394				
Related Species					
Notes	CDC Reportable.				
Photo Credit	Centers for Disease Control, PHIL# 1351, Dr. Martin Hicklin.				
References	Braude 1981, Freeman 1985, Kundsin 1968, Maniloff 1992, Mitscherlich 1984, Murray 1999, Prescott 1996, Ryan 1994, Canada 2001, Madoff 1971, Mandell 2000				

Nocardia asteroides

GROUP	Bacterial Spore
TYPE	Nocardiaceae
GENUS	Nocardia
FAMILY	Actinomycetes
DISEASE GROUP	Non-communicable
BIOSAFETY LEVEL	Risk Group 2

Infectious Dose (ID$_{50}$)	-
Lethal Dose (LD$_{50}$)	-
Infection Rate	none
Incubation Period	-
Peak Infection	na
Annual Cases	uncommon
Annual Fatalities	rare

Considered pathogenic, this Gram+ bacteria is classified as a pathogenic actinomycetes, this microorganism is a bacterium barely distinguishable from fungi. It can be found in some soils. It is an opportunistic pathogen and primarily affects patients who have been rendered susceptible by other diseases, especially those involving immunodeficiency.

Disease or Infection	nocardiosis, pneumonia				
Natural Source	Environmental, soils, sewage, nosocomial.				
Toxins	none				
Point of Infection	Upper Respiratory Tract				
Symptoms	Fever, cough, chest pain, CNS disease, headache, lethargy, confusion, seizures.				
Treatment	Surgical drainage, sulfonilamides (TMP-SMX, sulfisoxazole, sulfadiazine)				
Untreated Fatality Rate	10%	**Prophylaxis:** none		**Vaccine:** none	
Shape	ovoid spore				
Mean Diameter, μm	1.118	**Size Range:**	1-1.25 x 3-5 microns		
Growth Temperature	20-30 C	**Survival Outside Host:**		indefinitely in soil, water	
Inactivation	Moist heat: 121 C for 15 min. Dry heat: 170 for 1 hr.				
Disinfectants	1% sodium hypochlorite, 2% glutaraldehyde, formaldehyde.				
Nominal Filter Rating	MERV 6	MERV 8	MERV 11	MERV 13	MERV 15
Estimated Removal, %	18.8	40.4	57.1	92.5	100.0
UVGI Rate Constant	0.000123	cm^2/μW-s	**Media**	Plates (estimated)	
Dose for 90% Kill	18720	μW-s/cm^2	**Ref.**	Chick 1963	
Suggested Indoor Limit	0 CFU/cu.m.				
Genome Base Pairs					
Related Species	Gram+ bacteria. N. caviae., N. brasiliensis				
Notes	CDC Reportable.				
Photo Credit	Photo courtesy of Littleton / Englewood Wastewater Treatment Plant				
References	Austin 1991, Freeman 1985, Lacey 1988, Murray 1999, Slack 1975, Ryan 1994, Schaal 1979, Sikes 1973, Grigoriu 1987, Al-Doory 1987, Miyaji 1987, Canada 2001				

Paracoccidioides

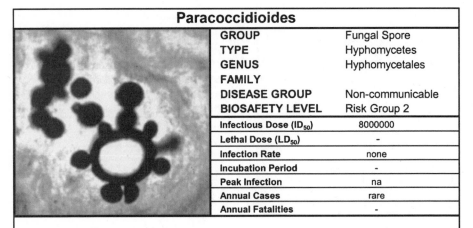

GROUP	Fungal Spore
TYPE	Hyphomycetes
GENUS	Hyphomycetales
FAMILY	
DISEASE GROUP	Non-communicable
BIOSAFETY LEVEL	Risk Group 2

Infectious Dose (ID$_{50}$)	8000000
Lethal Dose (LD$_{50}$)	-
Infection Rate	none
Incubation Period	-
Peak Infection	na
Annual Cases	rare
Annual Fatalities	-

Most common in South America, where an estimated 10 million people have been infected. Males are definitively more susceptible than females, representing 90% of all cases. One of only four fungi that cause true systemic infections. Can spread to the lymphatics and the lips and nose.

Disease or Infection	paracoccidioidomycosis, paracoccidioidal granuloma, Lutz's disease, South American blastomycosis
Natural Source	Environmental
Toxins	none
Point of Infection	Upper Respiratory Tract
Symptoms	Chronic mucocutaneous or cutaneous ulcers. Lymph nodes may become infected.
Treatment	Sulfonilamides, amphotericin B, azole compounds
Untreated Fatality Rate	- Prophylaxis: none Vaccine: none
Shape	spherical spore
Mean Diameter, μm	4.472 Size Range: 2-10 microns
Growth Temperature	37 C Survival Outside Host: indefinitely in soil, water
Inactivation	-
Disinfectants	-

Nominal Filter Rating	MERV 6	MERV 8	MERV 11	MERV 13	MERV 15
Estimated Removal, %	48.1	77.1	93.5	96.0	100.0
UVGI Rate Constant	(unknown)	cm^2/μW-s	Media		
Dose for 90% Kill	-	μW-s/cm^2	Ref.		
Suggested Indoor Limit	150-500 CFU/cu.m.				
Genome Base Pairs					
Related Species	P. brasiliensis				
Notes	CDC Reportable.				
Photo Credit	Centers for Disease Control, PHIL 527, Dr. Lucille K. Georg.				
References	Collins 1973, Freeman 1985, Howard 1983, Lacey 1988, Murray 1999, Ryan 1994, Smith 1989, Tuder 1985, Kashino 1985, Miyaji 1987				

Rickettsia prowazeki

GROUP	Bacteria
TYPE	Gram-
GENUS	Rickettsiae
FAMILY	Rickettsiaceae
DISEASE GROUP	Vector-borne
BIOSAFETY LEVEL	Risk Group 2-3

Infectious Dose (ID_{50})	10
Lethal Dose (LD_{50})	-
Infection Rate	-
Incubation Period	1-2 weeks
Peak Infection	-
Annual Cases	-
Annual Fatalities	-

Obligate intracellular bacterium that occurs in louse-infested areas. Infected lice excrete rickettsiae in their feces and this can infect bites or abrasions. Also caused by inhalation of infective louse feces or the bites of squirrel flea. The cause of epidemic typhus. Endemic in mountainous regions of Central and South America, Africa, and Asia.

Disease or Infection	Epidemic typhus, Louse-borne typhus fever, Brill-Zinsser disease				
Natural Source	Body louse, humans, squirrels, squirrel flea.				
Toxins	none				
Point of Infection	skin				
Symptoms	Onset can be sudden with fever, chills, headache, prostration, general pain, macular eruption, toxemia,.				
Treatment	Tetracycline, chloramphenicol, doxycycline				
Untreated Fatality Rate	10-40%	**Prophylaxis:** none		**Vaccine:**	Possible
Shape	pleomorphic				
Mean Diameter, μm	0.283	**Size Range:**	0.3-1.2 microns		
Growth Temperature	32-35 C	**Survival Outside Host:**		weeks	
Inactivation	Moist heat: 121 C for 15 min. Dry heat: 170 for 1 hr.				
Disinfectants	1% sodium hypochlorite, 70% ethanol, glutaraldehyde, formaldehyde.				
Nominal Filter Rating	**MERV 6**	**MERV 8**	**MERV 11**	**MERV 13**	**MERV 15**
Estimated Removal, %	4.4	9.2	15.5	42.6	74.9
UVGI Rate Constant	0.000292	$cm^2/\mu W\text{-}s$	**Media**	Plates	
Dose for 90% Kill	7886	$\mu W\text{-}s/cm^2$	**Ref.**	Allen 1954	
Suggested Indoor Limit	na				
Genome Base Pairs					
Related Species	R. canadensis, R. rickettsi, R. tsutsugamushi				
Notes	CDC Reportable.				
Photo Credit	Centers for Disease Control (R. tsutsugamushi shown), PHIL# 929, Dr. Edwin P. Ewing.				
References	Braude 1981, Freeman 1985, McCaul 1981, Mitscherlich 1984, Murray 1999, Prescott 1996, Ryan 1994, Canada 2001, Mandell 2000				

Rickettsia rickettsii

GROUP	Bacteria
TYPE	Gram-
GENUS	Rickettsiae
FAMILY	Rickettsiaceae
DISEASE GROUP	Vector-borne
BIOSAFETY LEVEL	Risk Group 2-3

Infectious Dose (ID_{50})	<10
Lethal Dose (LD_{50})	-
Infection Rate	-
Incubation Period	2-14 days
Peak Infection	-
Annual Cases	-
Annual Fatalities	-

Non-airborne, non-respiratory. Tick-borne bacterium. The cause of Rocky Mountain Spotted Fever. Occurs in U.S. during spring, summer, and fall. Can be confused with measles. Vectored by dog ticks, wood ticks, Lone Star ticks. Transmitted to humans, dogs, rodents, and various other animals.

Disease or Infection	Rocky Mountain Spotted Fever (RMSF), New World spotted fever, Tick-borne typhus, Sao Paulo fever.
Natural Source	Humans, dogs, rodents.
Toxins	none
Point of Infection	skin
Symptoms	Sudden onset with moderate to high fever for 2-3 weeks, severe muscle pain & headache, chills, maculopapular rash, hemorrhages can occur.
Treatment	Tetracyclines, chloramphenicol.
Untreated Fatality Rate	15-20% **Prophylaxis:** Antibiotics **Vaccine:** Available
Shape	pleomorphic
Mean Diameter, μm	0.85 **Size Range:** 0.6-1.2 microns
Growth Temperature	32-35 C **Survival Outside Host:** up to 1 year
Inactivation	Moist heat: 121 C for 15 min. Dry heat: 170 C for 1 hr.
Disinfectants	1% sodium hypochlorite, 70% ethanol, glutaraldehyde, formaldehyde.

Nominal Filter Rating	MERV 6	MERV 8	MERV 11	MERV 13	MERV 15
Estimated Removal, %	13.4	29.6	44.6	86.2	99.7

UVGI Rate Constant	0.000292	cm^2/μW-s	**Media**	Plates
Dose for 90% Kill	7886	μW-s/cm^2	**Ref.**	Allen 1954
Suggested Indoor Limit	na			
Genome Base Pairs				
Related Species	R. canadensis, R. prowazeki, R. tsutsugamushi			
Notes	CDC Reportable.			
Photo Credit	Centers for Disease Control (R. tsutsugamushi shown), PHIL# 930, Dr. Edwin P. Ewing.			
References	Braude 1981, Freeman 1985, McCaul 1981, Walker 1988, Murray 1999, Prescott 1996, Salvato 1993, Canada 2001, Mandell 2000			

Rift Valley fever

	GROUP	Virus
	TYPE	RNA
	GENUS	
	FAMILY	Bunyaviridae
	DISEASE GROUP	Vector-borne
	BIOSAFETY LEVEL	Risk Group 4

Infectious Dose (ID_{50})	unknown
Lethal Dose (LD_{50})	unknown
Infection Rate	na
Incubation Period	2-4 days
Peak Infection	-
Annual Cases	rare
Annual Fatalities	-

Non-respiratory and non-airborne. Rift Valley fever virus is a vector-borne pathogen that is naturally disseminated by mosquitoes in parts of Africa, particulary in the Rift Valley. It is an important cause of disease in sheep and cattle. It is an Arbovirus and has been mentioned by various sources as a potential biological weapon.

Disease or Infection	Rift Valley fever				
Natural Source	Mosquitoes, cattle, sheep, humans.				
Toxins	none				
Point of Infection	blood				
Symptoms	Febrile illness, induces a fever.				
Treatment	Ribavarin				
Untreated Fatality Rate	unknown	**Prophylaxis:**		**Vaccine:**	
Shape	spherical enveloped virion				
Mean Diameter, μm	0.0922	**Size Range:**	0.095-0.105 microns		
Growth Temperature	na	**Survival Outside Host:**			
Inactivation	Moist heat: 56 C for 30 minutes.				
Disinfectants	1% sodium hypochlorite, 2% glutaraldehyde.				
Nominal Filter Rating	MERV 6	MERV 8	MERV 11	MERV 13	MERV 15
Estimated Removal, %	7.5	13.6	20.4	45.9	72.2
UVGI Rate Constant	unknown	$cm^2/\mu W\text{-}s$	**Media**		
Dose for 90% Kill		$\mu W\text{-}s/cm^2$	**Ref.**		
Suggested Indoor Limit	na				
Genome Base Pairs	10,500-22,700		**Notes**	CDC Reportable.	
Related Species	Crimean-Congo hemorrhagic fever				
Photo Credit	(Rift Valley fever in Vervet cells) Reprinted w/ permission from Malherbe and Strickland 1980, Viral Cytopathology, Copyright CRC Press, Boca Raton, FL				
References	Dalton 1973, Fraenkel-Conrat 1985, Freeman 1985, Mahy 1975, Murray 1999, Ryan 1994, WHO 1982, Malherbe 1980				

Salmonella typhi

GROUP	Bacteria
TYPE	Gram-
GENUS	Salmonella
FAMILY	Enterobacteriaceae
DISEASE GROUP	Food-borne
BIOSAFETY LEVEL	Risk Group 2

Infectious Dose (ID_{50})	100000
Lethal Dose (LD_{50})	-
Infection Rate	-
Incubation Period	5-21 days
Peak Infection	-
Annual Cases	2,000,000
Annual Fatalities	-

Non-respiratory and non-airborne. A foodborne or ingested pathogen. The cause of typhoid fever. Can be transmitted person-to-person or by contaminated food handled by infected individuals. Infected flies can cause infection of uncovered foods. Once on food, bacteria can multiply until an infectious dose is reached. Endemic in some areas of the world. Non-typhoidal Salmonella causes gastroenteritis with a short incubation (6-48 hours), while typhiod fever has a longer incubation period (5-21 days).

Disease or Infection	Typhoid fever, Enteric fever, Typhus abdominalis.
Natural Source	Humans
Toxins	Enterotoxin
Point of Infection	Ingested.
Symptoms	Enteric fever, headache, malaise, anorexia, enlarged spleen, constipation, also nausea, vomiting, diarrhea, dehydration.
Treatment	Chloramphenicol, ampicillin, amoxicillin, TMP-SMX, fluoroquinolones
Untreated Fatality Rate	10-40% **Prophylaxis:** Antibiotics **Vaccine:** Available
Shape	Rods
Mean Diameter, μm	0.81 **Size Range:**
Growth Temperature	37 C **Survival Outside Host:** 17-130 days
Inactivation	Moist heat: 121 C for 15 min. Dry heat: 170 for 1 hr.
Disinfectants	1% sodium hypochlorite, 2% glutaraldehyde, formaldehyde, iodines, phenolics, 70% ethanol.

Nominal Filter Rating	MERV 6	MERV 8	MERV 11	MERV 13	MERV 15
Estimated Removal, %	12.5	27.8	42.4	84.6	99.6

UVGI Rate Constant	0.00223	$cm^2/\mu W\text{-}s$	**Media**	Plates
Dose for 90% Kill	1033	$\mu W\text{-}s/cm^2$	**Ref.**	Collins 1971
Genome Base Pairs				
Related Species	S. choleraesuis, S. enterica, S. paratyphi (paratyphoid fever)			
Notes	CDC Reportable.			
Photo Credit	Image of Salmonella Chester Reprinted with permission from Journal of Food Protection (Liao and Sapers 2000, U.S. Dept. of Agriculture) Copyright held by the International Association for Food Protection, Des Moines, IA, U.S.A.			
References	Braude 1981, Freeman 1985, Mitscherlich 1984, Murray 1999, Prescott 1996, Ryan 1994, Canada 2001, Mandell 2000, Fatah 2001			

Shigella

GROUP	Bacteria
TYPE	Gram-
GENUS	Shigella
FAMILY	Enterobacteriaceae
DISEASE GROUP	Food-borne
BIOSAFETY LEVEL	Risk Group 2
Infectious Dose (ID$_{50}$)	10-200
Lethal Dose (LD$_{50}$)	-
Infection Rate	0.2-0.4
Incubation Period	1-7 days
Peak Infection	-
Annual Cases	
Annual Fatalities	

Non-respiratory and non-airborne. A food borne pathogen. Four serogroups exist, A, B, C, & D. The cause of dysentery. Exists worldwide and most deaths are children under 10 years. Outbreaks occur in conditions of poor sanitation and overcrowding. Endemic in tropical climates.

Disease or Infection	Dysentery, Shigellosis, Bacillary dysentery.
Natural Source	Humans, primates.
Toxins	Shiga toxin
Point of Infection	Ingested.
Symptoms	Diarrhea, nausea, fever, sometimes toxemia, vomiting, cramps, and tenesmus, bloody stool.
Treatment	TMP-SMX, ampicillin, chloramphenicol, ciprofloxacin, ofloxacin
Untreated Fatality Rate	20% Prophylaxis: Antibiotics Vaccine: none
Shape	Rods
Mean Diameter, µm	0.80 Size Range:
Growth Temperature	37 C Survival Outside Host: 2-11 days
Inactivation	Moist heat: 121 C for 15 min. Dry heat: 170 for 1 hr.
Disinfectants	1% sodium hypochlorite, 2% glutaraldehyde, formaldehyde, iodines, phenolics, 70% ethanol.

Nominal Filter Rating	MERV 6	MERV 8	MERV 11	MERV 13	MERV 15
Estimated Removal, %	12.4	27.6	42.1	84.3	99.5

UVGI Rate Constant	0.000688	cm^2/µW-s	Media Water
Dose for 90% Kill	3347	µW-s/cm^2	Ref. Sharp 1940

Suggested Indoor Limit	na
Genome Base Pairs	
Related Species	S. dysenteriae, S. flexneri, S. boydii, S. sonnei
Notes	CDC Reportable.
Photo Credit	Photo reprinted with permission from K. Niebuhr & P.J. Sansonetti, 2000, Subcellular Biochemistry 33:251-287. Copyright by Kluwer Academic/Plenum Press.
References	Braude 1981, Freeman 1985, Mitscherlich 1984, Murray 1999, Prescott 1996, Ryan 1994, Canada 2001

Stachybotrys chartarum

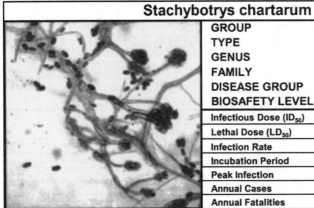

GROUP	Fungal Spore
TYPE	Hyphomycetes
GENUS	Stachybotris
FAMILY	Hyphomycetes
DISEASE GROUP	Non-communicable
BIOSAFETY LEVEL	Risk Group 1-2

Infectious Dose (ID_{50})	-
Lethal Dose (LD_{50})	-
Infection Rate	none
Incubation Period	-
Peak Infection	na
Annual Cases	-
Annual Fatalities	-

Produces a toxin that can be fatal to infants. Sometimes grows indoors. A recently identified fungal allergen, it has been identified as a contributing factor in some cases of Sick Building Syndrome.

Disease or Infection	bleeding lung disease, allergic reactions, stachybotritoxicosis, irritation, toxic reactions
Natural Source	Environmental, indoor growth on building materials & in humidifiers.
Toxins	trichothecenes, verrucarin J, roridin E, satratoxin F, satratoxin G, satratoxin H, sporidesmin G, trichoverrol, cyclosporins, stachybotryolactone
Point of Infection	Upper Respiratory Tract
Symptoms	Lung mycotoxicosis, cough, rhinitis, buring sensation in mouth and nasal passages. Bleeding lung disease reported in infants.
Treatment	

Untreated Fatality Rate	-	**Prophylaxis:**	none	**Vaccine:**	none

Shape	spherical spore

Mean Diameter, μm	5.623	**Size Range:**	5.1-6.2 microns

Growth Temperature	37 C	**Survival Outside Host:**	indefinitely

Inactivation	
Disinfectants	-

Nominal Filter Rating	MERV 6	MERV 8	MERV 11	MERV 13	MERV 15
Estimated Removal, %	49.3	77.4	93.9	96.0	100.0

UVGI Rate Constant	(unknown)	$cm^2/\mu W\text{-}s$	**Media**		
Dose for 90% Kill	-	$\mu W\text{-}s/cm^2$	**Ref.**		

Suggested Indoor Limit	0 CFU/cu.m.
Genome Base Pairs	
Related Species	(previously S. atra)
Notes	
Photo Credit	University of Minnesota, Dept. of Environmental Health & Safety.
References	Freeman 1985, Howard 1983, Lacey 1988, Montana 1988, Murray 1999, Nikulin 1996, Ryan 1994, Smith 1989

Streptococcus pneumoniae

GROUP	Bacteria
TYPE	Gram+
GENUS	Streptococcus
FAMILY	
DISEASE GROUP	Communicable
BIOSAFETY LEVEL	Risk Group 2

Infectious Dose (ID$_{50}$)	-
Lethal Dose (LD$_{50}$)	-
Infection Rate	0.1-0.3
Incubation Period	1-5 days
Peak Infection	2-10 days
Annual Cases	500,000
Annual Fatalities	50,000

The leading cause of death in the world. This microorganism is commonly known as pneumococcus, is is the prime agent of lobar pneumonia, which predominantly affects children. It is commonly carried asymptomatically in healthy individuals. Carriage rates among children are high -- about 30% for children and 10% for adolescents.

Disease or Infection	lobar pneumonia, sinusitis, meningitis, otitis media, toxic reactions
Natural Source	Humans, nosocomial.
Toxins	none
Point of Infection	Upper Respiratory Tract
Symptoms	Sudden onset with fever, shaking chills, pleural pain, dyspnea, coughing, leukocytosis, pneumonia, bacteremia, meningitis.
Treatment	Penicillin, erythromycin
Untreated Fatality Rate	5-40% **Prophylaxis:** Antibiotics **Vaccine:** Available
Shape	coccoid
Mean Diameter, μm	0.707 **Size Range:** 0.5-1 microns
Growth Temperature	25-42 C **Survival Outside Host:** 1-25 days
Inactivation	Moist heat: 121 C for 15 min. Dry heat: 170 for 1 hr.
Disinfectants	1% sodium hypochlorite, 2% glutaraldehyde, formaldehyde, iodines, 70% ethanol.

Nominal Filter Rating	MERV 6	MERV 8	MERV 11	MERV 13	MERV 15
Estimated Removal, %	10.6	23.6	37.1	79.8	98.9
UVGI Rate Constant	0.006161	cm^2/μW-s	**Media**	Air	
Dose for 90% Kill	374	μW-s/cm^2	**Ref.**	Lidwell 1950	
Suggested Indoor Limit	0 CFU/cu.m.				
Genome Base Pairs					
Related Species	aka pnemococcus				
Notes	CDC Reportable. Rate constant assumed same as S. pyogenes.				
Photo Credit	Centers for Disease Control, PHIL# 263, Dr. Richard Facklam.				
References	Braude 1981, Freeman 1985, Austin 1991, Mitscherlich 1984, Murray 1999, Prescott 1996, Ryan 1994, Canada 2001, Mandell 2000				

Variola (smallpox)

GROUP	Virus
TYPE	DNA
GENUS	Variola
FAMILY	Poxviridae
DISEASE GROUP	Communicable
BIOSAFETY LEVEL	Risk Group 4
Infectious Dose (ID$_{50}$)	10-100 (?)
Lethal Dose (LD$_{50}$)	-
Infection Rate	-
Incubation Period	12-14 days
Peak Infection	14-17 days
Annual Cases	none
Annual Fatalities	none

The cause of smallpox, previously a worldwide epidemic until eradicated. Exists in laboratory samples only, but has reportedly been developed as a biological weapon. A related species, Camelpox, may also have been weaponized.

Disease or Infection	Smallpox			
Natural Source	Humans			
Toxins	none			
Point of Infection	Upper Respiratory Tract			
Symptoms	Pustular skin rash.			
Treatment	Rifamycin, Isatin B-thiosemicarbazone			
Untreated Fatality Rate	10-40% **Prophylaxis:** - **Vaccine:** Available			
Shape	complex capsid			
Mean Diameter, μm	0.224 **Size Range:** 0.23-0.40 microns			
Growth Temperature	na **Survival Outside Host:** (18 months at 5 C)			
Inactivation	Moist heat: 121 C for 15 min.			
Disinfectants	1% sodium hypochlorite, 2% glutaraldehyde, formaldehyde.			

Nominal Filter Rating	MERV 6	MERV 8	MERV 11	MERV 13	MERV 15
Estimated Removal, %	4.3	8.6	14.1	38.0	68.0

UVGI Rate Constant	0.001528	cm^2/μW-s	**Media**	Air	
Dose for 90% Kill	1507	μW-s/cm^2	**Ref.**	Collier 1955 (Vaccinia)	
Suggested Indoor Limit	0 CFU/cu.m.				
Genome Base Pairs	130000-375000 nt		**Notes**	CDC Reportable.	
Related Species	Monkeypox, Camelpox, Mousepox, Vaccinia (cowpox).				
Photo Credit	(Variola minor in monkey cells) Reprinted w/ permission from Malherbe and Strickland 1980, Viral Cytopathology, Copyright CRC Press, Boca Raton, FL				
References	Dalton 1973, Fraenkel-Conrat 1985, Freeman 1985, Mahy 1975, Murray 1999, Ryan 1994, Henderson 1999, Malherbe 1980, Canada 2001, Franz 1997, Breman 1998				

VEE

GROUP	Virus
TYPE	ssRNA
GENUS	Alphavirus
FAMILY	Togaviridae
DISEASE GROUP	Vector-borne
BIOSAFETY LEVEL	Risk Group 3

Infectious Dose (ID$_{50}$)	1
Lethal Dose (LD$_{50}$)	-
Infection Rate	-
Incubation Period	1-15 days
Peak Infection	-
Annual Cases	-
Annual Fatalities	-

Non-respiratory and non-airborne. Eastern Equine Encephalitis is a virus vectored by mosquitoes to horses through bites. VEE occurs in South America. Virus overwinters in birds. Has reportedly been succesfully weaponized for aerosol dissemination.

Disease or Infection	Encephalitis, meningitis
Natural Source	Birds, horses, rodents, humans.
Toxins	none
Point of Infection	Blood
Symptoms	Acute inflammatory disease with rapid onset, high fever, headache, disorientation, stupor, convulsions, paralysis.
Treatment	None.
Untreated Fatality Rate	1-60% Prophylaxis: Possible Vaccine: Available
Shape	enveloped virus
Mean Diameter, μm	0.06 Size Range: .06-.07 microns
Growth Temperature	na Survival Outside Host: dies outside host
Inactivation	-
Disinfectants	1% sodium hypochlorite, 2% glutaraldehyde, formaldehyde, 70% ethanol.

Nominal Filter Rating	MERV 6	MERV 8	MERV 11	MERV 13	MERV 15
Estimated Removal, %	10.1	18.1	26.8	56.1	81.7
UVGI Rate Constant	unknown	cm^2/μW-s	Media		
Dose for 90% Kill		μW-s/cm^2	Ref.		
Suggested Indoor Limit					

Genome Base Pairs	9700-11800 nt, 11700 bp
Related Species	Japanese encephalitis, Russian spr.-sum. encephalitis, WEE, EEE
Notes	CDC Reportable.
Photo Credit	Photo shows cells infected with the Togavirus Hepatitus C. Image reprinted with permission from V.Agnello, G.Abel, M.Elfahal, G.B.Knight, and Q.-X.Zhang 1999, PNAS 96(22):12766-12771, Copyright 1999 National Academy of Sciences, U.S.A.
References	Dalton 1973, Fraenkel-Conrat 1985, Freeman 1985, Mahy 1975, Murray 1999, Fatah 2000, NATO 1996, Malherbe 1980, Canada 2001, Franz 1997

Vibrio cholerae

GROUP	Bacteria
TYPE	Gram-
GENUS	Vibrio
FAMILY	
DISEASE GROUP	Food-borne
BIOSAFETY LEVEL	Risk Group 2

Infectious Dose (ID$_{50}$)	106-1011
Lethal Dose (LD$_{50}$)	-
Infection Rate	-
Incubation Period	1-5 days
Peak Infection	-
Annual Cases	common
Annual Fatalities	rare

Non-respiratory and non-airborne. The cause of cholera, once a pandemic disease that hailed from India. Now confined to periodic outbreaks in diverse locations around the world. Has reportedly been used as a biological weapon.

Disease or Infection	Cholera
Natural Source	Humans, environmental.
Toxins	enterotoxin
Point of Infection	Ingested.
Symptoms	Sudden onset of profuse watery stools, some vomiting, rapid dehydration, acidosis,, circulatory collapse.
Treatment	Tetracycline, other antibiotics if not resistant, hydration therapy,
Untreated Fatality Rate	50% **Prophylaxis:** Antibiotics **Vaccine:** Possible
Shape	Rods
Mean Diameter, μm	2.12 **Size Range:** 1.5-3 x 0.5 microns
Growth Temperature	37 C **Survival Outside Host:** 3-50 days
Inactivation	Moist heat: 121 C for 15 min. Dry heat: 170 for 1 hr.
Disinfectants	70% ethanol, iodines, 2% glutaraldehyde, 8% formaldehyde.

Nominal Filter Rating	MERV 6	MERV 8	MERV 11	MERV 13	MERV 15
Estimated Removal, %	35.1	66.3	83.1	95.9	100.0
UVGI Rate Constant	-	cm^2/μW-s	Media		
Dose for 90% Kill	-	μW-s/cm^2	Ref.		
Suggested Indoor Limit	na				
Genome Base Pairs					
Related Species					
Notes	CDC Reportable.				
Photo Credit	Centers for Disease Control, PHIL# 1034, Dr. William A. Clark.				
References	Braude 1981, Freeman 1985, Mitscherlich 1984, Murray 1999, Prescott 1996, Ryan 1994, Canada 2001, NATO 1996, Mandell 2000				

WEE, EEE

GROUP	Virus
TYPE	ssRNA
GENUS	Alphavirus
FAMILY	Togaviridae
DISEASE GROUP	Vector-borne
BIOSAFETY LEVEL	Risk Group 3

Infectious Dose (ID$_{50}$)	10-100
Lethal Dose (LD$_{50}$)	-
Infection Rate	-
Incubation Period	1-5 days
Peak Infection	-
Annual Cases	-
Annual Fatalities	-

Non-respiratory and non-airborne. Eastern Equine Encephalitis is a virus vectored by mosquitoes to horses through bites. WEE occurs in the western and central USA. Virus overwinters in birds. Has potential for weaponization.

Disease or Infection	Encephalitis, meningitis				
Natural Source	Birds, horses, rodents, humans.				
Toxins	none				
Point of Infection	Blood				
Symptoms	Acute inflammatory disease with slow onset, high fever, headache, disorientation, stupor, convulsions, paralysis.				
Treatment	None.				
Untreated Fatality Rate	EEE 60%, WEE 3%		**Vaccine:**		Available
Shape	enveloped virus		**Prophylaxis:**		Possible
Mean Diameter, µm	0.07	**Size Range:**	.06-.07 microns		
Growth Temperature	na	**Survival Outside Host:**		dies outside host	
Inactivation	-				
Disinfectants	1% sodium hypochl., 2% glutaraldehyde, formaldehyde, 70% ethanol				
Nominal Filter Rating	MERV 6	MERV 8	MERV 11	MERV 13	MERV 15
Estimated Removal, %	10.1	18.1	26.7	56.0	81.6
UVGI Rate Constant	unknown	cm²/µW-s	**Media**		
Dose for 90% Kill		µW-s/cm²	**Ref.**		
Suggested Indoor Limit					
Genome Base Pairs	9700-11800 nt, 11700 bp				
Related Species	Japanese encephalitis, Russian spring-sum. encephalitis, VEE				
Notes	CDC Reportable.				
Photo Credit	Photo shows EEE lymphocytes migating through cell wall. Photo reprinted with permission from Williams et al 2000, Avian Diseases v44(4):1012-16.				
References	Dalton 1973, Fraenkel-Conrat 1985, Freeman 1985, Mahy 1975, Murray 1999, Ryan 1994, Canada 2001, Franz 1997				

Yellow Fever virus

GROUP	Virus
TYPE	ssRNA
GENUS	Arbovirus
FAMILY	Flaviviridae
DISEASE GROUP	Vector-borne
BIOSAFETY LEVEL	Risk Group 3

Infectious Dose (ID$_{50}$)	unknown
Lethal Dose (LD$_{50}$)	-
Infection Rate	-
Incubation Period	3-6 days
Peak Infection	-
Annual Cases	rare
Annual Fatalities	rare

Non-respiratory and non-airborne. Transmitted by the bite of infected mosquitoes. Host become infective for mosquitoes for first 3-5 days of illness. Enzootic in South America. Unknown in Asia and hasn't been seen in the U.S. for decades. Monkeys are the primary reservoir but other animals can be hosted.

Disease or Infection	Yellow fever, YF, Arbovirus			
Natural Source	Humans, monkeys, other primates.			
Toxins	none			
Point of Infection	Insect bite.			
Symptoms	Sudden onset of fever, aches, prostration, nausea, vomiting, leukopenia, weak pulse, albuminuria, anuria, hemorrhagic symptoms.			
Treatment	No treatment.			
Untreated Fatality Rate	<5% **Prophylaxis:** Vaccine **Vaccine:** Possible			
Shape	enveloped			
Mean Diameter, μm	0.04472136 **Size Range:** 0.04-0.05 microns			
Growth Temperature	na **Survival Outside Host:** does not survive			
Inactivation	Moist heat: 60 C for 10 min.			
Disinfectants	70% ethanol, iodines, 2% glutaraldehyde, 8% formaldehyde, 3% hydrogen peroxide, 1% iodine.			

Nominal Filter Rating	MERV 6	MERV 8	MERV 11	MERV 13	MERV 15
Estimated Removal, %	13.5	24.0	35.0	67.5	90.3

UVGI Rate Constant	-	cm^2/μW-s	**Media**
Dose for 90% Kill	-	μW-s/cm^2	**Ref.**
Suggested Indoor Limit	na		
Genome Base Pairs	9500-12500 nt, 10500 bp		
Related Species			
Notes	CDC Reportable.		
Photo Credit	Image shows flaviviridae in cells. Centers for Disease Control, Atlanta.		
References	Dalton 1973, Fraenkel-Conrat 1985, Freeman 1985, Mahy 1975, Murray 1999, Ryan 1994, WHO 1982, Malherbe 1980, Canada 2001, Franz 1997		

Yersinia pestis

GROUP	Bacteria
TYPE	Gram-
GENUS	Yersinia
FAMILY	Enterobacteriaceae
DISEASE GROUP	Communicable
BIOSAFETY LEVEL	Risk Group 2-3

Infectious Dose (ID$_{50}$)	100
Lethal Dose (LD$_{50}$)	-
Infection Rate	varies
Incubation Period	2-6 days
Peak Infection	-
Annual Cases	4
Annual Fatalities	-

The ancient cause of plague. Normally transmitted by flea bites. Aerosol transmission can become epidemic. Primarily a zoonotic disease with rodents as the natural reservoir. Occasionally causes infections in the Americas, Africa, the Near East and Middel East, Asia, and Indonesia. Common in Burma and Vietnam.

Disease or Infection	Plague, bubonic plague, pneumonic plague, sylvatic plague.
Natural Source	Rodents, fleas, humans
Toxins	none
Point of Infection	Upper Respiratory Tract, skin via flea bites
Symptoms	Bubonic plague with lymphadenitis around area of flea bite, fever. Pneumonic plague results in pneumonia, mediastinitis, pleural effusion.
Treatment	Streptomycin, tetracycline, chloramphenicol, kanamycin 8-24 hours after onset

Untreated Fatality Rate	50%	Prophylaxis: Antibiotics	Vaccine:	Possible
Shape	rods			
Mean Diameter, μm	0.707	Size Range:	0.5-1 x 1-2 microns	
Growth Temperature	25-30 C	Survival Outside Host:	100-270 days in bodies	
Inactivation	Moist heat: 121 C for 15 min. Dry heat: 170 for 1 hr.			
Disinfectants	1% sodium hypochlorite, 2% glutaraldehyde, formaldehyde, 70% ethanol.			

Nominal Filter Rating	MERV 6	MERV 8	MERV 11	MERV 13	MERV 15
Estimated Removal, %	10.6	23.6	37.1	79.8	98.9
UVGI Rate Constant	(unknown)	cm^2/μW-s	Media		
Dose for 90% Kill	-	μW-s/cm^2	Ref.		
Suggested Indoor Limit	0 CFU/cu.m.				
Genome Base Pairs					
Related Species	(previously Pasteurella pestis)				
Notes	CDC Reportable.				
Photo Credit	Centers for Disease Control, PHIL# 741, Dr. Marshall Fox.				
References	Braude 1981, Ewald 1994, Freeman 1985, Linton 1982, Mitscherlich 1984, Murray 1999, Prescott 1996, Ryan 1994, Canada 2001, NATO 1996, Mandell 2000				

Database of Pathogen Disease and Lethal Dose Curves

Pathogen	Disease group	Disease curve	Dose curve
Bacillus anthracis	Noncommunicable	Yes	Yes
Blastomyces dermatitidis	Noncommunicable	Yes	Yes
Bordetella pertussis	Communicable	Yes	Yes
Brucella	Noncommunicable	Yes	Yes
Burkholderia mallei	Noncommunicable	Yes	Yes
Burkholderia pseudomallei	Noncommunicable	Yes	No
Chikungunya virus	Vector borne	Yes	No
Chlamydia psittaci	Noncommunicable	Yes	No
Clostridium botulinum	Noncommunicable	Yes	No
Clostridium perfringens	Noncommunicable	Yes	Yes
Coccidioides immitis	Noncommunicable	Yes	Yes
Corynebacterium diphtheriae	Communicable	Yes	No
Coxiella burnetii	Noncommunicable	Yes	Yes
Crimean-Congo hemorrhagic fever	Communicable	Yes	No
Dengue fever virus	Vector borne	Yes	No
Ebola	Communicable	Yes	Yes
Francisella tularensis	Noncommunicable	Yes	Yes
Hantaan virus	Noncommunicable	Yes	No
Hepatitis A	Communicable	Yes	Yes
Histoplasma capsulatum	Noncommunicable	Yes	Yes
Influenza A virus	Communicable	Yes	Yes
Japanese encephalitis	Vector borne	Yes	No
Junin virus	Noncommunicable	Yes	Yes
Lassa fever virus	Communicable	Yes	Yes
Legionella pneumophila	Noncommunicable	Yes	Yes
Lymphocytic choriomeningitis	Noncommunicable	Yes	Yes
Machupo	Noncommunicable	Yes	Yes
Marburg virus	Communicable	Yes	No
Mycobacterium tuberculosis	Communicable	Yes	Yes

Pathogen	Disease group	Disease curve	Dose curve
Mycoplasma pneumoniae	Endogenous	Yes	Yes
Nocardia asteroides	Noncommunicable	Yes	No
Paracoccidioides	Noncommunicable	Yes	Yes
Rickettsia prowazeki	Vector borne	Yes	Yes
Rickettsia rickettsii	Vector borne	Yes	Yes
Rift Valley fever	Vector borne	Yes	No
Salmonella typhi	Food borne	Yes	Yes
Shigella	Food borne	Yes	Yes
Stachybotrys chartarum	Noncommunicable	No	No
Streptococcus pneumoniae	Communicable	Yes	No
Variola (smallpox)	Communicable	Yes	No
VEE	Vector borne	Yes	Yes
Vibrio cholerae	Food borne	Yes	Yes
WEE, EEE	Vector borne	Yes	Yes
Yellow fever virus	Vector borne	Yes	No
Yersinia pestis	Communicable	Yes	Yes

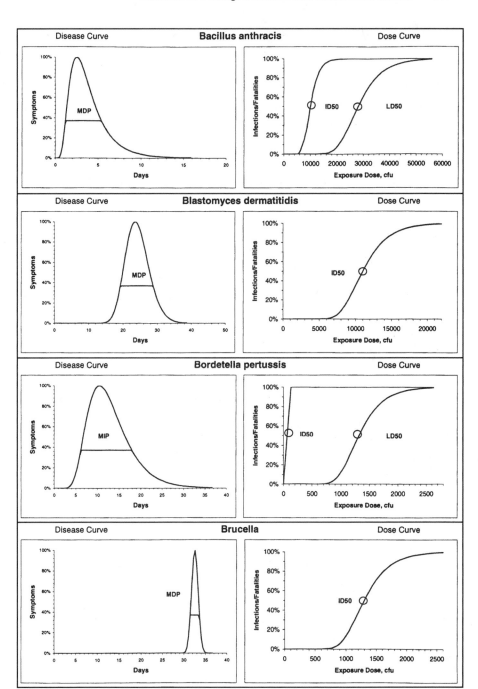

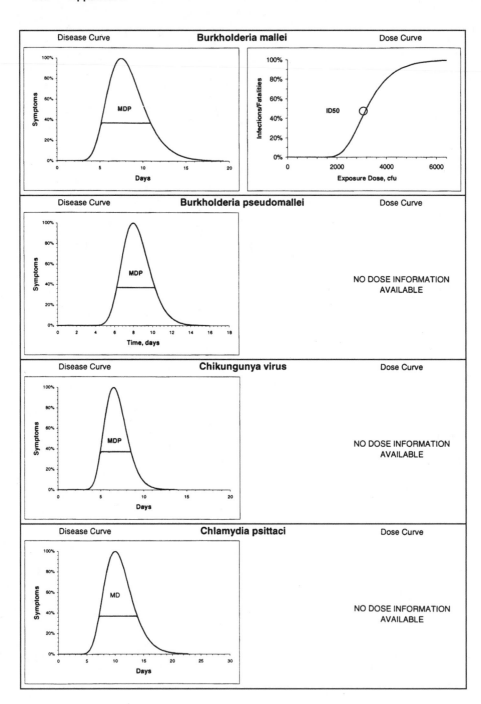

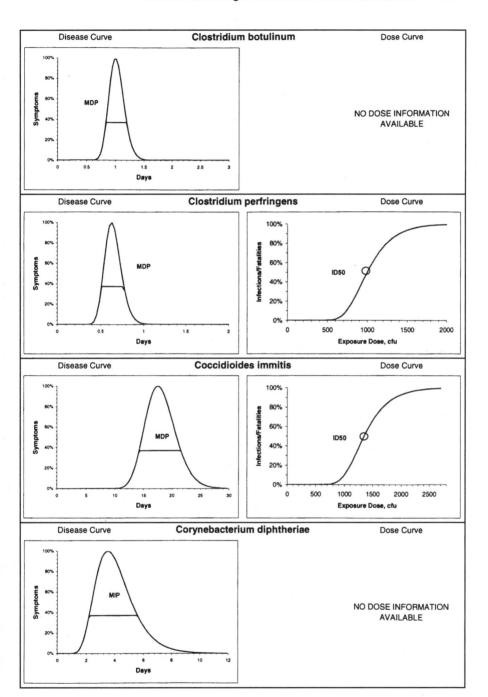

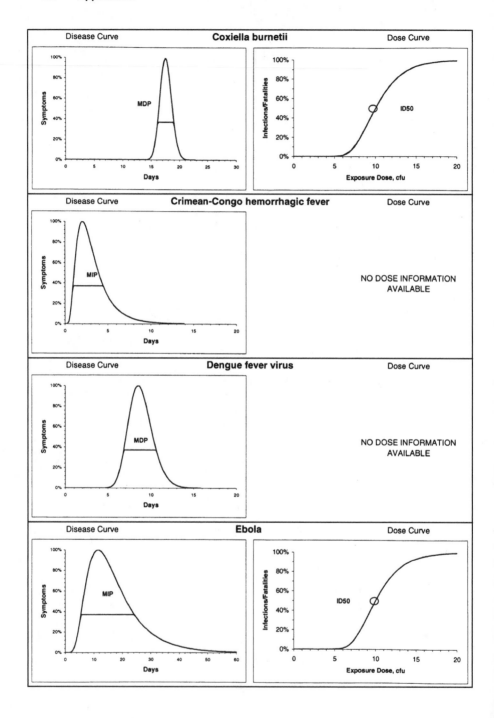

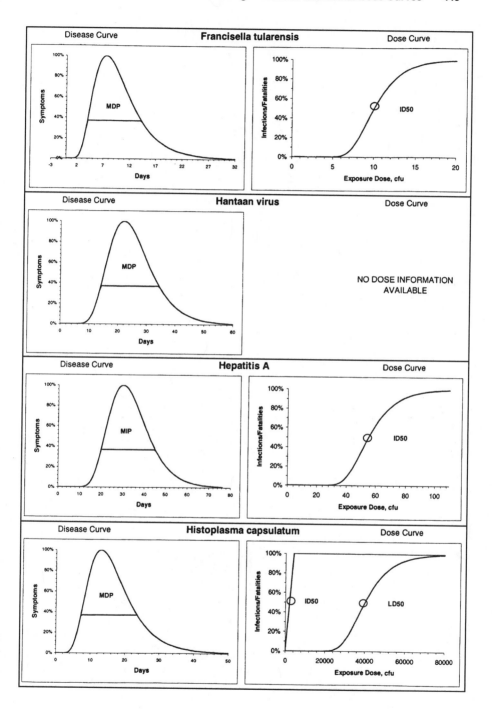

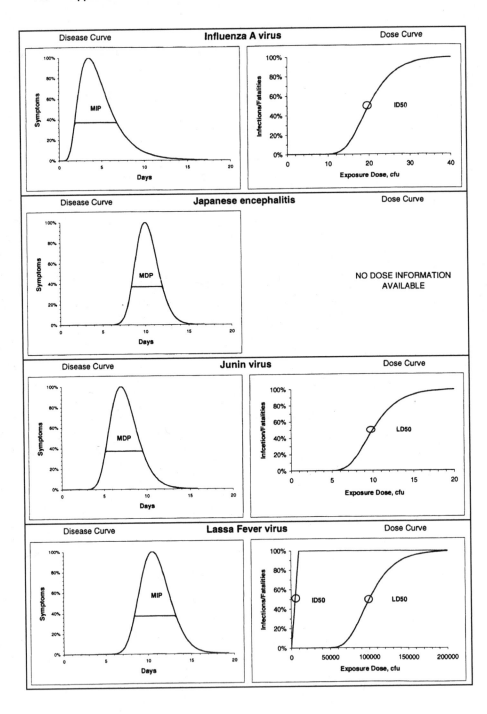

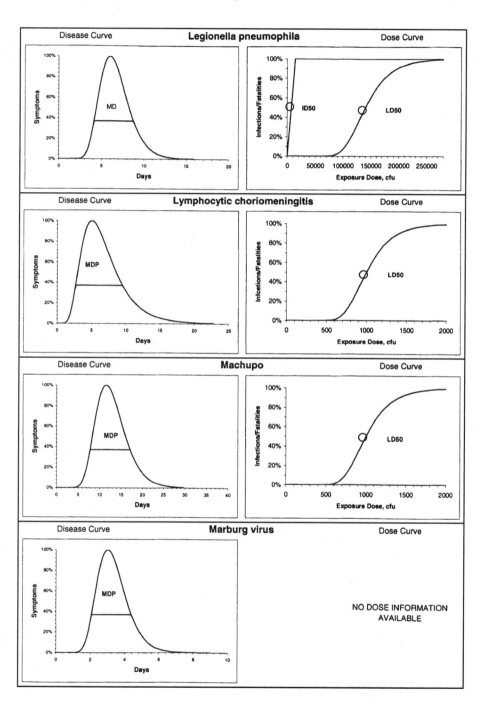

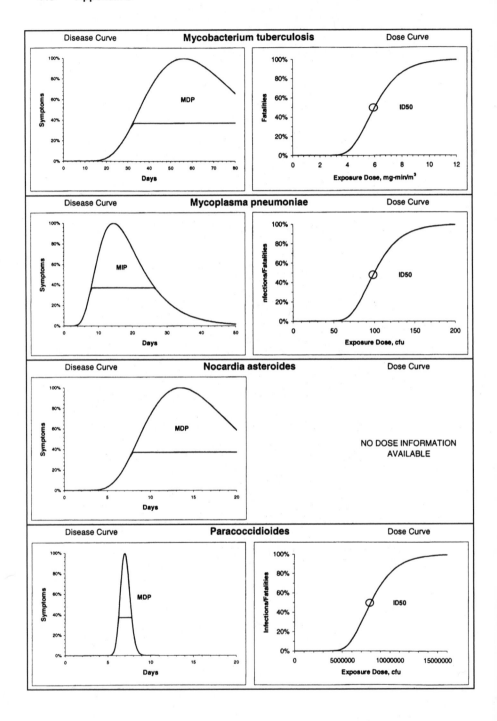

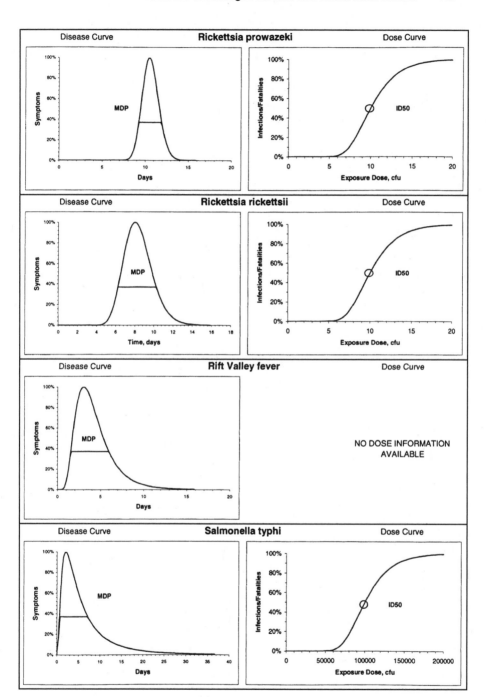

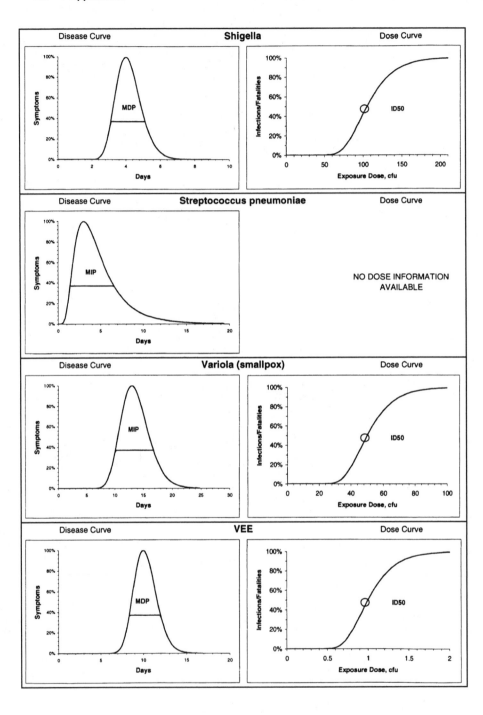

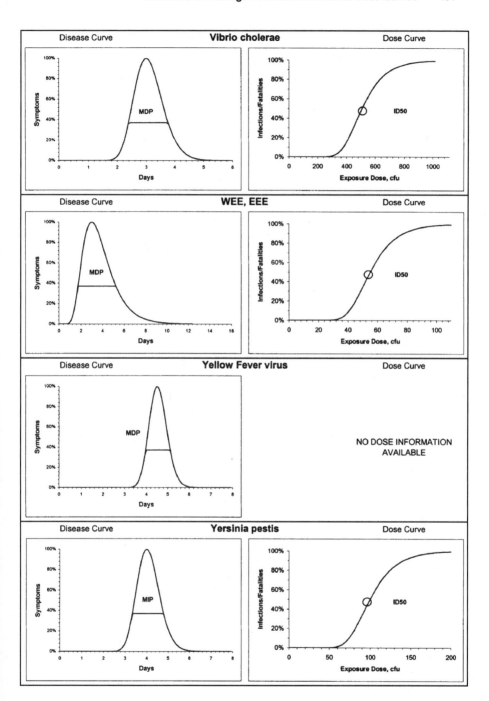

Database of Toxins and Dose Curves

Toxin	Type	Toxin	Type
Abrin	Plant	Maitotoxin	Neurotoxin
Aconitine	Plant	Microcystin	Peptide
Aflatoxin	Mycotoxin	Neuropeptide	Bioreglator
α-Latrotoxin	Neurotoxin	Neurotensin	Bioregulator
Anatoxin A	Neurotoxin	(NT)	
Angiotensin	Bioregulator	Notexin	Venom
Apamin	Venom	Oxytocin	Hormone
Atrial natriuretic	Bioregulator	Palytoxin	Neurotoxin
peptide		Ricin	Plant
Batrachatoxin	Neurotoxin	Sarafotoxin	Venom
β-Bungarotoxin	Neurotoxin	Saxitoxin	Neurotoxin
Bombesin (BN)	Bioregulator	Sea wasp toxin	Venom
Botulinum	Neurotoxin	Shiga toxin	Exotoxin
Bradykinin	Bioregulator	Somatostatin	Bioregulator
Brevetoxin	Neurotoxin	(SS)	
Cholecystokinin	Bioregulator	Staphylococcal	Enterotoxin
Ciguatoxin	Neurotoxin	enterotoxin A	
Citrinin	Mycotoxin	Staphylococcal	Enterotoxin
C. perfringens	Enterotoxin	enterotoxin B	
toxin		Substance P	Bioregulator
Cobrotoxin	Cytotoxin	T-2 toxin	Mycotoxin
Conotoxin	Venom	Taipoxin	Neurotoxin
Curare	Plant	Tetanus toxin	Neurotoxin
Diamphotoxin	Venom	Tetrodoxin (TTX)	Neurotoxin
Diphtheria toxin	Exotoxin	Textilotoxin	Neurotoxin
Dynorphin	Bioregulator	Thyroliberin	Bioregulator
Endothelin	Bioreglator	(TRF)	
Enkephalin	Bioregulator	Trichothecene	Mycotoxin
Gastrin	Bioregulator	toxins	
Gonadoliberin	Bioregulator	Vasopressin	Bioregulator
(LRF)		Warfarin	Anticoagulant

Agent	**Abrin**
Type	Plant
Source	Rosary Pea
LD_{50}, mg/kg	0.00004
Action	penetrates cell membranes
Formula	
Notes	a toxic lectin
References	NATO 1996, Stephen 1981, Harris 1986

Agent	**Aconitine**
Type	Plant
Source	Monkshod
LD_{50}, mg/kg	0.1
Action	highly toxic alkaloid
Formula	C34H49NO11
Notes	CAS# 302-27-2, acetyl benzoyl aconine
References	NATO 1996, Lewis 1993

Agent	**Aflatoxin**
Type	Mycotoxin
Source	Aspergillus spp.
LD_{50}, mg/kg	0.3
Action	carcinogen
Formula	
Notes	a group of polynuclear molds, primarily from Aspergillus
References	Lewis 1993, Kurata 1984

Agent	**α-latrotoxin**
Type	Neurotoxin
Source	Black widow spider
LD_{50}, mg/kg	0.01
Action	causes release of neurotransmitters from all synapses
Formula	
Notes	from Latrodectus spp.
References	Middlebrook 1986, Lackie 1995

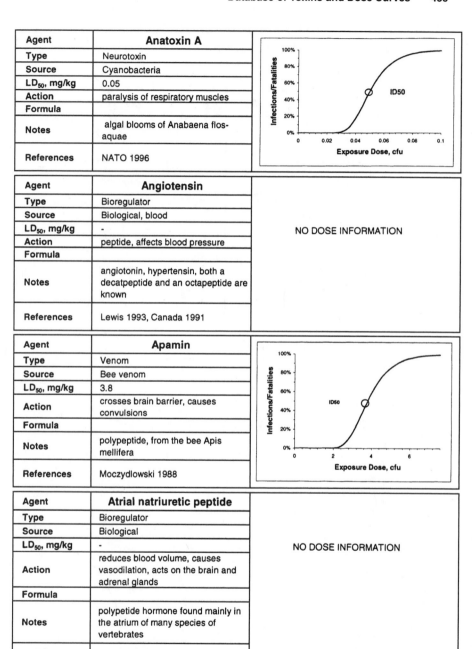

Agent	**Anatoxin A**
Type	Neurotoxin
Source	Cyanobacteria
LD$_{50}$, mg/kg	0.05
Action	paralysis of respiratory muscles
Formula	
Notes	algal blooms of Anabaena flos-aquae
References	NATO 1996

Agent	**Angiotensin**
Type	Bioregulator
Source	Biological, blood
LD$_{50}$, mg/kg	-
Action	peptide, affects blood pressure
Formula	
Notes	angiotonin, hypertensin, both a decatpeptide and an octapeptide are known
References	Lewis 1993, Canada 1991

NO DOSE INFORMATION

Agent	**Apamin**
Type	Venom
Source	Bee venom
LD$_{50}$, mg/kg	3.8
Action	crosses brain barrier, causes convulsions
Formula	
Notes	polypeptide, from the bee Apis mellifera
References	Moczydlowski 1988

Agent	**Atrial natriuretic peptide**
Type	Bioregulator
Source	Biological
LD$_{50}$, mg/kg	-
Action	reduces blood volume, causes vasodilation, acts on the brain and adrenal glands
Formula	
Notes	polypetide hormone found mainly in the atrium of many species of vertebrates
References	Lackie 1995

NO DOSE INFORMATION

Agent	**Batrachatoxin**
Type	Neurotoxin
Source	Phyllobates, poison dart frog
LD_{50}, mg/kg	0.002
Action	the strongest neurotoxin among venoms
Formula	C24H35NO5
Notes	Batrachatoxin A is an isomeric component, a steroidal alkaloid, as toxic as strychnine
References	NATO 1996, Wright 1990, Middlebrook 1986, Lewis 1993

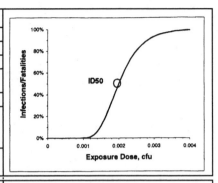

Agent	**β-bungarotoxin**
Type	Neurotoxin
Source	Venom
LD_{50}, mg/kg	0.014
Action	Inhibits release of ACh at neuromuscular junction and blocks potassium channels
Formula	
Notes	found in the venom of Bungarus multicinctus
References	Middlebrook 1986, Lackie 1995

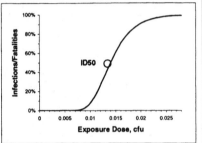

Agent	**Bombesin (BN)**
Type	Bioregulator
Source	Biological
LD_{50}, mg/kg	-
Action	cross reacts with gastride releasing peptides
Formula	
Notes	Tetradecapeptide neurohormone with both paracrine and autocrine effects
References	Lackie 1995, Canada 1991

NO DOSE INFORMATION

Agent	**Botulinum**
Type	Neurotoxin
Source	Clostridium botulinum
LD_{50}, mg/kg	0.000001
Action	Blocks acetylcholine release, respiratory paralysis
Formula	
Notes	destroyed by heating to 80 C for 0.5 hours
References	NATO 1996, Wright 1990, Middlebrook 1986, Lewis 1993

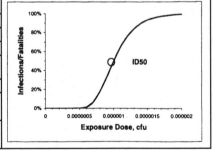

Agent	**Bradykinin**	
Type	Bioregulator	
Source	Biological	
LD$_{50}$, mg/kg	-	NO DOSE INFORMATION
Action	potent vasodilator, spasmogenic for some smooth muscle	
Formula		
Notes	Vasoactive nonapeptide	
References	Lackie 1995	

Agent	**Brevetoxin**	
Type	Neurotoxin	
Source	Red Tide dinoflagellate	
LD$_{50}$, mg/kg	35 nM*	NO DOSE INFORMATION
Action	Activates sodium channels, embryotoxic	
Formula	C50H74O15	
Notes	produced by Ptychodiscus brevis Davis or Gymnodinium breve Davis	
References	Harris 1986	

Agent	**Cholecystokinin**	
Type	Bioregulator	
Source	Biological	
LD$_{50}$, mg/kg	-	NO DOSE INFORMATION
Action	mediates action of trypsin on the pancreas	
Formula		
Notes	a hormone	
References	Harris 1986, Canada 1991	

Agent	**Ciguatoxin**	
Type	Neurotoxin	
Source	Gambierdiscus, dinoflagellate	
LD$_{50}$, mg/kg	0.0004	
Action	Opens sodium channels	
Formula	C28H52NO5Cl	
Notes	a type of quaternary ammonium compound, one part anticholinesterase	
References	NATO 1996, Lewis 1993	

Agent	**Citrinin**
Type	Mycotoxin
Source	Penicillium spp.
LD$_{50}$, mg/kg	35
Action	carcinogenic
Formula	
Notes	
References	Kurata 1984

Agent	**C. perfringens toxin**
Type	Enterotoxin
Source	Clostridium perfringens
LD$_{50}$, mg/kg	0.0003
Action	damages mucosa, facilitates infection
Formula	
Notes	
References	NATO 1996, Stephen 1981, Cliver 1990

Agent	**Cobrotoxin**
Type	Cytotoxin
Source	Chinese cobra
LD$_{50}$, mg/kg	0.075
Action	Blocks nicotinic receptors
Formula	
Notes	
References	Middlebrook 1986

Agent	**Conotoxin**
Type	Venom
Source	Marine Cone snail
LD$_{50}$, mg/kg	-
Action	Blocks calcium channels, or sodium channels, causing paralysis
Formula	
Notes	from cone shells (Conus spp).
References	Middlebrook 1986, Lackie 1995

NO DOSE INFORMATION

Agent	Curare
Type	Plant
Source	South American tree species
LD$_{50}$, mg/kg	0.5
Action	acts on the central nervous system
Formula	
Notes	CAS# 8063-06-7, a mixture of numerous toxic alkaloids
References	Middlebrook 1986, Lewis 1993

Agent	Diamphotoxin
Type	Venom
Source	African Beetle pupa
LD$_{50}$, mg/kg	-
Action	ion transfer across membranes
Formula	
Notes	an ionophore, used by the Kung for hunting
References	BWC 1995

NO DOSE INFORMATION

Agent	Diphtheria toxin
Type	Exotoxin
Source	C. diphtheria
LD$_{50}$, mg/kg	0.0001
Action	intracellular damage
Formula	
Notes	an enzyme that inactivates many molecules in cells
References	NATO 1996, Middlebrook 1986, Stephen 1981, Cliver 1990

Agent	Dynorphin
Type	Bioregulator
Source	Biological
LD$_{50}$, mg/kg	-
Action	opiate peptide
Formula	
Notes	peptide derived from the hypothalamic precursor pro-dynorphin
References	Lackie 1995, Canada 1991

NO DOSE INFORMATION

Agent	**Endothelin**	NO DOSE INFORMATION
Type	Bioregulator	
Source	Biological	
LD_{50}, mg/kg	-	
Action	potent vasoconstrictor hormone	
Notes	peptide hormones released by endothelial cells	
References	Lackie 1995	

Agent	**Enkephalin**	NO DOSE INFORMATION
Type	Bioregulator	
LD_{50}, mg/kg	Biological	
LD_{50}, mg/kg	-	
Action	binds to d -type opiate receptors	
Notes	natural opiate pentapeptides	
References	Lackie 1995, Canada 1991	

Agent	**Gastrin**	NO DOSE INFORMATION
Type	Bioregulator	
Source	Biological	
LD_{50}, mg/kg	-	
Action	stimulate secretion of protons and pancreatic enzymes	
Notes	secreted by the mucosal gut lining of some mammals	
References	Lackie 1995, Canada 1991	

Agent	**Gonadoliberin (LRF)**	NO DOSE INFORMATION
Type	Bioregulator	
Source	Biological	
LD_{50}, mg/kg	-	
Action	hormonal stimulation	
Notes	a protein hormone, or neuropeptide, (GnRH)	
References	Canada 1991	

Agent	**Maitotoxin**	
Type	Neurotoxin	
Source	Herbivorous ciguateric fish	
LD_{50}, mg/kg	0.0001	
Action	activates calcium channels and mobilizes intracellular calcium stores	
Notes	Acts on L-type voltage-sensitive calcium channels	
References	NATO 1996, Harris 1986, Lackie 1995	

Agent	**Microcystin**
Type	Peptide
Source	Blue-green algae
LD$_{50}$, mg/kg	0.05
Action	hepatoxic
Formula	
Notes	produced by cyanobacteria in waterblooms
References	NATO 1996

Agent	**Neuropeptide**	
Type	Bioregulator	
Source	Biological	
LD$_{50}$, mg/kg	-	NO DOSE INFORMATION
Action	indirectly modulates the nervous system	
Notes	peptide neurotransmitters	
References	Lackie 1995	

Agent	**Neurotensin (NT)**	
Type	Bioregulator	
Source	Biological	
LD$_{50}$, mg/kg	-	NO DOSE INFORMATION
Action	general vascular and neuroendocrine actions	
Notes	tridecapeptide hormone of gastrointestinal tract	
References	Lackie 1995, Canada 1991	

Agent	**Notexin**	
Type	Venom	
Source	Australian snakes	
LD$_{50}$, mg/kg	-	NO DOSE INFORMATION
Action	enzymatic action on synapses	
Notes	occurs in several Australian proteroglyphs	
References	Harris 1986	

Agent	**Oxytocin**	
Type	Hormone	
Source	Humans, pituitary gland	
LD$_{50}$, mg/kg	-	NO DOSE INFORMATION
Action	stimulates uterus muscles	
Formula	C43H66N12O12S2	
Notes	CAS#50-56-6, a-hypophamine, contains 8 amino acids	
References	Lewis 1993, Canada 1991	

Agent	**Palytoxin**
Type	Neurotoxin
Source	Marine soft coral
LD_{50}, mg/kg	0.00015
Action	Activates sodium channels
Formula	
Notes	Linear peptide from corals of Palythoa spp.
References	NATO 1996, Lackie 1995

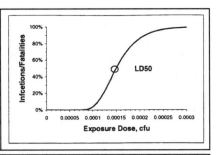

Agent	**Ricin**
Type	Plant
Source	Castor bean oil albumin
LD_{50}, mg/kg	0.003
Action	highly toxic by ingestion, in eyes or nose
Formula	
Notes	appears as white powder
References	NATO 1996, Wright 1990, Middlebrook 1986, Richardson 1992, Lewis 1993

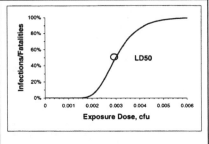

Agent	**Sarafotoxin**
Type	Venom
Source	Snakes
LD_{50}, mg/kg	-
Action	cardiotoxic
Notes	related to endothelins
References	Lackie 1995

NO DOSE INFORMATION

Agent	**Saxitoxin**
Type	Neurotoxin
Source	Dinoflagellate, shellfish
LD_{50}, mg/kg	0.002
Action	Blocks sodium channels
Formula	C10H17N7O4 2HCl
Notes	CAS# 35523-89-8, attacks central nervous system, muscular nerve block
References	NATO 1996, Wright 1990, Middlebrook 1986, Lewis 1993

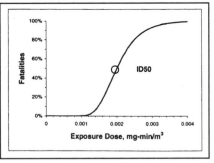

Agent	Sea wasp toxin	
Type	Venom	NO DOSE INFORMATION
Source	Jellyfish	
LD_{50}, mg/kg	-	
Action	paralysis of the central nervous system	
Formula		
Notes	from Chironex fleckeri and other jellyfish	
References		

Agent	Shiga toxin	
Type	Exotoxin	
Source	Shigella spp.	
LD_{50}, mg/kg	0.000002	
Action	inflammation and necrosis of mucosa in colon	
Formula		
Notes		
References	NATO 1996, Cliver 1990	

Agent	Somatostatin (SS)	
Type	Bioregulator	NO DOSE INFORMATION
Source	Biological	
LD_{50}, mg/kg	-	
Action	Inhibits gastric secretion, inhibits somatotropin release	
Formula		
Notes	Gastrointestinal and hypothalamic peptide hormone	
References	Lackie 1995, Canada 1991	

Agent	Staphylococcal enterotoxin A	
Type	Enterotoxin	
Source	Staphylococcus aureus	
LD_{50}, mg/kg	0.00005	
Action	interacts with membranes, emetic	
Formula		
Notes	SEA, contain numerous amino acids	
References	Wright 1990, Stephen 1981, Cliver 1990	

Agent	**Staphylococcal enterotoxin B**
Type	Enterotoxin
Source	Staphylococcus aureus
LD$_{50}$, mg/kg	0.027
Action	emetic action on intestines
Formula	
Notes	SEB, contain numerous amino acids
References	NATO 1996, Wright 1990, Cliver 1990

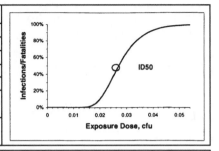

Agent	**Substance P**
Type	Bioregulator
Source	Biological
LD$_{50}$, mg/kg	0.000014
Action	induces vasodilation, salivation and increases capillary permeability
Formula	
Notes	vasoactive intestinal peptide
References	Lackie 1995, Canada 1991

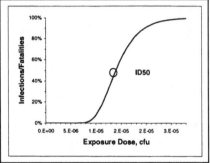

Agent	**T-2 toxin**
Type	Mycotoxin
Source	Fusarium spp.
LD$_{50}$, mg/kg	1.21
Action	digestive disorder, hemorrhage of stomach, heart, intestines
Formula	
Notes	a trichothecene
References	NATO 1996, CAST 1989

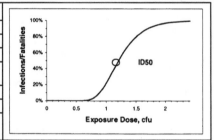

Agent	**Taipoxin**
Type	Neurotoxin
Source	Taipin snake
LD$_{50}$, mg/kg	0.005
Action	blocks transmission at the neuromuscular junction
Formula	
Notes	heterotrimeric toxin from Oxyuranus scutelatus scutelatus
References	NATO 1996, Middlebrook 1986, Harris 1986, Lackie 1995

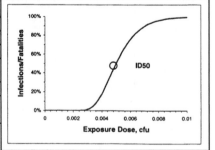

Agent	**Tetanus toxin**
Type	Neurotoxin
Source	Clostridium tetani
LD$_{50}$, mg/kg	0.000002
Action	acts on the central reflex apparatus of the spinal cord
Formula	
Notes	tetanospasmin
References	NATO 1996, Wright 1990, Middlebrook 1986, Stephen 1981

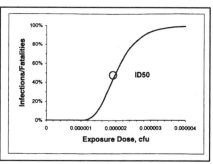

Agent	**Tetrodoxin (TTX)**
Type	Neurotoxin
Source	Japanese Pufferfish
LD$_{50}$, mg/kg	0.008
Action	blocks sodium channels
Notes	binds to the sodium channel, blocking the passage of action potentials
References	NATO 1996, Middlebrook 1986, Harris 1986

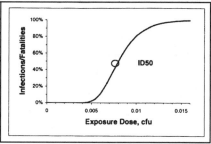

Agent	**Textilotoxin**
Type	Neurotoxin
Source	Elapid snake, Aus. brown snake
LD$_{50}$, mg/kg	0.0006
Action	blocks release of acetylcholine
Formula	
Notes	from venom of Pseudonaja textilis textilis
References	NATO 1996, Lackie 1995

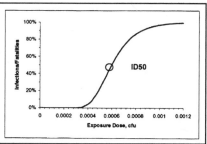

Agent	**Thyroliberin (TRF)**
Type	Bioregulator
Source	Biological
LD$_{50}$, mg/kg	-
Action	releases thyrotropin from the anterior pituitary
Notes	thyrotropic-releasing hormone
References	Lackie 1995, Canada 1991

NO DOSE INFORMATION

Agent	Trichothecene toxins	
Type	Mycotoxin	DOSES VARY FOR TRICHOTHECENE TOXINS
Source	various fungi	
LD_{50}, mg/kg	varies	
Action	block protein translation	
Notes	T-2 toxin is the most potent trichothecene	
References	CAST 1989	

Agent	Vasopressin	
Type	Bioregulator	NO DOSE INFORMATION
Source	Biological, pituitary glands	
Action	increases blood pressure and kidney water retention	
Notes	an octapeptide with 8 amino acids	
References	Lewis 1993, Canada 1991	

Agent	Warfarin	
Type	Anticoagulant	NO DOSE INFORMATION
Source	Biological	
Action	inhibits blood clotting	
Notes	Synthetic inhibitor of prothrombin activation	
References	Lackie 1995	

Additional Toxins Cited in CBW Literature

These toxins either have limited or dubious use as bioweapons, have no additional information, or play a minor role in associated disease infections.

Toxin	Reference
Aerolysin	BWC 1995
Amagassa venom	Ellis 1999
Amilyn	BWC 1995
Ammodytoxin	BWC 1995
Angiotensin-forming enzyme (bioregulator)	Canada 1991
Charybdotoxin	BWC 1995
Cholera toxin	BWC 1995
Delta sleep-inducing peptide (DSIP) (bioregulator)	Canada 1991
Dermophin	BWC 1995
Diacetoxyscirpenol (DAS)	BWC 1995
Erabutoxin	BWC 1995
Habu venom	Ellis 1999
Modeccin	BWC 1995
Nivalenol	BWC 1995
Noxiustoxin	BWC 1995
Pertussigen	BWC 1995
Renin (bioregulator)	Canada 1991
Roridin	BWC 1995
Trikabuto poison	Ellis 1999
Vasoactive intestinal polypeptide (VIP) (bioregulator)	Canada 1991
Verotoxin	BWC 1995
Verrucologen	BWC 1995

Database of Chemical Weapon Agents

CW agent	Code	Dose info	CW agent	Code	Dose info
0-Alkyl aminoethyl alkyl phosphonites		No	Cyclohexyl Sarin	GF	Yes
0-Alkyl phosphonochloridates		No	DC		No
			Dialkyl phosphoramidates		No
Adamsite	DM	Yes	Diethyl phosphate		Yes
Alkyl phosphoramidocyanidates		No	Dimethyl methylphosponate		No
			Dimethyl phosphate		Yes
Alkyl alkylphosphonofluorodates		No	Diphenylchloroarsine	DA	Yes
Alkyl phosphonothiolates		No	Diphenylcyanoarsine	DC	Yes
			Diphosgene	DP	No
Alkyl phosponyldifluorides		No	Ethyldiethanolamine		No
Amiton		Yes	Hydrogen chloride	HCL	No
Arsenic trichloride		Yes	Hydrogen cyanide	AC	Yes
Arsine	SA	No	Lewisite	L	Yes
Benzilic acid		No	Methyldiethanolamine		Yes
BZ	BZ	No	Methylphosphonyl dichloride	DF	No
Camite	CA	Yes	Mustard gas	H	Yes
Chlorine	CL	Yes	Mustard T-mixture	HT	No
Chloropicrin	PS	No	N,N-dialkylaminoethyl-2-chlorides		No
Chlorosarin		Yes	Nerve gas DFP	DFP	Yes
Chlorosoman		Yes	Nitrogen dioxide	NO2	No
CN	CN	Yes	Nitrogen mustard	HN-1	Yes
CS	CS	Yes	O-mustard	T	Yes
Cyanogen chloride	CK	Yes	Perfluoroisobutylene	PFIB	Yes
			Phosgene	CG	Yes

CW agent	Code	Dose info	CW agent	Code	Dose info
Phosgene oxime	CX	Yes	Sulphur dichloride		No
Phosphorus oxychloride		Yes	Sulphur monochloride		No
Phosphorus pentachloride		Yes	Sulphur trioxide	FS	No
Phosphorus trichloride		Yes	Tabun	GA	Yes
Pinacolyl alcohol		No	Thiodiglycol		Yes
QL	QL	No	Thionyl chloride		No
Quinuclidine-3-OL		No	Titanium tetrachloride	FM	No
Sarin	GB	Yes	Triethanolamine	TEA	Yes
Sesqui mustard	Q	Yes	Triethyl phosphate		Yes
Soman	GD	Yes	Triethylamine		No
Sulfur dioxide	SO2	No	Trimethyl phosphate		Yes
Sulfur mustard (distilled)	HD	Yes	VX	VX	Yes

CW Agents with Dose Information

CW Agent	Adamsite		DM
Type	Vomiting agent	CWC Schedule	
Formula	$C_6H_4(AsCl)(NH)C_6H_4$	Molecular Weight	277
Chem. Name	diphenylamine chloroarsine	HAZMAT ID#	1698
CAS #	578-94-9	Boiling Point, °C	410
LC_{50}		mg/l Ref.	
CD_{50}		mg-min/m³ Ref.	
$L(Ct)_{50}$	30000	mg-min/m³ Ref. Cookson 1969	
LD_{50}		mg/kg Ref.	
Appearance	yellow/brown crystals, liquid or powder, gas is heavier than air		
Odor	none		
Symptoms	irritant, headache, sneezing, coughing, chest pains, nausea, vomiting		
Antidote			
Decon.			
Solubility			
Removal	May be slightly or partially removable by high efficiency filters		
First Aid	Symptoms are temporary and may dissipate within minutes or hours. No specific first aid needed.		
Remediation	Use sand or inert absorbent to contain or absorb spills. Control vapor cloud with fine spray of water.		
Notes	noncombustible		
References	Hersh 1968, Cookson 1969, Wright 1990, Lewis 1993, Clarke 1968, Somani 1992, DOT 2000		
Dose Curve			

Dose Curve

CW Agent	Amiton		
Type		CWC Schedule	II
Formula	$C_{10}H_{24}NO_3PS$	Molecular Weight	269
Chem. Name	phosphorothoic acid	HAZMAT ID#	
CAS #	78-53-5	Boiling Point, °C	160-165
LC_{50}		mg/l Ref.	
CD_{50}		mg-min/m^3 Ref.	
$L(Ct)_{50}$		mg-min/m^3 Ref.	
LD_{50}	0.5	mg/kg Ref.	
Appearance	gas is heavier than air		
Odor			
Symptoms			
Antidote			
Decon.	SNL Foam		
Solubility			
Removal			
First Aid			
Remediation	Use sand or inert absorbent to contain or absorb spills. Control vapor cloud with fine spray of water.		
Notes			
References	Canada 1993, Thomas 1970, Canada 1992, OPCW 1997		

Dose Curve

CW Agent	Arsenic trichloride		
Type	Vesicant	CWC Schedule	II
Formula	$AsCl_3$	Molecular Weight	181.28
Chem. Name	arsenic chloride	HAZMAT ID#	1560
CAS #	7784-34-1	Boiling Point, °C	130.5
LC_{50}	30	mg/l	Ref.
CD_{50}		$mg\text{-}min/m^3$	Ref.
$L(Ct)_{50}$		$mg\text{-}min/m^3$	Ref.
LD_{50}	138	mg/kg	Ref. Richardson 1992
Appearance	colorless or pale yellow oily liquid, gas is heavier than air		
Odor			
Symptoms	irritant to eyes and skin		
Antidote	Dimercaprol (British anti-lewisite (BAL)) ointment		
Decon.	Water, SNL Foam		
Solubility			
Removal	May be slightly or partially removable by high efficiency filters		
First Aid	move victim to fresh air, provide oxygen if breathing difficulty, flush skin and eyes with water for 20 mins		
Remediation	Use sand or inert absorbent to contain or absorb spills. Control vapor cloud with fine spray of water.		
Notes	noncombustible		
References	Richardson 1992, Canada 1993, Lewis 1993, SIPRI 1971, Canada 1992, OPCW 1997, DOT 2000, Ellison 2000		
Dose Curve			

CW Agent	Camite		CA
Type	Tear Gas	CWC Schedule	
Formula	$C_6H_5CHBrCN$	Molecular Weight	196.05
Chem. Name	brombenzyl cyanide	HAZMAT ID#	2188
CAS #	16532-79-9	Boiling Point, °C	225
LC_{50}		mg/l Ref.	
CD_{50}	3500	mg-min/m^3 Ref.	Clarke 1968
$L(Ct)_{50}$	3500	mg-min/m^3 Ref.	Clarke 1968
LD_{50}	100	mg/kg Ref.	
Appearance	pink crystals, yellow liquid, gas is heavier than air		
Odor	sour or rotting fruit		
Symptoms	burning throat		
Antidote			
Decon.			
Solubility			
Removal	May be slightly or partially removable by high efficiency filters		
First Aid	Symptoms are temporary and may dissipate within minutes or hours. No specific first aid needed.		
Remediation	absorb or cover with dry earth, sand or other non-combustible material and transfer to approved containers		
Notes			
References	Richardson 1992, Cookson 1969, Clarke 1968, Somani 1992, CDC 2001, DOT 2000, Ellison 2000		
Dose Curve			

CW Agent	Chlorine		CL	
Type	Choking agent	**CWC Schedule**		
Formula	Cl_2	**Molecular Weight**		70.91
Chem. Name	chlorine	**HAZMAT ID#**		1017
CAS #	7782-50-5	**Boiling Point, °C**		
LC$_{50}$	5600	mg/l	**Ref.**	
CD$_{50}$		mg-min/m^3	**Ref.**	
L(Ct)$_{50}$	52740	mg-min/m^3	**Ref.**	Richardson 1992
LD$_{50}$		mg/kg	**Ref.**	
Appearance	greenish-yellow gas, gas is heavier than air			
Odor	pungent odor			
Symptoms	Corrosive burning, pain, cough			
Antidote				
Decon.				
Solubility	slightly soluble in cold water			
Removal	Granular Activated Carbon removes with 10-25% efficiency			
First Aid	Inhalation exposure: Provide oxygen if breathing difficulty. Give artificial respiration if breathing has stopped. Give mouth to mouth resuscitation only if no facial contamination. Skin exposure: Immediately wash skin and clothes with water. Remove clothing and flush skin with water. Flush eyes with water.			
Remediation	Use sand or inert absorbent to contain or absorb spills. Control vapor cloud with fine spray of water.			
Notes	noncombustible			
References	Richardson 1992, CDC 2001, Lewis 1993, Haber 1986, Wright 1990, Miller 1999, DOT 2000			
Dose Curve				

CW Agent	Chlorosarin		
Type		CWC Schedule	I
Formula	$C_4H_{10}ClO_2P$	Molecular Weight	156.5
Chem. Name		HAZMAT ID#	
CAS #	1445-76-7	Boiling Point, °C	
$L(Ct)_{50}$		mg-min/m^3 Ref.	
LD_{50}	4.5	mg/kg Ref.	
Appearance	gas is heavier than air		
Odor			
Symptoms			
Antidote	atropine, diazepam (CANA), pralidoxime chloride (2 PAM Cl)		
Decon.	Bleaching powder, Supertropical bleach, SNL Foam		
First Aid	Use all three nerve agent antidote kits if symptoms are severe. Use Mark I injectors or atropine. If symptoms progress, use injectors at 5-20 minute intervals (maximum of three injections). Give artificial respiration if breathing stops. Use oxygen delivery if face is contaminated or if breathing is labored. Skin exposure: Immediately wash skin and clothes with solution of 5% sodium hypochlorite or common household liquid bleach. Remove clothing and flush skin with same solution, then wash areas with soap and water. Flush eyes with water for 10-15 minutes.		
Remediation	Cover with fine sand, diatomaceous earth, sponges, paper towels, cloth, vermiculite, or clay. Neutralize with large quantities of 5.25% solution of sodium hypochlorite. Place all material in suitable sealed containers. Decontaminate exterior of containers. Label and dispose according to applicable regulations. Alternate solution: calcium hypochlorite, DS2, supertropical bleach slurry (STB), in that order.		
Notes	a precursor agent		
References	Canada 1993, OPCW 1997		
Dose Curve			

CW Agent	CN		CN
Type	Tear Gas	CWC Schedule	
Formula	various	Molecular Weight	154.6
Chem. Name	chloroacetophenone	HAZMAT ID#	1697
CAS #	532-27-4, 78-92-2, 57-55-6, 5989-27-5, 34590-94-8		
LC_{50}	1400	Boiling Point, °C	247
CD_{50}	5	mg-min/m^3 Ref.	
$L(Ct)_{50}$	8500	mg-min/m^3 Ref.	Clarke 1968
LD_{50}		mg/kg Ref.	
Appearance	white crystals, liquid, clear amber liquid		
Odor	apple blossom, aromatic odor		
Symptoms	burning skin, irritation, dizziness		
Antidote			
Decon.	CO2/water spray		
Solubility	slightly soluble in water		
Removal	May be slightly or partially removable by high efficiency filters		
First Aid	Symptoms are temporary and may dissipate within minutes or hours. Provide fresh air. Give oxygen if breathing difficulty.		
Remediation	absorb or cover with dry earth, sand or other non-combustible material and transfer to approved containers		
Notes			
References	Hersh 1968, Cookson 1969, Wright 1990, Clarke 1968, Somani 1992, Wright 1990, Miller 1999, CDC 2001, DOT 2000, Ellison 2000		
Dose Curve			

Dose Curve: Fatalities (%) vs Exposure Dose, mg-min/m^3. CD50 marked at approximately 8500.

CW Agent	CS		CS
Type	Tear Gas	CWC Schedule	
Formula	$ClC_6H_4CH=C(CN)_2$	Molecular Weight	189
Chem. Name	o-chlorobenzylidene malononitrile	HAZMAT ID#	2810
CAS #	2698-41-1	Boiling Point, °C	500
LC_{50}		mg/l Ref.	
CD_{50}	43500	mg-min/m^3 Ref.	
$L(Ct)_{50}$		mg-min/m^3 Ref.	
LD_{50}	8	mg/kg Ref.	
Appearance	white crystals, liquid, gas is heavier than air		
Odor	peppery		
Symptoms	stinging skin		
Antidote			
Decon.			
Solubility	insoluble in water		
Removal	May be slightly or partially removable by high efficiency filters		
First Aid	Symptoms are temporary and may dissipate within minutes or hours. No specific first aid needed.		
Remediation	absorb or cover with dry earth, sand or other non-combustible material and transfer to approved containers		
Notes			
References	Hersh 1968, Cookson 1969, Wright 1990, Lewis 1993, Clarke 1968, Miller 1999, DOT 2000		
Dose Curve			

CW Agent	Cyanogen chloride		CK
Type	Blood agent	CWC Schedule	III
Formula	CNCl	Molecular Weight	61.47
Chem. Name	chlorine cyanide, Mauguinite	HAZMAT ID#	1589
CAS #	506-77-4	Boiling Point, °C	12.5
LC_{50}		mg/l Ref.	
CD_{50}		mg-min/m^3 Ref.	
$L(Ct)_{50}$	3800	mg-min/m^3 Ref. Richardson 1992	
LD_{50}	20	mg/kg Ref. Richardson 1992	
Appearance	colorless compressed gas, gas is heavier than air		
Odor	pungent odor		
Symptoms	eye irritation		
Antidote	amyl nitrite, sodium nitrite, sodium thiosulfate		
Decon.	SNL Foam		
Solubility	soluble in water, alcohol, and ether		
Removal	Granular Activated Carbon removes with unknown efficiency		
First Aid	Inhalation: Provide fresh air and place victims in half-upright position. Skin exposure: Rinse skin with plenty of water or place victim in shower. Rinse eyes with plenty of water for several minutes.		
Remediation	Use sand or inert absorbent to contain or absorb spills. Control vapor cloud with fine spray of water.		
Notes	Reacts violently with oxidants forming toxic chlorine gas.		
References	Richardson 1992, CDC 2001, Lewis 1993, Paddle 1996, Miller 1999, Canada 1992, OPCW 1997, DOT 2000		

Dose Curve

CW Agent	Cyclohexyl Sarin		GF
Type	Nerve agent	CWC Schedule	
Formula	$C_7H_{14}FO_2P$	Molecular Weight	180.16
Chem. Name	cyclohexyl methylphosphonofluoridate	HAZMAT ID#	2810
CAS #		Boiling Point, $^\circ$C	
$L(Ct)_{50}$	35	mg-min/m^3 Ref.	NAS 1997
LD_{50}	5	mg/kg Ref.	NAS 997
Appearance	colorless liquid, gas is heavier than air		
Odor	odorless		
Symptoms	Breathing difficulty, convulsions, coma		
Antidote	atropine, diazepam (CANA), pralidoxime chloride (2 PAM Cl)		
Decon.	NaOH bleaches, DS2, phenol, ethanol, SNL Foam, Bleaching powder, Supertropical bleach, C8		
Removal			
First Aid	Use all three nerve agent antidote kits if symptoms are severe. Use Mark I injectors or atropine. If symptoms progress, use injectors at 5-20 minute intervals (maximum of three injections). Give artificial respiration if breathing stops. Use oxygen delivery if face is contaminated or if breathing is labored. Skin exposure: Immediately wash skin and clothes with solution of 5% sodium hypochlorite or common household liquid bleach. Remove clothing and flush skin with same solution, then wash areas with soap and water. Flush eyes with water for 10-15 minutes.		
Remediation	Cover with fine sand, diatomaceous earth, vermiculite, or clay. Neutralize with large quantities of aqueous sodium hydroxide solution. Place all material in suitable sealed containers. Decontaminate exterior of containers. Label and dispose according to applicable regulations. Alternate solution: calcium hypochlorite, DS2, supertropical bleach slurry (STB), in that order.		
Notes	aka CMPF		
References	CDC 2001, Cookson 1969, Miller 1999, Clarke 1968, DOT 2000		
Dose Curve			

CW Agent	Diethyl phosphite		
Type		CWC Schedule	III
Formula	$(C_2H_5O)_2HPO$	Molecular Weight	138.12
Chem. Name		HAZMAT ID#	
CAS #	762-04-9	Boiling Point, °C	187-188
LC_{50}		mg/l	Ref.
CD_{50}		mg-min/m^3	Ref.
$L(Ct)_{50}$		mg-min/m^3	Ref.
LD_{50}	3900	mg/kg Ref.	Canada 1993
Appearance	water-white liquid, gas is heavier than air		
Odor			
Symptoms			
Antidote			
Decon.			
Solubility	soluble in water & most organic solvents		
Removal	May be slightly or partially removable by high efficiency filters		
First Aid			
Remediation	Use sand or inert absorbent to contain or absorb spills. Control vapor cloud with fine spray of water.		
Notes	combustible, a precursor agent		
References	Canada 1993, Lewis 1993, Canada 1992, OPCW 1997		

Dose

Curve

CW Agent	Dimethyl phosphite		
Type		CWC Schedule	III
Formula	$(CH_3O)_2P(O)H$	Molecular Weight	110.05
Chem. Name		HAZMAT ID#	
CAS #	868-85-9	Boiling Point, °C	170-171
LC_{50}		mg/l Ref.	
CD_{50}		mg-min/m³ Ref.	
$L(Ct)_{50}$		mg-min/m³ Ref.	
LD_{50}	3050	mg/kg Ref.	Canada 1993
Appearance	colorless liquid, gas is heavier than air		
Odor	mild odor		
Symptoms			
Antidote			
Decon.			
Solubility	soluble in water, many organic solvents		
Removal	May be slightly or partially removable by high efficiency filters		
First Aid			
Remediation	Use sand or inert absorbent to contain or absorb spills. Control vapor cloud with fine spray of water.		
Notes	combustible, a precursor agent		
References	Canada 1993, Canada 1992, OPCW 1997		
Dose Curve			

CW Agent	Diphenylchloroarsine		DA
Type	Vomiting agent	**CWC Schedule**	
Formula	$C_7H_{14}FO_2P$	**Molecular Weight**	264.5
Chem. Name	diphenylchloroarsine	**HAZMAT ID#**	2810
CAS #	712-48-1	**Boiling Point, °C**	246
LC_{50}	1500	mg/l **Ref.**	Ellison 2000
CD_{50}		mg-min/m^3 **Ref.**	
$L(Ct)_{50}$	15000	mg-min/m^3 **Ref.**	Somani 1992
LD_{50}		mg/kg **Ref.**	
Appearance	colorless crystals/dark brown liquid		
Odor	none		
Symptoms	strong irritant to skin and eyes, vomiting agent		
Antidote			
Decon.			
Removal	May be slightly or partially removable by high efficiency filters		
First Aid	Use all three nerve agent antidote kits if symptoms are severe. Use Mark I injectors or atropine. If symptoms progress, use injectors at 5-20 minute intervals (maximum of three injections). Give artificial respiration if breathing stops. Use oxygen delivery if face is contaminated or if breathing is labored. Skin exposure: Immediately wash skin and clothes with solution of 5% sodium hypochlorite or common household liquid bleach. Remove clothing and flush skin with same solution, then wash areas with soap and water. Flush eyes with water for 10-15 minutes.		
Remediation	Use sand or inert absorbent to contain or absorb spills. Control vapor cloud with fine spray of water.		
Notes			
References	Lewis 1993, DOT 2000, Ellison 2000		

Dose Curve

CW Agent	Diphenylcyanoarsine		DC
Type	Vomiting agent	CWC Schedule	
Formula	$(C_6H_5)_2AsCN$	Molecular Weight	255
Chem. Name		HAZMAT ID#	2810
CAS #	23525-22-6	Boiling Point, °C	333
LC_{50}	30	mg/l Ref.	
CD_{50}		mg-min/m³ Ref.	
$L(Ct)_{50}$	30	mg-min/m³ Ref.	Somani 1992
LD_{50}		mg/kg Ref.	
Appearance	gas is heavier than air		
Odor	similar to garlic or bitter almonds		
Symptoms	vomiting, nausea		
Antidote			
Decon.			
Solubility	soluble in CCl4, almost insoluble in water		
Removal	May be slightly or partially removable by high efficiency filters		
First Aid	move victim to fresh air, provide oxygen if breathing difficulty, flush skin and eyes with water for 20 mins		
Remediation	Use sand or inert absorbent to contain or absorb spills. Control vapor cloud with fine spray of water.		
Notes			
References	Cookson 1969, Somani 1992, CDC 2001, DOT 2000, Ellison 2000		
Dose Curve			

CW Agent	Hydrogen cyanide		AC
Type	Blood agent	CWC Schedule	III
Formula	HCN	Molecular Weight	27.03
Chem. Name	prussic acid	HAZMAT ID#	1614, 3294
CAS #	74-90-8	Boiling Point, °C	25.7
LC_{50}	200	mg/l **Ref.**	
CD_{50}		mg-min/m^3 **Ref.**	
$L(Ct)_{50}$	5070	mg-min/m^3 **Ref.**	Richardson 1992
LD_{50}	50	mg/kg **Ref.**	Canada 1993
Appearance	colorless liquid		
Odor	bitter almonds		
Symptoms	Confusion, nausea, shortness of breath		
Antidote			
Decon.			
Solubility	soluble in ether, miscible in water, alcohol		
Removal	Granular Activated Carbon removes with zinc oxide impregnation		
First Aid	Inhalation exposure: Provide oxygen if breathing difficulty. Give artificial respiration if breathing has stopped. Give mouth to mouth resuscitation only if no facial contamination. Skin exposure: Immediately wash skin and clothes with water. Remove clothing and flush skin with water. Flush eyes with water.		
Remediation	Use sand or inert absorbent to contain or absorb spills. Control vapor cloud with fine spray of water.		
Notes	flammable		
References	Richardson 1992, CDC 2001, Lewis 1993, Clarke 1968, Paddle 1996, Wright 1990, Miller 1999, Canada 1992, OPCW 1997, DOT 2000		
Dose Curve			

CW Agent	Lewisite		L
Type	Vesicant/Blister agent	CWC Schedule	I
Formula	$C_2H_2AsCl_3$	Molecular Weight	207.32
Chem. Name	2-chlorovinyldichloroarsine	HAZMAT ID#	2810
CAS #	54-25-3, 40334-69-8, 40334-70-1	Boiling Point, °C	190
$L(Ct)_{50}$	1200	mg-min/m^3 Ref.	Somani 1992
LD_{50}	38	mg/kg Ref.	Richardson 1992
Appearance	amber to dark brown liquid, gas is heavier than air		
Odor	geraniums		
Symptoms	blindness, destroys lung tissue		
Antidote			
Decon.	Neutralized by British anti-Lewisite		
Solubility	soluble in organic solvents, oils, and alcohol		
Removal	May be slightly or partially removable by high efficiency filters		
First Aid	Inhalation exposure: Provide oxygen if breathing difficulty. Give artificial respiration if breathing has stopped. Give mouth to mouth resuscitation only if no facial contamination. Skin exposure: Immediately wash skin and clothes with solution of 5% sodium hypochlorite or common household liquid bleach. Remove clothing and flush skin with same solution, then wash areas with soap and water. Flush eyes with water.		
Remediation	Cover with fine sand, diatomaceous earth, vermiculite, or clay. Neutralize with large quantities of 5.25% solution of sodium hypochlorite. Place all material in suitable sealed containers. Decontaminate exterior of containers. Label and dispose according to applicable regulations. Alternate solution: calcium hypochlorite, DS2, supertropical bleach slurry (STB), in that order.		
Notes	L-1, L-2, L-3, HL (when mixed with Mustard)		
References	Richardson 1992, CDC 2001, NRC 1999, Somani 1992, Thomas 1970, Wright 1990, Miller 1999, Canada 1992, OPCW 1997, DOT 2000		
Dose Curve			

CW Agent	Methyldiethanolamine		
Type		CWC Schedule	III
Formula	$CH_3N(C_2H_4OH)_2$	Molecular Weight	119.2
Chem. Name	MDEA	HAZMAT ID#	
CAS #	105-59-9	Boiling Point, °C	247.2
LC_{50}		mg/l Ref.	
CD_{50}		mg-min/m³ Ref.	
$L(Ct)_{50}$		mg-min/m³ Ref.	
LD_{50}	4780	mg/kg Ref.	Canada 1993
Appearance	colorless liquid, gas is heavier than air		
Odor	amine-like odor		
Symptoms			
Antidote			
Decon.			
Solubility	miscible with benzene and water		
Removal	May be slightly or partially removable by high efficiency filters		
First Aid			
Remediation	Use sand or inert absorbent to contain or absorb spills. Control vapor cloud with fine spray of water.		
Notes	combustible, a precursor agent		
References	Canada 1993, Lewis 1993, OPCW 1997		

Dose Curve

CW Agent	Mustard gas		H
Type	Vesicant/Blister agent	CWC Schedule	
Formula	$S(CH_2CH_2Cl)_2$	Molecular Weight	207.32
Chem. Name	dichlorodiethyl sulfide	HAZMAT ID#	2810
CAS #	505-60-2	Boiling Point, °C	190
$L(Ct)_{50}$	1500	mg-min/m^3 Ref.	Wald 1970
LD_{50}	40	mg/kg Ref.	Richardson 1992
Appearance	yellow liquid, gas is heavier than air		
Odor	burning garlic		
Symptoms	severe irritation, coughing, stinging eyes		
Antidote	Dimercaprol (British anti-lewisite (BAL)) ointment		
Decon.	chlorines or bleach, chloramine, DS2		
Solubility	soluble in fats, oils, gasoline, kerosene, acetone, CCl4, alcohol		
Removal	May be slightly or partially removable by high efficiency filters		
First Aid	Inhalation exposure: Provide oxygen if breathing difficulty. Give artificial respiration if breathing has stopped. Give mouth to mouth resuscitation only if no facial contamination. Skin exposure: Immediately wash skin and clothes with solution of 5% sodium hypochlorite or common household liquid bleach. Remove clothing and flush skin with same solution, then wash areas with soap and water. Flush eyes with water.		
Remediation	Cover with fine sand, diatomaceous earth, vermiculite, or clay. Neutralize with large quantities of 5.25% solution of sodium hypochlorite. Place all material in suitable sealed containers. Decontaminate exterior of containers. Label and dispose according to applicable regulations. Alternate solution: calcium hypochlorite, DS2, supertropical bleach slurry (STB), in that order.		
Notes	Also HS, a sulfur mustard		
References	Richardson 1992, CDC 2001, Somani 1992, Thomas 1970, Wright 1990, Miller 1999, OPCW 1997, DOT 2000		
Dose Curve			

Dose curve: Mustard Gas. Fatalities (%) vs Exposure Dose, mg-min/m^3. LD50 marked at approximately 1500.

CW Agent	Nerve Gas DFP		DFP
Type	Vesicant/Blister agent	CWC Schedule	
Formula	[(CH)$_2$CHO]POF	Molecular Weight	121
Chem. Name	di-isopropyl fluorophosphate	HAZMAT ID#	
CAS #	55-91-4	Boiling Point, °C	
L(Ct)$_{50}$		mg-min/m^3 Ref.	
LD$_{50}$	1	mg/kg Ref.	
Appearance	colorless liquid		
Odor	odorless		
Symptoms	blurred vision, sweating, vomiting, convulsions		
Antidote	atropine sulfate, pralidoxime iodide, Dimercaprol (BAL) ointment		
Decon.	NaOH bleaches, DS2, phenol, ethanol, SNL Foam, Bleaching powder, Supertropical bleach		
Solubility	soluble in organic solvents		
Removal	Granular Activated Carbon removes with heavy metal salt impregnation		
First Aid	Inhalation exposure: Provide oxygen if breathing difficulty. Give artificial respiration if breathing has stopped. Give mouth to mouth resuscitation only if no facial contamination. Skin exposure: Immediately wash skin and clothes with solution of 5% sodium hypochlorite or common household liquid bleach. Remove clothing and flush skin with same solution, then wash areas with soap and water. Flush eyes with water.		
Remediation	Cover with fine sand, diatomaceous earth, vermiculite, or clay. Neutralize with large quantities of 5.25% solution of sodium hypochlorite. Place all material in suitable sealed containers. Decontaminate exterior of containers. Label and dispose according to applicable regulations. Alternate solution: calcium hypochlorite, DS2, supertropical bleach slurry (STB), in that order.		
Notes	aka isofluorophate, also used as an insecticide, a cholinesterase inhibitor		
References	Clarke 1968, Lewis 1993		
Dose Curve			

Dose Curve:

Fatalities (%) vs Exposure Dose, mg-min/m^3. LD50 marked near 2200 mg-min/m^3 at ~48% fatalities.

CW Agent	Nitrogen mustard		HN-1
Type	Vesicant/Blister agent	CWC Schedule	I
Formula	$C_6H_{13}Cl_2N$	Molecular Weight	170.08
Chem. Name	N,N-dialkyl phosphoramidic dihalides	HAZMAT ID#	2810
CAS #	538-07-8	Boiling Point, °C	194
$L(Ct)_{50}$	1500	mg-min/m^3 Ref.	Canada 1993
LD_{50}	15	mg/kg Ref.	
Appearance	colorless to pale yellow liquid, gas is heavier than air		
Odor	fish		
Symptoms	eye irritation, rashes, blisters		
Antidote	Dimercaprol (British anti-lewisite (BAL)) ointment		
Decon.	Bleaching powder, Supertropical bleach, SNL Foam, chloramine, DS2		
Solubility	soluble in fats, oils, gasoline, kerosene, acetone, CCl4, alcohol		
Removal	May be slightly or partially removable by high efficiency filters		
First Aid	Inhalation exposure: Provide oxygen if breathing difficulty. Give artificial respiration if breathing has stopped. Give mouth to mouth resuscitation only if no facial contamination. Skin exposure: Immediately wash skin and clothes with solution of 5% sodium hypochlorite or common household liquid bleach. Remove clothing and flush skin with same solution, then wash areas with soap and water. Flush eyes with water.		
Remediation	Cover with fine sand, diatomaceous earth, vermiculite, or clay. Neutralize with large quantities of 5.25% solution of sodium hypochlorite. Place all material in suitable sealed containers. Decontaminate exterior of containers. Label and dispose according to applicable regulations. Alternate solution: calcium hypochlorite, DS2, supertropical bleach slurry (STB), in that order.		
Notes	Hydrolizes slowly in water		
References	Richardson 1992, CDC 2001, Somani 1992, Paddle 1996, Wright 1990, Canada 1992, OPCW 1997, DOT 2000		
Dose Curve			

CW Agent	Nitrogen mustard		HN-2
Type	Vesicant/Blister agent	CWC Schedule	I
Formula	$C_5H_{11}Cl_2N$	Molecular Weight	156.07
Chem. Name	N,N-dialkylaminoethane-2-OLS	HAZMAT ID#	2810
CAS #	51-75-2	Boiling Point, °C	87
$L(Ct)_{50}$	3000	mg-min/m^3 Ref.	Canada 1993
LD_{50}	2	mg/kg Ref.	
Appearance	amber liquid, gas is heavier than air		
Odor	herring		
Symptoms	eye irritation, rashes, blisters		
Antidote	Dimercaprol (British anti-lewisite (BAL)) ointment		
Decon.	Bleaching powder, Supertropical bleach, SNL Foam, chloramine, DS3		
Solubility	soluble in fats, oils, gasoline, kerosene, acetone, CCl4, alcohol		
Removal	May be slightly or partially removable by high efficiency filters		
First Aid	Inhalation exposure: Provide oxygen if breathing difficulty. Give artificial respiration if breathing has stopped. Give mouth to mouth resuscitation only if no facial contamination. Skin exposure: Immediately wash skin and clothes with solution of 5% sodium hypochlorite or common household liquid bleach. Remove clothing and flush skin with same solution, then wash areas with soap and water. Flush eyes with water.		
Remediation	Cover with fine sand, diatomaceous earth, vermiculite, or clay. Neutralize with large quantities of 5.25% solution of sodium hypochlorite. Place all material in suitable sealed containers. Decontaminate exterior of containers. Label and dispose according to applicable regulations. Alternate solution: calcium hypochlorite, DS2, supertropical bleach slurry (STB), in that order.		
Notes	Hydrolizes slowly in water		
References	Richardson 1992, CDC 2001, Somani 1992, Thomas 1970, Wright 1990, Canada 1992, OPCW 1997, DOT 2000		

Dose Curve

CW Agent	Nitrogen mustard		HN-3
Type	Vesicant/Blister agent	**CWC Schedule**	I
Formula	$C_6H_{12}Cl_3N$	**Molecular Weight**	204.54
Chem. Name	N,N-dialkylaminoethane-2-THIOLS	**HAZMAT ID#**	2810
CAS #	555-77-1	**Boiling Point, °C**	256
$L(Ct)_{50}$	1000	mg-min/m³ **Ref.**	Clarke 1968
LD_{50}		mg/kg **Ref.**	
Appearance	colorless to pale yellow liquid, gas is heavier than air		
Odor	fish		
Symptoms	eye irritation, lung damage		
Antidote	Dimercaprol (British anti-lewisite (BAL)) ointment		
Decon.	Bleaching powder, Supertropical bleach, SNL Foam, chloramine, DS4		
Solubility	soluble in fats, oils, gasoline, kerosene, acetone, CCl4, alcohol		
Removal	May be slightly or partially removable by high efficiency filters		
First Aid	Inhalation exposure: Provide oxygen if breathing difficulty. Give artificial respiration if breathing has stopped. Give mouth to mouth resuscitation only if no facial contamination. Skin exposure: Immediately wash skin and clothes with solution of 5% sodium hypochlorite or common household liquid bleach. Remove clothing and flush skin with same solution, then wash areas with soap and water. Flush eyes with water.		
Remediation	Cover with fine sand, diatomaceous earth, vermiculite, or clay. Neutralize with large quantities of 5.25% solution of sodium hypochlorite. Place all material in suitable sealed containers. Decontaminate exterior of containers. Label and dispose according to applicable regulations. Alternate solution: calcium hypochlorite, DS2, supertropical bleach slurry (STB), in that order.		
Notes	Hydrolizes slowly in water		
References	Richardson 1992, CDC 2001, Clarke 1968, Somani 1992, Paddle 1996, Wright 1990, Canada 1992, OPCW 1997, DOT 2000		
Dose Curve			

CW Agent	O-Mustard			T
Type	Vesicant/Blister agent	**CWC Schedule**		I
Formula	$(ClCH_2CH_2SCH_2CH_2)O_2$	**Molecular Weight**		263.3
Chem. Name	bis(2-chloroethylthioethyl)ether	**HAZMAT ID#**		2810
CAS #	63918-89-8	**Boiling Point, °C**		120
L(Ct)$_{50}$	400	mg-min/m^3	**Ref.**	Clarke 1968
LD$_{50}$		mg/kg	**Ref.**	
Appearance	oily liquid, gas is heavier than air			
Odor	none			
Symptoms	blisters, irritation, redness			
Antidote	Dimercaprol (British anti-lewisite (BAL)) ointment			
Decon.	Bleaching powder, Supertropical bleach, SNL Foam, chloramine			
Solubility	soluble in fats, oils, gasoline, kerosene, acetone, CCl4, alcohol			
Removal	May be slightly or partially removable by high efficiency filters			
First Aid	Inhalation exposure: Provide oxygen if breathing difficulty. Give artificial respiration if breathing has stopped. Give mouth to mouth resuscitation only if no facial contamination. Skin exposure: Immediately wash skin and clothes with solution of 5% sodium hypochlorite or common household liquid bleach. Remove clothing and flush skin with same solution, then wash areas with soap and water. Flush eyes with water.			
Remediation	Cover with fine sand, diatomaceous earth, vermiculite, or clay. Neutralize with large quantities of 5.25% solution of sodium hypochlorite. Place all material in suitable sealed containers. Decontaminate exterior of containers. Label and dispose according to applicable regulations. Alternate solution: calcium hypochlorite, DS2, supertropical bleach slurry (STB), in that order.			
Notes	a sulfur mustard			
References	Cookson 1969, Clarke 1968, OPCW 1997, DOT 2000, Ellison 2000			
Dose Curve				

CW Agent	Perfluoroisobutylene		PFIB
Type	Choking agent	CWC Schedule	II
Formula	C_4F_8	Molecular Weight	200
Chem. Name	Octafluoroisobutylene	HAZMAT ID#	
CAS #	382-21-8	Boiling Point, °C	
LC_{50}		mg/l Ref.	
CD_{50}		mg-min/m^3 Ref.	
$L(Ct)_{50}$	5000	mg-min/m^3 Ref.	
LD_{50}	50	mg/kg Ref.	
Appearance	colorless gas, gas is heavier than air		
Odor			
Symptoms	cough, shortness of breath, pain		
Antidote			
Decon.			
Solubility			
Removal			
First Aid	Inhalation: Provide fresh air and place victims in half-upright position. Skin exposure: Rinse skin with plenty of water or place victim in shower. Rinse eyes with plenty of water for several minutes.		
Remediation	Use sand or inert absorbent to contain or absorb spills. Control vapor cloud with fine spray of water.		
Notes			
References	CDC 2001, Paddle 1996, Canada 1992, OPCW 1997		

Dose Curve

CW Agent	Phosgene		CG
Type	Choking agent	CWC Schedule	III
Formula	$COCl_2$	Molecular Weight	98.92
Chem. Name	carbonyl chloride	HAZMAT ID#	1076
CAS #	75-4-5	Boiling Point, °C	8.3
LC_{50}	500	mg/l	Ref.
CD_{50}		mg-min/m³	Ref.
$L(Ct)_{50}$	3200	mg-min/m³ Ref.	Richardson 1992
LD_{50}	660	mg/kg	Ref.
Appearance	colorless gas, gas is heavier than air		
Odor	fresh hay		
Symptoms	asphyxia		
Antidote			
Decon.	SNL Foam, sodium bicarbonate, soda sh and slaked lime		
Solubility	soluble in benzene and toulene		
Removal	Granular Activated Carbon removes with 10-25% efficiency		
First Aid	Inhalation: Provide fresh air and place victims in half-upright position. Skin exposure: Rinse skin with plenty of water or place victim in shower. Rinse eyes with plenty of water for several minutes.		
Remediation	Neutralize spilled liquid with sodium bicarbonate or equal mixture of soda ash and slaked lime.		
Notes	non-combustible, decomposes on heating to produce the toxic gases hydrogen chloride, carbon monoxide, and chlorine fumes. Reacts violently with strong oxidants, including amines and aluminum.		
References	Richardson 1992, CDC 2001, Lewis 1993, Somani 1992, Paddle 1996, Wright 1990, Miller 1999, Canada 1992, OPCW 1997, DOT 2000		

Dose Curve

CW Agent	Phosgene oxime		CX
Type	Vesicant/Blister agent	CWC Schedule	
Formula	CHCl$_2$NO	Molecular Weight	113.93
Chem. Name	dichloroformoxime	HAZMAT ID#	2811
CAS #	1794-86-1	Boiling Point, °C	-
L(Ct)$_{50}$	1500	mg-min/m^3 Ref.	
LD$_{50}$	25	mg/kg Ref.	
Appearance	yellow-brown liquid, crystal powder, gas is heavier than air		
Odor	disagreeable, penetrating odor		
Symptoms	pain and irritation, inflammation of skin		
Antidote	Dimercaprol (BAL) ointment , M291 Skin Decontaminating Kit		
Decon.			
Solubility	70% soluble in water, alcohol, ether, and benzene		
Removal			
First Aid	Inhalation exposure: Provide oxygen if breathing difficulty. Give artificial respiration if breathing has stopped. Give mouth to mouth resuscitation only if no facial contamination. Skin exposure: Immediately wash skin and clothes with solution of 5% sodium hypochlorite or common household liquid bleach. Remove clothing and flush skin with same solution, then wash areas with soap and water. Flush eyes with water.		
Remediation	Cover with fine sand, diatomaceous earth, vermiculite, or clay. Neutralize with large quantities of 5.25% solution of sodium hypochlorite. Place all material in suitable sealed containers. Decontaminate exterior of containers. Label and dispose according to applicable regulations. Alternate solution: calcium hypochlorite, DS2, supertropical bleach slurry (STB), in that order.		
Notes			
References	Miller 1999, Thomas 1970, CDC 2001, DOT 2000		
Dose Curve			

CW Agent	Phosphorus oxychloride		
Type		CWC Schedule	III
Formula	$POCl_3$	Molecular Weight	153.33
Chem. Name	phosphoryl chloride	HAZMAT ID#	1810
CAS #	10025-87-3	Boiling Point, °C	107.2
$L(Ct)_{50}$		mg-min/m^3 Ref.	
LD_{50}	380	mg/kg Ref.	Canada 1993
Appearance	colorless liquid, gas is heavier than air		
Odor	pungent odor		
Symptoms	eye and lung irritant		
Antidote	Dimercaprol (British anti-lewisite (BAL)) ointment		
Decon.	SNL Foam, decomposed by water and alcohol in heat		
Solubility	hot water and alcohol		
Removal	May be slightly or partially removable by high efficiency filters		
First Aid	Inhalation exposure: Provide oxygen if breathing difficulty. Give artificial respiration if breathing has stopped. Give mouth to mouth resuscitation only if no facial contamination. Skin exposure: Immediately wash skin and clothes with solution of 5% sodium hypochlorite or common household liquid bleach. Remove clothing and flush skin with same solution, then wash areas with soap and water. Flush eyes with water.		
Remediation	Cover with fine sand, diatomaceous earth, vermiculite, or clay. Neutralize with large quantities of 5.25% solution of sodium hypochlorite. Place all material in suitable sealed containers. Decontaminate exterior of containers. Label and dispose according to applicable regulations. Alternate solution: calcium hypochlorite, DS2, supertropical bleach slurry (STB), in that order.		
Notes	precursor agent, reacts with water to produce toxic HCl gas		
References	Richardson 1992, Lewis 1993, Canada 1992, OPCW 1997, DOT 2000		
Dose

Curve | | | |

Dose Curve figure: Fatalities (%) vs Exposure Dose, mg-min/m^3, with LD50 marked.

CW Agent	Phosphorus pentachloride		
Type		CWC Schedule	III
Formula	PCl$_5$	Molecular Weight	208.24
Chem. Name	phosphoric chloride	HAZMAT ID#	1806
CAS #	1026-13-8	Boiling Point, °C	160-165
L(Ct)$_{50}$		mg-min/m^3 Ref.	
LD$_{50}$	660	mg/kg Ref.	Canada 1993
Appearance	slightly yellow & crystalline, liquid/solid, gas is heavier than air		
Odor	irritating odor		
Symptoms	eye and lung irritant		
Antidote	Dimercaprol (British anti-lewisite (BAL)) ointment		
Decon.	SNL Foam		
Solubility	soluble in carbon disulfide and CCl5		
Removal	May be slightly or partially removable by high efficiency filters		
First Aid	Inhalation exposure: Provide oxygen if breathing difficulty. Give artificial respiration if breathing has stopped. Give mouth to mouth resuscitation only if no facial contamination. Skin exposure: Immediately wash skin and clothes with solution of 5% sodium hypochlorite or common household liquid bleach. Remove clothing and flush skin with same solution, then wash areas with soap and water. Flush eyes with water.		
Remediation	Cover with fine sand, diatomaceous earth, vermiculite, or clay. Neutralize with large quantities of 5.25% solution of sodium hypochlorite. Place all material in suitable sealed containers. Decontaminate exterior of containers. Label and dispose according to applicable regulations. Alternate solution: calcium hypochlorite, DS2, supertropical bleach slurry (STB), in that order.		
Notes	flammable, reacts with water to produce HCl gas, a precursor agent		
References	Richardson 1992, Lewis 1993, Canada 1992, OPCW 1997, DOT 2000		
Dose Curve			

CW Agent	Phosphorus trichloride		
Type		CWC Schedule	III
Formula	PCl_3	Molecular Weight	137.33
Chem. Name	phosphorus chloride	HAZMAT ID#	1809
CAS #	7719-12-2	Boiling Point, °C	76
$L(Ct)_{50}$		mg-min/m^3 Ref.	
LD_{50}	550	mg/kg Ref.	Canada 1993
Appearance	colorless fuming liquid, gas is heavier than air, odorless		
Symptoms	eye and lung irritant		
Antidote	Dimercaprol (British anti-lewisite (BAL)) ointment		
Decon.	SNL Foam, decomposes in moist air		
Solubility	soluble in ether, benzene, carbon disulfide, & CCl5		
Removal	May be slightly or partially removable by high efficiency filters		
First Aid	Inhalation exposure: Provide oxygen if breathing difficulty. Give artificial respiration if breathing has stopped. Give mouth to mouth resuscitation only if no facial contamination. Skin exposure: Immediately wash skin and clothes with solution of 5% sodium hypochlorite or common household liquid bleach. Remove clothing and flush skin with same solution, then wash areas with soap and water. Flush eyes with water.		
Remediation	Cover with fine sand, diatomaceous earth, vermiculite, or clay. Neutralize with large quantities of 5.25% solution of sodium hypochlorite. Place all material in suitable sealed containers. Decontaminate exterior of containers. Label and dispose according to applicable regulations. Alternate solution: calcium hypochlorite, DS2, supertropical bleach slurry (STB), in that order.		
Notes	a precursor agent, reacts with water to produce toxic HCl gas		
References	Richardson 1992, Lewis 1993, SIPRI 1971, Canada 1992, OPCW 1997, DOT 2000		
Dose Curve			

CW Agent	Sarin		GB
Type	Nerve agent	CWC Schedule	I
Formula	$C_4H_{10}FO_2P$	Molecular Weight	140.11
Chem. Name	methylphosphonofluoride acid	HAZMAT ID#	2810
CAS #	107-44-8	Boiling Point, °C	147
CD_{50}	35	mg-min/m^3 Ref.	Canada 1993
$L(Ct)_{50}$	100	mg-min/m^3 Ref.	Richardson 1992
LD_{50}	0.108	mg/kg Ref.	Richardson 1992
Appearance	colorless liquid, gas is heavier than air		
Odor	odorless		
Symptoms	blurred vision, sweating, vomiting, convulsions		
Antidote	atropine, diazepam (CANA), pralidoxime chloride (2 PAM Cl)		
Decon.	NaOH bleaches, DS2, phenol, ethanol, SNL Foam, Bleaching powder, Supertropical bleach, C8		
Solubility	soluble in organic solvents, half life in water 24-46 hours at 20 C		
First Aid	Use all three nerve agent antidote kits if symptoms are severe. Use Mark I injectors or atropine. If symptoms progress, use injectors at 5-20 minute intervals (maximum of three injections). Give artificial respiration if breathing stops. Use oxygen delivery if face is contaminated or if breathing is labored. Skin exposure: Immediately wash skin and clothes with solution of 5% sodium hypochlorite or common household liquid bleach. Remove clothing and flush skin with same solution, then wash areas with soap and water. Flush eyes		
Remediation	Cover with fine sand, diatomaceous earth,sponges, paper towels, cloth, vermiculite, or clay. Neutralize with large quantities of 5.25% solution of sodium hypochlorite. Place all material in suitable sealed containers. Decontaminate exterior of containers. Label and dispose according to applicable regulations. Alternate solution: calcium hypochlorite, DS2, supertropical bleach slurry (STB), in that order.		
Notes	nerve agent absorbed by skin, aka G-agent		
References	Richardson 1992, CDC 2001, Lewis 1993, NRC 1999, Somani 1992, NATO 1996, Miller 1999, OPCW 1997, DOT 2000		
Dose Curve			

CW Agent	Sesqui mustard		Q
Type	Vesicant/Blister agent	CWC Schedule	I
Formula	$C_4H_8Cl_2S$	Molecular Weight	159
Chem. Name	1,2-bis(2-chloroethylthio)ethane	HAZMAT ID#	2810
CAS #	3563-36-8	Boiling Point, °C	353
$L(Ct)_{50}$	300	mg-min/m^3 **Ref.**	Clarke 1968
LD_{50}		mg/kg **Ref.**	
Appearance	colorless, oily liquid, gas is heavier than air		
Odor	garlic		
Symptoms	blindness, ulceration of skin		
Antidote	Dimercaprol (British anti-lewisite (BAL)) ointment		
Decon.	Bleaching powder, Supertropical bleach, SNL Foam		
Solubility			
Removal	May be slightly or partially removable by high efficiency filters		
First Aid	Inhalation exposure: Provide oxygen if breathing difficulty. Give artificial respiration if breathing has stopped. Give mouth to mouth resuscitation only if no facial contamination. Skin exposure: Immediately wash skin and clothes with solution of 5% sodium hypochlorite or common household liquid bleach. Remove clothing and flush skin with same solution, then wash areas with soap and water. Flush eyes with water.		
Remediation	Cover with fine sand, diatomaceous earth,sponges, paper towels, cloth, vermiculite, or clay. Neutralize with large quantities of 5.25% solution of sodium hypochlorite. Place all material in suitable sealed containers. Decontaminate exterior of containers. Label and dispose according to applicable regulations. Alternate solution: calcium hypochlorite, DS2, supertropical bleach slurry (STB), in that order.		
Notes	a sulfur mustard		
References	Cookson 1969, Clarke 1968, CDC 2001, OPCW 1997, DOT 2000		

Dose Curve

CW Agent	Soman		GD
Type	Nerve agent	CWC Schedule	I
Formula	$C_7H_{16}FO_2P$	Molecular Weight	182
Chem. Name	methylphosphonofluoridic acid 1,2,2-trimethylpropyl ester		
CAS #	96-64-0	Boiling Point, °C	167
CD_{50}	35	mg-min/m^3 **Ref.**	Canada 1993
$L(Ct)_{50}$	70	mg-min/m^3 **Ref.**	NAS 1997
LD_{50}	0.01	mg/kg **Ref.**	Lewis 1993
Appearance	colorless liquid, gas is heavier than air		
Odor	odorless	HAZMAT ID#	2810
Symptoms	severe congestion, sweating, twitching		
Antidote	atropine, diazepam (CANA), pralidoxime chloride (2 PAM Cl)		
Decon.	Bleaches, (NaOCl of Ca(OCl)2), DS2, SNL Foam, Bleaching powder, Supertropical bleach, C8		
Solubility	soluble in gasoline, alcohols, fats, and oils, half-life in water 82 hours		
First Aid	Use all three nerve agent antidote kits if symptoms are severe. Use Mark I injectors or atropine. If symptoms progress, use injectors at 5-20 minute intervals (maximum of three injections). Give artificial respiration if breathing stops. Use oxygen delivery if face is contaminated or if breathing is labored. Skin exposure: Immediately wash skin and clothes with solution of 5% sodium hypochlorite or common household liquid bleach. Remove clothing and flush skin with same solution, then wash areas with soap and water. Flush eyes with water for 10-15 minutes.		
Remediation	Cover with fine sand, diatomaceous earth, sponges, paper towels, cloth, vermiculite, or clay. Neutralize with large quantities of 5.25% solution of sodium hypochlorite. Place all material in suitable sealed containers. Decontaminate exterior of containers. Label and dispose according to applicable regulations. Alternate solution: calcium hypochlorite, DS2, supertropical bleach slurry (STB), in that order.		
Notes	absorbed by skin, aka G-agent		
References	CDC 2001, Lewis 1993, NRC 1999, Somani 1992, Paddle 1996, Wright 1990, NATO 1996, Miller 1999, OPCW 1997, DOT 2000		
Dose Curve			

CW Agent	Sulfur Mustard (Distilled)		HD
Type	Vesicant/Blister agent	CWC Schedule	I
Formula	$C_4H_8Cl_2S$	Molecular Weight	159.08
Chem. Name	distilled mustard	HAZMAT ID#	2810
CAS #	39472-40-7	Boiling Point, °C	218
$L(Ct)_{50}$	1000	mg-min/m³ Ref.	Clarke 1968
LD_{50}	20	mg/kg Ref.	NAS 1997
Appearance	colorless, oily liquid, gas is heavier than air		
Odor	faint garlic		
Symptoms	blindness, ulceration of skin		
Antidote	Dimercaprol (British anti-lewisite (BAL)) ointment		
Decon.	Common bleach and chloramine, SNL Foam, Bleaching powder, Supertropical bleach, DS2, C8		
Solubility	soluble in fats, oils, gasoline, kerosene, acetone, CCl4, alcohol		
First Aid	Inhalation exposure: Provide oxygen if breathing difficulty. Give artificial respiration if breathing has stopped. Give mouth to mouth resuscitation only if no facial contamination. Skin exposure: Immediately wash skin and clothes with solution of 5% sodium hypochlorite or common household liquid bleach. Remove clothing and flush skin with same solution, then wash areas with soap and water. Flush eyes with water.		
Remediation	Cover with fine sand, diatomaceous earth, sponges, paper towels, cloth, vermiculite, or clay. Neutralize with large quantities of 5.25% solution of sodium hypochlorite. Place all material in suitable sealed containers. Decontaminate exterior of containers. Label and dispose according to applicable regulations. Alternate solution: calcium hypochlorite, DS2, supertropical bleach slurry (STB), in that order.		
Notes	Half-life in water 7.5-110 minutes at 25 C		
References	CDC 2001, Lewis 1993, NRC 1999, Paddle 1996, DOT 2000		

Dose

Curve

CW Agent	Tabun		GA
Type	Nerve agent	CWC Schedule	I
Formula	$C_5H_{11}N_2O_2P$	Molecular Weight	162.13
Chem. Name	dimethylphosphoramidocyanidic acid, ethyl ester		
CAS #	7-81-6	Boiling Point, °C	246
CD_{50}	300	mg-min/m³ Ref.	Canada 1993
$L(Ct)_{50}$	135	mg-min/m³ Ref.	NAS 1997
LD_{50}	0.01	mg/kg Ref.	Lewis 1993
Appearance	colorless/dark brown liquid, gas is heavier than air		
Odor	bitter almonds	HAZMAT ID#	2810
Symptoms	vision blurs, severe congestion, convulsions		
Antidote	atropine, diazepam (CANA), pralidoxime chloride (2 PAM Cl)		
Decon.	Bleaching powder, DS2, phenol, aqueous NaOH, SNL Foam, STB, C8		
Solubility	partially soluble in water, half-life 7-8.5 hours in water at 20 C		
First Aid	Use all three nerve agent antidote kits if symptoms are severe. Use Mark I injectors or atropine. If symptoms progress, use injectors at 5-20 minute intervals (maximum of three injections). Give artificial respiration if breathing stops. Use oxygen delivery if face is contaminated or if breathing is labored. Skin exposure: Immediately wash skin and clothes with solution of 5% sodium hypochlorite or common household liquid bleach. Remove clothing and flush skin with same solution, then wash areas with soap and water. Flush eyes with water for 10-15 minutes.		
Remediation	Cover with fine sand, diatomaceous earth, sponges, paper towels, cloth, vermiculite, or clay. Neutralize with large quantities of 5.25% solution of sodium hypochlorite. Place all material in suitable sealed containers. Decontaminate exterior of containers. Label and dispose according to applicable regulations. Alternate solution: calcium hypochlorite, DS2, supertropical bleach slurry (STB), in that order.		
References	Richardson 1992, CDC 2001, Lewis 1993, NRC 1999, Somani 1992, Wright 1990, Miller 1999, OPCW 1997, DOT 2000		
Dose Curve			

CW Agent	Thiodiglycol		
Type		CWC Schedule	II
Formula	$(CH_2CH_2OH)_2S$	Molecular Weight	122.19
Chem. Name	thiodiethylene glycol	HAZMAT ID#	
CAS #	111-48-8	Boiling Point, °C	283
LC_{50}		mg/l Ref.	
CD_{50}		$mg\text{-}min/m^3$ Ref.	
$L(Ct)_{50}$		$mg\text{-}min/m^3$ Ref.	
LD_{50}	4500	mg/kg Ref.	
Appearance	syrupy colorless liquid, gas is heavier than air		
Odor	characteristic odor		
Symptoms	irritates eyes and lungs		
Antidote			
Decon.	SNL Foam		
Solubility	soluble in acetone, alcohol, chloroform, water		
Removal	May be slightly or partially removable by high efficiency filters		
First Aid			
Remediation	Use sand or inert absorbent to contain or absorb spills. Control vapor cloud with fine spray of water.		
Notes	combustible, precursor agent		
References	Richardson 1992, SIPRI 1971, Canada 1992, OPCW 1997		

Dose
Curve

CW Agent	Triethanolamine		TEA
Type		CWC Schedule	III
Formula	$(HOCH_2CH_2)_3N$	Molecular Weight	149.19
Chem. Name	tri[2-hydroxyethyl]-amine	HAZMAT ID#	
CAS #	102-71-6	Boiling Point, °C	336
LC_{50}		mg/l Ref.	
CD_{50}		mg-min/m^3 Ref.	
$L(Ct)_{50}$		mg-min/m^3 Ref.	
LD_{50}	7200	mg/kg Ref.	Canada 1993
Appearance	colorless viscous liquid, gas is heavier than air		
Odor	slight ammonia-like odor		
Symptoms			
Antidote			
Decon.			
Solubility	soluble in chloroform, miscible in water, alcohol		
Removal	May be slightly or partially removable by high efficiency filters		
First Aid			
Remediation	Use sand or inert absorbent to contain or absorb spills. Control vapor cloud with fine spray of water.		
Notes	combustible, a precursor agent		
References	Richardson 1992, Lewis 1993, SIPRI 1971, OPCW 1997		

Dose Curve

CW Agent	Triethyl phosphite		
Type		CWC Schedule	III
Formula	$(C_2H_5)_3PO_3$	Molecular Weight	166.18
Chem. Name		HAZMAT ID#	2323
CAS #	122-52-1	Boiling Point, °C	156.6
LC_{50}		mg/l Ref.	
CD_{50}		mg-min/m^3 Ref.	
$L(Ct)_{50}$		mg-min/m^3 Ref.	
LD_{50}	3200	mg/kg Ref.	Canada 1993
Appearance	colorless liquid, gas is heavier than air		
Odor			
Symptoms	skin and eye irritant		
Antidote			
Decon.			
Solubility	soluble in alcohol and ether		
Removal	May be slightly or partially removable by high efficiency filters		
First Aid	move victim to fresh air, provide oxygen if breathing difficulty, flush skin and eyes with water for 20 mins, wash with soap and water		
Remediation	Use sand or inert absorbent to contain or absorb spills. Control vapor cloud with fine spray of water.		
Notes	combustible, a precursor agent, flammable liquid		
References	Richardson 1992, Lewis 1993, OPCW 1997, DOT 2000		
Dose Curve			

CW Agent	Trimethyl phosphite		
Type		CWC Schedule	III
Formula	$(CH_3O)P$	Molecular Weight	125.08
Chem. Name		HAZMAT ID#	2329
CAS #	121-45-9	Boiling Point, °C	112
LC_{50}		mg/l Ref.	
CD_{50}		mg-min/m^3 Ref.	
$L(Ct)_{50}$		mg-min/m^3 Ref.	
LD_{50}	1600	mg/kg Ref.	Richardson 1992
Appearance	colorless liquid, gas is heavier than air		
Odor			
Symptoms	eye and skin irritant		
Antidote			
Decon.			
Solubility	soluble in hexane, benzene, acetone, alcohol, ether		
Removal	May be slightly or partially removable by high efficiency filters		
First Aid	move victim to fresh air, provide oxygen if breathing difficulty, flush skin and eyes with water for 20 mins, wash with soap and water		
Remediation	Use sand or inert absorbent to contain or absorb spills. Control vapor cloud with fine spray of water.		
Notes	flammable, a precursor agent		
References	Canada 1993, Lewis 1993, Canada 1992, OPCW 1997, DOT 2000		

Dose

Curve

CW Agent	VX		VX
Type	Nerve agent	CWC Schedule	I
Formula	$C_{11}H_{26}NO_2PS$	Molecular Weight	267.37
Chem. Name	methylphosphonothioic acid	HAZMAT ID#	2810
CAS #	50782-69-9	Boiling Point, °C	298
CD_{50}	5	mg-min/m^3 Ref.	Canada 1993
$L(Ct)_{50}$	10	mg-min/m^3 Ref.	Wald 1970
LD_{50}	6	mg/kg Ref.	Wald 1970
Appearance	amber colored liquid, gas is heavier than air, odorless gas		
Symptoms	eye irritation, confusion, shortness of breath		
Antidote	atropine, diazepam (CANA), pralidoxime chloride (2 PAM Cl)		
Decon.	Bleaching powder, Supertropical bleach, SNL Foam, C8		
Solubility	partially soluble in water, half-life 7-8.5 hours in water at 20 C		
First Aid	Use all three nerve agent antidote kits if symptoms are severe. Use Mark I injectors or atropine. If symptoms progress, use injectors at 5-20 minute intervals (maximum of three injections). Give artificial respiration if breathing stops. Use oxygen delivery if face is contaminated or if breathing is labored. Skin exposure: Immediately wash skin and clothes with solution of 5% sodium hypochlorite or common household liquid bleach. Remove clothing and flush skin with same solution, then wash areas with soap and water. Flush eyes with water for 10-15 minutes.		
Remediation	Cover with fine sand, diatomaceous earth,sponges, paper towels, cloth, vermiculite, or clay. Use alcoholic HTH mixture (add 10 ml of denatured ethanol to a 900 ml slurry of 10% HTH in water. Mix with VX and agitate for 1 hour. Neutralize with large quantities of 5.25% solution of sodium hypochlorite. Place all material in suitable sealed containers. Decontaminate exterior of containers. Label and dispose according to applicable regulations. Alternate solution: calcium hypochlorite, DS2, supertropical bleach slurry		
References	CDC 2001, NRC 1999, Clarke 1968, Somani 1992, Paddle 1996, Wright 1990, NATO 1996, Miller 1999, OPCW 1997, DOT 2000		
Dose Curve			

CW Agents with Incomplete Dose Information

CW Agent	Arsine		SA
Type	Blood agent	CWC Schedule	
Formula	AsH_3	Molecular Weight	77.9
Chem. Name	Arsenic trihydride, Hydrogen arsenide		
CAS #	7784-42-1	Boiling Point, °C	-62
Appearance	colorless liquefied gas, gas is heavier than air		
Odor	has a characteristic odor	HAZMAT ID#	2188
Symptoms	Abdominal pain, short breath, vomiting		
Antidote			
Decon.	decomposes on heating		
Solubility	soluble in water		
Removal	Granular Activated Carbon removes with heavy metal salt impregnation		
First Aid	Inhalation: Provide fresh air and place victims in half-upright position. Skin exposure: Rinse skin with plenty of water or place victim in shower. Rinse eyes with plenty of water for several minutes.		
Remediation	Contact with oxidative compounds may produce violent reactions and explosion hazards.		
Notes	decomposition in heat, moisture, and light produces toxic fumes		
References	CDC 2001, Lewis 1993, DOT 2000, Ellison 2000		

CW Agent	BZ		BZ
Type	Incapacitating agent	CWC Schedule	II
Formula	$C_{21}H_{23}NO_3$	Molecular Weight	337.5
Chem. Name	3-quinuclidinyl benzilate	HAZMAT ID#	2810
CAS #	6581-06-2	Boiling Point, °C	
Appearance	gas is heavier than air		
Odor			
Symptoms	disables temporarily		
Solubility			
Removal			
First Aid	Symptoms are temporary and may dissipate within minutes or hours. No specific first aid needed.		
Notes	nonlethal		
References	Hersh 1968, Cookson 1969, Wright 1990, Clarke 1968, CDC 2001, OPCW 1997, DOT 2000		

CW Agent	Chloropicrin		PS
Type	Tear Gas	**CWC Schedule**	
Formula	CCl_3NO_2	**Molecular Weight**	164.38
Chem. Name	trichloronitromethane	**HAZMAT ID#**	1580
CAS #	76-06-2	**Boiling Point, °C**	112.3
LC$_{50}$	2000	mg/l **Ref.**	Richardson 1992
Appearance	colorless refractive liquid, gas is heavier than air		
Odor	stinging, pungent odor		
Symptoms	strong irritant		
Decon.	SNL Foam		
Solubility	slightly soluble in water		
Removal	Granular Activated Carbon removes with 20-40% efficiency		
First Aid	Symptoms are temporary and may dissipate within minutes or hours. No specific first aid needed.		
Remediation	absorb or cover with dry earth, sand or other non-combustible material and transfer to approved containers		
Notes	non-flammable		
References	Richardson 1992, Canada 1992, Lewis 1993, Haber 1986, Wright 1990, Miller 1999, OPCW 1997, DOT 2000, Ellison 2000		

CW Agent	Chlorosoman	
Type	precursor	**CWC Schedule** I
Formula	$C_7H_{16}ClO_2P$	**Molecular Weight** 198.5
Chem. Name		**HAZMAT ID#**
CAS #	7040-57-5	**Boiling Point, °C**
Appearance	gas is heavier than air	
Decon.	Bleaching powder, Supertropical bleach, SNL Foam	
First Aid	Use all three nerve agent antidote kits if symptoms are severe. Use Mark I injectors or atropine. If symptoms progress, use injectors at 5-20 minute intervals (maximum of three injections). Give artificial respiration if breathing stops. Use oxygen delivery if face is contaminated or if breathing is labored. Skin exposure: Immediately wash skin and clothes with solution of 5% sodium hypochlorite or common household liquid bleach. Remove clothing and flush skin with same solution, then wash areas with soap and water. Flush eyes with water for 10-15 minutes.	
Remediation	Cover with fine sand, diatomaceous earth,sponges, paper towels, cloth, vermiculite, or clay. Neutralize with large quantities of 5.25% solution of sodium hypochlorite. Place all material in suitable sealed containers. Decontaminate exterior of containers. Label and dispose according to applicable regulations. Alternate solution: calcium hypochlorite, DS2, supertropical bleach slurry (STB), in that order.	
Notes	a precursor agent	
References	Canada 1993, OPCW 1997	

CW Agent	Diphosgene		DP
Type	Choking agent	CWC Schedule	
Formula	ClCOOCCl$_3$	Molecular Weight	198
Chem. Name	trichloromethyl chloroformate	HAZMAT ID#	1076
CAS #	503-38-8	Boiling Point, °C	127
Appearance	colorless liquid, gas is heavier than air		
Odor	new mown hay		
Symptoms	strong irritant to tissue		
Decon.	decomposed by heat, alkalies, hot water		
Solubility	alkalies, hot water		
Removal	May be slightly or partially removable by high efficiency filters		
First Aid	Inhalation exposure: Provide oxygen if breathing difficulty. Give artificial respiration if breathing has stopped. Give mouth to mouth resuscitation only if no facial contamination. Skin exposure: Immediately wash skin and clothes with water. Remove clothing and flush skin with water. Flush eyes with water.		
Remediation	Do not direct water jet on liquid. Control vapor cloud with fine spray of water.		
Notes	activated carbon converts to phosgene		
References	Paddle 1996, Lewis 1993, DOT 2000, Ellison 2000		

CW Agent	GE		GE
Type	Nerve agent	CWC Schedule	
Formula		Molecular Weight	
Chem. Name		HAZMAT ID#	
CAS #		Boiling Point, °C	
Appearance	vapor, liquid, aerosol		
Symptoms	drooling, sweating. nausea, vomiting, cramps, involuntary defecation and unrination, twitching, jerking, staggering, headache, confusion, drowsiness, coma, convulsions, asphyxia		
Antidote	atropine, diazepam (CANA), pralidoxime chloride (2 PAM Cl)		
First Aid	Use all three nerve agent antidote kits if symptoms are severe. Use Mark I injectors or atropine. If symptoms progress, use injectors at 5-20 minute intervals (maximum of three injections). Give artificial respiration if breathing stops. Use oxygen delivery if face is contaminated or if breathing is labored. Skin exposure: Immediately wash skin and clothes with solution of 5% sodium hypochlorite or common household liquid bleach. Remove clothing and flush skin with same solution, then wash areas with soap and water. Flush eyes with water for 10-15 minutes.		
Remediation	Cover with fine sand, diatomaceous earth, sponges, paper towels, cloth, vermiculite, or clay. Neutralize with large quantities of 5.25% solution of sodium hypochlorite. Place all material in suitable sealed containers. Decontaminate exterior of containers. Label and dispose according to applicable regulations. Alternate solution: calcium hypochlorite, DS2, supertropical bleach slurry (STB), in that order.		
References	Cookson 1969, Clarke 1968, Ellison 2000		

CW Agent	Hydrogen chloride		HCL	
Type	Blood agent	CWC Schedule		
Formula	HCl	Molecular Weight		36.5
Chem. Name	anhydrous hydrogen chloride	HAZMAT ID#		1050
CAS #	7647-01-0	Boiling Point, °C		-85
Appearance	colorless gas			
Odor	pungent odor			
Symptoms	strong irritant to eyes and skin			
Decon.				
Solubility	very soluble in water, soluble in ether and alcohol			
Removal	Granular Activated Carbon removes with negligible efficiency			
First Aid	Inhalation: Provide fresh air and place victims in half-upright position. Skin exposure: Rinse skin with plenty of water or place victim in shower. Rinse eyes with plenty of water for several minutes.			
Remediation	Use sand or inert absorbent to contain or absorb spills. Control vapor cloud with fine spray of water.			
References	CDC 2001, Lewis 1993, DOT 2000			

CW Agent	Methylphosphonyl dichloride		DF	
Type		CWC Schedule		I
Formula	CH_3Cl_2OP	Molecular Weight		100
Chem. Name	methylphosphonyl dichloride	HAZMAT ID#		
CAS #	676-99-3, 676-97-1, 676-83-5	Boiling Point, °C		178
Appearance	gas is heavier than air			
Odor				
Symptoms				
Antidote				
Decon.	SNL Foam			
Solubility				
Removal				
First Aid				
Remediation	Use sand or inert absorbent to contain or absorb spills. Control vapor cloud with fine spray of water.			
Notes	a precursor agent, dose: 0.141 mg/liter/4h			
References	Canada 1993, Canada 1992, OPCW 1997, Ellison 2000			

CW Agent	Mustard T-mixture		HT
Type	Vesicant/Blister agent	CWC Schedule	
Formula	na	Molecular Weight	-
Chem. Name		HAZMAT ID#	2810
CAS #	na	Boiling Point, °C	190
Appearance			
Symptoms			
Antidote	Dimercaprol (British anti-lewisite (BAL)) ointment		
Decon.	Bleaching powder, Supertropical bleach, Ds2		
Solubility	soluble in fats, oils, gasoline, kerosene, acetone, CCl4, alcohol		
Removal	May be slightly or partially removable by high efficiency filters		
First Aid	Inhalation exposure: Provide oxygen if breathing difficulty. Give artificial respiration if breathing has stopped. Give mouth to mouth resuscitation only if no facial contamination. Skin exposure: Immediately wash skin and clothes with solution of 5% sodium hypochlorite or common household liquid bleach. Remove clothing and flush skin with same solution, then wash areas with soap and water. Flush eyes with water.		
Remediation	Cover with fine sand, diatomaceous earth, vermiculite, or clay. Neutralize with large quantities of 5.25% solution of sodium hypochlorite. Place all material in suitable sealed containers. Decontaminate exterior of containers. Label and dispose according to applicable regulations. Alternate solution: calcium hypochlorite, DS2, supertropical bleach slurry (STB), in that order.		
Notes	A mixture of two or more sulfur mustards and other agents		
References	Richardson 1992, CDC 2001, Somani 1992, DOT 2000		

CW Agent	Nitrogen dioxide		NO2
Type	Choking agent	CWC Schedule	
Formula	NO_2	Molecular Weight	46.01
Chem. Name	Nitrogen oxide	HAZMAT ID#	1067
CAS #	10102-44-0	Boiling Point, °C	
Appearance	reddish-brown gas		
Odor	pungent odor		
Symptoms	cough, dizziness, pain		
Removal	Granular Activated Carbon removes with negligible efficiency		
First Aid	Inhalation: Provide fresh air and place victims in half-upright position. Skin exposure: Rinse skin with plenty of water or place victim in shower. Rinse eyes with plenty of water for several minutes.		
Remediation	Use sand or inert absorbent to contain or absorb spills. Control vapor cloud with fine spray of water.		
Notes	non-combustible, reacts with water to form nitric acid and nitric oxide		
References	Lewis 1993, CDC 2001, DOT 2000		

CW Agent	QL		QL
Type		CWC Schedule	I
Formula	$C_{11}H_{26}NO_2P$	Molecular Weight	235
Chem. Name		HAZMAT ID#	
CAS #	57856-11-8	Boiling Point, °C	
Appearance	gas is heavier than air		
Odor	strong fishy odor		
Symptoms			
Antidote			
Decon.			
First Aid			
Remediation	Use sand or inert absorbent to contain or absorb spills. Control vapor cloud with fine spray of water.		
Notes	a binary precursor agent for VX2		
References	Canada 1993, OPCW 1997, Ellison 2000		

CW Agent	Sulfur dioxide		SO2
Type	Choking agent	CWC Schedule	
Formula	SO_2	Molecular Weight	64.1
Chem. Name	Sulfurous oxide	HAZMAT ID#	1079
CAS #	7449-09-5	Boiling Point, °C	-10
Appearance	colorless gas, gas is heavier than air		
Odor	pungent odor		
Symptoms	cough, shortness of breath, eye irritant		
Antidote			
Decon.			
Solubility	soluble in water, alcohol, ether		
Removal	Granular Activated Carbon removes with 15% efficiency		
First Aid	Inhalation: Provide fresh air and place victims in half-upright position. Skin exposure: Rinse skin with plenty of water or place victim in shower. Rinse eyes with plenty of water for several minutes.		
Remediation	Do not direct water jet on liquid. Control vapor cloud with fine spray of water.		
Notes	Delayed effects, Reacts violently with ammonia, acetylene, chlorine, and amines. Reacts with water or steam. Attacks metals in the presence of water.		
References	CDC 2001, Lewis 1993, DOT 2000		

CW Agent	Sulphur monochloride		
Type		CWC Schedule	III
Formula	S_2Cl_2	Molecular Weight	135.03
Chem. Name	sulfur chloride	HAZMAT ID#	1834
CAS #	10025-67-9	Boiling Point, °C	138
Appearance	amber to yellowish red oily liquid, gas is heavier than air		
Odor	penetrating odor		
Symptoms	strong irritant to tissue		
Antidote			
Decon.	SNL Foam, decomposes on contact with water		
Solubility	water		
First Aid	move victim to fresh air, provide oxygen if breathing difficulty, flush skin and eyes with water for 20 mins		
Remediation	Use sand or inert absorbent to contain or absorb spills. Control vapor cloud with fine spray of water.		
Notes	combustible, a precursor agent		
References	Canada 1993, Lewis 1993, Canada 1992, OPCW 1997, DOT 2000		

CW Agent	Sulphur trioxide		FS
Type	Choking agent	CWC Schedule	
Formula	SO_3	Molecular Weight	80.1
Chem. Name	sulfuric anhydride	HAZMAT ID#	2810
CAS #	7449-11-9	Boiling Point, °C	
Appearance	sublimating solid		
Symptoms	strong irritant to tissue		
Antidote			
Decon.	combines with water to form sulfuric acid		
Solubility	water		
Removal	Granular Activated Carbon removes with 10-25% efficiency		
First Aid	Inhalation exposure: Provide oxygen if breathing difficulty. Give artificial respiration if breathing has stopped. Give mouth to mouth resuscitation only if no facial contamination. Skin exposure: Immediately wash skin and clothes with water. Remove clothing and flush skin with water. Flush eyes with water.		
Remediation	Do not direct water jet on liquid. Control vapor cloud with fine spray of water.		
References	Lewis 1993, CDC 2001, DOT 2000		

CW Agent	Titanium tetrachloride		FM
Type	Choking agent	CWC Schedule	
Formula	TiCl$_4$	Molecular Weight	189.69
Chem. Name	titanic chloride	HAZMAT ID#	1838
CAS #	7750-45-0	Boiling Point, °C	136.4
Appearance	colorless liquid, white cloud		
Symptoms	strong irritant to skin and tissue		
Antidote			
Decon.			
Solubility	soluble in water with heat		
Removal	May be slightly or partially removable by high efficiency filters		
First Aid	Inhalation exposure: Provide oxygen if breathing difficulty. Give artificial respiration if breathing has stopped. Give mouth to mouth resuscitation only if no facial contamination. Skin exposure: Immediately wash skin and clothes with water. Remove clothing and flush skin with water. Flush eyes with water.		
Remediation	Do not direct water jet on liquid. Control vapor cloud with fine spray of water.		
References	Lewis 1993, DOT 2000		

(Appendix D continues on p. 518.)

Additional CW Agents and Precursors

CW agent	Case no.	CWC	HAZMAT ID no.
0-Alkyl aminoethyl alkyl phosphonites			
0-Alkyl phosphonochloridates			
1,3-bis(2-chloroethylthio)n-butane	142868-83-7	I	
1,3-bis(2-chloroethylthio)-n-pentane	142868-94-8	I	
1,3-bis(2-chloroethylthio)-n-propane	63905-10-2	I	
2,2-diphenyl-2-hydroxyacetic acid	76-93-7	II	
2-chloroethylchloromethylsulfide	2625-76-5	I	
3,3-dimethyl-2-butanol	464-07-3	II	
Alkyl phosphoramidocyanidates			
Alkyl alkylphosphonofluorodates			
Alkyl phosponothiolates			
Alkyl phosponyldifluorides			
Benzilic acid			
Bis(2-chloroethylthio)methane	63869-13-6	I	
Bis(2-chloroethylthiomethyl)ether	63918-90-1	I	
DC			2810
Dialkyl phosphoramidates		II	
Dimethyl methylphosphonate	756-79-6	II	
Ethyldichloroarsine (ED)			1892
Ethyldiethanolamine	139-87-7	III	
Methyldichloroarsine (MD)			
N,N-dialkyl phosphoramidic dihalides		II	1556
N,N-dialkylaminoethane-2-ols		II	
N,N-dialkylaminoethyl-2-chlorides		II	
N,N-diethylaminoethanol	100-37-8	II	2686
N,N-dimethylaminoethanol	108-01-0	II	2051
Phenyldichloroarsine (PD)			1556
Pinacolyl alcohol	464-07-3	II	
Quinuclidine-3-OL	1619-34-7	II	
Sulphur dichloride	10545-99-0	III	1828
Thionyl chloride	7719-09-7	III	1836

References: Canada 1993, CDC 2001, DOT 2000, OPCW 1997.

UVGI System Sizes and Kill Rates

Width, cm	Height, cm	Length, cm	UV power, W	Reflectivity, %	Kill rate, %		I_{avg}, μW/cm^2	AvgDir, μW/cm^2
					Anthrax	Smallpox		
10	10	20	3	50	4	32	3,225	
				75	7	43	4,599	
				85	8	48	5,384	1,785
			11	50	17	76	11,825	
				75	25	87	16,863	
				85	30	91	19,740	6,545
	20	40	3	50	4	32	1, 598	
				75	7	43	2,321	
				85	8	48	2,719	830
			11	50	17	76	5,860	
				75	25	87	8,511	
				85	32	97	9,970	3,042
	40	80	3	50	4	30	732	
				75	6	40	1,053	
				85	7	44	1,219	371
			11	50	15	73	2,685	
				75	23	84	3,860	
				85	27	88	4,469	1,359
20	10	20	6	50	9	54	6,346	
				75	13	65	8,676	
				85	15	70	9,961	3,757
			22	50	32	94	23,270	
				75	44	98	31,812	
				85	50	99	36,524	13,775
	20	40	6	50	9	55	3,294	
				75	14	68	4,714	
				85	17	74	5,524	1,806
			22	50	33	95	12,079	
				75	47	98	17,284	
				85	54	99	20,254	6,623

Width, cm	Height, cm	Length, cm	UV power, W	Reflectivity, %	Kill rate, %		I_{avg}, μW/cm²	AvgDir, μW/cm²
					Anthrax	Smallpox		
20	40	80	6	50	9	53	1,577	
				75	14	67	2,290	
				85	16	73	2,683	820
			22	50	32	94	5,784	
				75	47	98	8,397	
				85	53	99	9,836	3,006
	80	160	6	50	7	48	668	
				75	11	60	960	
				85	13	66	1,111	339
			22	50	28	91	2,449	
				75	41	97	3,519	
				85	46	98	4,072	1,243
40	10	20	11	50	16	74	11,266	
				75	21	83	14,770	
				85	24	87	16,594	7,065
			45	50	54	100	46,090	
				75	66	100	60,421	
				85	72	100	67,884	28,900
	20	40	11	50	16	76	5,863	
				75	23	85	8,024	
				85	27	89	9,216	3,462
			45	50	56	100	23,985	
				75	70	100	32,824	
				85	76	100	37,701	14,163
	40	80	11	50	16	75	2,846	
				75	24	86	4,045	
				85	29	90	4,729	1,588
			45	50	55	100	11,641	
				75	71	100	16,548	
				85	78	100	19,346	6,497
	80	160	11	50	14	69	1,228	
				75	21	82	1,767	
				85	25	86	2,063	655
			45	50	50	99	5,024	
				75	67	100	7,229	
				85	73	100	8,440	2,681
	160	320	11	50	11	59	468	
				75	16	72	667	
				85	19	77	770	244

Width, cm	Height, cm	Length, cm	UV power, W	Reflectivity, %	Kill rate, %		I_{avg}, μW/cm²	AvgDir, μW/cm²
					Anthrax	Smallpox		
40	160	320	45	50	41	98	1,914	
				75	56	99	2,728	
				85	62	100	3,149	998
	320	640	11	50	7	46	158	
				75	10	58	222	
				85	12	62	253	82
			45	50	29	92	647	
				75	41	97	908	
				85	47	98	1,037	337
	640	1280	11	50	4	32	50	
				75	6	41	69	
				85	7	45	78	26
			45	50	19	79	204	
				75	27	89	282	
				85	31	92	320	106
80	10	20	22	50	19	81	13,928	
				75	26	89	18,106	
				85	29	91	20,223	8,841
			90	50	61	100	56,978	
				75	74	100	74,069	
				85	78	100	82,729	36,167
	20	40	22	50	20	82	7,154	
				75	28	90	9,553	
				85	32	93	10,825	4,416
			90	50	62	100	29,268	
				75	76	100	39,080	
				85	81	100	44,285	18,066
	40	80	22	50	19	81	3,488	
				75	28	90	4,770	
				85	32	93	5,488	2,118
			90	50	62	100	14,268	
				75	76	100	19,514	
				85	82	100	22,451	8,664
	80	160	22	50	18	78	1,573	
				75	26	88	2,189	
				85	30	91	2,542	935
			90	50	58	100	6,437	
				75	73	100	8,957	
				85	79	100	10,399	3,825

Width, cm	Height, cm	Length, cm	UV power, W	Reflectivity, %	Kill rate, % Anthrax	Kill rate, % Smallpox	I_{avg}, $\mu W/cm^2$	AvgDir, $\mu W/cm^2$
80	160	320	22	50	15	71	650	
				75	22	83	909	
				85	25	87	1,052	373
			90	50	52	99	2,658	
				75	67	100	3,720	
				85	73	100	4,304	1,527
	320	640	22	50	11	60	240	
				75	16	72	334	
				85	18	77	382	134
			90	50	41	98	981	
				75	56	99	1,365	
				85	62	100	1,564	547
	640	1280	22	50	7	46	81	
				75	10	58	111	
				85	12	62	126	44
			90	50	29	92	329	
				75	41	97	454	
				85	46	98	516	181
160	20	40	45	50	21	84	7,564	
				75	29	91	10,183	
				85	34	94	11,549	4,543
			180	50	64	100	30,254	
				75	77	100	40,731	
				85	82	100	46,194	18,172
	40	80	45	50	21	84	3,820	
				75	31	92	5,302	
				85	36	95	6,109	2,210
			180	50	65	100	15,278	
				75	79	100	21,208	
				85	85	100	24,438	8,842
	80	160	45	50	21	83	1,847	
				75	31	92	2,639	
				85	37	95	3,087	1,020
			180	50	64	100	7,389	
				75	80	100	10,558	
				85	85	100	12,347	4,081
	160	320	45	50	19	80	839	
				75	29	91	1,223	
				85	35	94	1,443	444

Width, cm	Height, cm	Length, cm	UV power, W	Reflectivity, %	Kill rate, % Anthrax	Kill rate, % Smallpox	I_{avg}, $\mu W/cm^2$	AvgDir, $\mu W/cm^2$
160	160	320	180	50	61	100	3,355	
				75	78	100	4,890	
				85	84	100	5,771	1,776
	320	640	45	50	16	73	343	
				75	24	85	500	
				85	29	90	587	177
			180	50	54	99	1,372	
				75	71	100	2,001	
				85	77	100	2,347	706
	640	1280	45	50	11	62	126	
				75	17	75	180	
				85	20	80	209	64
			180	50	43	98	503	
				75	58	100	722	
				85	65	100	835	256
320	80	160	90	50	23	85	1,994	
				75	34	94	2,879	
				85	40	96	3,364	1,040
			359	50	67	100	7,953	
				75	83	100	11,484	
				85	88	100	13,417	4,150
	160	320	90	50	22	85	971	
				75	35	94	1,447	
				85	41	96	1,714	467
			359	50	67	100	3,874	
				75	83	100	5,773	
				85	89	100	6,837	1,864
	320	640	90	50	21	82	442	
				75	33	93	675	
				85	39	96	809	201
			359	50	64	100	1,762	
				75	82	100	2,693	
				85	87	100	3,225	802
	640	1280	90	50	17	75	181	
				75	27	88	275	
				85	32	92	327	81
			359	50	57	100	721	
				75	75	100	1,097	
				85	81	100	1,304	323

Width, cm	Height, cm	Length, cm	UV power, W	Reflectivity, %	Kill rate, %		I_{avg}, μW/cm²	AvgDir, μW/cm²
					Anthrax	Smallpox		
640	80	160	180	50	24	87	2,124	
				75	36	95	3,068	
				85	42	97	3,568	1,060
			719	50	70	100	8,485	
				75	85	100	12,255	
				85	89	100	14,252	4,236
	160	320	180	50	25	87	1,065	
				75	38	95	1,587	
				85	44	97	1,868	481
			719	50	71	100	4,254	
				75	86	100	6,341	
				85	91	100	7,462	1,920
	320	640	180	50	25	87	520	
				75	39	95	803	
				85	45	98	959	213
			719	50	71	100	2,077	
				75	87	100	3,207	
				85	92	100	3,832	849

Source Code for Direct UVGI Field Average Intensity*

```
// Global variables and arrays
double Average;                      // Average UVGI Intensity
double DirectField[51][51][101];     // Direct Intensity Field
double DistanceMtx[51][51][101];     // matrix of distances to lamp
axis
double PositionMtx[51][51][101];     // matrix of position along
                                     // lamp axis
int NUMLAMPS = 1;              // number of lamps in system (set this
                                  to 1 or more)
double SurfInt[1] = 10000.0;   // Lamp surface intensity for lamp #1,
                                  µW/cm2, (set value)
double arclength[1] = 35.0;        // Lamp arclength for lamp #1,
                                      cm (set value)
double radius[1] = 1.0;            // radius of lamp # 1, cm (set
                                      value)
                               // add additional lamps here as neces-
                                  sary


brun(){       // this is the first routine that calls the computa-
                 tional subroutines in C++
              // Compute the direct intensity field for all lamps in
                 enclosure
              // and fill the matrix DirectField[][][] with the
                 intensity values at each point
    DirectIntField();
              // Compute the average intensity
    AverageDirect();
              // the result, Average, can be printed at this point
                 or output to a file
}
```

*Note: For the complete source code implementation, including the reflectivity subroutines, see Kowalski (2001).

```
void DirectIntField()
{      // Computes 50x50x100 UV Intensity Matrix of Direct Intensity
          Field
    int i, j, k, l;
    double tempsum = 0.0, x, paxis, db;
    for (i=0; i<=50; i++){
        db=0.0;
        for (j=0; j<=50; j++){
            for (k=0; k<=100; k++){
                for (l=0; l<NUMLAMPS; l=l+1) {
                    // Compute distance x to lamp axis
                    x = Distance(i,j,k,l);
                    DistanceMtx[i][j][k]=x;
                    // Compute position on lamp axis
                    paxis = fabs(Position(i,j,k,l));
                    PositionMtx[i][j][k]=paxis;
                    // Is it within lamp arclength?
                    if (paxis < arclength[l]){
                        // Compute Intensity within Lamp arclength
                        tempsum = Intensity(SurfInt[l],arclength[l],
                                radius[l],x,paxis);
                    }
                    else {      // Compute Intensity beyond Lamp end
                        db = paxis-arclength[l];
                        tempsum = IBeyondEnds(SurfInt[l],arclength
                                [l],radius[l],x,db);
                    }
                    // Add intensities for all lamps
                    DirectField[i][j][k] = DirectField[i][j][k] +
                                        tempsum;
                    tempsum = 0.0;
                }
            }
        }
    }
}


double Distance(int i, int j, int k, int l)
{       //  Compute shortest Distance to Lamp Axis
    double xi, yj, zk;
    xi = i;
    yj = j;
    zk = k;
    double x = xi*xincr;
    double y = yj*yincr;
    double z = zk*zincr;
    double dist = PointLine(x, y, z, l);
    return dist;
}


double PointLine(double x, double y, double z, int l)
{          //  Compute Distance from a Point to a Line (lamp axis)
        double x1=x-lampx1[l];
        double y1=y-lampy1[l];
        double z1=z-lampz1[l];
```

```
            double x2=lampx2[l]-lampx1[l];
            double y2=lampy2[l]-lampy1[l];
            double z2=lampz2[l]-lampz1[l];
            double dist, DotProd, a;
            double p1=x1*x1+y1*y1+z1*z1;
            double p2=x2*x2+y2*y2+z2*z2;
            if (p1*p2>0){
                DotProd = (x1*x2+y1*y2+z1*z2)/sqrt(p1*p2);
                a = acos(DotProd);
                dist=fabs(sin(a))*sqrt(p1);
            }
            else {
                dist = 0;
            }
        return dist;
}


double Position(int i, int j, int k, int l)
{           // Compute Position along lamp axis
            double xi, yj, zk, p1, p2, posit, p3, p4, a, x;
            double DotProd, y, z, x1, y1, z1, x2, y2, z2;
            double pc, pa, pd, p5, posit1, posit2;
            xi = i;
            yj = j;
            zk = k;
            x = xi*xincr;
            y = yj*yincr;
            z = zk*zincr;
            posit = 1;
            x1=x-lampx1[l];
            y1=y-lampy1[l];
            z1=z-lampz1[l];
            x2 = lampx2[l]-lampx1[l];
            y2 = lampy2[l]-lampy1[l];
            z2 = lampz2[l]-lampz1[l];
            p1 = x1*x1+y1*y1+z1*z1;
            p2 = x2*x2+y2*y2+z2*z2;
            pc = x1*x2+y1*y2+z1*z2;
            pa =p1*p2;
            if (pa>0){
                DotProd = pc/sqrt(pa);
                a = acos(DotProd);
                posit1 = cos(a)*sqrt(p1);
            }
            else {
                posit1 = 0.000001;
            }
            x1=x-lampx2[l];
            y1=y-lampy2[l];
            z1=z-lampz2[l];
            x2 = lampx1[l]-lampx2[l];
            y2 = lampy1[l]-lampy2[l];
            z2 = lampz1[l]-lampz2[l];
            p3 = x1*x1+y1*y1+z1*z1;
            p4 = x2*x2+y2*y2+z2*z2;
            pd = x1*x2+y1*y2+z1*z2;
            p5 =p3*p4;
```

```
        if (p5>0){
            DotProd = pd/sqrt(p5);
            a = acos(DotProd);
            posit2 = cos(a)*sqrt(p3);
        }
        else {
            posit2 = 0.000001;
        }
        posit = max(posit1,posit2);
    return posit;
}




double Intensity(double IS, double arcl, double r, double x, double l)
{           // Compute Intensity Field
            // IS=Surface Intensity, arcl=arclength, r=radius,
            // x=distance from axis, l = distance along axis
        double intense;
        double VF, VF1, VF2;
            // Compute VF Lamp segment 1
        VF1 = VFCylinder(l,r,x);
            // Compute Lamp segment 2
        VF2 = VFCylinder(arcl-l,r,x);
            // Total VF for Lamp
        VF = VF1 + VF2;
            // Compute intensity at the point
        intense = IS*VF;
        return intense;
}




double IBeyondEnds(double IS, double arcl, double r, double x, double
db )
{           // Compute Intensity field beyond the ends of the lamp
            // IS=Surface Intensity, arcl=arclength, r=radius
            // x=distance from axis, db=distance beyond lamp end
        double intense;
        double VF, VF1, VF2;
            // Compute Lamp + Ghost Lamp segment
        VF1 = VFCylinder(arcl+db,r,x);
            // Compute Ghost Lamp segment
        VF2 = VFCylinder(db,r,x);
            // Compute Lamp VF
        VF = VF1 - VF2;
            // Compute intensity at the point
            // use absolute value since near zero values may be
            //     negative
        intense = fabs(IS*VF);
        return intense;
}




double VFCylinder(double l, double r, double h)     // View Factor #15
per Modest 1993
{           // l=length, r=radius, h=height above axis
        double H, L, X, Y, p1, p2, p3, VF;
```

```
    if (h<r) h=r+0.000001;  // Not inside lamp
    H = h/r;
    L = 1/r;
    if (L==0) L=0.000001;
    if (H==1) H=H+0.000001;
    X = (1+H)*(1+H)+L*L;
    Y = (1-H)*(1-H)+L*L;
            // Compute Parts of View Factor
    p1 = atan( L/sqrt(H*H-1) )/L;
    p2 = (X-2*H)*atan( sqrt( (X/Y)*(H-1)/(H+1) ))/sqrt(X*Y);
    p3 =  atan( sqrt((H-1)/(H+1)) );
    VF = L*(p1+p2-p3)/(Pi*H);
    return VF;



}



double AverageDirect()
{    // compute average intensity field
    double total = 0;
    double Avg = 0;
    for (int i=0; i<=50; i++)
        for (int j=0; j<=50; j++)
            for (int k=0; k<=100; k++)
                total = total + DirectField[i][j][k];
    Average = total/(51*51*101);
    return Avg;
```

Glossary

ACH Air changes per hour. The number of times per hour the room volume is exchanged.

Actinomycetes A group of gram-positive bacteria that grow in filaments.

Adsorption The process by which molecules will adhere to surfaces as the result of van der Waal's forces, as in carbon adsorption.

Aerobiological engineering The engineering of the indoor air environment to control the aerobiology or indoor microbial air quality, through the use of air disinfection or air-cleaning technologies.

Aerobiology The study of the biology of the air.

Aerosol Fine liquid or solid particles suspended in a gas such as air, for example, fog or smoke.

Aerosolizer A device that can generate a fine suspension of liquid or solid particles in air.

Aerosol toxicity The toxicity of an aerosolized agent in air.

Aflatoxin Fungal metabolite with toxic properties.

After-hours UVGI system Ultraviolet germicidal irradiation (UVGI) area disinfection systems that operate only when personnel are not present.

Agglomeration The accumulation of particles in air. Usually the result of static charge or molecular forces.

AHU Air-handling unit; usually including fans, filters, cooling coils, heating coils, and control dampers. May also include air disinfection equipment.

Air sampling The process of drawing an air sample for identification of airborne chemicals or microorganisms.

Amphotericin B An antibiotic used for fungal infections.

Amplifier A building that amplifies the microbial content of the air, due to moisture, mold growth, or other problems.

Anthrax An infectious disease caused by *Bacillus anthracis* spores.

Antibiotic A substance that inhibits the growth of or kills microorganisms.

Antibody An immunoglobulin released by the immune system in response to the presence of an antigen.

Antimicrobial coatings Any one of a number of compounds that can be used to inhibit microbial growth on surfaces.

Antiseptic A chemical used for disinfection.

Antisera   The liquid part of blood containing antibodies that react against disease-causing agents.

Antitoxin   An antibody that combines with and neutralizes a microbial toxin.

Arclength   The length of the filament of a tubular or cylindrical lamp, as opposed to the full physical lamp length.

Asymptomatic   The condition when an infection is present without apparent symptoms.

Atomization   Reducing a liquid or solid to an aerosol.

Atropine   A compound used as an antidote for nerve agents.

Average intensity   The average UV irradiance on a surface, or the average UV fluence rate through a volume.

Axial flow   The UVGI configuration in which the lamp is oriented with its axis parallel to the direction of airflow.

BACnet   Internet protocol for building automation systems.

Bacteria   Single-celled organisms that multiply by cell division and that can cause disease in humans, plants, or animals.

Bactericidal   Having the ability to kill bacteria.

Bacteriostatic   Having the ability to prevent bacterial growth.

BAL   British anti-Lewisite, an antidote for the blister agent Lewisite.

BAS   Building automation systems.

Bell curve   A normal, or Gaussian, distribution.

Biaxial   Having an axis in two separate locations, as in biaxial lamps that have two conjoined cylindrical sections, each with its own axis.

Binary chemicals   Two precursor chemicals that, when mixed, form a chemical weapon agent.

Biochemicals   The chemicals that make up or are produced by living things.

Biocidal   Having the ability to kill living organisms.

Biodetection   The process of detecting biological agents such as microbes or toxins.

Biological warfare   The intentional use of biological agents as weapons to kill or injure humans, animals, or plants, or to damage facilities.

Biological weapon agents   Living organisms or the materials derived from them that cause disease in or harm to humans, animals, or plants, or cause deterioration of material.

Bioregulators   Naturally occurring biochemicals that regulate bodily functions. Bioregulators that are produced by the body are termed "endogenous." Some of these same bioregulators can be chemically synthesized.

Bioremediation   The removal or degradation of biological agents.

Biosafety level   Four biosafety levels have been established, ranked from the least dangerous (I) to the most dangerous (IV). They carry specific requirements for protective clothing, equipment, and facilities, and for handling microorganisms.

Biosensor   Sensor for biological agents in which a sensing agent is fixed to a solid substrate and monitored for changes that indicate a reaction has taken place.

Blastomycosis   A systemic infection caused by *Blastomyces dermatitidis.*

Bleaching agents   Solutions of bleaches, like sodium hydroxide, that may be used to decontaminate surfaces.

Bleaching powders   Dry bleaching agents usually used for soaking up chemical spills.

Blister agents   Substances that cause blistering of the skin. Exposure is through liquid or vapor contact with any exposed tissue (eyes, skin, lungs).

Blood agents   Substances that injure a person by interfering with cell respiration (the exchange of oxygen and carbon dioxide between blood and tissues).

Botulism   A form of food poisoning caused by botulinum toxin from *Clostridium botulinum.*

BRI   Building-related illness.

Brownian motion   The random motion of airborne particles subject to buffeting by air molecules.

BSL   Biosafety level.

Bubonic plague   See *Plague.*

BW agent   Biological weapon agent.

Capsule   A layer of organized material that surrounds many bacteria or viruses.

Carbon adsorber   Granulated activated carbon (GAC) used as a filter for gases and vapors.

Casualty   The incapacitation due to a CW agent or the infection due to a BW agent. Not necessarily the same as a fatality.

Casualty agents   Toxic agents that produce incapacitation, serious injury, or death. They can be used to incapacitate or kill victims.

Catalyst   A substance that accelerates a chemical reaction without being consumed.

CDC   Abbreviation for the Centers for Disease Control and Prevention, located in Atlanta, GA.

Central nervous system depressants   Compounds that have the predominant effect of depressing or blocking the activity of the central nervous system.

The primary mental effects include the disruption of the ability to think, sedation, and lack of motivation.

Central nervous system stimulants   Compounds that have the predominant effect of flooding the brain with too much information. The primary mental effect is loss of concentration, causing indecisiveness and the inability to act in a sustained, purposeful manner.

CFD   Computational fluid dynamics.

Cfu   Colony-forming unit. Essentially synonymous with a bacterium or a spore, depending on the context. In the culturing of a plate of bacteria or fungi, a colony forms as the result of multiplication from a single microbe; therefore, 1 cfu indicates that a single original microbe was present in the culture. Similar to pfu (plaque-forming unit), used for viruses.

Chemical agent   A chemical substance that is intended for use in military operations to kill, seriously injure, or incapacitate people through its physiological effects.

Chloramphenicol   A commonly used broad-spectrum antibiotic.

Choking agents   Substances that cause physical injury to the lungs. Exposure is through inhalation. In extreme cases, membranes swell and the lungs become filled with liquid. Death results from lack of oxygen; hence, the victim is asphyxiated.

Cholera   An acute infectious enteritis caused by *Vibrio cholerae.*

Clumping   The tendency of micron-sized particles to agglomerate.

CNS   Central nervous system.

Coccidioidomycosis   A fungal disease caused by *Coccidioides immitis.*

Coefficient of variation (COV)   A constant that defines the distribution of lognormally distributed sizes of particles. The standard deviation divided by the mean.

Cold zone   Area where the command post and support functions are located to control a hazardous spill or incident. Also known as a "clean zone" or "support zone."

Combustible   Having a flash point greater than 60.5°C (141°F).

Commensal   Any microorganism that coexists with another organism, to the mutual benefit of both or to the harm of neither. Bacteria in the human stomach that protect us against pathogens and aid in digestion are one example.

Communicable   Capable of being transmitted from one infected host to another, usually by contact, airborne spread, or other mechanisms.

Constant volume system   Ventilation at a constant airflow rate.

Contagious   Communicable. When a disease is capable of being transmitted from one person to another.

Control dampers Mechanical dampers in a ventilation system used for isolation or control of airflow.

Control system architecture The logic behind a control system design.

Control systems Electrical, electronic, or digital systems used to control mechanical systems like ventilation systems.

Crossflow The UVGI system configuration in which the airflow direction is perpendicular to the axis of the UV lamp.

Culture A population of microorganisms grown in a medium such as a petri dish.

Cutaneous Of or pertaining to the skin.

CW agent Chemical weapon agent.

CWC Chemical Weapons Convention.

CWC Schedule A classification of chemical weapon agents, in which they are ranked from most dangerous (I) to least dangerous (III).

Cyanobacteria A large group of photosynthetic bacteria.

Cytotoxic Causing damage to or death of a cell.

Cytotoxin A toxin that has a specific action upon cells, whether human or bacterial.

Dalton Atomic mass unit that gives the same number as the atomic weight.

Death curve Decay curve or survival curve of a microbe exposed to biocidal factors.

Decay curve The decay of a microbial population under any biocidal factor.

Decontamination The process of removing harmful agents by absorbing, destroying, neutralizing, making harmless, or removing the hazardous material from the contaminated area or people.

Dedicated outdoor air systems Ventilation systems that use 100 percent outside air to meet the minimum outside air requirements.

Dehumidification Drying the air or reducing the relative humidity by various means.

Desiccant Materials that absorb moisture from the air, lowering humidity.

Desiccant cooling The combination of a desiccant and an evaporative cooler to provide cooling or reduce enthalpy in an airstream.

Detect-to-alarm A control system architecture that alarms after a positive detection signal.

Detect-to-isolate A control system architecture that isolates a building or ventilation system after a positive detection signal.

Detect-to-treat A control system architecture that is set up to detect biological agents in sufficient time to allow for treatment of exposed occupants.

Diffusion The motion of molecules through air or solid materials.

Dilution ventilation Purging of the air by mixing it with clean air or outdoor air. The normal action of any ventilation system using outside air for dilution. Also called "purge ventilation."

Dioctyl phthalate (DOP) An oily liquid aerosolized to test filter penetration.

Disease progression curve Graph of the severity of disease symptoms over time.

Disinfection Having a bactericidal or viricidal effect. Total disinfection is sterilization.

DOAS Dedicated outdoor air systems.

DOP Dioctyl phthalate, a test aerosol or simulant.

Dose-response curve The relationship between the dose and the fatalities or casualties, or between the dose and the infections, of any particular agent.

Dosimetry The science of measuring dose.

Droplet nuclei Small particles in the 1- to 4-μm size range that remain airborne after the evaporation of droplets.

DS2 Decontamination solution number 2.

DSP Dust spot efficiency. A rating system for filters.

Dysentery Pain and frequent defecation resulting from an inflammation of the colon or other intestines.

E. Coli *Escherichia coli*. A common test microorganism or pathogen simulant.

Edema The accumulation of an excessive amount of fluid in cells or tissue.

EEE Eastern equine encephalitis virus.

Electret A filter using static or electrical charge to enhance efficiency.

Electrostatic filter A filter that maintains a static charge to enhance filtration effficiency.

Emergency systems Any system engaged during an emergency, such as purge systems or secondary trains of air-cleaning equipment.

Endemic disease A disease that is common in a particular population or geographic area.

Endogenous Native to an individual's own body.

Endogenous infection An infection caused by an individual's own microbiota.

Endotoxin A toxin produced by certain gram-negative bacteria.

Energy recovery   The use of heat exchangers to transfer heat energy.

Enterotoxin   A toxin that specifically affects the cells of the intestinal mucosa.

Epidemic   An outbreak of disease.

Epidemiology   The study of the factors influencing the spread of disease in a population.

Epizootic   Sudden outbreak of a disease in an animal population.

ePTFE   Expanded polytetrafluoroethylene.

Erythema   Redness of skin due to cell damage.

Exotoxin   A toxin produced by certain bacteria and released into the surroundings.

Exposure time   The time of exposure of a microbe to a biocidal factor like UVGI or ozone.

Fine water spray   Water sprays used to move clouds of vapor.

Fixed-point gas monitoring   Placement of gas sensors at fixed locations.

Fomites   Particles or droplets left on surfaces or parts of the body that contain viable infectious microorganisms and can transmit infections.

Food-borne pathogens   Pathogenic microorganisms that may incubate or survive in foods.

Food poisoning   Usually refers to the ingestion of foods containing toxin-producing microbes or food-borne pathogens.

Fuller's earth   A material containing kaolin and other compounds used for containing chemical spills.

Fungi   Any of a group of eukaryotic organisms mainly characterized by the absence of chlorophyll and vascular tissue. Fungi range from microscopic single-celled microbes such as molds and mildews to large plantlike forms such as mushrooms.

GAC   Granulated activated carbon. Used in carbon adsorbers.

Gamma irradiation   The use of ionizing gamma radiation to sterilize materials.

Gas phase filtration   Equipment, such as carbon adsorbers, used to remove gases from the air.

Gastroenteritis   An acute inflammation of the lining of the stomach.

Gastritis   Inflammation of the stomach.

GE   Genetically engineered.

Genome   The full set of genes present in a cell or virus.

Genus   A well-defined group of one or more species.

Germination   Growth of a fungus from spores.

Gompertz curve   An S-shaped curve common in natural and electronic processes.

Gram stain   A differential staining procedure that divides bacteria into gram-positive or gram-negative groups.

G-series nerve agents   Chemical agents of moderate to high toxicity developed in the 1930s. Examples are tabun (GA), sarin (GB), soman (GD), and GF.

Hazard zones   U.S. Department of Transportation (DOT) classification for hazardous chemical releases.

Hazard Zone A LC_{50} less than or equal to 200 ppm.

Hazard Zone B LC_{50} between 200 and 1000 ppm.

Hazard Zone C LC_{50} between 1000 and 3000 ppm.

Hazard Zone D LC_{50} greater than 3000 ppm.

HAZMAT teams   Specialists trained in the control and decontamination of hazardous chemical spills. "HAZMAT" is a contraction of the words "hazardous materials."

Hemolysis   The disruption of red blood cells.

HEPA filter   High-efficiency particulate air filters that have removal efficiencies of 99.97 percent or higher at 0.3 micron. Rated by methods (DOP) that differ from high-efficiency filters.

Herd immunity   The resistance of a population to infection and spread of an infectious organism due to the immunity of a high percentage of the population.

High-efficiency filters   Filters with rated efficiencies of 20 percent DSP or higher. This category usually excludes dust filters, prefilters, HEPA filters, and ultra-low-penetration air (ULPA) filters. May be rated MERV 6 to 16.

Histoplasmosis   A systemic fungal infection caused by *Histoplasma capsulatum*.

Host   An animal or plant that is used by another organism, typically a microorganism, for shelter and sustenance. Sometimes this may be a commensal relationship, but often it is a parasitic relationship in which the host can only lose.

Hot zone   Area immediately surrounding a hazardous materials spill or release incident. Also called an "exclusion zone" or "restricted zone."

IAQ   Indoor air quality.

ID_{50} mean infectious dose   The dose or number of microorganisms that will cause infections in 50 percent of an exposed population. Applies only to microorganisms, not toxins or chemical weapons. Units are always in terms of number of microorganisms, or, more correctly, the number of colony-forming units per cubic meter (cfu/m^3).

Immiscible   A material that does not mix readily with water.

Immune building   A building that has the engineered ability to prevent or protect against the spread of disease or the dissemination of chemical agents.

Immunocompromised   The condition in which a person's immunity to disease has been reduced, whether due to health or injury.

Immunodeficiency   The state of being unable to protect against disease. The inability to produce normal antibodies in response to antigens.

Immunosuppression   The reduction of immune system response that leaves a person susceptible to diseases.

Impaction   The collision of a particle with a filter fiber that results in attachment.

Impregnated carbon   Carbon adsorber material that has compounds added to increase adsorption efficiency for specific chemicals.

Incapacitating agent   Agents that produce physical or psychological effects, or both, that may persist for hours or days after exposure, rendering victims incapable of performing normal physical and mental tasks.

Incubation   The time during which a pathogen grows and multiplies to cause an infection. The process of maintaining a bacterial culture in ideal conditions to induce growth of colonies.

Infectious agents   Biological agents capable of causing disease in a susceptible host.

Infectious period   The period during which an infected person is contagious.

Infectivity   (1) The ability of an organism to spread. (2) The number of organisms required to cause an infection in secondary hosts. (3) The capability of an organism to spread out from the site of an infection and cause disease in the host organism. Infectivity also can be viewed as the number of organisms required to cause an infection.

Inoculation   The use of subcutaneous vaccines. Also, the process of initiating a culture on a petri dish.

Interception   The process by which particles attach to filter fibers when they pass close enough to come into contact.

Inverse square law   A simplistic description of the decrease of light or radiation intensity as a function of the inverse of the square of the distance from the source.

Ionization   The process of stripping electrons from atoms to produce ions.

Ionizing radiation   Radiation of short wavelengths or high energy that causes atoms to lose their electrons, or become ionized.

Irradiance   The density of radiation incident upon a flat surface.

Isolation room   Rooms in which the pressure is controlled to prevent microbial pathogens from entering or exiting. May be positively or negatively pressurized, depending on the function.

Kill zone   Areas or volumes in which pathogens are destroyed. Usually relates to UVGI exposure.

LC$_{50}$ Mean lethal concentration. The concentration of chemical agents in the air that will cause 50 percent fatalities in the exposed population. Does not apply to microorganisms and is not used for toxins. The units are either mg·min/m^3 or mg/kg. The former unit represents an exposure while the latter represents an absorbed, inhaled, injected, or ingested dose.

L(Ct)$_{50}$ Mean lethal exposure. Dose specified in terms of a concentration and an exposure time.

Line-source delivery system   A delivery system in which the biological agent is dispersed from a linear array of release cylinders, or from a moving ground or air vehicle in a line perpendicular to the direction of the prevailing wind. Used for spreading out the contaminated zone.

LD$_{50}$ Mean lethal dose. The dose or number of microorganisms that will cause fatalities in 50 percent of an exposed population. Applies to microorganisms, toxins, or chemical weapons. For microorganisms, the units are number of microbes (or cfu/m^3). For toxins and chemicals, the units are either mg·min/m^3 or mg/kg. The former unit represents an exposure while the latter represents an absorbed, inhaled, injected, or ingested dose.

LD$_{Lo}$ Usually refers to the lowest lethal dose.

Legionnaires' disease   A pulmonary infection caused by *Legionella pneumophila*.

Lethality curve   The dose-response curve for fatalities.

Light pipes   Flexible tubes used to deliver light or radiation.

Liquid agent   A chemical agent in the form of a liquid. Sometimes these are compressed gases in liquid form.

Logarithmic decay   The classical shape of the decay curve of a population exposed to biocidal factors.

Logmean   The average of the logarithms of any quantity, such as microbe diameters.

Lognormal   The distribution in which the logarithm of some factor such as size is distributed normally (i.e., as a bell curve). The characteristic size distribution for all microorganisms and micron-sized particles in which larger numbers of particles exist at the smaller sizes. Has a standard deviation which is not constant but is a function of the coefficient of variation.

Mean diameter   The actual median diameter for a population of microbes or particles.

Mean disease period   The mean period for any disease during which symptoms are manifest.

Mean infectious period   The mean period for any contagious disease during which the disease is communicable.

Media velocity The actual velocity through pleated filter media, as opposed to the face velocity through the filter cartridge.

MERV Minimum efficiency reporting value. Rating system for filters based on ASHRAE Standard 52.2-1999. Filters are rated from1 (lowest) to 16 (highest).

Microbe Synonym for microorganism.

Microemulsions Mixtures of water, oil, surfactants, and cosurfactants that behave as a homogenous mixture and can dissolve chemical agents.

Micron One-millionth of a meter; also μm.

Microorganism Any organism, such as bacteria, viruses, fungi, and protozoa, that can be seen only with a microscope.

Mildew A term that refers to mold when it grow on fabrics or textiles.

Miscible A material that readily mixes with water.

Mist A suspension of liquid droplets. Usually created by atomization or vapor condensation.

Mold The form in which fungi will grow in dry or moist conditions. Called "mildew" when growth occurs on fabrics or textiles. The alternative growth form, yeast, occurs in solution or in a host during an infection.

Morbidity Reduced health or lethargy due to disease.

Multihit model A model of the decay of microbial populations that accounts for shoulder curves.

Multizone Being divided into separate ventilation zones.

Mycoplasma Bacteria of the class Mollicutes that lack a cell wall.

Mycosis Any disease caused by fungi.

Mycotoxin A toxin produced by fungi, such as T-2 mycotoxin, aflatoxin, or ochratoxin.

Natural ventilation Ventilation without the use of powered equipment.

Nebulizer A device for producing a fine spray or aerosol.

Nerve agents Substances that interfere with the central nervous system. Exposure is primarily through contact with the liquid (skin and eyes) and secondarily through inhalation of the vapor. Symptoms of nerve agents include pin-point pupils, extreme headache, and severe tightness in the chest.

Neurotoxin A toxin that destroys nerve tissue.

Nonpersistent agent An agent that upon release loses its ability to cause casualties after 10 to 15 min.

Normal distribution The Gaussian distribution of data in a bell curve.

Nosocomial Refers to infections that occur in hospitals.

Open-air lethality The lethality of an agent in open air.

Organism   Any individual living thing, whether animal or plant.

Organophosphorous compound   A compound, containing the elements phosphorus and carbon, whose physiological effects include the inhibition of acetylcholinesterase.

Overload attack   Using much more of an agent than is necessary to cause 100 percent fatalities.

Ozonation   The use of ozone to disinfect air or water.

Ozone   A corrosive form of oxygen consisting of three bound oxygen molecules.

Parasite   Any organism that lives in or exploits another organism without providing any benefit in return.

Passive release   Release of an agent by spilling, dusting, or other means without the use of dispersion equipment.

Passive solar exposure   The use of solar exposure to disinfect surfaces or materials.

Pathogen   Any organism capable of producing disease or infection.

Pathogenic agents   Biological agents capable of causing serious disease.

Pathogenicity   The disease-causing ability of a pathogen. The degree of health hazard posed by such microbes.

Pathology   The study of disease or the course of a particular disease.

PCO   Photocatalytic oxidation.

Peak infection period   The mean period for any contagious disease during which the host is infectious.

Penetration   The degree to which particles or microbes penetrate a filter. The complement of removal efficiency. %Penetration = 100 − %Efficiency.

Penicillins   A group of antibiotics that are active against gram-negative bacteria.

Percutaneous agent   Able to be absorbed by the body through the skin.

Persistent agent   An agent that upon release retains its casualty-producing effects for an extended period of time. Considered to be a long-term hazard.

Pfu   Plaque-forming unit. Used to measure the presence of viruses in culture. Similar to cfu.

Photocatalytic oxidation   A technology in which material coated with titanium dioxide is exposed to visible or ultraviolet light to produce an oxidative effect on both volatile organic compounds (VOCs) and biological agents.

Physical security   Mechanical devices, such as locks or fences, that provide security.

Plague   An acute, febrile, infectious disease caused by *Yersinia pestis*.

Plug flow   The type of airflow pattern in a room in which air moves from one side to the other in piston fashion.

Pneumonia   The presence of excessive fluid in the lungs. May be caused by any one of a number of pathogens.

Point-source delivery system   A delivery system in which the biological agent is dispersed from a stationary position. This delivery method results in coverage over a smaller area than with the line-source system.

Potable water   Clean drinking water.

Prefilters   Filters used before other equipment, including UVGI systems, carbon adsorbers, and high-efficiency filters. Prefilters may typically be dust filters or 25 percent high-efficiency filters.

Protective clothing   Respiratory and physical protection. Four levels are defined: Level A—self-contained breathing apparatus (SCBA) with totally encapsulating nonpermeable clothing; Level B—SCBA with hooded nonpermeable clothing; Level C—full or half-face respirator with hooded nonpermeable clothing; Level D—coverall with no respiratory protection.

Protozoa   A multicelled microscopic animal. These microorganisms may sometimes be involved in human disease.

Pulmonary edema   The accumulation of excessive fluid in the lungs.

Pulsed filtered light   Pulsed white light with the UVC spectrum removed.

Pulsed-light disintegration   The disintegrating effect on bacterial cells from excess power levels of pulsed light. A non-UV-dependent phenomenon.

PWL   Pulsed white light.

Q-fever   An acute zoonotic disease caused by *Coxiella burnetii*.

Radiation view factor   A quantity that defines the amount of radiation transmitted from a radiating surface and absorbed by another surface.

Radiological   Involving radioactive materials.

Recirculation   The return of airflow, processed or otherwise, to the building volume by the ventilation system.

Relative risk   The risk of a CBW attack on a building in relation to other buildings.

Remediation   The cleanup or decontamination of an area that has been exposed to CBW agents, molds, or other hazards.

Retrofit   To add a new component to an existing system.

Route of exposure   The path by which a person comes into contact with an agent or organism, for example, through breathing, digestion, or skin contact.

Second stage   The portion of a microbial population decay curve after the first stage. Usually results from a fraction of the population being resistant to some biocidal factor.

Sensitive information   Any information which may compromise the security of an immune building.

Septicemia   A disease associated with the presence of pathogens in the blood.

Sheltering in place   The use of an area of a building for protection against CBW attacks.

Simulant   A normally harmless microbe used to simulate a pathogenic microbe for testing, or a harmless chemical used to simulate a CW agent for testing.

Single-cell protein   Protein-rich material obtained from cultured algae, fungi, protein, and bacteria, and often used as food or animal feed.

Sorbent   Any material used to absorb chemical agents.

Spore   A seed-like form that some microorganisms will adopt as a reproductive mechanism, or as a hedge against adverse environmental conditions. Spores are dormant, with no biological activity, and tend to be resistant to biocidal factors. They reactivate, or germinate, when moisture and temperature conditions are appropriate.

Standard deviation   A parameter of a normal or bell curve that defines how far out the curve spreads.

Streptomycin   An antibiotic produced by *Streptomyces griseus*.

Sudden release   The release of a quantity of a CBW agent all at once, or in a short time period.

Supertropical bleach   A bleaching agent designed for the decontamination of chemical agents, especially G agents. A mixture of 93 percent calcium hypochlorite and 7 percent sodium hydroxide.

Surfactants   Materials that solubilize chemical agents and convert them into a solution that can be detoxified.

Tear gas agents   These produce irritating or disabling effects that rapidly disappear within minutes after exposure ceases.

Tetanus   An often fatal disease caused by *Clostridium tetani*.

TIH   Toxic inhalation hazard.

Total arrestance   A rating system for filters that measures the total removal efficiency over a broad range of particle sizes.

Toxicity   A measure of the harmful effect produced by a given amount of a toxin on a living organism. The relative toxicity of an agent can be expressed in milligrams of toxin needed per kilogram of body weight to kill experimental animals.

Toxigenic   Containing or capable of producing toxins.

Toxins   Biologically generated poisonous substances. Includes venoms, paralytic agents, enterotoxins, endotoxins, neurotoxins, and other toxins created by plants, insects, animals, fish, bacteria, or fungi.

Tracer gas   A harmless gas used for testing ventilation systems and equipment.

Trichothecenes   A class of compounds produced by certain fungi that includes some toxins.

Tuberculosis   An infectious disease caused by *Mycobacterium tuberculosis.*

Tularemia   A plaguelike disease of animals caused by *Francisella tularemia.*

Typhoid fever   A bacterial infection transmitted by contaminated food or water. Caused by *Salmonella typhi,* which is present in human feces.

ULPA filters   Ultra-low-penetration air filters. Filters that have efficiencies exceeding those of HEPA filters. May be 99.999 percent or higher at 0.3 micron.

Ultrasonic atomizers   Aerosolizers that use ultrasonics to generate vapors or mists.

Ultraviolet light   Light in the range of approximately 200 to 400 nm.

Upper air irradiation   The use of UVGI to irradiate the air in rooms above the level of occupancy so as to disinfect the air and limit human exposure.

UR   Upper respiratory.

URD   Upper respiratory disease.

URV   UVGI rating value.

UV   Common abbreviation for ultraviolet.

UVA   Ultraviolet light in the A band region, approximately 315 to 400 nm. Has minor biocidal effects.

UVB   Ultraviolet light in the B band region, approximately 280 to 315 nm. Has minor biocidal effects.

UVC   Ultraviolet light in the C band region, approximately 200 to 280 nm. Responsible for the greatest part of the biocidal effects of UV light.

UVGI   Ultraviolet germicidal irradiation. The term defined by the CDC to describe the technology of using ultraviolet light to disinfect air and surfaces.

Vaccine   A preparation of killed or weakened microorganism products used to artificially induce immunity against a disease.

Vapor   Aerosolized form of a chemical agent or liquid that exists in airborne droplet form.

Vapor pressure   The pressure at which a liquid and its vapor are in equilibrium at a given temperature.

Variable air volume   Ventilation systems that adjust the amount of outside air based on outside air conditions to save energy.

Variolation   An early form of vaccination against smallpox.

VAV   Variable air volume.

Vector   An agent, such as a mosquito, tick, or rat, serving to transfer a pathogen from one organism to another.

VEE   Venezuelan equine encephalitis virus.

Venom   A poisonous toxin produced in the glands of certain animals like snakes, scorpions, wasps, or bees.

Ventilation effectiveness   A measure of how well the air is purged from a room.

Vesicant   A blister agent or chemical agent that is absorbed through the skin.

View factor   Radiation view factor.

Virion   A single virus particle.

Virulence   The quality that defines how pathogenic a microbe is, or how deadly it is, or how incapacitating a disease it causes, or how rapidly it causes an infection.

Virus   An infectious microorganism that exists as a particle rather than as a complete cell. Particle sizes range from 20 to 400 nanometers (one-billionth of a meter). Viruses are not capable of reproducing outside of a host cell and may consist of little more than a strand of DNA or RNA.

VOCs   Volatile organic compounds.

Volatile organic compounds   Airborne compounds that come from microbial sources such as bacteria or fungi.

Volatility   A measure of how readily a substance will vaporize.

Vomiting agents   These produce nausea and vomiting effects; they can also cause coughing, sneezing, pain in the nose and throat, nasal discharge, and tears.

V-series nerve agents   Chemical agents of moderate to high toxicity developed in the 1950s. They are generally persistent. Examples are VE, VG, VM, VS, and VX.

Warm zone   Area between a hot zone and a cold zone where decontamination functions take place.

Water-borne pathogens   Pathogenic microorganisms that can survive or grow in water and which can transmit to humans who drink the water.

Water spray   A method of spraying water in a fine aerosol. Can be used to move gases.

Weaponization   The process of converting a chemical or biological agent into a form that can be easily delivered or that can survive delivery mechanisms. For toxins this involves grinding to a size that permits aerosolization. For microorganisms this usually involves selecting virulent strains that are highly infectious, and hardy strains that can survive aerosolization or other delivery mechanisms.

Weathering   The use of the natural elements of sunlight and temperature extremes to decontaminate or disinfect surfaces or materials.

WEE   Western equine encephalitis virus.

Yellow fever   An acute infectious disease caused by a flavivirus and transmitted by mosquitoes.

Zoonosis   Disease of animals that can be transmitted to humans.

References

Abshire, R. L., and Dunton, H. (1981). "Resistance of selected strains of *Pseudomonas aeruginosa* to low-intensity ultraviolet radiation." *Appl. Envir. Microb.* 41(6), 1419–1423.

ACGIH (1973). *Threshold Limit Values of Physical Agents*. The American Conference of Government Industrial Hygienists, Cincinnati, OH.

Aerotech (2001). *IQ Sampling Guide*. Aerotech Laboratories, Phoenix, AZ.

Agnello, V., Abel, G., Elfahal, M., Knight, G. B., and Zhang, Q.-X. (1999). "Hepatitis C virus and other Flaviviridae viruses enter cells via low density lipoprotein receptor." *PNAS* 96(22), 12766–12771.

AIA (1993). *Guidelines for Construction and Equipment of Hospital and Medical Facilities*. Mechanical Standards/American Institute of Architects, Washington, DC.

AL (2002). *Eikon Version 3.02*. Automated Logic. http://www.automatedlogic.com/eikon.

Al-Doory, Y., and Ramsey, S. (1987). *Moulds and Health: Who Is at Risk?* Charles C. Thomas, Springfield, IL.

Alibek, K., and Handelman, S. (1999). *Biohazard*. Random House, New York, NY.

Alimov, A. S., Knapp, E. A., Shvedunov, V. I., and Trower, W. P. (2000). "High-power CW LINAC for food irradiation." *Applied Radiation and Isotopes* 53(4-5), 815–820.

Allegra, L., Blasi, F., Tarsia, P., Arosio, C., Fagetti, L., and Gazzano, M. (1997). "A novel device for the prevention of airborne infections." *J. Clin. Microbiol.* 35(7), 1918–1919.

Allen, E. G., Bovarnick, M. R., and Snyder, J. C. (1954). "The effect of irradiation with ultraviolet light on various properties of *Typhus rickettsiae*." *J. Bacteriol.* 67, 718–723.

Alpen, E. L. (1990). *Radiation Biophysics*. Prentice-Hall, Englewood Cliffs, NJ.

Alper, T. (1979). *Cellular Radiobiology*. Cambridge University Press, Cambridge, U.K.

Andrews, L. S., Park, D. L., and Chen, Y. P. (2000). "Low temperature pasteurization to reduce the risk of vibrio infections from raw shell-stock oysters." *Food Addit. Contam.* 17(9), 787–791.

Antopol, S. C., and Ellner, P. D. (1979). "Susceptibility of *Legionella pneumophila* to ultraviolet radiation." *Appl. Environ. Microbiol.* 38(2), 347–348.

Arrieta, R. T., and Huebner, J. S. (2000). "Photo-electric chemical and biological sensors." Proceedings of the SPIE 2000, *Chemical and Biological Sensing*, Orlando, FL, 132–142.

ASD (2002). *American School of Defense Home Page*. American School of Defense. http://www.asod.org/index.htm.

Ashford, D. A., Hajjeh, R. A., Kelley, M. F., Kaufman, L., Hutwagner, L., and McNeil, M. M. (1999). "Outbreak of histoplasmosis among cavers attending the National Speleological Society Annual Convention, Texas, 1994." *American Journal of Tropical Medicine and Hygiene* 60(6), 899–903.

ASHRAE (1985). *Handbook of Fundamentals*. American Society of Heating, Refrigeration, and Air Conditioning Engineers, Atlanta, GA.

ASHRAE (1991). "Health facilities." In *ASHRAE Handbook of Applications*. American Society of Heating, Refrigeration, and Air Conditioning Engineers, Atlanta, GA.

ASHRAE (1992). "Air cleaners for particulate contaminants." Chapter 25 in *ASHRAE Systems Handbook*. American Society of Heating, Refrigeration, and Air Conditioning Engineers, Atlanta, GA.

ASHRAE (1995). *Standard 135-1995*. American Society of Heating, Refrigeration, and Air Conditioning Engineers, Atlanta, GA.

ASHRAE (1999a). *ASHRAE Standard 52.2-1999*. American Society of Heating, Refrigeration, and Air Conditioning Engineers, Atlanta, GA.

ASHRAE (1999b). "Fire and smoke management." Chapter 51 in *HVAC Applications Handbook*. American Society of Heating, Refrigeration, and Air Conditioning Engineers, Atlanta, GA.

ASHRAE (1999c). "Control of gaseous indoor air contaminants." Chapter 44 in *HVAC Applications Handbook*. American Society of Heating, Refrigeration, and Air Conditioning Engineers, Atlanta, GA.

ASHRAE (2002). *Risk Management Guidance for Health and Safety under Extraordinary Incidents*. American Society of Heating, Refrigeration, and Air Conditioning Engineers, Atlanta, GA.

Atlas, R. M. (2001). "Bioterrorism before and after September 11." *Crit. Rev. Microbiol.* 27(4), 335–379.

ATSDR (2002). *Agency for Toxic Substances and Disease Registry*. http://atsdr1.atsdr.cdc.gov/atsdrhome.html.

Austin, B. (1991). *Pathogens in the Environment*. The Society for Applied Bacteriology/Blackwell Scientific Publications, Oxford, U.K.

AWWA (1971). *Water Quality and Treatment*. The American Water Works Association/McGraw-Hill, New York, NY.

Bakken, L. R., and Olsen, R. A. (1983). "Buoyant densities and dry-matter contents of microorganisms: Conversion of a measured biovolume into biomass." *Appl. Environ. Microbiol.* 45.

Bardi, J. (1999). "Aftermath of a hypothetical smallpox disaster." *Emerging Infectious Diseases* 5(4), 547–551.

Barnaby, W. (2000). *The Plague Makers: The Secret World of Biological Warfare*. Continuum, New York, NY.

Bashor, M. M. (1998). "International terrorism and weapons of mass destruction." *Risk Analysis* 18(6), 675.

Baskerville, A., and Hambleton, P. (1976). "Pathogenesis and pathology of respiratory tularemia in the rabbit." *Br. J. Exp. Pathol.* 57, 339–347.

Beebe, J. M. (1958). "Stability of disseminated aerosols of *Pasteurella tularensis* subjected to simulated solar radiations at various humidities." *Journal of Bacteriology* 78, 18–24.

Belgrader, P., Benett, W., Begman, W., Langlois, R. R., Mariella, J., Milanovich, F., Miles, R., Venkateswaran, K., Long, G., and Nelson, W. (1998). "Autonomous system for pathogen detection and identification." Proceedings of the SPIE, *Air Monitoring and Detection of Chemical and Biological Agents*, Boston, MA, 198–206.

Bell, A. A., Jr. (2000). *HVAC Equations, Data, and Rules of Thumb*. McGraw-Hill, New York, NY.

Beltran, F. J. (1995). "Theoretical aspects of the kinetics of competitive ozone reactions in water." *Ozone Sci. Eng.* 17, 163–181.

Berendt, R. F., Young, H. W., Allen, R. G., and Knutsen, G. L. (1980). "Dose-response of guinea pigs experimentally infected with aerosols of *Legionella pneumophila*." *Journal of Infectious Diseases* 141(2), 186–192.

BioChem (2002). Personal communication with W. J. Kowalski from BioChem Technologies, LLC, Toronto, Canada.

Block, S. S., and Goswami, D. Y. (1995). "Chemically enhanced sunlight for killing bacteria." *Transactions of the ASME—Solar Engineering* 1, 431–437.

Boelter, K. J., and Davidson, J. H. (1997). "Ozone generation by indoor electrostatic air cleaners." *Aerosol Science and Technology* 27(6), 689–708.

Bolton, J. R. (2001). *Ultraviolet Applications Handbook*. Bolton Photosciences, Inc., Ayr, Ontario, Canada.

Bosche, H. V., Odds, F., and Kerridge, D. (1993). *Dimorphic Fungi in Biology and Medicine*. Plenum Publishers, New York, NY.

Boss, M. J., and Day, D. W. (2001). *Air Sampling and Industrial Hygiene Engineering*. Lewis Publishers, Boca Raton, FL.

Boyce, W. E., and DiPrima, R. C. (1997). *Elementary Differential Equations and Boundary Value Problems*. John Wiley & Sons, New York, NY.

Brachman, P. S., Kaufmann, A. F., and Dalldorf, F. G. (1966). "Industrial inhalation anthrax." *Bacteriological Reviews* 30(3), 646–657.

Bradley, D., Burdett, G. J., Griffiths, W. D., and Lyons, C. P. (1992). "Design and performance of size selective microbiological samplers." *Journal of Aerosol Science* 23(S1), s659–s662.

Bratbak, G., and Dundas, I. (1984). "Bacterial dry matter content and biomass estimations." *Appl. Environ. Microbiol.* 48, 755–757.

Braude, A. I., Davis, C. E., and Fierer, J. (1981). *Infectious Diseases and Medical Microbiology,* 2nd ed. W. B. Saunders Company, Philadelphia, PA.

Breman, J. G., and Henderson, D. A. (1998). "Poxvirus dilemmas—monkeypox, smallpox, and biologic terrorism." *New England Journal of Medicine* 339(8), 556–559.

Brletich, N. R., Waters, M. J., Bowen, G. W., and Tracy, M. F. (1995). *Worldwide Chemical Detection Equipment Handbook.* Defense Technical Information Center, Chemical and Biological Defense Information Analysis Center, Gunpowder Br. APG, MD.

Brown, R. C. (1993). *Air Filtration.* Pergamon Press, Oxford, U.K.

Brown, R. C., and Wake, D. (1991). "Air filtration by interception—theory and experiment." *Journal of Aerosol Science* 22(2), 181–186.

Burmester, B. R., and Witter, R. L. (1972). "Efficiency of commercial air filters against Marek's disease virus." *Appl. Microbiol.* 23(3), 505–508.

Burroughs, H. E. B. (1998). "Improved filtration in residential environments." *ASHRAE J.* 40(6), 47–51.

Bushby, S. T. (1999). *GSA Guide to Specifying Interoperable Building Automation and Control Systems Using ANSI / ASHRAE Standard 135-1995.* NIST GPO, U.S. Dept. of Commerce, Technology Admin., NIST, Gaithersburg, MD.

Bushnell, A., Clark, W., Dunn, J., and Salisbury, K. (1997). "Pulsed light sterilization of products packaged by blow-fill-seal techniques." *Pharmaceutical Engineering* 17(5), 74–84.

Bushnell, A., Cooper, J. R., Dunn, J., Leo, F., and May, R. (1998). "Pulsed light sterilization tunnels and sterile-pass-throughs." *Pharmaceutical Engineering.* March/April, 48–58.

BWC (1995). *Ad Hoc Group of the States Parties to the Convention on the Prohibition of the Development, Production and Stockpiling of Bacteriological (Biological) and Toxin Weapons and on their Destruction: List of Agents."* Biological Weapons Convention, Geneva, Switzerland.

Calder, K. L. (1957). *Mathematical Models for Dosage and Casualty Coverage Resulting from Single Point and Line Source Release of Aerosol near Ground Level.* BWL Tech. Study 3, AD310-361. Defense Technical Information Center, Ft. Belvoir, VA.

Canada (1991a). *Novel Toxins and Bioregulators: The Emerging Scientific and Technology Issues Relating to Verification and the Biological and Toxin Weapons Convention.* External Affairs and International Trade Canada, Ottawa, Ontario, Canada.

Canada (1991b). *Collateral Analysis and Verification of Biological and Toxin Research in Iraq.* External Affairs and International Trade Canada, Ottawa, Ontario, Canada.

Canada (1992). *The Chemical Weapons Convention and the Control of Scheduled Chemicals in Canada.* External Affairs and International Trade Canada, Ottawa, Ontario, Canada.

Canada (1993). *Chemical Weapons Convention Verification: Handbook on Scheduled Chemicals.* External Affairs and International Trade Canada, Ottawa, Ontario, Canada.

Canada (2001). *Office of Laboratory Security Material Safety Data Sheets.* Canada Population and Public Health Branch, Ottawa, Ontario, Canada. http://www.hc-sc.gc.ca/pphb-dgspsp/msds-ftss/index.html.

Cannaliato, V. J., Jezek, B. W., Hyttinen, L., Strawbridge, J., and Ginley, W. J. (2000). "Short range biological standoff detection system (SR-BSDS)." Proceedings of the SPIE, *Chemical and Biological Sensing,* Orlando, FL, 219–223.

Cao, K. L., Anderson, G. P., Ligler, F. S., and Ezzell, J. (1995). "Detection of *Yersinia pestis* Fraction 1 antigen with a fiber optic biosensor." *J. Clin. Microbiol.* 33, 336–341.

Cardosi, M., Birch, S., Talbot, J., and Phillips, A. (1991). "An electrochemical immunoassay for *Clostridium pefringens* phospholipase C." *Electroanalysis* 3, 1–8.

Carlson, M. A., Bargeron, C. B., Benson, R. C., Fraser, A. B., Groopman, J. D., Ko, H. W., Phillips, T. E., Strickland, P. T., and Velky, J. T. (1999). "Development of an automated handheld imunoaffinity fluorometric biosensor." *Johns Hopkins APL Technical Review* 20(3), 372–380.

Carpenter, G. A., Cooper, A. W., and Wheeler, G. E. (1986a). "The effect of air filtration on air hygiene and pig performance in early-weaner accommodation." *Animal Production* 43, 505–515.

Carpenter, G. A., Smith, W. K., MacLaren, A. P. C., and Spackman, D. (1986b). "Effect of internal air filtration on the performance of broilers and the aerial concentrations of dust and bacteria." *British Poultry Science* 27, 471–480.

Carrier (1993). *Energy Analysis User's Manual for the Hourly Analysis Program.* Carrier Corporation, Farmington, CT.

Carrier (1994). *Design Load User's Manual for the Hourly Analysis Program and the System Design Load Program.* Carrier Corporation, Farmington, CT.

Carus, W. S. (1998). "Biological Warfare Threats in Perspective." *Crit. Rev. Microbiol.* 27(3), 149–155.

Casarett, A. P. (1968). *Radiation Biology.* Prentice-Hall, Englewood Cliffs, NJ.

CAST (1989). *Mycotoxins: Economic and Health Risks.* Center for the Advancement of Science and Technology, Council for Agricultural Science and Technology, Ames, IA.

Castle, M., and Ajemian, E. (1987). *Hospital Infection Control.* John Wiley & Sons, New York, NY.

Cater, R. M., Jacobs, M. B., Lubrano, G. J., and Guillbault, G. G. (1995). "Piezoelectric detection of ricin and affinity-purified goat anti-ricin antibody." *Anal. Lett.* 28, 1379–1386.

Cavalcante, M. J. B., and Muchovej, J. J. (1993). "Microwave irradiation of seeds and selected fungal spores." *Seed Sci. Technol.* 21, 247–253.

CDC (2001). *Counter-Terrorism Cards (CDC/NIOSH).* Centers for Disease Control, Atlanta, GA. http://www.bt.cdc.gv/Agent/Agentlist.asp.

CDC (2002a). *Bioterrorism Info for Healthcare.* Centers for Disease Control, Atlanta, GA. http://www.cdc.gov/ncidod/hip/Bio/bio.htm.

CDC (2002b). *Emergency Response Website.* Centers for Disease Control, Atlanta, GA. http://www.bt.cdc.gov/EmContact/index.asp.

Cerf, O. (1977). "A review: Tailing of survival curves of bacterial spores." *J. Appl. Bacteriol.* 42, 1–19.

Chang, C. Y., Chiu, C. Y., Lee, S. J., Huang, W. H., Yu, Y. H., Liou, H. T., Ku, Y., and Chen, J. N. (1996). "Combined self-absorption and self-decomposition of ozone in aqueous solutions with interfacial resistance." *Ozone Sci. Eng.* 18, 183–194.

Chang, I. (1997). *The Rape of Nanking—The Forgotten Holocaust of World War II.* Basic Books, New York, NY.

Chen, C. C., and Huang, S. H. (1998). "The effects of particle charge on the performance of a filtering facepiece." *American Industrial Hygiene Association J.* 59, 227–233.

Cheng, Y. S., Yeh, H. C., and Kanapilly, G. M. (1981). "Collection efficiencies of a point-on-plane electrostatic precipitator." *Am. Ind. Hyg. Assoc. J.* 42, 605–610.

Cheremisinoff, P. N., and Ellerbusch, F. (1978). *Carbon Adsorption Handbook.* Ann Arbor Science, Ann Arbor, MI.

Chick, E. W., A. B. Hudnell, J., and Sharp, D. G. (1963). "Ultraviolet sensitivity of fungi associated with mycotic keratitis and other mycoses." *Sabouviad* 2(4), 195–200.

Chick, H. (1908). "An investigation into the laws of disinfection." *J. Hyg.* 8, 92.

Christopher, G. W., Cieslak, T. J., Pavlin, J. A., and Eitzen, E. M. (1997). "Biological warfare: A historical perspective." *JAMA* 278(5), 412–417.

Cieslak, T. J., and E. M. Eitzen, J. (1999). "Clinical and epidemiological principles of anthrax." *Emerg. Infect. Dis.* 5(4), 552–555.

Clark, W., Bushnell, A., Dunn, J., and Ott, T. (1997). *Pulsed Light and Pulsed Electric Fields for Food Preservation.* Annual Meeting Abstract. American Institute of Chemical Engineers (AIChE), New York, NY.

Clarke, R. (1968). *We All Fall Down.* Allen Lane The Penguin Press, London, U.K.

Clarke, R. A. (1999). "Finding the right balance against bioterrorism." *Emerging Infectious Diseases* 5(4), 497. http://www.cdc.gov/ncidod/EID/vol5no4/clarke.htm.

Cliver, D. O. (1990). *Foodborne Diseases.* Academic Press, San Diego, CA.

CNS (2002). *"Agro-Terrorism: Chronology of CBW Attacks Targeting Crops & Livestock 1915–2000."* Center for Non-Proliferation Studies, Monterey, CA. http://cns.miis.edu/research/cbw/agchron.htm#N_8_.

Coggle, J. E. (1971). *Biological Effects of Radiation.* Wykeham Publ., London, U.K.

Cole, L. A. (1996). "The specter of biological weapons." *Sci. Amer.* Dec., 60–65.

Collier, L. H., McClean, D., and Vallet, L. (1955). "The antigenicity of ultra-violet irradiated vaccinia virus." *J. Hyg.* 53(4), 513–534.

Collins, C. H. (1993). *Laboratory-Acquired Infections.* Butterworth-Heinemann, Oxford, U.K.

Collins, F. M. (1971). "Relative susceptibility of acid-fast and non-acid fast bacteria to ultraviolet light." *Appl. Microbiol.* 21, 411–413.

Cooke, J. R. (2001). "News article: Researcher gets anti-semitic e-mail." *The Daily Collegian* 11-29-01 (Thursday). Pennsylvania State University, State College, PA.

Cookson, J., and Nottingham, J. (1969). *A Survey of Chemical and Biological Warfare.* Monthly Review Press, New York, NY.

Cornell (2001). *Material Safety Data Sheets.* Cornell University, Ithaca, NY. http://msds.pdc.cornell.edu/msdssrch.asp.

Cornish, T. J., and Bryden, W. A. (1999). "Miniature time-of-flight mass spectrometer for a field-portable biodetection system." *Johns Hopkins APL Technical Digest* 20(3), 335–342.

CSIS (1995). *The Threat of Chemical / Biological Terrorism.* Canadian Security Intelligence Service, Ottawa, Ontario, Canada. http://www.csis-scrs.gc.ca/eng/comment/com60e.html.

Cullis, C. F., and Firth, J. G. (1981). *Detection and Measurement of Hazardous Gases.* Heineman, London, U.K.

Dalton, A. J., and Haguenau, F. (1973). *Ultrastructure of Animal Viruses and Bacteriophages: An Atlas.* Academic Press, New York, NY.

Darlington, A., Dixon, M. A., and Pilger, C. (1998). "The use of biofilters to improve indoor air quality: The removal of toluene, TCE, and formaldehyde." *Life Support Biosph. Sci.* 5(1), 63–69.

Darlington, A. B., Dat, J. F., and Dixon, M. A. (2001). "The biofiltration of indoor air: Air flux and temperature influences the removal of toluene, ethylbenzene, and xylene." *Environ. Sci. Technol.* 35(1), 240–246.

David, H. L. (1973). "Response of mycobacteria to ultraviolet radiation." *Am. Rev. Resp. Dis.* 108, 1175–1184.

Davidovich, I. A., and Kishchenko, G. P. (1991). "The shape of the survival curves in the inactivation of viruses." *Mol. Gen. Microbiol. Virol.* 6, 13–16.

Davies, C. N. (1973). *Air Filtration.* Academic Press, London, U.K.

Davies, J. C. (1982). "A major epidemic of anthrax in Zimbabwe, Part 1." *Cent. Afr. J. Med.* 28(12), 291–298.

Davies, J. C. (1983). "A major epidemic of anthrax in Zimbabwe, Part 2." *Cent. Afr. J. Med.* 29(1), 8–12.

Davies, J. C. (1985). "A major epidemic of anthrax in Zimbabwe: The experience at the Beatrice Road Infectious Diseases Hospital, Harare." *Centr. Afr. J. Med.* 31(9), 176–180.

Davis, J., and Johnson-Winegar, A. (2000). *The Anthrax Terror: DOD's Number-One Biological Threat.* USAF, Maxwell AFB, AL. http://www.airpower.maxwell.af.mil/airchronicles/apj/apj00/win00/davis.htm.

Dept. of the Army (1989). *Field Manual FM 8-285: Treatment of Chemical Agent Casualties and Conventional Military Chemical Injuries.* Government Printing Office, Washington, DC. http://www.nbc-med.org/SiteContent/MedRef/OnlineRef/ FieldManuals/ fm8_285/about.htm.

Dept. of the Army (1996). *Field Manual 3-4: NBC Protection.* U.S. Marine Corps, Washington DC. http://155.217.58.58/cgi-bin/atdl.dll/fm/3-4/toc.htm.

Di Salvo, A. F. (1983). *Occupational Mycoses.* Lea & Febiger, Philadelphia, PA.

Diaz-Cinco, M., and Martinelli, S. (1991). "The use of microwaves in sterilization." *Dairy Food Environ. Sanit.* 11(12), 722–724.

Dietz, P., Bohm, R., and Strauch, D. (1980). "Investigations on disinfection and sterilization of surfaces by ultraviolet radiation." *Zbl. Bakt. Mikrobiol. Hyg.* 171(2-3), 158–167.

D-Mark, Inc. (1995). *Concept of Molecular Screening in Micropores.* D-Mark, Chesterfield, MI.

Dobyns, H. F. (1966). "Estimating aboriginal American population: An appraisal of techniques with a new hemisphere estimate." *Current Anthropology* 7, 415.

DOE (1994). *BLAST: Building Loads Analysis and System Thermodynamics.* Department of Energy, Washington, DC.

Donlon, M., and Jackman, J. (1999). "DARPA integrated chemical and biological detection system." *Johns Hopkins APL Technical Digest* 20(3), 320–325.

Dorgan, C. B., Dorgan, C. E., Kanarek, M. S., and Willman, A. J. (1998). "Health and productivity benefits of improved air quality." *ASHRAE Transactions* 104(1), 658–666.

DOT (2000). *Emergency Response Guidebook.* Transportation Canada/Secretariat of Transport and Communications of Mexico/U.S. Department of Transportation, Washington, DC. http://hazmat.dot.gov/erg2000/erg2000.pdf.

Drell, S. D., Sofaer, A. D., and Wilson, G. D. (1999). *The New Terror: Facing the Threat of Biological and Chemical Weapons.* Hoover National Security Forum Series/Hoover Institution Press, Stanford, CA.

Drielak, S. C., and Brandon, T. R. (2000). *Weapons of Mass Destruction: Response and Investigation.* Charles C. Thomas, Springfield, IL.

Druett, H. A., Robinson, J. M., Henderson, D. W., Packman, L., and Peacock, S. (1956). "Studies on respiratory infection, II & III." *J. Hygiene* 54, 37–57.

Dumyahn, T., and First, M. (1999). "Characterization of ultraviolet upper room air disinfection devices." *Am. Ind. Hyg. Assoc. J.* 60(2), 219–227.

Dunn, J. (2000). *Pulsed Light Disinfection of Water and Sterilization of Blow/Fill/Seal Manufactured Aseptic Pharmaceutical Products.* Automatic Liquid Packaging, Woodstock, IL.

Dunn, J., Burgess, D., and Leo, F. (1997). "Investigation of pulsed light for terminal sterilization of WFI filled blow/fill/seal polyethylene containers." *Parenteral Drug Association J. of Pharm. Sci. & Tech.* 51(3), 111–115.

Dyas, A., Boughton, B. J., and Das, B. C. (1983). "Ozone killing action against bacterial and fungal species; microbiological testing of a domestic ozone generator." *J. Clin. Pathol.* 36, 1102–1104.

Ehrt, D., Carl, M., Kittel, T., Muller, M., and Seeber, W. (1994). "High performance glass for the deep ultraviolet range." *Journal of Non-Crystalline Solids* 177, 405–419.

EIA (1989). *Commercial Buildings Characteristics.* U.S. Dept. of Energy, Energy Information Administration, Washington, DC.

El-Adhami, W., Daly, S., and Stewart, P. R. (1994). "Biochemical studies on the lethal effects of solar and artificial ultraviolet radiation on *Staphylococcus aureus*." *Arch. Microbiol.* 161, 82–87.

Elford, W. J., and van den Eude, J. (1942). "An investigation of the merits of ozone as an aerial disinfectant." *J. Hygiene* 42, 240–265.

Ellis, J. W. (1999). *Police Analysis and Planning for Chemical, Biological and Radiological Attacks: Prevention, Defense and Response.* Charles C. Thomas, Springfield, IL.

Ellison, D. H. (2000). *Handbook of Chemical and Biological Warfare Agents.* CRC Press, Boca Raton, FL.

Endicott, S., and Hagerman, E. (1998). *The United States and Biological Warfare: Secrets from the Early Cold War and Korea.* Indiana University Press, Bloomington, IN.

Ensor, D. S., Hanley, J. T., and Sparks, L. E. (1991). "Particle-Size-Dependent Efficiency of Air Cleaners" *IAQ '91,* 334–336. Healthy Buildings/IAQ, Washington, DC.

Estola, T., Makela, P., and Hovi, T. (1979). "The effect of air ionization on the air-borne transmission of experimental Newcastle disease virus infections in chickens." *J. Hyg.* 83, 59–67.

Eversole, J. D., Roselle, D., and Seaver, M. E. (1998). "Monitoring biological aerosols using UV fluorescence." Proceedings of the SPIE, *Air Monitoring and Detection of Chemical and Biological Agents,* Boston, MA, 34–42.

Ewald, P. W. (1994). *Evolution of Infectious Disease.* Oxford University Press, Oxford, U.K.

Farquharson, S., and Smith, W. W. (1998). "Biological agent identification by nucleic acid base-pair analysis using surface-enhanced Raman spectroscopy." Proceedings of the SPIE, *Air Monitoring and Detection of Chemical and Biological Agents,* Boston, MA, 207–214.

Fatah, A. A., Barrett, J. A., Arcilesi, R. D., Ewing, K. J., Lattin, C. H., Helinski, M. S., and Baig, I. A. (2001). *Guide for the Selection of Chemical and Biological Decontamination*

Equipment for Emergency First Responders, vol. 1. National Institute of Standards and Technology, Office of Law Enforcement Standards, U.S. Department of Justice, Office of Justice Program, National Institute of Justice, Washington, DC. http://www.ncjrs.org/pdffiles1/nij/189724.pdf   (v.   1); http://www.ncjrs.org/pdffiles1/nij/189725.pdf (v. 2).

FDI (1998). *FIDAP 8.* Fluid Dynamics International, Inc. Lebanon, NH.

Fellman, G. (1999). *ASHRAE Study Shows Mixed Results for Antimicrobial Filters.* Indoor Environment Connections Online. http://www.ieconnections.com/archive/nov_99/1199_article2.htm.

Fernandez, R. O. (1996). "Lethal effect induced in *Pseudomonas aeruginosa* exposed to ultraviolet-A radiation." *Photochem. Photobiol.* 64(2), 334–339.

Fetner, R. H., and Ingols, R. S. (1956). "A comparison of the bactericidal activity of ozone and chlorine against *Escherichia coli* at 1 C." *Journal of General Microbiology* 15, 381–385.

Fields, B. N., and Knipe, D. M. (1991). *Fundamental Virology.* Raven Press, New York, NY.

First, M. W., Nardell, E. A., Chaisson, W., and Riley, R. (1999). "Guidelines for the application of upper-room ultraviolet germicidal irradiation for preventing transmission of airborne contagion." *ASHRAE Transactions* 105(1), 869–876.

Fisk, W. (1994). "The California healthy buildings study." *Center for Building Science News* Spring 1994, 7, 13.

Fisk, W., and Rosenfeld, A. (1997). "Improved productivity and health from better indoor environments." *Center for Building Science News,* Summer, 5.

Flannigan, B., Samson, R. A., and Miller, J. D. (2001). *Microorganisms in Home and Indoor Work Environments.* Taylor and Francis, Andover, Hants, U.K.

Foarde, K. K., Hanley, J. T., and Veeck, A. C. (2000). "Efficacy of antimicrobial filter treatments." *ASHRAE J.* Dec., 52–58.

Folmsbee, T. W., and Ganatra, C. P. (1996). "Benefits of membrane surface filtration." *World Cement* 27(10), 59–61.

Fraenkel-Conrat, H. (1985). *The Viruses: Catalogue, Characterization, and Classification.* Plenum Press, New York, NY.

Franklin, M. W. (1914). *Transactions of the Fourth International Congress School of Hygiene,* New York, NY, 346.

Franz, D. R. (1997). "Defense against Toxin Weapons." Chapter 30 in *Medical Aspects of Chemical and Biological Warfare,* Office of the Surgeon General at TMM Publications, Washington, DC, 603–619.

Franz, D. R., Jahrling, P. B., Friedlander, A. M., McClain, D. J., Hoover, D. L., Bryne, W. R., Pavlin, J. A., Christopher, G. W., and Eitzen, E. M. (1997). "Clinical recognition and management of patients exposed to biological warfare agents." *JAMA* 278(5), 399–411.

Freeman, B. A., ed. (1985). *Burrows Textbook of Microbiology.* W. B. Saunders, Philadelphia, PA.

Freeman, D. S. (1951). *George Washington: A Biography,* vol. 4. Charles Scribner's Sons, New York, NY.

Fujikawa, H., and Itoh, T. (1996). "Tailing of thermal inactivation curve of *Aspergillus niger* spores." *Appl. Microbiol.* 62(10), 3745–3749.

Fung, D. Y. C., and Cunningham, F. E. (1980). "Effect of microwaves on microorganisms in foods." *J. Food Prot.* 43, 641–650.

Fuortes, L., and Hayes, T. (1988). "An outbreak of acute histoplasmosis in a family." *American Family Physician* 37(5), 128–132.

Futter, B. V., and Richardson, G. (1967). "Inactivation of bacterial spores by visible radiation." *J. Appl. Bacteriol.* 30(2), 347–353.

Gabbay, J. (1990). "Effect of ionization on microbial air pollution in the dental clinic." *Environ. Res.* 52(1), 99.

Garate, E., Evans, K., Gornostaeva, O., Alexeff, I., Kang, W. L., Rader, M., and Wood, T. K. (1998). "Atmospheric plasma induced sterilization and chemical neutralization." *Proceedings of the 1998 IEEE International Conference on Plasma Science,* Raleigh, NC.

Gard, S. (1960). "Theoretical considerations in the inactivation of viruses by chemical means." *Annals of the New York Academy of Sciences* 82, 638–648.

Gardner, P. J. (2000). *Chemical and Biological Sensing.* Volume 4036 of *Proceedings of the SPIE.* The International Society for Optical Engineering, Orlando, FL.

Garrett, L. (1994). *The Coming Plague*. Penguin Books, New York, NY.

Gates, F. L. (1929). "A study of the bactericidal action of ultraviolet light." *J. Gen. Physiol.* 13, 231–260.

Geissler, E. (1986). *Biological and Toxin Weapons Today*. SIPRI/Oxford University Press, New York, NY.

Geissler, E., and van Courtland Moon, J. E. (1999). *Biological and Toxin Weapons: Research, Development and Use from the Middle Ages to 1945*. SIPRI/Oxford University Press, New York, NY. http://editors.sipri.se/pubs/cbw18.html.

Gibbon, E. (1995). *The Decline and Fall of the Roman Empire*. Modern Library, New York, NY.

Gibson, J. E. (1937). *Dr. Bodo Otto and the Medical Background of the American Revolution*. George Banta Publishing Co., Baltimore, MD.

Gilpin, R. W. (1984). "Laboratory and Field Applications of UV Light Disinfection on Six Species of *Legionella* and Other Bacteria in Water." *Legionella: Proceedings of the 2nd International Symposium*, C. Thornsberry, ed. American Society for Microbiology, Washington, DC.

Glaze, W. H., Payton, G. R., Huang, F. Y., Burleson, J. L., and Jones, P. C. (1980). "Oxidation of water supply refractory species by ozone with ultraviolet radiation." *600*, U.S. EPA, Washington, DC.

Goldblith, S. A., and Wang, D.I.C. (1967). "Effect of microwaves on *Escherichia coli* and *Bacillus subtilis*." *Appl. Microbiol.* 15, 1371–1375.

Gorman, O. T., Donis, R. O., Kawaoka, Y., and Webster, R. G. (1990). "Evolution of *Influenza A* virus PB2 genes: Implications for evolution of the ribonucleoprotein complex and origin of human *Influenza A* virus." *J. Virol.* 64, 4893–4902.

Goswami, D. Y., Trivedi, D. M., and Block, S. S. (1997). "Photocatalytic disinfection of indoor air." *Journal of Solar Energy Engineering* 119(1), 92–96.

Griego, V. M., and Spence, K. D. (1978). "Inactivation of *Bacillus thuringiensis* spores by ultraviolet and visible light." *Appl. Environ. Microbiol.* 35(5), 906–910.

Griffiths, W. D., Upton, S. L., and Mark, D. (1993). "An investigation into the collection efficiency & bioefficiencies of a number of aerosol samplers." *Journal of Aerosol Science* 24(S1), s541–s542.

Grigoriu, D., Delacretaz, J., and Borelli, D. (1987). *Medical Mycology*. Hans Huber Publishers, Toronto, Ontario, Canada.

Haar, B. T. (1991). *The Future of Biological Weapons*. Praeger, New York, NY.

Haas, C. N. (2002). "On the risk of mortality to primates exposed to anthrax spores." *Risk Analysis* 118(4), 573–582.

Haas, C. N. (1983). "Estimation of risk due to low doses of microorganisms." *Am. J. Epidemiol.* 118(4), 573–582.

Haber, L. F. (1986). *The Poisonous Cloud: Chemical Warfare in the First World War*. Clarendon Press, Oxford, U.K.

Hamre, D. (1949). "The effect of ultrasonic waves upon *Klebsiella pneumoniae, Saccharomyces cerevisiae, Miyaga wanella felis*, and *Influenza virus A*." *J. Bacteriol.* 57, 279–295.

Han, R., Wu, J. R., and Gentry, J. W. (1993). "The development of a sampling train and test chamber for sampling biological aerosols." *Journal of Aerosol Science* 24(S1), s543–s544.

Happ, J. W., Harstad, J. B., and Buchanan, L. M. (1966). "Effect of air ions on submicron T1 bacteriophage aerosols." *Appl. Microbiol.* 14, 888–891.

Harakeh, M. S., and Butler, M. (1985). "Factors influencing the ozone inactivation of enteric viruses in effluent." *Ozone Sci. Eng.* 6, 235–243.

Harm, W. (1980). *Biological Effects of Ultraviolet Radiation*. Cambridge University Press, New York, NY.

Harper, G. J. (1961). "Airborne micro-organisms: Survival tests with four viruses." *Journal of Hygiene* 59, 479–486.

Harris, J. B. (1986). *Natural Toxins: Animal, Plant, and Microbial*. Oxford Science Publications/Clarendon Press. Oxford, U.K.

Harris, S. H. (1994). *Factories of Death: Japanese Biological Warfare 1932–45 and the American Cover-Up*. Routledge, London, U.K.

Harris (2002). *Emergency Decontamination Triage and Treatment.* Harris County, Texas, Fire and Emergency Services, Houston, TX. http://www.co.harris.tx.us/fmarshal/ Training%20Programs/OG-emer_decon_triage_treatment.pdf.

Harstad, J. B., Decker, H. M., Buchanan, L. M., and Filler, M. E. (1967). "Air filtration of submicron virus aerosols." *American Journal of Public Health* 57(12), 2186–2193.

Harstad, J. B., and Filler, M. E. (1969). "Evaluation of air filters with submicron viral aerosols and bacterial aerosols." *American Industrial Hygiene Association Journal* 30, 280–290.

Hart, J., Walker, I., and Armstrong, D. C. (1995). "The use of high concentration ozone for water treatment." *Ozone Sci. Eng.* 17, 485–497.

Hartman (1925). *J. Am. Soc. Heat. Vent. Engrs.* 31, 33.

Hartman, T. B. (1993). *Direct Digital Control for HVAC Systems.* McGraw-Hill, New York, NY.

H.A.S.C. (2000). *Terrorist Threats to the U.S.* Document 106-52. House Armed Services Committee, Washington, DC.

Hautanen, J. H., Janka, K., Lehtimaki, M., and Kivisto, T. (1986). "Optimization of filtration efficiency and ozone production of the electrostatic precipitator." *Journal of Aerosol Science and Technology* 17(3), 622–626.

Hayek, C. S., Pineda, F. J., Doss III, O. W., and Lin, J. S. (1999). "Computer-assisted interpretation of mass spectra." *Johns Hopkins APL Technical Digest* 20(3), 363–371.

Heinsohn, R. J. (1991). *Industrial Ventilation: Principles and Practice.* John Wiley & Sons, New York, NY.

Henderson, D. A. (1999a). "The looming threat of bioterrorism." *Science* 283, 1279–1282.

Henderson, D. A. (1999b). "Smallpox: Clinical and epidemiological features." *Emerging Infectious Diseases* 5(4), 537–539.

Henschel, D. B. (1998). "Cost analysis of activated carbon versus photocatalytic oxidation for removing organic compounds from indoor air." *J. Air Waste Mgt.* 48(10), 985–994.

Hersh, S. J. (1968). *Chemical & Biological Warfare: America's Hidden Arsenal.* Doubleday & Co., Garden City, NY.

Hess-Kosa, K. (2002). *Indoor Air Quality.* Lewis Publishers, Boca Raton, FL.

Hickman, D. C. (1999). *A Chemical and Biological Warfare Threat: USAF Water Systems at Risk.* USAF Counterproliferation Center, Maxwell AFB, AL.

Hicks, R. E., Sengun, M. Z., and Fine, B. C. (1996). "Effectiveness of local filters for infection control in medical examination rooms." *HVAC&R Research* 2(3), 173–194.

Higgins, I. J., and Burns, R. G. (1975). *The Chemistry and Microbiology of Pollution.* Academic Press, London, U.K.

Hill, W. F., Hamblet, F. E., Benton, W. H., and Akin, E. W. (1970). "Ultraviolet devitalization of eight selected enteric viruses in estuarine water." *Appl. Microbiol.* 19(5), 805–812.

Hines, A. L., Ghosh, T. K., Loyalka, S. K., and R. C. Warder, J. (1992). *A Summary of Pollutant Removal Capacities of Solid and Liquid Desiccants from Indoor Air.* NTIS Document No. PB95-104683. Gas Research Institute, Chicago, IL.

Hirai, K., Nagata, Y., and Maeda, Y. (1996). "Decomposition of chlorofluorocarbons and hydrofluorocarbons in water by ultrasonic irradiation." *Ultrasonics Sonochemistry* 3, S205–S207.

Hiromoto, A. (1984). Method of Sterilization. U.S. Patent No. 4464336.

Hobson, N. S., Tothill, I., and Turner, A. P. F. (1996). "Microbial detection." *Biosensors and Bioelectronics* 11(5), 455–477.

Hoffman, L., and Coutu, P. R. (2001). "Did Peter Pond participate in the ethnic cleansing of Western Canada's Indigenous Peoples?" (A chapter from the upcoming book *Inkonze*). The School of Native Studies, University of Alberta, Edmonton, Alberta, Canada. http://www.ualberta.ca/~nativest/pim/ClashofWorlds.html.

Hoffman, M. R., Hua, I., and Hochemer, R. (1996). "Application of ultrasonic irradiation for the degradation of chemical contaminants in water." *Ultrasonics Sonochemistry* 3, S163–S172.

Hollaender, A. (1943). "Effect of long ultraviolet and short visible radiation (3500 to 4900) on *Escherichia coli*." *J. Bacteriol.* 46, 531–541.

Holler, S., Pan, Y., Bottiger, J. R., Hill, S. C., Hillis, D. B., and Chang, R. K. (1998). "Two-dimensional angular scattering measurements of single airborne micro-particles." Proceedings of the SPIE, *Air Monitoring and Detection of Chemical and Biological Agents,* Boston, MA, 64–72.

Holloway, H. C., Norwood, A. E., Fullerton, C. S., Engel, C. C., and Ursano, R. J. (1997). "The threat of biological weapons." *JAMA* 278(5), 425–427.

Honeywell (2001). *HUV: Honeywell Ultraviolet Air Treatment Application Guide.* Software Package, Honeywell, Inc., Golden Valley, MN.

Horvath, M., Bilitzky, L., and Huttnerr, J. (1985). *Ozone.* Elsevier, Amsterdam, The Netherlands.

Howard, D. H., and Howard, L. F. (1983). *Fungi Pathogenic for Humans and Animals.* Marcel Dekker, New York, NY.

Howell, J. R. (1982). *A Catalog of Radiation View Factors.* McGraw-Hill, New York, NY.

Huang, S.-H., and Chen, C. C. (2001). "Filtration characteristics of a miniature electrostatic precipitator." *Aerosol Science and Technology* 35, 792–804.

Hughes, J. M. (1999). "The emerging threat of bioterrorism." *Emerging Infectious Diseases* 5(4), 494–495.

IITRI (1999). *Laboratory Tests for Kill of Anthrax Spores in SNL Foam.* IITRI Project No. 1084, Study Nos. 1–4, Illinois Institute of Technology Research Institute, Chicago, IL.

Inglesby, T. V. (1999). "Anthrax: A possible case history." *Emerging Infectious Diseases* 5(4), 556–560.

Inglesby, T. V., Henderson, D. A., Bartlett, J. G., Ascher, M. S., Eitzen, E., Friedlander, A. M., Hauer, J., McDade, J., Osterholm, M. T., O'Toole, T., Parker, G., Perl, T. M., Russell, P. K., and Tonat, K. (1999). "Anthrax as a biological weapon: Medical and public health management." *JAMA* 281(18), 1735–1745. http://jama.ama-assn.org/issues/v281n18/ffull/jst80027.html.

Irvine, R. L. (1998). *Chemical/Biological/Radiological Incident Handbook.* Interagency Intelligence Committee on Terrorism/CIA Unclassified Document, Washington, DC.

Ishizaki, K., Shinriki, N., and Matsuyama, H. (1986). "Inactivation of Bacillus spores by gaseous ozone." *J. Appl. Bacteriol.* 60, 67–72.

Jaax, N., Davis, K., and Geisbert, T. (1996). "Lethal experimental infection of Rhesus monkeys with *Ebola-Zaire* (Mayinga) virus by the oral and conjunctival route of exposure." *Arch. Pathol. Lab. Med.* 120, 140–155.

Jaax, N., Jahrling, P., and Geisbert, T. (1995). "Transmission of Ebola virus (Zaire strain) to uninfected control monkeys in a biocontaminant laboratory." *Lancet* 346, 1669–1671.

Jackson, R. H. (1994). *Indian Population Decline: The Mission of Northwest Spain, 1687–1840.* University of New Mexico Press, Albuquerque, NM.

Jacoby, W. A., Blake, D. M., Fennell, J. A., Boulter, J. E., and Vargo, L. M. (1996). "Heterogeneous photocatalysis for control of volatile organic compounds in indoor air." *J. Air Waste Mgt.* 46, 891–898.

Jacoby, W. A., Maness, P. C., and Wolfrum, E. J. (1998). "Mineralization of bacterial cell mass on a photocatalytic surface in air." *Environ. Sci. Technol.* 32(17), 2650–2653.

Jaisinghani, R. (1999). "Bactericidal properties of electrically enhanced HEPA filtration and a bioburden case study." *InterPhex Conference,* New York, NY. http://www.cleanroomsys.com/downloads.htm.

Jaisinghani, R. A., and Bugli, N. J. (1991). "Conventional air cleaners versus electrostatic air cleaners in HEPA and indoor applications." *Particulate Science & Technology* (9), 1–18.

Janney, C., Janus, M., Saubier, L. F., and Widder, J. (2000). *Test Report: System Effectiveness Test of Home/Commercial Portable Room Air Cleaners.* Contract No. SPO900-94-D-0002, Task No. 491. U.S. Army Soldier and Biological Chemical Command, Washington, DC.

Jensen, M. (1967). "Bacteriophage aerosol challenge of installed air contamination control systems." *Appl. Microbiol.* 15(6), 1447–1449.

Jensen, M. M. (1964). "Inactivation of airborne viruses by ultraviolet irradiation." *Applied Microbiology* 12(5), 418–420.

Jensen, P. A., Todd, W. F., Davis, G. N., and Scarpino, P. Y. (1992). "Evaluation of eight bioaerosol samplers challenged with aerosols of free bacteria." *Am. Ind. Hyg. Assoc. J.* 53(10), 660–667.

Johnson, E., Jaax, N., White, J., and Jahrling, P. (1995). "Lethal experimental infections of Rhesus monkeys by aerosolized Ebola virus." *Int. J. Exp. Pathol.* 76, 227–236.

Johnson, T. (1982). "Flashblast: The light that cleans." *Popular Science,* July, 82–84.

Jordan, E. O., and Carlson, A. J. (1913). *JAMA* 61, 1007.

Kakita, Y., Kashige, N., Murata, N., Kuroiwa, A., Funatsu, M., and Watanabe, K. (1995). "Inactivation of *Lactobacillus* bacteriophage PL-1 by microwave irradiation." *Microbiol. Immunol.* 39, 571–576.

Kapur, V., Whittam, T. S., and Musser., J. M. (1994). "Is *Mycobacterium tuberculosis* 15,000 years old?" *J. Infect. Dis.* 170, 1348–1349.

Kashino, S. S., Calich, V. L. G., Burger, E., and Singer-Vermes, L. M. (1985). "In vivo and in vitro characteristics of six *Paracoccidioides brasiliensis* strains." *Mycopathologia* 92, 173–178.

Katzenelson, E., and Shuval, H. I. (1973). In *Studies on the Disinfection of Water by Ozone: Viruses and Bacteria.* Edited by R. G. Rice and M. E. Browning. Hampson Press, Washington, DC.

Kaufman, A. F., Meltzer, M. I., and Schmid, G. P. (1997). "The economic impact of a bioterrorist attack: Are prevention and postattack intervention programs justifiable?" *Emerging Infectious Diseases* 3(2), 83–94. http://www.cdc.gov/ncidod/eid/vol3no2/Kaufman.htm.

Kelle, A., Dando, M. R., and Nixdorff, K. (2001). *The Role of Biotechnology in Countering BTW Agents.* NATO Science Series/Kluwer Academic Publishers, Norwell, MA.

Kemp, S. J.,Kuehn, T. H., Pui, D. Y. H., Vesley, D., and Streifel, A. J. (1995). "Filter collection efficiency and growth of microorganisms on filters loaded with outdoor air." *ASHRAE Transactions* 101(1), 228.

Kennedy, R. (2002). *Anti-Microbial Control by Coatings through the Selection of Raw Materials.* European Coatings Net, Hanover, Germany. http://www.coatings.de/articles/ecs01papers/Kennedy/kennedy.htm.

Kenyon, R. H., Green, D. E., Maiztegui, J. I., and Peters, C. J. (1988). "Viral strain dependent differences in experimental Argentine Hemorrhagic Fever (Junin virus) infection of guinea pigs." *Intervirology* 29, 133–143.

Kiel, J. L., Seaman, R. L., Marthur, S. P., Parker, J. E., Wright, J. R., Alls, J. L., and Morales, P. J. (1999). "Pulsed microwave induced light, sound, and electrical discharge enhanced by a biopolymer." *Bioelectromagnetics* 20(4), 216–223.

Kim, C. K., Kim, S. S., Kim, D. W., Lim, J. C., and Kim, J. J. (1998). "Removal of aromatic compounds in the aqueous solution via micellar enhanced ultrafiltration: Part 1. Behavior of nonionic surfactants." *Journal of Membrane Science* 147(1), 13–22.

Klens, P. F., and Yoho, J. R. (1984). "Occurrence of Alternaria species on latex paint." *The 6th International Biodeterioration Symposium,* Washington, DC.

Knudson, G. B. (1986). "Photoreactivation of ultraviolet-irradiated, plasmid-bearing, and plasmid-free strains of *Bacillus anthracis.*" *Appl. Environ. Microbiol.* 52(3), 444–449.

Koch, A. L. (1961). "Some calculations on the turbidity of mitochondria and bacteria." *Biochim. Biophys. Acta.* 51, 429–441.

Koch, A. L. (1966). "The logarithm in biology: Mechanisms generating the lognormal distribution exactly." *J. Theor. Biol.* 12, 276–290.

Koch, A. L. (1969). "The logarithm in biology: Distributions simulating the log-normal." *J. Theor. Biol.* 23, 251–268.

Koch, A. L. (1995). *Bacterial Growth and Form.* Chapman & Hall, New York, NY.

Kolavic, S. A., Kimura, A., Simons, S. L., Slutsker, L., Barth, S., and Haley, C. E. (1997). "An outbreak of *Shigella dysenteriae* Type 2 among laboratory workers due to intentional food contamination." *JAMA* 278, 396–398.

Koller, L. R. (1952). *Ultraviolet Radiation.* John Wiley & Sons, New York, NY.

Konig, B., and Gratzel, M. (1993). "Detection of viruses and bacteria with piezoelectric immunosensors." *Anal. Lett.* 26, 1567–1585.

Korobeinichev, O. P., Shvartsberg, V. M., and Chernov, A. A. (1999). "Destruction chemistry of organophosphorus compounds in flames—II: Structure of a hydrogen-oxygen flame doped with trimethyl phosphate." *Combustion and Flame* 118(4), 727–732.

Kortepeter, M. G., and Parker, G. W. (1999). "Potential biological weapons threats." *Emerging Infectious Diseases* 5(4), 523–527.

Kovak, B., Heiman, P. R., and Hammel, J. (1997). "The sanitizing effects of desiccant-based cooling." *ASHRAE J.* 39(4), 60–64.

Kowalski, W., and Bahnfleth, W. P. (1998). "Airborne respiratory diseases and technologies for control of microbes." *HPAC* 70(6)34–48. http://www.engr.psu.edu/ae/wjk/ardtie.html.

Kowalski, W. J. (2001). *Design and optimization of UVGI air disinfection systems*. Ph.D. dissertation. Pennsylvania State University, University Park, PA. http://etda.libraries.psu.edu/theses/available/etd-0622101-204046/.

Kowalski, W. J., and Bahnfleth, W. P. (2000a). "UVGI design basics for air and surface disinfection." *HPAC* 72(1), 100–110. http://www.engr.psu.edu/ae/wjk/uvhpac.html.

Kowalski, W. J., and Bahnfleth, W. P. (2000b). "Effective UVGI system design through improved modeling." *ASHRAE Transactions* 106(2), 4–15. http://www.engr.psu.edu/ae/wjk/uvmodel.html.

Kowalski, W. J., and Bahnfleth, W. P. (2002a). "Airborne-microbe filtration in indoor environments." *HPAC Engineering* 74(1), 57–69. http://www.bio.psu.edu/people/faculty/whittam/research/amf.pdf.

Kowalski, W. J., and Bahnfleth, W. P. (2002b). *Aerobiological Engineering Website*. Pennsylvania State University Department of Architectural Engineering and Biology Department, University Park, PA. http://www.engr.psu.edu/ae/wjkaerob.html.

Kowalski, W. J., Bahnfleth, W. P., and Whittam, T. S. (1998). "Bactericidal effects of high airborne ozone concentrations on *Escherichia coli* and *Staphylococcus aureus*." *Ozone Sci. Eng.* 20(3), 205–221. http://www.bio.psu.edu/people/faculty/whittam/research/ozone.html.

Kowalski, W. J., W. P. Bahnfleth, and T. S. Whittam (1999). "Filtration of airborne microorganisms: Modeling and prediction." *ASHRAE Transactions* 105(2), 4–17. http://www.engr.psu.edu/ae/wjk/fom.html.

Kowalski, W. J., Bahnfleth, W. P., and Whittam, T. S. (2003). "Surface disinfection with airborne ozone and catalytic ozone removal." *AIHA J.* (accepted).

Kowalski, W. J., Bahnfleth, W. P., Witham, D., Severin, B. F., and Whittam, T. S. (2000). "Mathematical modeling of UVGI for air disinfection." *Quantitative Microbiology*, 2(3), 249–270; http://www.bio.psu.edu/people/faculty/whittam/research/mmuad.pdf.

Kowalski, W. J., and Burnett, E. (2001). *Mold and Buildings*. Builder Brief BB0301. Pennsylvania Housing Research Center, University Park, PA. http://www.bio.psu.edu/people/faculty/whittam/research/B0301.pdf.

Kroll (2001). *Mailroom Protocols*. Kroll Associates, New York, NY. http://www.krollassociates.com/mailroom_protocols.cfm.

Kundsin, R. B. (1966). "Characterization of Mycoplasma aerosols as to viability, particle size, and lethality of ultraviolet radiation." *J. Bacteriol.* 91(3), 942–944.

Kundsin, R. B. (1968). "Aerosols of Mycoplasmas, L forms, and bacteria: Comparison of particle size, viability, and lethality of ultraviolet radiation." *Applied Microbiology* 16(1), 143–146.

Kurata, H., and Ueno, Y. (1984). *Toxigenic Fungi—Their Toxins and Health Hazard*. Elsevier, Amsterdam, The Netherlands.

Lacey, J., and Crook, B. (1988). "Fungal and actinomycete spores as pollutants of the workplace and occupational illness." *Ann. Occup. Hyg.* 32, 515–533.

Lackie, J. M., and Dow, J. A. T. (1995). *Dictionary of Cell and Molecular Biology*. Academic Press, London, U.K.

Lancaster-Brooks, R. (2000). "Water terrorism: An overview of water & wastewater security Problems and Solutions." *J. Homeland Security* Feb. http://www.homelandsecurity.org/journal/Articles/article.cfm?article=33.

Latham, R. E. (1983). *Lucretius: On the Nature of the Universe*. Penguin Press, New York, NY.

Latimer, J. M., and Matson, J. M. (1977). "Microwave oven irradiation as a method for bacterial decontamination in a clin. microbiology laboratory." *J. Clin. Microbiol.* 4, 340–342.

Lee, J.-K., Kim, S. C., Shin, J. H., Lee, J. E., Ku, J. H., and Shin, H. S. (2001). "Performance evaluation of electrostatically augmented air filters coupled with a corona precharger." *Aerosol Science and Technology* 35, 785–791.

Lee, J. Y., and Deininger, R. A. (2000). "Survival of bacteria after ozonation." *Ozone Sci. Eng.* 22, 65–75.

Lee, W. E., and Hall, J. G. (1992). *Biosensor-Based Detection of Soman*. Suffield Memorandum No. 1380. Defence Research Establishment, Suffield, Canada.

Leitenberg, M. (2001). "Biological weapons in the twentieth century: A review and analysis." *Crit. Rev. Microbiol.* 27(4), 267–320.

Leonelli, J., and Althouse, M. L. (1998). *Air Monitoring and Detection of Chemical and Biological Agents.* Volume 3533 of *Proceedings of the SPIE.* The International Society for Optical Engineering, Boston, MA.

Letenberg, M. (1993). "The biological weapons program of the former Soviet Union." *Biologicals* 21, 187–191.

Levetin, E., Shaughnessy, R., Rogers, C. A., and Scheir, R. (2001). "Effectiveness of germicidal UV radiation for reducing fungal contamination within air-handling units." *Appl. Environ. Microbiol.* 67(8), 3712–3715.

Lewis, R. J. (1993). *Hawley's Condensed Chemical Dictionary.* Van Nostrand Reinhold, New York, NY.

Li, C.-S., Hao, M. L., Lin, W. H., Chang, C. W., and Wang, C. S. (1999). "Evaluation of microbial samplers for bacterial microorganisms." *Aerosol Sci. Technol.* 30, 100–108.

Liao, C.-H., and Sapers, G. M. (2000). "Attachment and growth of *Salmonella chester* on apple fruits and in vivo response of attached bacteria to sanitizer treatments." *J. Food Protection* 63(7), 876–883.

Lidwell, O. M., and Lowbury, E. J. (1950). "The survival of bacteria in dust." *Annual Review of Microbiology* 14, 38–43.

Linton, A. H. (1982). *Microbes, Man, and Animals: The Natural History of Microbial Interactions.* John Wiley & Sons, New York, NY.

Little, J. S., Kishimoto, R. A., and Canonico, P. G. (1980). "In vitro studies of interaction of *Rickettsia* and macrophages: Effect of ultraviolet light on *Coxiella burnetii* inactivation and macrophage enzymes." *Infect. Immun.* 27(3), 837–841.

Livy (24 B.C.E.). *History of Rome.* Concordance.com, http://www.concordance.com/.

Luciano, J. R. (1977). *Air Contamination Control in Hospitals.* Plenum Press, New York, NY.

Luckiesh, M. (1946). *Applications of Germicidal, Erythemal and Infrared Energy.* D. Van Nostrand Co., New York, NY.

Madoff, S. (1971). *Mycoplasma and the L Forms of Bacteria.* Gordon and Breach Science Publishers, New York, NY.

Mahy, B. W. J., and Barry, R. D. (1975). *Negative Strand Viruses.* Academic Press, London, U.K.

Makela, P., Ojajarvi, J., Graeffe, G., and Lehtimaki, M. (1979). "Studies on the effects of ionization on bacterial aerosols in a burns and plastic surgery unit." *J. Hyg.* 83, 199–206.

Malherbe, H. H., and Strickland-Cholmley, M. (1980). *Viral Cytopathology.* CRC Press, Boca Raton, FL.

Mandell, G. L., Gerald, L., Bennett, J. E., and Dolin, R. (2000). *Principles and Practice of Infectious Diseases.* Churchill Livingstone, Philadelphia, PA.

Maniloff, J., McElhaney, R. N., Finch, L. R., and Baseman, J. B. (1992). *Mycoplasmas: Molecular Biology and Pathogenesis.* ASM Press, Washington, DC.

Marsden, C. (2001). Personal communication with W. J. Kowalski. Trio3 Industries, Inc., Fort Pierce, FL.

Martin, R. J., and Shackleton, R. C. (1990). "Comparison of two partially activated carbon fabrics for the removal of chlorine and other impurities from water." *Water Res.* 24(4), 477–484.

Martini, G. A., and Siegert, R. (1971). *Marburg Virus Disease.* Springer-Verlag, New York, NY.

Masaoka, T., Kubota, Y., Namiuchi, S., Takubo, T., Ueda, T., Shibata, H., Nakamura, H., Yoshitake, J., Yamayoshi, T., Doi, H., and Kamiki, T. (1982). "Ozone decontamination of bioclean rooms." *Appl. Environ. Microbiol.* 43(3), 509–513.

Matson, J. V. (2001). Personal communication with W. J. Kowalski, December, ARL meeting on biodefense.

Matteson, M. J., and Orr, C. (1987). *Filtration: Principles and Practices.* Chemical Industries/Marcel Dekker, New York, NY.

Maus, R., Goppelsroder, A., and Umhauer, H. (1997). "Viability of bacteria in unused air filter media." *Atmos. Environ.* 31(15), 2305–2310.

Maus, R., Goppelsroder, A., and Umhauer, H. (2001). "Survival of bacterial and mold spores in air filter media." *Atmos. Environ.* 35, 105–113.

Mayes, S. E., and Ruisinger, T. (1998). "Using ozone for water treatment." *ASHRAE Journal* 40(5), 39–42.

McCaa, R. (1995). *Spanish and Nahuatl Views on Smallpox and Demographic Catastrophe in the Conquest of Mexico.* University of Minnesota, Minneapolis, MN. http://www.hist.umn.edu/~rmccaa/vircatas/vir6.htm.

McCarthy, J. J., and Smith, C. H. (1974). "A review of ozone and its application to domestic wastewater treatment." *J. Am. Water Works Assoc.* 74, 718–729.

McCathy, R. D. (1969). *The Ultimate Folly: War by Pestilence, Asphyxiation, and Defoliation.* Alfred A. Knopf, New York, NY.

McCaul, T. F., and Williams, J. C. (1981). "Developmental cycle of *Coxiella burnetii*: Structure and morphogenesis of vegetative and sporogenic differentiations." *J. Bacteriol.* 147(3), 1063–1076.

McCormick, J. B. (1987). "Epidemiology and control of Lassa fever." *Arenaviruses.* M. B. A. Oldstone, ed. Springer-Verlag, New York, NY.

McDermott, H. J. (1985). *Handbook of Ventilation for Contaminant Control.* Butterworth Publishers, Boston, MA.

McGowan, J. J. (1995). *Direct Digital Control: A Guide to Distributed Building Automation.* Fairmont Press, Lilburn, GA.

McLean, K. J. (1988). "Electrostatic precipitators." *IEEE Proceedings* 135(6), 347–362.

McLoughlin, M. P., Allmon, W. R., Anderson, C. W., Carlson, M. A., DeCicco, D. J., and Evancich, N. H. (1999). "Development of a field-portable time-of-flight mass spectrometer system." *Johns Hopkins APL Technical Digest* 20(3), 326–334.

McQuiston, F. C., and Parker, J. D. (1994). *Heating, Ventilating, and Air Conditioning Analysis and Design.* John Wiley & Sons, New York, NY.

Meselson, M., Guillemin, J., Hugh-Jones, M., Langmuir, A., Popova, I., Shelokov, A., and Yampolskaya, O. (1994). "The Sverdlovsk anthrax outbreak of 1979." *Science* 266(18), 1202–1208.

Middlebrook, J. L. (1986). "Cellular mechanism of action of botulinum neurotoxin." *Journal of Toxicology and Toxin Reviews* 5(2), 177–190.

Miller, K. (1999). "Chemical agents as weapons: Medical implications." *Fire Engineering* 152(2), 61–69.

Mitchell, T. J., Godfree, F. F., and Stewart-Tull, D. E. S. (1998). *Toxins.* Society for Applied Microbiology, Symposium Series Number 27. Blackwell Science, Glasgow, U.K.

Mitscherlich, E., and Marth, E. H. (1984). *Microbial Survival in the Environment.* Springer-Verlag, Berlin, Germany.

Miyaji, M. (1987). *Animal Models in Medical Mycology.* CRC Press, Boca Raton, FL.

Moats, W. A., Dabbah, R., and Edwards, V. M. (1971). "Interpretation of nonlogarithmic survivor curves of heated bacteria." *J. Food Sci.* 36, 523–526.

Moczydlowski, E., Lucchesi, K., and Ravindran, A. (1988). "An emerging pharmacology of peptide toxins targeted against potassium channels." *J. Membr. Biol.* 105(2), 95–111.

Modec, Inc. (2001). *Formulations for the Decontamination and Mitigation of CB Warfare Agents, Toxic Hazardous Materials, Viruses, Bacteria and Bacterial Spores.* Technical Report No. MDF2001-1002. Denver, CO.

Modest, M. F. (1993). *Radiative Heat Transfer.* McGraw-Hill, New York, NY.

Mongold, J. (1992). "DNA repair and the evolution of transformation in *Haemophilus influenzae.*" *Genetics* 132, 893–898.

Montana, E., Etzel, R., Sorenson, W., Kullman, G., Allan, T., and Dearborn, D. (1998). "Acute pulmonary hemorrhage in infants associated with exposure to *Stachybotris atra* and other fungi." *Arch. Pediatr. Adolesc. Med.* 152, 757–762.

Morgan, J. S., Bryden, W. A., Vertes, R. F., and Bauer, S. (1999). "Detection of chemical agents in water by membrane-introduction mass spectrometry." *Johns Hopkins APL Technical Digest* 20(3), 381–387.

Morrissey, R. F., and Phillips, G. B. (1993). *Sterilization Technology.* Van Nostrand Reinhold, New York, NY.

Mumma, S. A. (2001a). "Dedicated Outside Air Systems." *ASHRAE IAQ Applications* 2(1), 20–22. http://doas.psu.edu/IAQ_Winter2001pgs20-22.pdf.

Mumma, S. A. (2001b). "Designing dedicated outdoor air systems." *ASHRAE J.* 43(5), 28–31. http://doas.psu.edu/journal_01_doas.pdf.

Mumma, S. A. (2002). "Safety and Comfort Using DOAS: Radiant Cooling Panel Systems." *IAQ Applications* Winter, 20–21. http://doas.psu.edu/IAQ_winter_02.pdf.

Munakata, N., Saito, M., and Hieda, K. (1991). "Inactivation action spectra of *Bacillus subtilis* spores in extended ultraviolet wavelengths (50–300 nm) obtained with synchrotron radiation." *Photochem. Photobiol.* 54(5), 761–768.

Murray, P. R. (1999). *Manual of Clinical Microbiology.* ASM Press, Washington, DC.

Musser, A. (2000). "Multizone modeling as an indoor air quality design tool." *Healthy Buildings 2000,* Espoo, Finland, August 6–10.

Musser, A. (2001). "An analysis of combined CFD and multizone IAQ model assembly issues." *ASHRAE Transactions* 106(1), 371–382.

Musser, A., Palmer, J., and McGrattan, K. (2001). *Evaluation of a Fast, Simplified Computational Fluid Dynamics Model for Solving Room Airflow Problems.* NIST, Gaithersburg, MD.

NAS (1997). *Review of Acute Human-Toxicity Estimates for Selected Chemical-Warfare Agents.* National Academy of Sciences, Washington, DC. http://www.nap.edu/readingroom/books/toxicity/.

Nass, M. (1992a). "Anthrax Epizootic in Zimbabwe, 1978–1980: Due to Deliberate Spread?" *Physicians for Social Responsibility Quarterly.* http://www.anthraxvaccine.org/zimbabwe.html.

Nass, M. (1992b). "Chemical and biological war: Zimbabwe, South Africa and anthrax." *Covert Action* 43(Winter).

NATO (1996). *Handbook on the Medical Aspects of NBC Defensive Operations FM 8-9.* Dept. of the Army, Washington, DC.

NBC (2002). *NBC Industry Group Home Page.* NBC Industry Group, Springfield, VA. http://www.nbcindustrygroup.com.

Newton, H. N. (1994). *Direct Digital Control of Building Systems.* John Wiley & Sons, New York, NY.

NFPA (1991). *National Fire Protection Association Standard—Life Safety Code,* NFPA 101. Quincy, MA.

Nicas, M., and Miller, S. L. (1999). "A multi-zone model evaluation of the efficacy of upper-room air ultraviolet germicidal irradiation." *Appl. Environ. Occup. Hyg. J.* 14, 317–328.

Niebuhr, K., and Sansonetti, P. J. (2000). "Invasion of cells by bacterial pathogens." *Subcell. Biochem.* 33, 251–287.

Nikulin, M., Reijula, K., Jarvis, B. B., and Hintikka, E. (1996). "Experimental lung mycotoxicosis in mice induced by *Stachybotris atra*." *Int. J. Exp. Pathol.* 77, 213–218.

NIST (1992). "Photoinitiated ozone-water reaction." *J. Res. NIST* 97(4), 499.

NIST (2002). *CONTAMW: Multizone Airflow and Contaminant Transport Analysis Software.* National Institute of Standards and Technology, Gaithersburg, MD. http://www.bfrl.nist.gov/IAQanalysis/CONTAMWdownload1.htm.

NRC (1999). *National Research Council Review of the U.S. Army's Health Risk Assessments for Oral Exposure to Six Chemical-Warfare Agents.* National Academy Press, Washington, DC.

NRC (1999). *Chemical and Biological Terrorism.* National Research Council/National Academy Press, Washington, DC. http://bob.nap.edu/html/terrorism/index.html.

Oeuderni, A., Limvorapituk, Q., Bes, R., and Mora, J. C. (1996). "Ozone decomposition on glass and silica." *Ozone Sci. Eng.* 18, 385–415.

Offerman, F. J., Loiselle, S. A., and Sextro, R. G. (1992). "Performance of air cleaners in a residential forced air system." *ASHRAE J.* 34(7), 51–57.

Ogert, R. A., Brown, E. J., Singh, B. R., Shriver-Lake, L. C., and Ligler, F. S. (1992). "Detection of *Clostridium botulinum* toxin A using a fibre optic-based biosensor." *Anal. Biochem.* 205, 306–312.

Olsen, J. C., and Ulrich, W. H. (1914). *Journal of Industrial Engineering & Chemistry* 6, 619.

Olson, K. B. (1999). "Aum Shinrikyo: Once and future threat?" *Emerging Infectious Diseases* 5(4), 513–516.

OMH (1997). *Medical Aspects of Chemical and Biological Warfare.* Office of the Surgeon General at TMM Publications, Washington, DC.

OPCW (1997). *A List of Schedule 1 Chemicals.* Organization for the Prohibition of Chemical Weapons, The Hague, Netherlands. http://www.opcw.nl/cwc/schedul1.htm.

OPCW (1997). *A List of Schedule 2 Chemicals.* Organization for the Prohibition of Chemical Weapons, The Hague, Netherlands. http://www.opcw.nl/cwc/schedul2.htm.

OPCW (1997). *A List of Schedule 3 Chemicals.* Organization for the Prohibition of Chemical Weapons, The Hague, Netherlands. www.opcw.nl/cwc/schedul3.htm.

Ornberg, S. C. (1978). *Design, Construction and Testing of Nuclear Air Cleaning Systems.* Sargent & Lundy Engineers, Chicago, IL.

OSHA (1999). *OSHA Technical Manual.* Occupational Health and Safety Administration, U.S. Department of Labor, Washington, DC. http://www.osha-slc.gov/dts/osta/otm/otm_toc.html.

O'Toole, T. (1999). "Smallpox: An attack scenario." *Emerging Infectious Diseases* 5(4), 540–546.

Paddle, B. M. (1996). "Biosensors for chemical and biological agents of defence interest." *Biosensors & Bioelectronics* 11(11), 1079–1113.

Paula, G. (1998). "Crime-fighting sensors." *Mechanical Engineering* 120(1), 66–72.

Pavlin, J. A. (1999). "Epidemiology of bioterrorism." *Emerging Infectious Diseases* 5(4), 528–530.

Peccia, J., Werth, H. M., Miller, S., and Hernandez, M. (2001). "Effects of relative humidity on the ultraviolet induced inactivation of airborne bacteria." *Aerosol Sci. Technol.* 35, 728–740.

Peters, C. J., Jahrling, P. B., Liu, C. T., Kenyon, R. H., Jr., K. T. M., and Oro, J. G. B. (1987). "Experimental studies of Arenaviral hemorrhagic fevers." *Arenaviruses.* M. B. A. Oldstone, ed. Springer-Verlag, New York, NY.

Pethig, R. (1979). *Dielectric and Electronic Properties of Biological Materials.* John Wiley & Sons, Chichester, U.K.

Philips (1985). *UVGI Catalog and Design Guide.* Catalog No. U.D.C. 628.9. Netherlands.

Phillips, G., Harris, G. J., and Jones, M. V. (1964). "Effect of air ions on bacterial aerosols." *Int. J. Biometeorol.* 8, 27–37.

Pile, J. C., Malone, J. D., Eitzen, E. M., and Friedlander, A. M. (1998). "Anthrax as a potential biological warfare agent." *Arch. Intern. Med.* 158, 429–434.

Plomer, M., Guilbault, G. G., and Hock, B. (1992). "Development of a piezoelectric immunosensor for the detection of enterobacteria." *Enzyme Microbiol. Technol.* 14, 230–235.

Poindexter, J. S., and Leadbetter, E. R. (1985). *Bacteria in Nature.* Plenum Press, New York, NY.

Poupard, J. A., and Miller, L. A. (1992). *History of Biological Warfare: Catapults to Capsomeres.* R. A. Zilinskas, ed. The Microbiologist and Biological Defense Research/ New York Academy of Sciences, New York, NY.

Prescott, L. M., Harley, J. P., and Klein, D. A. (1996). *Microbiology.* Wm. C. Brown Publishers, Dubuque, IA.

Pruitt, K. M., and Kamau, D. N. (1993). "Mathematical models of bacterial growth, inhibition and death under combined stress conditions." *J. Ind. Microbiol.* 12, 221–231.

Prusak-Sochaczewski, E., Luong, J. H. T., and Guilbault, G. G. (1990). "Development of a piezoelectric immunosensor for the detection of *Salmonella typhimurium.*" *Enzyme Microbiol. Technol.* 12, 173–177.

Qualls, R. G., and Johnson, J. D. (1983). "Bioassay and dose measurement in UV disinfection." *Appl. Microbiol.* 45(3), 872–877.

Qualls, R. G., and Johnson, J. D. (1985). "Modeling and efficiency of ultraviolet disinfection systems." *Water Res.* 19(8), 1039–1046.

Raber, R. R. (1986). "Fluid Filtration: Gas." *Symposium on Gas and Liquid Filtration,* Philadelphia, PA.

Rahn, R. O., Xu, P., and Miller, S. L. (1999). "Dosimetry of room-air germicidal (254 nm) radiation using spherical actinometry." *Photochem. Photobiol.* 70(3), 314–318.

Rainbow, A. J., and Mak, S. (1973). "DNA damage and biological function of human adenovirus after U.V. irradiation." *Int. J. Radiat. Biol.* 24(1), 59–72.

Rauth, A. M. (1965). "The physical state of viral nucleic acid and the sensitivity of viruses to ultraviolet light." *Biophysical Journal* 5, 257–273.

Raynor, P. C., and Leith, D. (1999). "Evaporation of accumulated multicomponent liquids from fibrous filters." *Annals of Occupational Hygiene* 43(3), 181–192.

Raynor, P. C., and Leith, D. (2000). "Influence of accumulated liquid on fibrous filter performance." *Journal of Aerosol Science* 31(1), 19–34.

Reiger, I. H., Feucht, G., and Schonfeld, A. (1995). "Selective adsorption of noxon for the detection of ozone." *Odours & VOC's Journal* (Dec.), 39–44.

Reijula, K., Nikulin, M., Jarvis, B. B., and Hintikka, E.-L. (1996). "*Stachybotris atra*-induced lung injury." *The 7th International Conference on IAQ and Climate,* Nagoya, Japan, 639–643.

Rentschler, H. C., and Nagy, R. (1942). "Bactericidal action of ultraviolet radiation on airborne microorganisms." *J. Bacteriol.* 44, 85–94.

Rentschler, H. C., Nagy, R., and Mouromseff, G. (1941). "Bactericidal effect of ultraviolet radiation." *J. Bacteriol.* 42, 745–774.

Rice, R. G. (1997). "Applications of ozone for industrial wastewater treatment—a review." *Ozone Sci. Eng.* 18, 477–515.

Richardson, M. I., and Gangolli, S. (1992). *The Dictionary of Substances and Their Effects.* Clays Ltd., Bugbrooke, Northamptonshire, U.K.

Riley, R. L., and O'Grady, F. (1961). *Airborne Infection.* The Macmillan Company, New York, NY.

Riley, R. L., and Kaufman, J. E. (1972). "Effect of relative humidity on the inactivation of airborne *Serratia marcescens* by ultraviolet radiation." *Applied Microbiology* 23(6), 1113–1120.

Riley, R. L., and Nardell, E. A. (1989). "Clearing the air: The theory and application of ultraviolet disinfection." *Am. Rev. Resp. Dis.* 139, 1286–1294.

Roelants, P., Boon, B., and Lhoest, W. (1968). "Evaluation of a commercial air filter for removal of viruses from the air." *Appl. Microbiol.* 16(10), 1465–1467.

Rogers, K. R., Foley, M., Alter, S., Koga, P., and Eldefrawi, M. (1991). "Light addressable potentiometric biosensor for the detection of anticholinerase." *Anal. Lett.* 24, 191–198.

Rosaspina, S., Anzanel, D., and Salvatorelli, G. (1993). "Microwave sterilization of enterobacteria." *Microbios* 76, 263–270.

Rose, S. (1968). *CBW: Chemical and Biological Warfare.* Beacon Press, Boston, MA.

Rowan, N. J., MacGregor, S. J., Anderson, J. G., Fouracre, R. A., McIlvaney, L., and Farish, O. (1999). "Pulsed-light inactivation of food-related microorganisms." *Appl. Environ. Microbiol.* 65(3), 1312–1315.

Russell, A. D. (1982). *The Destruction of Bacterial Spores.* Academic Press, New York, NY.

Ryan, K. J., ed. (1994). *Sherris Medical Microbiology.* Appleton & Lange, Norwalk, CT.

Sakuma, S., and Abe, K. (1996). "Prevention of fungal growth on a panel cooling system by intermittent operation." *The 7th International Conference on IAQ and Climate,* Nagoya, Japan, 179–184.

Salvato, M. S. (1993). *The Arenaviridae.* Plenum Press, New York, NY.

Sander, D. M. (2002). *All the Virology on the WWW: Biological Weapons and Warfare.* http://www.virology.net/garryfavwebbw.html.

Sanz, B., Palacios, P., Lopez, P., and Ordonez, J. A. (1985). *Effect of Ultrasonic Waves on the Heat Resistance of Bacillus Stearothermophilus Spores.* Fundamental and Applied Aspects of Bacterial Spores/Academic Press, London, U.K.

Sartin, J. S. (1993). "Infectious Diseases during the Civil War: The Triumph of the 'Third Army.'" *Clinical Infectious Diseases* 16, 580–584. http://www.imsdocs.com/civilwar.htm.

Sawyer, R. D., and M.-C. Sawyer, translators (1994). *The Art of War by Sun-tzu.* Westview Press, Boulder, CO.

SBCCOM (2000). *Guidelines for Mass Decontamination during a Terrorist Chemical Agent Incident.* U.S. Army Soldier and Biological Chemical Command, Washington, DC. http://ww2.sbccom.army.mil/hld.

Schaal, K. P., and Pulverer, G. (1979). *Actinomycetes.* Proceedings of the Fourth International Symposium on Actinomycete Biology, Cologne, Germany.

Scherba, G., Weigel, R. M., and W. D. O'Brien, J. (1991). "Quantitative assessment of the germicidal efficacy of ultrasonic energy." *Appl. Microbiol.* 57, 2079–2084.

Scholl, P. F., Leonardo, M. A., Rule, A. M., Carlson, M. A., Antoine, M. D., and Buckley, T. J. (1999). "The development of a matrix-assisted laser desorption/ionization time-of-flight mass spectrometry for the detection of biological warfare agent aerosols." *Johns Hopkins APL Technical Digest* 20(3), 343–362.

Severin, B. F. (1986). "Ultraviolet disinfection for municipal wastewater." *Chemical Engineering Progress* 81, 37–44.

Severin, B. F., Suidan, M. T., and Englebrecht, R. S. (1983). "Kinetic modeling of U.V. disinfection of water." *Water Res.* 17(11), 1669–1678.

Severin, B. F., Suidan, M. T., and Englebrecht, R. S. (1984). "Mixing effects in UV disinfection." *J. Water Pollution Control Federation* 56(7), 881–888.

Sharp, G. (1939). "The lethal action of short ultraviolet rays on several common pathogenic bacteria." *J. Bacteriol.* 37, 447–459.

Sharp, G. (1940). "The effects of ultraviolet light on bacteria suspended in air." *J. Bacteriol.* 38, 535–547.

Shaughnessy, R., Levetin, E., and Rogers, C. (1999). "The effects of UV-C on biological contamination of AHUs in a commercial office building: Preliminary results." *Indoor Environment '99,* 195–202.

Shepard, G. S., Stockenstrom, S., de Villiers, D., Englebrecht, W. J., Sydenham, E. W., and Wessels, G. F. (1998). "Photocatalytic degradation of cyanobacterial microcystin toxins in water." *Water Research* 36(12), 1895–1901.

Sidell, F. R., Takafuji, E. T., and Franz, D. R. (1997). *Medical Aspects of Chemical and Biological Warfare.* Office of the Surgeon General at TMM Publications, Washington, DC. http://www.nbcmed.org/sitecontent/homepage/whatsnew/medaspects/contents.html.

Seigel, J. A., and Walker, I. S. (2001). *Deposition of Biological Aerosols on HVAC Heat Exchangers.* LBNL-47476. Lawrence Berkley National Laboratory, Berkeley, CA.

Siegrist, D. W. (1999). "The threat of biological attack: Why concern now?" *Emerging Infectious Diseases* 5(4), 505–508.

Siegrist, D. W., and Graham, J. M. (1999). *Countering Biological Terrorism in the U.S.: An Understanding of Issues and Status.* Oceana Publ., Dobbs Ferry, NY.

Sikes, G., and Skinner, F. A. (1973). *Actinomycetes: Characteristics and Practical Importance.* Academic Press, London, U.K.

Simon, J. D. (1997). "Biological terrorism: Preparing to meet the threat." *JAMA* 278(5), 428–430.

Singh, A., Kuhad, R. C., Sahai, V., and Ghosh, P. (1994). "Evaluation of biomass." *Adv. Biochem. Eng. Biotechnol.* 51, 48–66.

SIPRI (1971). *The Prevention of CBW.* SIPRI/Humanities Press, New York, NY.

Skistad, H. (1994). *Displacement Ventilation.* John Wiley & Sons, New York, NY.

Slack, J. M., and Gerencser, M. A. (1975). *Actinomycetes, Filamentous Bacteria: Biology and Pathogenicity.* Burgess Publishing Co., Minneapolis, MN.

Smart, J. K. (1997). "History of Chemical and Biological Warfare: An American Perspective." Chapter 2 in *Medical Aspects of Chemical and Biological Warfare.* Office of Surgeon General at TMM Publications, Washington, DC, pp. 9–21. http://www.nbc-med.org/SiteContent/HomePage/WhatsNew/MedAspects/contents.html.

Smerage, G. H., and Teixeira, A. A. (1993). "Dynamics of heat destruction of spores: A new view." *J. Ind. Microbiol.* 12, 211–220.

Smit, M. H., and Cass, A. E. G. (1990). "Cyanide detection using a substrate-regenerating, peroxidase-based biosensor." *Analytical Chemistry* 62(22), 24–29.

Smith, J. M. B. (1989). *Opportunistic Mycoses of Man and Other Animals.* BPCC Wheatons, Exeter, U.K.

Smith, R. A., Ingels, J., Lochemes, J. J., Dutkowsky, J. P., and Pifer, L. L. (2001). "Gamma irradiation of HIV-1." *J. Orthop. Res.* 19(5), 815–819.

Somani, S. M. (1992). *Chemical Warfare Agents.* Academic Press, New York, NY.

Sorensen, K. N., Clemons, K. V., and Stevens, D. A. (1999). "Murine models of blastomycosis, coccidioidomycosis, and histoplasmosis." *Mycopathologia* 146, 53–65.

Spendlove, J. C., and Fannin, K. F. (1983). "Source, significance, and control of indoor microbial aerosols: Human health aspects." *Public Health Reports* 98(3), 229–244.

Stafford, R. G., and Ettinger, H. J. (1972). "Filter efficiency as a function of particle size and velocity." *Atmospheric Environment* 6, 353–362.

Stearn, E. W., and Stearn, A. E. (1945). *The Effect of Smallpox on the Destiny of the Amerindians.* Bruce Humphries, Boston, MA.

Stephen, J., and Pietrowski, R. A. (1981). *Bacterial Toxins.* American Society for Microbiology, Washington, DC.

Stern, J. (1999). "The prospect of domestic bioterrorism." *Emerging Infectious Diseases* 5(4), 517–522.

Stoecker, W. F., and Stoecker, P. A. (1989). *Microcomputer Control of Thermal and Mechanical Systems.* Van Nostrand Reinhold, New York, NY.

Storz, J. (1971). *Chlamydia and Chlamydia-Induced Diseases.* Charles C. Thomas, Springfield, IL.

Straja, S., and Leonard, R. T. (1996). "Statistical analysis of indoor bacterial air concentration and comparison of four RCS biotest samplers." *Environment International* 22(4), 389.

Suneja, S. K., and Lee, C. H. (1974). "Aerosol filtration by fibrous filters at intermediate Reynolds numbers (<100)." *Atmos. Environ.* 8, 1081–1094.

Sylvania (1981). *Germicidal and Short-Wave Ultraviolet Radiation.* Sylvania Engineering Bulletin 0-342. GTE Products Corp.

Takafuji, E. T., Johnson-Winegar, A., and Zajtchuk, R. (1997). Chapter 35 in *Medical Aspects of Chemical and Biological Warfare.* Office of the Surgeon General at TMM Publications, Washington, DC.

Takeuchi, Y., and Itoh, T. (1993). "Removal of ozone from air by activated carbon treatment." *Sep. Technol.* 3, 168–175.

Tamai, H., Kakii, T., Hirota, Y., and Kumamoto, T. (1996). "Synthesis of extremely large mesoporous activated carbon and its unique adsorption for giant molecules." *Chemistry of Materials* 8(2), 454.

Tanner, H. H., et al. (1986). *Atlas of Great Lakes Indian History.* University of Oklahoma Press, Norman, OK.

Tennal, K. B., Mazumder, M. K., Siag, A., and Reddy, R. N. (1991). "Effect of loading with an oil aerosol on the collection efficiency of an electret filter." *Particulate Science & Technology* (9), 19–29.

Thomas, A. V. W. (1970). *Legal Limits on the Use of Chemical and Biological Weapons.* Southern Methodist University Press, Dallas, TX.

Thomas, W. (1998). *Bringing the War Home.* Earthpulse Press, Anchorage, AK.

Thompson, H. G., and Lee, W. E. (1992). "Rapid immunofiltration assay of *Francisella tularensis.*" *1992 Suffield Memorandum No. 1376,* Canada, 1–17.

Thorne, H. V., and T.M.Burrows (1960). "Aerosol sampling methods for the virus of foot-and-mouth disease and the measurement of virus penetration through aerosol filters." *J. Hyg.* 58, 409–417.

Thornsberry, C., Balows, A., Feeley, J., and Jakubowski, W. (1984). *Legionella: Proceedings of the 2nd International Symposium.* Atlanta, GA.

Thurman, R., and Gerba, C. (1989). "The molecular mechanisms of copper and silver ion disinfection of bacteria and viruses." *CRC Crit. Rev. Environ. Control* 18, 295–315.

Tolley, G., Kenkel, D., and Fabian, R. (1994). *Valuing Health for Policy.* University of Chicago Press, Chicago, IL.

Torok, T. J., Tauxe, R. V., Wise, R. P., Livengood, J. R., Sokolow, R., Mauvais, S., Birkness, K. A., Skeels, M. R., Horan, J. M., and Foster, L. R. (1997). "A large community outbreak of Salmonellosis caused by intentional contamination of restaurant salad bars." *JAMA* 278, 389–395.

Tucker, J. B. (1999). "Historical trends related to bioterrorism: An empirical analysis." *Emerging Infectious Diseases* 5(4), 498–504.

Tuder, R. M., Ibrahim, E., Godoy, C. E., and Brito, T. D. (1985). "Pathology of the human paracoccidioidomycosis." *Mycopathologia* 92, 179–188.

Turner, M. (1980). "Anthrax in humans in Zimbabwe." *Centr. Afr. J. Med.* 26(7), 160–161.

USAMRIID (2000). *History of Biological Warfare.* American War College, Washington, DC.

USAMRIID (2001). *Medical Management Of Biological Casualties Handbook,* 4th ed. U.S. Army Medical Research Institute of Infectious Diseases, Fort Detrick, MD. http://www.nbc-med.org/SiteContent/HomePage/WhatsNew/MedManual/Feb01/handbook.htm.

U. S. Army Corps of Engineers (2001). *Protecting Buildings and Their Occupants from Airborne Hazards.* Document TI 853-01. U. S. Army Corps of Engineers, Washington, DC. http://www.state.nd.us/dem/Documents/Building%20Protection.pdf.

U. S. Congress (2000). "Terrorist threats to the United States: Hearing before the Special Oversight Panel on Terrorism of the Committee on Armed Services, House of Representatives, 106th Congress." *Y 4.AR 5/2 A, 999-2000/52,* U.S. GPO, Washington, DC.

U. S. Congress (2001). *Patterns of Global Terrorism and Threats to the United States.* House Armed Services Committee, Washington, DC.

USPS (2002). *The United States Post Office Website.* U.S. Post Office. http://www.usps.com/.

UVDI (1999). *Report on Lamp Photosensor Data for UV Lamps.* Ultraviolet Devices, Valencia, CA.

UVDI (2001). *UVD: Ultraviolet Air Disinfection Design Program.* Ultraviolet Devices, Valencia, CA.

vanLancker, M., and Bastiaansen, L. (2000). "Electron-beam sterilization. Trends and developments." *Med Device Technol.* 11(4), 18–21.

VanOsdell, D. W. (1994). "Evaluation of test methods for determining the effectiveness and capacity of gas-phase air filtration equipment for indoor air applications." *ASHRAE Transactions* 100(2), 511.

VanOsdell, D. W., and Sparks, L. E. (1995). "Carbon adsorption for indoor air cleaning." *ASHRAE J.* 37(2), 34–40.

van Veen, J. A., and Paul, E. A. (1979). "Conversion of biovolume measurements of soil organisms, grown under various moisture tensions, to biomass and their nutrient content." *Appl. Environ. Microbiol.* 37, 686–692.

Vollmer, A. C., Kwakye, S., Halpern, M., and Everbach, F. C. (1998). "Bacterial stress response to 1-megahertz pulsed ultrasound in the presence of microbubbles." *Applied and Environmental Microbiology* 64(10), 3927–3931.

Wagner, F. S., Eddy, G. A., and Brand, O. M. (1977). "The African green monkey as an alternate primate host for studying Machupo virus infection." *The American Journal of Tropical Medicine and Hygiene* 26(1), 159–162.

Waid, D. E. (1969). "Incineration of organic materials by direct gas flame for air pollution control." *American Industrial Hygiene Association Journal* 30, 291–297.

Wake, D., Redmayne, A. C., Thorpe, A., Gould, J. R., Brown, R. C., and Crook, B. (1995). "Sizing and filtration of microbiological aerosols." *J. Aerosol Sci.* 26(S1), s529–s530.

Wald, G., and United Nations (1970). *Chemical and Bacteriological (Biological) Weapons and the Effects of Their Possible Use.* Report E.69.I.24. Ballantine, New York, NY.

Walker, D. H. (1988). *Biology of Rickettsial Diseases.* vols. 1 and 2. CRC Press, Boca Raton, FL.

Walker, P. L., and Thrower, P. A. (1975). *Chemistry and Physics of Carbon,* vol. 12. Marcel Dekker, New York, NY.

Walter, C. W. (1969). "Ventilation and air conditioning as bacteriologic engineering." *Anesthesiology* 31, 186–192.

Walton, G. N. (1989). "Airflow network models for element-based building airflow modeling." *ASHRAE Transactions* 1989, 611–620.

Washam, C. J. (1966). "Evaluation of filters for removal of bacteriophages from air." *Appl. Microbiol.* 14(4), 497–505.

Webb, A. R. (1991). "Solar ultraviolet radiation in southeast England: The case for spectral measurements." *Photochemistry and Photobiology* 54(5), 789–794.

Webb, S. J., and Booth, A. D. (1969). "Absorption of microwaves by microorganisms." *Nature* 222(June), 1199–1200.

Weinberger, S., Evenchick, Z., and Hertman, I. (1984). "Transitory UV resistance during germination of UV-sensitive spores produced by a mutant of *Bacillus cereus 569.*" *Photochem. Photobiol.* 39(6), 775–780.

Weinstein, R. A. (1991). "Epidemiology and control of nosocomial infections in adult intensive care units." *Am. J. Med.* 91(suppl 3B), 179S–184S.

Weissenbacher, M. C., Laguens, R. P., and Coro, C. E. (1987). "Argentine hemorrhagic fever." *Arenaviruses.* M. B. A. Oldstone, ed. Springer-Verlag, New York, NY.

Wekhof, A. (1991). "Treatment of contaminated water, air, and soil with UV flashlamps." *Environmental Progress* 10(4), 241–247.

Wekhof, A. (2000). "Disinfection with flashlamps." *PDA J. Pharmaceutical Science and Technology* 54(3), 264–267. http://www.wektec.com/.

Wekhof, A., Folsom, E. N., and Halpen, Y. (1992). "Treatment of groundwater with UV flashlamps—the third generation of UV systems." *Hazardous Materials Control* 5(6), 48–54.

Wekhof, A., Trompeter, I.-J., and O.Franken (2001). "Pulsed UV-disintegration, a new sterilization mechanism for broad packaging and medical-hospital applications." *Proceedings of the First International Congress on UV-Technologies,* Washington, DC. http://www.wektec.com/.

Westinghouse (1982). *Booklet A-8968.* Westinghouse Electric Corp., Lamp Div., Monroeville, PA.

White, H. J. (1977). "Electrostatic precipitation of fly ash." *Journal of Air Pollution Control Association* 27, 15–21.

Whitmore, T. M. (1992). *Disease and Death in Early Mexico.* Westview Press, Boulder, CO.

WHO (1970). *Health Aspects of Chemical and Biological Weapons.* World Health Organization, Geneva, Switzerland.

WHO (1982). *Rift Valley Fever: An Emerging Human and Animal Problem.* World Health Organization, Geneva, Switzerland.

Wilkinson, T. R. (1966). "Survival of bacteria on metal surfaces." *Applied Microbiology* 14, 303–307.

Williams, S. M., Fulton, R. M., Patterson, J. S., and Reed, W. M. (2000). "Diagnosis of eastern equine encephalitis by immunohistochemistry in two flocks of Michigan ring-necked pheasants." *Avian Diseases* 44(4), 1012–1016.

Woods, J. E., Grimsrud, D. T., and Boschi, N. (1997). *Healthy Buildings / IAQ '97.* ASHRAE, Washington, DC.

Wright, S. (1990). *Preventing a Biological Arms Race.* MIT Press, Cambridge, MA.

Yaghoubi, M. A., Knappmiller, K., and Kirkpatrick, A. (1995). "Numerical prediction of contaminant transport and indoor air quality in a ventilated office space." *Particulate Science and Technology* 13, 117–131.

Yang, C. S., Hung, L.-L., Lewis, F. A., and Zampiello, F. A. (1993). "Airborne fungal populations in non-residential buildings in the United States." *IAQ '93,* Helsinki, Finland, 219–224.

Youmans, G. P. (1979). *Tuberculosis.* W. B. Saunders, Philadelphia, PA.

Zeterberg, J. M. (1973). "A review of respiratory virology and the spread of virulent and possibly antigenic viruses via air conditioning systems." *Annals of Allergy* 31, 228–299.

Zhang, Y., Fox, J. G., Ho, Y., Zhang, L., Stills, H. F., and Smith, T. H. (1993). "Comparison of the major outer membrane protein gene of mouse pneumonitis and hamster SFPD strains of *Chlamydia trachomatis* with other *Chlamydia* strains." *Mol. Biol. Evol.* 10, 1327–1342.

Zilinskas, R. A. (1997). "Iraq's biological weapons." *JAMA* 278(5), 418–424.

Zilinskas, R. A. (1999a). "Cuban Allegations of Biological Warfare by the United States: Assessing the Evidence." *Crit. Rev. Microbiol.* 25(3), 173–227.

Zilinskas, R. A. (1999b). "Terrorism and Biological Weapons: Inevitable Alliance." *Persp. Biol. Med.* 34(1), 45–72.

Zilinskas, R. A. (1999c). *Assessing the Threat of Bioterrorism.* Monterey Institute of International Studies, October 20, 1999. http://www.house.gov/reform/ns/press/test3.htm.

Zilinskas, R. A. (2000). *Biological Warfare.* Lynne Rienner Publ., Boulder, CO.

Zink, S. (1999). *Understanding the Threat of Bioterrorism.* Johns Hopkins Center for Civilian Biodefense Studies, DHHS, IDSA, & ASM, Washington, DC. http://www.acl.lanl.gov/RAMBO/Minutes/Bioterrorism_conf_2.99.htm.

Zoon, K. C. (1999). "Vaccines, pharmaceutical products, and bioterrorism: Challenges for the U.S. Food and Drug Administration." *Emerging Infectious Diseases* 5(4), 534–536.

Index

abortion clinics, 129–130
abrin, 91, 453, **454**
absorption efficiency, chemical weapons agents and, 85
access control to security areas, 288–289
access points, commercial buildings and, 122
aconite, 3
aconitine, 91, 453, **454**
actinomycetes, 34
adamsite, 64, 82, 469, 471, **471**
Aedes aegypti (Yellow fever mosquito), **39**
aerolysin, **454**, 467
aerosol dispersal systems, 16
aerosolization, **103**, 146–147, **148**
  air intake release of, 143
  atomizers for, 104
  biological weapons agents and, 49
  combustion mechanisms or explosive devices for, 104
  condensation of supersaturated particles in, 104
  dispersion and delivery in, 101–109
  explosive devices in, 108
  filtration and, liquid types, 185–188, **187**
  generation of particulate clouds for, 104, **105**
  indoor release of, 146–147, **148**
  internal releases into ducts of, 147–150, **149**
  liquids in, 104, **106**
  nebulizers and compressors for, 105–106, **106, 107, 108**
  passive dissemination in, 107, 108, **109, 110**
  powders in, 104–106, **106**
  smoke generator for, **105**
  sprayers for, 104
  venturi style aerosol nozzles for, 104, **108**
aerosolizers, 97, 99, 101–109, 301, **302**
  disabling, 301–303, **302**
affinity class biological sensors, 263
aflatoxin, 43, 52, 453, **454**
  lethal doses for, 91
  open air aerosol toxicity of, 101
African hemorrhagic fever (*see* Ebola)
agro-terrorism, 18
airborne concentration
  biological weapons agents and, 76
  chemical weapons agents and, 85
  dispersion and delivery in, 98, 99

airborne disease, 25, 27–37, 388–389
air change rate (ACH), ventilation systems and, 155
air cleaning and disinfection systems, 165–209, **166**, 278–279
  alternative technologies for, 327–348, **328**
  carbon adsorption as, 165, **166**, 203–208, **205, 206, 207, 208**
  dilution ventilation in, 165, **166**, 167
  diminishing returns principle in, 234, 236, **236**
  effectiveness of, 165, **166**
  emergency systems for, 278–279
  filtration as, 165, **166**, 167–188, **167**
  gamma irradiation as, 166
  gas phase filtration as, 203–208, **204**
  granulated activated carbon (GAC) in, 204–207, **205**
  impregnated filtration as, 166
  ionization as, 166
  large-sized buildings and, 231–232, **235**
  medium-sized buildings and, 231, **234**
  microwaves as, 166
  overload attack and, 242–244, **245**
  ozone as, 166
  photocatalytic oxidation (PCO) as, 166
  plasma fields as, 166
  pulsed electric fields as, 166
  pulsed light as, 166
  simulation of building attack scenarios and, 214, 220, **221, 223, 224**, 230, 231, 233, 234
  sudden release attack simulation, 244–246, **245, 246**
  ultraviolet germicidal irradiation (UVGI) as, 165, **166**, 188–200
air flow rate and simulation of building attack scenarios and, 231
air handling units (AHU), security of, 288–289, **289**
air intake release, 141–143, **142, 143**
  in simulation of building attack scenarios, 236–238, **238, 239**
air mixing (ventilation effectiveness), 155–156, 159, 177
airports, 123, **124**
air returns, access to, 289, **290**
air sampling systems, 249, 257, 258–261, **258, 259, 261**, 270

Note: Pages shown in **boldface** have illustrations on them.

alarm systems, 270, 271, **271, 272, 273**
algal toxins, 43, 46
Alibek, Ken, 35
alkyl alkylphosphonofluorodates, 469, 518
alkyl aminoethyl alkyl phosphonites, 469, 518
alkyl phosphonochoridates, 469, 518
alkyl phosphonothiolates, 469, 518
alkyl phosphonyldifluorides, 469, 518
alkyl phosphoramidocyandiates, 518, 469
all-outside-air ventilation systems, 152, 154, 165
allergies (*see Stachybotrys chartarum*)
alpha Latrotoxin, 453, **454**
Alphaviruses, 34, 35
alternative disinfection technologies, 327–348, **328**, 385
aluminum–magnesium silicate, in decontamination, 313
amagassa venom, 467
American Nazis, 386
Amherst, Jeffrey, 7–8, **8**, 10
amilyn, 467
amiton, 64, **316**, 469, 472, **472**
ammodytoxin, 467
amyl nitrite, 306
Anabaena flosquae, 43
anatoxin A, 43, 91, **455**
angiotensin forming enzyme, 455
angiotensis/angiotonin, 48, 453, **455**
animals as "biosensors," 296
anionic surfactants, in decontamination, 313
annualized cost (AC) of UVGI, 359
anthrax (*see Bacillus anthracis*)
antibodies, 262
antidotes, 306
antigens, 262
anti-Lewisite, 306
antimicrobial coatings, in decontamination, **328**, 332–333
antimicrobial treatments for filters, 179
apamin, 91, 453, **453**
apartment building attack scenarios, 130–131, **131**, 231
Arbovirus, 35, 37, 40, 436
arbrin, open air aerosol toxicity of, 101
Arenaviruses, 33
Argentinean hemorrhagic fever (*see* Junin virus)
arsenic trichloride, 64, 469, 510
decontaminant/neutralizer for, **316**
arsenic, 12–13
arsine, 64
carbon adsorption filters vs., **208**
decontaminant/neutralizer for, **316**
Art of War, 3
Aryan Nations, 20, 386
ASHRAE standards
for filtration, 168, 179
for gas phase filtration, 206
Aspergillus niger, 345
microwave radiation vs., 378, **378**
ozone disinfection and, 318–321, **322**
pulsed filter light vs., 344–345, **345**
pulsed light treatment vs. UVGI in, 342–343, **344, 345**
Aspergillus spp., 43
(*see also* aflatoxin)
assembly buildings, 136–138
auditoriums as, 136–137
churches, places of worship, 137–138, **138**
size, occupancy, risk level for, **115**

assembly buildings (*cont.*):
stadiums as, 137, **137**
theaters as, 136
Atlanta Olympics, Centennial Park 1996 attack, 130, 288
atomizers, 104
atrial natriuretic peptide, 453, **455**
atropine, 306
attack scenarios (*see* buildings and attack scenarios)
auditorium attack scenarios, 136–137, 231
Aum Shinrikyo cult, 20, 98, 100, 140–141, 296, 308
Avernian place, 6
Aztec tapestry, showing smallpox epidemic, **7**

bacillary dysentery (*see Shigella*)
Bacillus anthracis (anthrax), 15, 16, 17, 20, 25, 27, 30, 35, 36, 97, 98, 138, 140–141, **230**, 296, 308, 372, 393, **393**
biosensors and assay times for, **264**
CONTAMW program modeling of release of, 227–229, **229, 231**
decontamination and remediation of, 311, 322–323
disease and dose curves for, 439, **441**
disease progression in, 94, **95**
dose value for, 80, 81
explosive dissemination devices for, 101
incubation time of, 276, **277**
mailroom contamination with, 367–382
microwave irradiation of, 378, **378**
open air aerosol toxicity of, **102**
overload attack simulation of, 242–244, **245**
ozone disinfection and, 321, **322**
post office attacks using, 125–126, 146
removal rates, 236, **237**, 238
simulation of building attack scenarios and, 214, 215, 224–226, **225**
SNL foam vs., 323, **324**
UVGI plus filtration vs., **202**
UVGI vs., 200, 374–377, **375, 376**
Bacillus cereus
ozone vs., 318–321, **322**, 377
pulsed electric field (PEF) vs., 346
Bacillus pumilus
pulsed light treatment vs. UVGI in, 342–343, **344, 345**
Bacillus stearothermophilus
pulsed light treatment vs. UVGI in, 342–343, **344, 345**
Bacillus subtilis, **344**
pulsed light treatment vs. UVGI in, 342–343, **344, 345**
BACnet, 270–271, 281–285
control systems and, 269
security and, 292
bacteria as disease organisms, 23
ultraviolet germicidal irradiation (UVGI) vs., 52–59, **54, 55, 57, 59**
bacterial toxins, 41–47, **44–45**
banana spider, **48**
banks, 122
Barbarossa, 6
baseline design conditions, 211–214
basic agents in, in decontamination, 314
batrachatoxin, 43, 453, **456**
lethal doses for, 91
open air aerosol toxicity of, 101
Belgrade, Battle of (1456), 6
benzilic acid, 469, 518

biaxial UVGI lamps, 192, **193**
binary chemicals, 303
biological weapons agents, 1–3, 23–59, 391–437
  aerosolization delivery for, 101–109
  airborne concentration vs., 76
  airborne diseases as, 27–37
  anthrax as, 27, 30, 27
  bioregulators as, 47–49
  breathing rates vs., 85–86, **86**, 219–220, 224–226
  carbon adsorption vs., 204
  chloramine B to decontaminate, 314
  database of, 391–437
  "detect to treat" principle in, 275–276
  detection systems and, 257–268
  disease progression curves in, 89–95, **91**
  dispersion and delivery systems for, 97–112
  division and classification of, 23–24, **24**
  dose and epidemiology of, 73–96, **76**
  dose response curve in, 76–77, **76, 77, 78, 80,** 439–451
  filtration systems for, 50–52, **51**
  food-borne pathogens as, 40–41
  genetic engineering and, 26–27
  growth or culture of, 35–36, **36,** 49, 257
  illnesses and absences as indication of biological attack, 295, 296
  incubation periods in, 90–91
  indoor dispersion and delivery systems for, 99–109
  infectious period in, 90–93
  ingestion dosimetry in, 87–89
  lethality vs. concentration of agents in, 226, **226, 233**
  list of pathogenic microorganisms for, **28–29**
  mean disease period (MDP) in, 93
  mean infectious dose (ID_{50}) in, 73, 76
  mean lethal dose ($L(Ct)_{50}$) in, 74
  mean lethal dose (LD_{50}) in, 74, 75–80
  microorganisms in, 24–41
  outdoor dispersion systems for, 97–99
  ozone disinfection and, 338–340, **338**
  particle detectors in, 263–267, **265.** 267
  Petri dish for culturing biological agents in, 258–259, **260**
  preparation of, 49
  simulation of building attack scenarios and, 214, 215
  size distribution of, 50–52, **51,** 168–169
  SNL foam vs., 323, **324**
  spore size vs. filtration efficiency in, 168–169, **180,** 181–182
  toxicity (relative) of, **28–29,** 49
  toxins as, 41–47, **42, 44–45,** 52
  transmission routes for, 25–26, **25**
  ultraviolet germicidal irradiation (UVGI) vs., 50, 52–59, **54, 55, 57, 59,** 188–200
  vector-borne pathogens as, 37–40
  water-borne pathogens as, 40–41
  weaponizing BW agents in, 49
Biopreparat program (USSR), 16–17, 19, 27, 35
bioregulators, 47–49, 453–466, **455–466**
  mean lethal dose (LD_{50}) in, 90, **91**
biosensors, 254, 261–263, **262, 264,** 273–274
bioterrorism, 19–21, **21,** 385–386
Blastomyces dermatitidis, 26, 30, 391, 394, **394**
  disease and dose curves for, 439, **441**
  disease progression in, 94
  dose values for, 81
blastomycosis (*see Blastomyces dermatitidis*)

bleach in decontamination and remediation, 311, 322–323
bleach-soaked towels, as gas mask substitute, 300
bleaching powder, in decontamination, 314
bleeding lung disease, 34, 430
  (*see also Stachybotrys chartarum*)
blister agent or vesicant, 63, 66–68, 81, 86, 299
blood agents, 62, 66, 68
blue-green algae, 43
blueprints, security and, 292–293
Boer War, 13
Bolivian hemorrhagic fever (*see* Machupo virus)
bombesin, 453, **456**
bombs (*see* explosive devices)
Bordetella pertussis (whooping cough), 30, 391, 395, **395**
  disease and dose curves for, 439, **441**
  disease progression in, 94
  dose values for, 81
  size of microbe, vs. filtration, **182**
Botkin's disease (*see* Hepatitis A virus)
botulinum toxin/botulism, 453, **456**
  (*see also Clostridium botulinum*)
Bouquet, Henry, 7, 9
box jellyfish (*Chironex fleckeri*), **47**
bradykinin, 453, **457**
breakbone fever (*see* Dengue Fever virus)
breathing rates vs. exposure, 85–86, **86,** 219–220, 224–226
brevetoxin B, 43, 453, **456**
Brickhill, Jeremy, 17
Brill–Zinsser disease (*see Rickettsia prowazeki*)
British anti-Lewisite (BAL), 68
British use of biological/chemical weapons, 13
Brownian motion detection, 266
Brucella spp. (brucellosis), 391, 396, **396**
  biosensors and assay times for, **264**
  disease and dose curves for, 439, **441**
  disease progression in, 94
  dose values for, 81
  open air aerosol toxicity of, **102**
  UVGI and filtration vs., **202**
brucellosis (*see Brucella*)
bubonic plague (*see Yersinia pestis*)
building automation system (BAS), 269, 270–271, 281–285, 292, 309
building input data/parameters for building attack scenarios, 215, **216**
building related illness (BRI), 387
buildings and attack scenarios, 113–150
  air intake releases in, 141–143, **142, 143**
  assembly buildings and, **115,** 136–138
  commercial buildings and, **115,** 119–123, **120, 121**
  education buildings and, **115,** 132–134
  filtration vs., 148
  food and entertainment buildings and, **115,** 127–129
  government buildings and, **115,** 123–127
  healthcare buildings and, **115,** 129–130, **130**
  immunity factors in, 118
  indoor aerosolization release in, 146–147, **148**
  indoor explosive release in, 144–146, **144, 145**
  indoor passive releases, 146
  internal releases into ducts in, 147–150, **149**
  location of buildings in, 113–114
  lodging buildings and, **115,** 130–132
  mercantile buildings and, **115,** 134–136

buildings and attack scenarios (*cont.*):
  military installations and, 117
  occupancy vs. risk in, 116–118
  outdoor releases in, 140–141, **142**
  profile (importance, visibility) and, 116–118
  relative risk assessment for, 118–119, **119**, **150**
  risk factors in, 116–119
  risk or vulnerability assessment for, 113–119
  size or floorspace of buildings in, 113–114
  special facilities and, 138–140
  types of buildings and relative risk of,
       115–140, **115**
  ventilation systems in, 113–114, 148–149
bungarotoxin, 43, 453, **466**
  biosensors and assay times for, **264**
  lethal doses for, 91
Bunyaviruses, 34, 37
Burkholderia mallei (glanders), 31, 35, 391,
     397, **397**
  disease and dose curves for, 439, **442**
  disease progression in, 94
  dose values for, 81
  open air aerosol toxicity of, **102**
Burkholderia pseudomallei (melioidosis), 31,
     35, 391, 398, **398**
  disease and dose curves for, 439, **442**
  disease progression in, 94
  size of microbe, vs. filtration, **182**
BZ, 64, 469, 510

C8 microemulsion, in decontamination, 315
calcium hydroxide, 314
calcium hypochlorite, 314
calculus methods for ventilation modeling,
     156–157
calon snake, **46**
calorimetric biosensors, 263
camelpox, **202**, 432
camite, 64, 82, 469, 474, **474**
carbon adsorption, 165, **166**, 203–208, **205–208**,
     327, 337, 387
  economics and optimization of, 363–364
  filtration and, 364
  UVGI and, 364
  metallic salt impregnated (MSI), 338
  simulation of building attack scenarios and,
       214
carbon dioxide, 250
Carulite, 338
castor bean oil, 20, 46
casualty distribution throughout building during
     attack, 295–296
catalytic and noncatalytic biosensors, 263
catalytic detectors, 252
cationic surfactants, in decontamination, 313
CD50 (*see* mean casualty dose)
Central Asian hemorrhagic fever (*see*
     Crimean–Congo Fever virus)
charybdotoxin, 467
chemical means of decontamination, 314–315,
     316–317, 327
chemical weapons agents, 1–3, 61–71, **62**,
     469–518
  absorption efficiency in, 85
  aerosolization delivery for, 101–109
  airborne concentration vs., 85
  animals as "biosensors," 296
  blister agents or vesicants in, 63, 66
  blood agents in, 62, 66, 68
  boiling points of, vs. filtration, 186–187
  breathing rates vs., 85–86, **86**

chemical weapons agents (*cont.*):
  carbon adsorption vs., 205–206, **208**
  choking agents in, 63, 66, 68
  classification of, 61–66, **63**
  decontamination and remediation in, 314
  design of, 66–70, **66**
  detection systems and, 250–257, **251**
  dispersion and delivery systems for, 97–112
  dose and epidemiology of, 73–96
  dose response curves in, 83–84, **83**, 471–518
  filtration and, 185–188, **187**
  HAZMAT ID numbers in, 66
  incapacitating agents in, 66
  indoor dispersion and delivery systems for,
       99–109
  ingestion dosimetry in, 87–89
  inhalation dosimetry for, 85–87, **87**
  lethality vs. concentration of agents in, 226,
       226, **233**
  mean casualty dose (CD50) in, 74, 81
  mean incapacitating dose, 73–74, 81
  mean lethal concentration (LC50) in, 74, 88
  mean lethal dose (L(Ct)50) in, 74
  mean lethal dose (LD50) in, 74, 82–84, 86–87
  nerve agents in, 14, 63, 64, 66, 67, 307
  nonlethal vs. lethal, 62
  open air toxicity of aerosolized, 99, **100**, **101**
  outdoor dispersion systems for, 97–99
  purging by outside air in, 246–247, **247**
  simulants of, 70, **70**
  simulation of building attack scenarios and,
       214
  source of precursor chemicals for, 70, **70**
  symptoms of, as warning of attack, 294–295
  toxicity (relative) of, 63, **64–65**, **66**, 66–70
  vomiting agents, 62, 63, 66
Chicago disease (*see Blastomyces dermatitidis*)
Chikungunya virus, 35, 37, 40, 391, 399, **399**
  size of microbe, vs. filtration, **182**, 182
  disease and dose curves for, 439, **442**
  disease progression in, 94
China, 3
Chlamydia pneumoniae, 52
Chlamydia psittaci, 31, 391, 400, **400**
  disease and dose curves for, 439, **442**
  disease progression in, 94
  size of microbe, vs. filtration, **182**
chloramine B, 314
chlorine gas, 12–13, 63, 64, 66, 69–70, 469, 475,
     475
  carbon adsorption filters vs., **208**
  decontaminant/neutralizer for, **316**
  dose values for, 82
  simulation of building attack scenarios and,
       214
chlorine dioxide as disinfectant, 311, 322–323
chlorinechloroethylchloromethylsulfide, 518
chloroethylthio n pentane, 518
chloroethylthio n propane, 518
chloroethylthio n butane, 518
chloroethylthiomethane, 518
chloroethylthiomethyl ether, 518
chloropicrin, 64, 469, 511
  carbon adsorption filters vs., **208**
  decontaminant/neutralizer for, **316**
chlorosarin, 64, 469, 476, **476**
  decontaminant/neutralizer for, **316**
chlorosoman, 64, 469, 511
  decontaminant/neutralizer for, **316**
choking agents, 63, 66, 68
cholecystokinin, 453, **457**

cholera toxin, 467
cholera (*see* Vibrio cholerae), 434
Christian Identity, 20, 386
churches, places of worship, 137–138, **138**
ciguatoxin, 91, 453, **457**
citrinin, 91, 453, **458**
City Hall, Boston, **126**
Civil War, 12–13
classification of chemical weapons agents and, 61–66, **63**
clinics, 129–130
  (*see also* healthcare buildings)
Clostridium botulinum (botulism/botulinum toxin), 14, 40, 43, 46, 49, 140–141, 308, 391, 401, **401**
  biosensors and assay times for, 262, **264**
  carbon adsorption vs., 204
  disease and dose curves for, 439, **443**
  disease progression in, 94
  lethal doses for, 91
  open air aerosol toxicity of, 101
  removal rates, 236, **237**, 238
Clostridium perfringens, 40, 46, 391, 402, **402**
  biosensors for, 262
  disease and dose curves for, 439, **443**
  disease progression in, 94
  dose values for, 81
  lethal doses for, 91
  UVGI and filtration vs., **202**
Clostridium perfringens toxin, 453, 467, **458**
clostridium, 17
CN, 64, 82, 469, 477, **477**
cobra (*Naja naja*), 43
cobrotoxin, 43, 91, 453, **458**
Coccidioides immitis, 31–32, **31**, 391, 403, **403**
  disease and dose curves for, 439, **443**
  disease progression in, 94
  dose values for, 81
  open air aerosol toxicity of, **102**
  UVGI and filtration vs., **202**
coccidioidomycosis (*see Coccidioides immitis*)
colleges and universities, 132–134, **134**
combustion mechanisms (*see* explosive devices)
commercial buildings, 119–123, **120, 121**
  access points in, 122
  airports as, 123, **124**
  banks as, 122
  constant volume airflow systems in, 120
  equipment rooms and access to, 122
  filtration in, 121
  manufacturing facilities as, 122–123, **123**
  occupancy levels in, 123
  risk assessment for, 119–120
  security issues in, 122
  simulation of building attack scenarios and, 231
  size, occupancy, risk level for, **115**
  variable air volume (VAV) systems in, 120, 149
  ventilation systems of, 120–122
compressors for aerosol delivery, 105–107, **106, 107, 108**
  disabling, 301–303, **302**
computational fluid dynamics (CFD), ventilation systems design, 154, 163
computational methods for ventilation modeling, 157–162
concentration camps, 14–15
concentration, lethality vs., 226, **226, 233**
concentration, ventilation modeling and, 156–162, **156, 158, 160, 161**

conotoxin, 453, **458**
condensation of supersaturated particles, 104
constant volume airflow systems, 120, 152
consumption (*see Mycobacterium tuberculosis*)
contaminants, 1
CONTAMW computer program, 161–163, 226–229, **227, 228, 231, 232, 233**
  multizone building attack simulation using, 238–242, **240, 241, 242**
  simulation of building attack scenarios and, 214–215
continuous release attack simulation, 214
control systems, 269–285, 384
  air sampling systems for, 270
  alarm systems for, 270, 271, **271, 272, 273**
  architectures for, 271–276, **271**
  building automation system (BAS) in, 269, 270–271, 281–285
  critical functions of, 269
  dampers in, 309
  "detect to alarm" principles in, 271, **271, 272, 273**
  "detect to isolate" principle in, 271, **271**, 273–275, **274, 275**
  "detect to treat" principle in, 271, **271**, 275–276
  detection systems for, 270
  emergency functions of, 269, 276–281
  logic of response in, **284**, 285
  network based (BACnet), 269, 270–271, 281–285
  purging systems in, 277–278, **278**
  secondary system operation and, 278–279, **279, 280**
  sheltering zone isolation and, 279–281, **281**, 304–306, **305**
  shutdown systems in, 274–275
  smoke detection/control in, 282–283, **282, 283**
  types of, 270–271
cooling towers, 140
Corynebacterium diphtheria, 46, 391, 404, **404**
  disease and dose curves for, 439, **443**
  disease progression in, 94
  lethal doses for, 91
  open air aerosol toxicity of, 101
  UVGI and filtration vs., **202**
cost of security/prevention, 2, 211, 212–213
Coulter counters, 266
courthouses as, 124–125, **125**
Coxiella burnetii (Q fever), 32, 35, 36, 39–40, 391, 405, **405**
  biosensors and assay times for, **264**
  disease and dose curves for, 439, **444**
  disease progression in, 94
  dose values for, 81
  open air aerosol toxicity of, **102**
  size of microbe, vs. filtration, **182**
Crimea, Crimean War, 6
Crimean–Congo hemorrhagic fever, 34, 37, 39, 391, 406, **406**
  disease and dose curves for, 439, **444**
  disease progression in, 94
  size of microbe, vs. filtration, **182**
  UVGI and filtration vs., **202**
critical functions of control systems and, 269
crop dusting planes to disperse toxins, 16, 97–98, **98**
cryogenic freezing, in decontamination, 328, 329
Cryptosporidium, pulsed electric field (PEF) vs., 346
CS, 64, 82, 469, 478, **478**

Cuba and bioterrorism, 18
Cunningham slip factor, filtration and, 174–175
curare, 46, 91, 453, **459**
cyanide, 112
cyanogen chloride, 64, 66, 68, 469, 479, **479**
   carbon adsorption filters vs., **208**
   decontaminant/neutralizer for, **316**
   dose values for, 82
   open air aerosol toxicity of, 100
cyclohexyl sarin (GF), 64, 66, 67, 82, 469, 480, **480**
cytotoxins, 453–466, **455–466**

DA, decontaminant/neutralizer for, **316**
dampers, 309
DC, 469, 518
Decline and Fall of the Roman Empire, 5
decontamination and remediation, 311–325, 385
   anthrax, 311, 322–323
   basic agents in, 314
   bleaching powder in, 314
   chemical means for, 314–315, **316–317**
   chloramine B in, 314
   chlorine dioxide in, 322–323
   Decontamination Solution 2 (DS2) in, 314
   delivery vehicles and, 379–380, **379**
   disinfectants and bleach in, 311, 314
   Ebola, 311
   forced spore germination in, 313
   Hart Senate Office Building anthrax and, 322–323
   hot air or steam, 313
   letters and packages, 373–378
   mail processing equipment and, 378–379
   microemulsions in, 315
   miscibility of decontamination agents in, 315
   oxidizing agents in, 314
   ozone as, 315–321, **319, 320, 322**
   physical means for (scrubbing, washing), 312–313
   SNL foam as, 323, **324**
   sunlight as, 313
   surfactants in, 313
   ultraviolet germicidal irradiation (UVGI) in, 312
   weathering as, 313
decontamination kits, 307
Decontamination Solution 2 (DS2), 314
dedicated outside air systems (DOAS), 152
delivery vehicles, decontamination of, 379–380, **379**
delta sleep inducing peptide (DSIP), 467
Dengue Fever virus, 39, 391, 407, **407**
   disease and dose curves for, 439, **444**
   disease progression in, 94
   size of microbe, vs. filtration, **182**, 182
department stores and malls, 134–135, **135**
dermophin, 467
desert rheumatism (*see Coccidioides immitis*)
desiccation, in decontamination, 328, 329–330
design of systems, 1
"detect to alarm" principle in, 258, 271, **271, 272, 273**
"detect to isolate" principle in, 255–256, **257, 258, 271, 271, 273–275, 274, 275**
"detect to treat" principle in, 258, 260, 271, **271, 275–276**
detection systems, 249–268, **250,** 270, 384
   affinity class biological sensors in, 263
   air sampling in, 249, 257, 258–261, **258, 259, 261, 270**

detection systems (*cont.*):
   antigens, antibodies and immune system, 262
   biological agents and, 257–268
   biosensors for, 254, 261–263, **262, 264**
   Brownian motion detectors in, 266
   calorimetric biosensors in, 263
   catalytic and noncatalytic biosensors in, 263
   catalytic detectors in, 252
   chemical agents and, 250–257, **251**
   cost of, 249
   Coulter counters in, 266
   culturing of biological agents for, 257
   "detect to alarm" principle in, 258
   "detect to isolate" principle in, 255–256, **257,** 258
   "detect to treat" principle in, 258, 260
   dynamic light scattering in, 266
   electrochemical sensors in, 262
   electrochemical techniques in, 251
   electron capture detectors in, 252
   evanescent wave (EW) biosensors in, 262
   flame photometry in, 254
   flammable materials and, 250, 252
   fluorescence biosensors in, 262, 267
   gas chromatography (GC) in, 251, 253
   ion mobility spectrometry (IMS) in, 253
   light detection and ranging (LIDAR) in, 258, 267
   limits to, 249
   luminescence biosensors in, 262
   mass spectrometry (MS) in, 253, 263, 267
   monitoring points for, 254–255
   nephelometry in, 266
   paper tape sensors (color change) in, 251
   particle detectors in, 263–267, **265, 267**
   Petri dish for culturing biological agents in, 258–259, **260**
   photoacoustic gas analyzer in, **255**
   photoionization detectors in, 252
   piezoelectric biosensors in, 261, 262
   portable gas detectors in, 252–253, **252, 253, 254**
   Raman spectroscopy, 253–254, 267
   response times of, 256–257
   smoke detection/control in, 282–283, **282, 283**
   solid state detectors in, 252
   spectroscopy in, 251, 253–254
   surface acoustic wave (SAW) detectors in, 254
   surface plasmon resonance (SPR) biosensors in, 262
   two-dimensional angular optical scattering (TAOS) in, 266
   ultraviolet laser induced fluorescence (UV LIF) in, 267
   UV resonance Raman spectroscopy in, 267
detergents, 312
diacetoxcirpenol (DAH), 467
dialkyl phosphoramidates, 470, 518
diamphotoxin, 453, **459**
diatomaceous earth, 312
diazepam, 306
dietheylenetriamine, 314
diethyl phosphite, 64, **316,** 470, 481, **481**
diffusion in filtration and, 171–172, **171, 173**
dilution ventilation, 154–155, 165, **166,** 167, 215, 327
dimethyl butanol, 518
dimethyl methylphosphonate, 470, 518
dimethyl phosphite, 64, **316,** 470, 482, **482**
diminishing returns principle, 234, 236, **236**

dioctyl phthalate (DOP), as filter test, 187
diphenyl hydroxyacetic acid, 518
diphenylchloroarsine, 82, 470, 483, **483**
diphenylcyanosarsine, 64, 82, 470, 484, **484**
diphosgene, 64, **316**, 470, 512
diphtheria (*see Corynebacterium diphtheria*)
diphtheria toxin, 453, **459**
disabling devices, 301–303, **302**
disease and dose curves, 439–451
disease progression curves, 89–95, **91, 95**
disease transmission, naturally occurring, 1
disinfectants and bleach, 311, 314
disinfection (*see* air cleaning and disinfection
    systems)
dispersion and delivery systems, 97–112
    aerosolization, 101–109, **103**
    aerosolizers in, 99
    airborne concentration and, 98, **99**
    crop dusting planes as, 97–98, **98**
    explosive devices for, 99, 100, 101, **103**, 108
    food-borne agents and, 110–112
    indoor, 99–109
    mean lethal dose (LD50) and, 99, **99**
    open air toxicity of aerosolized toxins, 99,
        100, 101
    outdoor, 97–99
    passive dissemination in, 99, 100, **103**,
        107–108, **109, 110**
    shells or cylinders for, 98
    toxins and, 99
    water-borne agents and, 110–112
dosage simulation, 219–220
dose and epidemiology, 73–96
    absorption efficiency in, 85
    airborne concentration vs., 76
    biological weapons agents and, 75–81, **76**
    breathing rates vs., 85–86, **86**, 219–220,
        224–226
    chemical weapons agents and, 81–85
    disease progression curves in, 89–95, **91, 95**
    dose response curve in, 74–77, **76, 77, 78, 80**,
        83–84, **83**
    dosimetry used in, 73–75, **75**
    ingestion dosimetry in, 87–89
    inhalation dosimetry for chemical agents,
        85–87, **87**
    mean casualty dose (CD50) in, 74, 81
    mean disease period (MDP) in, 93
    mean incapacitating dose (ID50) in, 73–74,
        81
    mean infectious dose (ID50) in, 73, 76
    mean lethal concentration (LC50) in, 74, 88
    mean lethal dose (L(Ct)50) in, 74
    mean lethal dose (LD50) in, 74–80, 82–84,
        86–87, 90, **91**
dose response curve, 74–75, 439–451
    biological weapons agents and, 76–77, **76, 77,
        78, 80**
    chemical weapons agents and, 83–84, **83**
    toxins, 453–466, **455–466**
dosimetry, 73–75, **75**
drawings, blueprints, security and, 292–293
duct systems
    access to, 289, **291**
    detector response time and, 256–257
    internal release into, 147–150, **149**
    for UVGI, cost of, 358
dust spot (DSP) efficiency rating, 52, 185, 168,
    199
dynamic light scattering, 266
dynorphin, 453, **459**

dysentery, 12, 26, 429
    (*see also Shigella*)

Eastern equine encephalitis, 35, 37, 392, 399,
        433, 435, **435**
    disease and dose curves for, 440, **451**
    disease progression in, 94
    dose values for, 81
    open air aerosol toxicity of, **102**
    size of microbe, vs. filtration, **182**
Ebola (EBO/EBOV) virus, 16, 26, 32, 35, 36,
        372, 391, 408, **408**, 420
    decontamination and remediation in, 311
    disease and dose curves for, 439, **444**
    disease progression in, 94
    dose values for, 81
    open air aerosol toxicity of, **102**
    size of microbe, 181, **182**
    UVGI and filtration vs., **202**
Echoviruses, dose–response curves for, 78
economics and optimization, 349–366
    carbon adsorber, 363–364
    energy analysis in, 365–366
    filtration, 350–354, **352, 354**
    performance criteria selection, 349–350, **350,
        351**
    ultraviolet germicidal irradiation (UVGI),
        354–363
education buildings, **115**, 132–134
effectiveness of ventilation (air mixing),
        155–156, 159
efficiency of filtration system, 353–354, **354**
Egypt, 16
EHF (*see* Ebola)
electrically enhanced filter (EEF), 333, **335**
electrochemical sensors, 251, 262
electron beams, in decontamination, **328**, 347
electron capture detectors, 252
electrostatic filters, **328**, 333, **334**
elevator shafts and stairwells, 161, **162**, 227
emergency functions of control systems and, 269
Emergency Response Guidebook 298, 307
emergency response teams, 297–301, **299**
emergency systems, 276–281
    purging systems in, 277–278, **278**
    secondary system operation and, 278–279,
        279, 280
    sheltering zone isolation and, 279–281, **281**,
        304–306, **305**
encephalitis (*see* Eastern equine encephalitis;
        Venezuelan equine encephalitis; Western
        equine encephalitis)
endothelin, 48, 453, **460**
endotoxins, 41–47, **44–45**
energy analysis, cost and optimization, 365–366
engineer's role in disease control, 388–390
enkephalin, 453, **460**
enteric fever (*see Salmonella typhi*)
enterotoxin, 47, 453–466, **455–466**
entertainment (*see* food and entertainment
        buildings)
ethyldichloroarsine, 518
ethyldiethanolamine, 470
epidemic jaundice (*see* Hepatitis A virus)
epidemic polyarthritis (*see* Chikungunya virus)
epidemic typhus (*see Rickettsia prowazeki*)
epidemics, 6, 7–12
epidemiology (*see* dose and epidemiology)
equine encephalitis viruses (*see* Eastern equine
        encephalitis; Venezuelan equine encephali-
        tis; Western equine encephalitis)

equipment rooms and access security, 122
erabutoxin, 467
Escherichia coli (*E. coli*), 55
  ozone disinfection and, 338–340, **338**
  pulsed electric field (PEF) vs., 346
ethanol, 314
Ethiopia, 13
ethylene glycol monomethyl ether, 314
EU standards for filtration and, 168
evacuation procedures, 294, 299–300, 303–304,
  309–310
evanescent wave (EW) biosensors, 262
exhaust hoods, 372–373, **374**
exotoxins, 453–466, **455–466**
explosive devices, 293–294
  aerosol agents and, 104, 108
  biological weapons agents and, 49
  disabling, 303
  dispersion and delivery using, 99, 100, 101,
    103
  indoor, 144–146, **144, 145**
extremist groups, 20

fans (*see* filtration; ventilation systems)
fatality rate computation, in simulation of
  building attack, 222, 233, 234
fiber diameter vs. efficiency in filters, 175–176,
  176
filtration systems, 112, 165–188, **166, 167,**
  278–279, 327, 383–384, 387
  age of filter vs. performance in, 177, 353
  air mixing and, 177
  antimicrobial treatments for filters in, 179,
    332–333
  applications for, 177–184
  ASHRAE standards for, 168, 179
  bacterial size vs., 168–169
  biological weapons agents and, 50–52, **51**
  buildings and attack scenarios and, 148
  carbon adsorbers and, 364
  classification of, 168, **169**
  commercial buildings and, 121
  Cunningham slip factor in, 174–175
  diffusion in, 171–172, **171, 173**
  diminishing returns principle in, 234, 236,
    236
  dust spot efficiency (DSP) and, 52, 185, 168
  economics of, 350–354, **352, 354**
  efficiency of, 173–177, 353–354, **354**
  electrically enhanced filter (EEF) in, 333, **335**
  electrostatic, **328,** 333, **334**
  emergency systems for, 278–279
  EU standards for, 168
  fiber diameter vs. efficiency in, 175–176, **176**
  fiber types in, 177
  food processing facilities and, 127
  gas molecule mean free path in, 175
  gas phase, 203–208, **204**
  glass fibers within, **170**
  high efficiency particulate air (HEPA), 168,
    176, 179, **179,** 181, 183, 333
  impaction in, 171–172, **171, 172**
  interception in,171–172, **171, 172,** 175
  Kuwabara hydrodynamic factor in, 174
  large-sized buildings and, 231–232, **235**
  life cycle costs of, 350–354, **352, 354**
  liquid aerosols and, 185–188, **187**
  loading of filter vs. performance in, 177, 353
  mailrooms and, 369
  maintenance costs of, 361–362
  mathematical modeling of, 172–177, **176**

filtration systems (*cont.*):
  medium-sized buildings and, 231, **234**
  minimum efficiency reporting value (MERV)
    rating for, 168, 178, 182
  overload attack vs., 242–244, **245**
  parameters for, multifiber MERV, **178**
  particle diffusion coefficient in, 174
  Peclet number for determining efficiency of,
    174
  performance criteria selection for, 349–350,
    350, 351
  performance curves for, 169–171, **171,** 177
  poison gases vs., 185–188, **187**
  polytetrafluoroethylene (ePTFE) fibers in,
    179, **180**
  prefilters in, 179
  pressure drop vs. load in, 177, 353
  recirculation of air through, 182–183, **186**
  removal rates in, 183–184, **184, 185**
  retrofitting, 177
  simulation of attack and, 215–216, **216,**
    222–226, **225, 226,** 230, 231, 233, 234
  size of filters in, 167–168
  spore size of fungi and bacteria vs., 168–169,
    180, 181–182
  sudden release attack simulation vs.,
    244–246, **245, 246**
  test results of, vs. microbes, 184–185, **185, 186**
  ultra low penetration air (ULPA), 168, 176
  ultraviolet germicidal irradiation (UVGI)
    and, 179, 200–203, **200, 201, 202,** 359
  velocity of airflow vs., 177, 181
  ventilation systems and, 154, 161–162
  viruses vs., 177
  water system, **111**
fire, 1
firefighters as emergency response team,
  297–301
first aid kits, 306–307
flame photometry, 254
flammable fluids as chemical weapons, 3, 5,
  12–13, 250, 252
Flaviviruses, 34, 37, 39
fluorescence biosensors, 262, 267
fomites, 26
food and entertainment buildings, 127–129
  food processing facilities as, 127–128
  nightclubs as, 128–129
  restaurants as, 128, **128**
  size, occupancy, risk level for, **115**
food-borne agents/diseases, 24–27, 40–41, 88,
  127–128, 135–136
  biosensors and, 261
  manufacturing facilities and, 122–123, **123**
  dispersion and delivery in, 110–112
food poisoning, 111, 402
  (*see also Clostridium perfringens;*
    Salmonella)
food processing facilities, 127–128
forced air ventilation systems, 153
France, 5, 13
Francisella tularensis (tularemia), 14, 32, 35,
  36, 372, 391, 409, **409**
  biosensors and assay times for, 262, **264**
  disease and dose curves for, 439, **445**
  disease progression in, 94
  dose values for, 81
  open air aerosol toxicity of, **102**
  size of microbe, 181, **182**
  UVGI and filtration vs., **202**
freezing (*see* cryogenic freezing)

French and Indian War, 7–9
full-body protection suit, **15**
Fuller's earth, 312–313
fumes (*see* poison gases)
fungi/fungal toxins, 23, 26, 30, 41–47, **44–45**,
    155, 387, 388
    spore size vs. filtration in, 168–169, **180**,
        181–182
    ultraviolet germicidal irradiation (UVGI) vs.,
        52–59, **54, 55, 57, 59**
Fusarium sporotrichoides (T2 toxin), 47
future of bioterrorism, 385–386

GA, 86
Galen, 3
gamma irradiation, in decontamination, 166,
    328, 340, 346–347, **347**, 377
gas chromatography (GC) detection systems
    and, 251, 253
gas masks, 300, **300**
gas molecule mean free path, filtration and,
    175
gas phase filtration, 203–208, **204**, 337
    ASHRAE standards for, 206
    carbon adsorption in, 203–208, **205, 206, 207,
        208**
    granulated active carbon (GAC) in, 204–207,
        205
    replacement time for, 206
    simulation of building attack scenarios and,
        214
    sizing of, 206–207
    van der Waal's forces and, 203
    (*see also* carbon adsorption)
gastrin, 453, **460**
GD, Decontamination Solution 2 (DS2) to
    neutralize, 314
GE, 64, 470, 512
generation of particulate clouds for aerosols,
    104, **105**
genetically engineered organisms, 16, 20,
    26–27, 33
Geneva protocol on chemical/biological
    weapons, 13
Germany, 5, 13, 14, 19, 314
ghost lamp calculation of view factor in UVGI
    systems and, 191–192, **191**
Gibbons, Edward, 5
Gilchrist's disease (*see Blastomyces dermatitidis*)
glanders, 31
    (*see also Burkholderia mallei; Pseudomonas
        mallei*)
glass fibers within filter, **170**
gloves, 372
gonadoliberin, 453, **460**
government building, 123–127
    courthouses as, 124–125, **125**
    military installations as, 126–127
    municipal facilities as, 125, **126**
    police stations as, 125
    post offices as, 125–126
    prisons as, 126
    security issues in, 124
    simulation of building attack scenarios and,
        231
    size, occupancy, risk level for, **115**
    ventilation systems of, 124
    (*see also* national monuments)
granulated active carbon (GAC), 204–207, **205**,
    363–364
Greek fire, 5, **5**

Greek use of chemical warfare, 5
Gulf War, 19
Gulf War Syndrome, 34, 422
    (*see also Mycoplasma pneumoniae*)

HA (*see* Hepatitis A virus)
habu venom, 467
Hanta virus (*see* Hantaan virus)
Hantaan virus, 32, **33**, 391, 410, **410**
    disease and dose curves for, 439, **445**
    disease progression in, 94
    size of microbe, vs. filtration, **182**
    UVGI and filtration vs., **202**
Harris, Larry Wayne, 20
Hart Senate Office Building, in decontamination,
    322–323
hate groups, 386
HAV (*see* Hepatitis A virus)
hazardous materials used as weapons, 18–19
HAZMAT ID numbers, 66
HAZMAT teams, 297–301, **297**
HD
    chloramine B to decontaminate, 314
    Decontamination Solution 2 (DS2) to
        neutralize, 314
    supertropical bleach (STB) to neutralize, 314
healthcare buildings, 129–130, **130**
    ventilation systems and, 152
    size, occupancy, risk level for, **115**
heat treatments, in decontamination, 313, 328,
    329
heating, ventilating, air conditioning (HVAC),
    carbon adsorption and, 204
hellebore, 3
hemlock, 3, 5
hemorrhagic fevers, 33–34, 37, 39, 181, **182**,
    407
    (*see also* Crimean–Congo Fever virus;
        Dengue Fever virus; Ebola; Hantaan
        virus; Junin virus; Machupo virus;
        Marburg virus)
Henry, Alexander, 10–11
hepatic toxicosis, 43
Hepatitis A virus (HAV), 41, 391, 411, **411**
    disease and dose curves for, 439, **445**
    disease progression in, 94
    dose values for, 81
    open air aerosol toxicity of, **102**
    size of microbe, vs. filtration, **182**
herd immunity, 388–390
high efficiency particulate air (HEPA),
    278–279, 333
    anthrax vs., 27
    filtration and, 168, 176, **179**
    life cycle and costs of, 351–354, **352, 354**
    prefilters for, 179
    recirculation of air vs. efficiency in, 183
    removal rates in, 183–184, **184, 185**
    simulation of building attack scenarios and,
        225
    velocity of airflow vs. efficiency in, 181
Hippocrates, 3
Histoplasma capsulatum, 26, 32, 391, 412, **412**
    disease and dose curves for, 439, **445**
    disease progression in, 94
    dose values for, 81
    open air aerosol toxicity of, **102**
    UVGI and filtration vs., **202**
histoplasmosis (*see Histoplasma capsulatum*)
history of chemical/biological warfare, 3–19, **4**
homes, 131–132

hormones, 48
hospitals, 129, 152
    (*see also* healthcare buildings)
hotels, 132, **133**
hydrogen chloride (HCL), 62, 64, 470, 513
    carbon adsorption filters vs., **208**
    decontaminant/neutralizer for, **316**
hydrogen cyanide, 15, 62, 64, 66, 68, 303, 470,
    485, **485**
    decontaminant/neutralizer for, **316**
    dose values for, 82
    ingestion dose for, **89**
    open air aerosol toxicity of, 100
hydrogen peroxide, 323
hydrolysis, in decontamination, 312, 314
hypertensin (*see* angiotensin)

ID50 (*see* mean incapacitating dose; mean
    infectious dose)
illnesses and absences as indication of biological
    attack, 295, 296, 310
immune buildings, 1–2, 118, 134
immune system response, 262
immunity of humans, 388–389
immunization and vaccination, 9, 26, 372, 389–390
impaction, filtration and, 171–172, **171, 172**
impregnated filtration as air purifier, 166
incapacitating agents, 62, 66
incident recognition and response in, 293–297
incubation periods, 90–91, 276, **277**
indicators of agent present (mist, oily films,
    dead insects, etc.), 294–295
indoor air quality (IAQ), 258, 387–388
indoor dispersion and delivery systems, 99–109
indoor explosive release in, 144–146, **144, 145**
indoor passive releases, 146
in-duct UVGI system, 188, **189**
infectious hepatitis (*see* Hepatitis A virus)
infectious periods, biological agents, 90–93
Influenza A virus, 12, 33, 388, 391, 413, **413**
    disease and dose curves for, 439, **446**
    disease progression in, 94
    dose values for, 81
    open air aerosol toxicity of, **102**
    size of microbe, vs. filtration, **182**
    UVGI and filtration vs., **202**
ingestion dosimetry, 87–89
inhalation dosimetry for chemical agents,
    85–87, **87**
inoculation, smallpox, 9
    (*see also* immunization and vaccination)
insect toxins, 47
inspections, 308–309
intensity field of UVGI systems, 188, 192–195,
    199–200
interception, in filtration and, 171–172, **171,
    172**, 175
Internet and sensitive information, 386
inverse square law (ISL) to compute intensity
    of UVGI, 188
ion mobility spectrometry (IMS) detection sys-
    tems and, 253
ionization, in decontamination, 166, **328**, 333
Iraq, bioterrorism and, 19–20
Islamic extremists, 19, 386
isolation (*see* "detect to isolate")
Israel, 15
Italy, 13

Japan, 14, 16, 19, 20
Japanese encephalitis (JBE, JE, JEV), 39, 391,
    414, **414**
    disease and dose curves for, 439, **446**
Japanese encephalitis (*cont.*):
    disease progression in, 94
    size of microbe, vs. filtration, **182**
jellyfish toxin, 47, **47**
Jewell, Richard, 288
Jewish Resistance during WWII, 14–15
John Hancock Building, Boston, **114**
Junin virus, 33, 391, 415, **415**, 416
    disease and dose curves for, 439, **446**
    disease progression in, 94
    dose values for, 81
    size of microbe, vs. filtration, **182**
    UVGI and filtration vs., **202**

Kaczynski, Ted, 132
Kaffa, 6
kaolin, in decontamination, 313
kill rate of UVGI, combined with filtration,
    201–202, **203**, 215, **216**, 222–226, **225, 226**
Korean hemorrhagic fever (*see* Hantaan virus)
Kosovo, 18–19, 18
krait (*Bungarus multicinctus*), 43
Ku Klux Klan (KKK), 138, 386
Kuwabara hydrodynamic factor, filter, 174

L(Ct)50 (*see* mean lethal dose)
Lactobacillus brevis, pulsed electric field (PEF)
    vs., 346
large-sized buildings in simulation of attack,
    231–232, **235**
Lassa Fever virus, 33, 36, 391, 416, **416**, 420
    disease and dose curves for, 439, **446**
    disease progression in, 94
    dose values for, 81
    size of microbe, vs. filtration, **182**
    UVGI and filtration vs., **202**
latex gloves, 372
latrotoxin, lethal doses for, 91
LC50 (*see* mean lethal concentration)
Legionella pneumophila (Legionnaire's disease),
    27, 33, 52, **53**, 140, 391, 417, **417**
    disease and dose curves for, 439, **447**
    disease progression in, 94
    dose values for, 81
    size of microbe, vs. filtration, **182**
    UVGI and filtration vs., **202**
lethality vs. concentration of agents in, 226,
    226, 233
Lewis, Andrew, 9
Lewisite, 63, 64, 66, 68–69, 470, 486, **486**
    decontaminant/neutralizer for, **316**
    dose values for, 82
    open air aerosol toxicity of, 100
libraries, 133
lice, 37–40
light detection and ranging (LIDAR) detection
    systems and, 258, 267
liquid aerosols, filtration vs., 185–188, **187**
liquid-splash protective suits, 299
liquid agents
    aerosolization and, 104, **106**
    air intake release of, 143
    indoor passive releases of, 146
Livy, mass murder by Roman matrons, 6
lobar pneumonia (*see* Streptococcus pneumoniae)
location of buildings in, 113–114
lodging buildings, 130–132
    apartment buildings as, 130–131, **131**
    hotels as, 132, **133**
    residences and homes as, 131–132
    size, occupancy, risk level for, **115**

louse-borne typhus fever (*see Rickettsia prowazeki*)
Lucretius, 3
luminescence biosensors, 262
Lutz's disease (*see Paracoccidioides*)
lyddite, 13
Lymphocytic choriomeningitis, 33, 391, 418, **418**
  disease and dose curves for, 439, **447**
  disease progression in, 94
  dose values for, 81
  size of microbe, vs. filtration, **182**

Machupo virus, 33, 35, 36, 391, 416, 419, **419**
  disease and dose curves for, 439, **447**
  disease progression in, 94
  dose values for, 81
  size of microbe, vs. filtration, **182**
mailrooms and CBW agents, 367–382, **368**
  anthrax contamination in, 367–368
  contaminated letters and packages in, decontamination of, 373–378
  delivery vehicles and, 379–380, **379**
  filtration in, 369
  gamma irradiation in, 377
  general areas of building and, 369–372
  human exposure limits to UVGI and, 369, **371**
  mail handling by employees and, 372–373
  mail processing equipment and, 378–379
  microwave irradiation and, 378, **378**
  ozone in, 377, 380
  protocol for security in, 380–382
  pulsed electric fields in, 377
  pulsed light in, 377
  ultraviolet germicidal irradiation (UVGI) in, 369–372, **371**, 374–377, **375, 376**
  ventilation systems in, 368–369
maintenance costs
  carbon adsorbers and, 364
  filtration, 361–362
  UVGI, 361–362
maitotoxin, 91, 453, **460**
malfunctions or changes in equipment operation, 295
malls, 134–135, **135**, 136
Manchuria, 16
manufacturing facilities, 122–123, **123**
Marburg virus, 32, 35, 391, 420, **420**
  disease and dose curves for, 439, **447**
  disease progression in, 94
  size of microbe, 181, **182**
  UVGI and filtration vs., **202**
Mark I Nerve Agent Antidote Injector Kit, 307, **307**
mass spectrometry (MS), 253, 263, 267
mathematical modeling of UVGI systems and, 189–196
mathematical modeling of filtration, 172–177, **176**
mean casualty dose (CD50) in, 74, 81
mean disease period (MDP), 93
mean incapacitating dose (ID50), 73–74, 81
mean infectious dose (ID50), 73, 76
mean lethal concentration (LC50) in, 74, 88
mean lethal dose (L(Ct)50) in, 74
  dispersion and delivery in, 99, **99**
  toxins and bioregulators, 90, **91**
mean lethal dose (LD50), 74
  biological weapons agents and, 75–80
  chemical weapons agents and, 82–84, 86–87
measles, 6
medical emergency response, 306–307

medium-sized buildings in simulation of attack, 231, **234**
melioidosis, 31
  (*see also Burkholderia pseudomallei; Pseudomonas pseudomallei*)
meningitis, 32
meningitis (*see* Eastern equine encephalitis; *Streptococcus pneumoniae*; Venezuelan equine encephalitis; Western equine encephalitis)
mercantile buildings, 134–136
  department stores and malls as, 134–135, **135**
  malls as, 136
  size, occupancy, risk level for, **115**
  supermarkets as, 135–136
mercury, as food poisoning, 112
metal detectors, 292, **293**
metallic salt impregnated (MSI) carbon adsorbers, 338
methyl salicylate, as filter test, 187–188, **187**
methyldiethanolamine, 64, **316**, 470, 487, **487**
methylphosphonyl dichloride, 64, 470, 513
Mexico, 7
microbial growth control (MGC), UVGI, 355
microcystin, 43, 52, 91, 453, **461**
microemulsions, in decontamination, 315
microorganisms for disease, 24–41, **28–29**
microwave irradiation, in decontamination, 166, **328**, 340–341, **341**, 378, **378**
Middle Ages and chemical/biological warfare, 6
military installations, 117, 126–127
minimum efficiency reporting value (MERV), 178, 182
  filtration and, 168
  UVGI systems and, 199
Minnesota Patriots Council, 20
minute-by-minute concentrations/calculations in attack scenarios, 216–220, **217, 219**
miscibility of decontamination agents, 315
modeccin, 467
modeling an attack scenario, 214–226
modeling filtration systems, 172–177, **176**
modeling UVGI systems and, 189–196
molds, 311
monitoring points for detection systems and, 254–255
mosquito-borne diseases, 24, 26, 34, 35, 37–40
mosquito, **39**
mosquito-borne encephalitis (*see* Japanese encephalitis)
Mozambique, 17
MS1 (*see* Hepatitis A virus)
multizone building attack simulation, 238–242, **240, 241, 242**
municipal facilities as, 125, **126**
Murray Valley fever, 39
museums, 133–134
mustard gas, 13, 63, 64, 67, 469, 488, **488**
  bleaching powder to neutralize, 314
  decontaminant/neutralizer for, **316**
  Decontamination Solution 2 (DS2) to neutralize, 314
  dose response curve for, 84, **85**
  dose values for, 82
  SNL foam vs., 323
mustard T mixture, 64, **316**, 469, 514
Mycobacterium tuberculosis (tuberculosis; TB), 91, 126, 388 392, 421, **421**
  disease and dose curves for, 440, **448**
  disease progression in, 94
  dose values for, 81
  open air aerosol toxicity of, **102**

Mycobacterium tuberculosis (cont.):
   removal rates, 236, **237**, 238
   simulation of building attack scenarios and,
      214, 215
   UVGI and filtration vs., **202**
Mycoplasma pneumoniae, 34, 392, 422, **422**
   disease and dose curves for, 440, **448**
   disease progression in, 94
   dose values for, 81
   open air aerosol toxicity of, **102**
   size of microbe, vs. filtration, **182**
   (*see also* Gulf War Syndrome)
mycotoxins, 34, 41–47, **44–45**, 453–466, **455–466**

Namibia, 18
naphtha, 5
national monuments, 140
   (*see also* government buildings)
Native Americans, smallpox and other
      epidemics used against, 6–12, **7, 8, 11**, 385
natural ventilation, 151, 153
nebulizers and compressors for aerosols,
      105–107, **106, 107, 108**
negative ionization, **328**, 333–334, **336**
neo-fascists, 386
neo-Nazis, 386
nephelometry detection systems and,
      266
nerve agents, 14, 63, 64, 66, 67, 307
nerve gas, 64
nerve gas DFP, 469, 489, **489**
neuropeptide, 453, **461**
neurotensin (NT), 453, **461**
neurotoxins, 43, 453–466, **456–466**
New World spotted fever (*see Rickettsia
      rickettsii*)
news media facilities, 140
nightclubs, 128–129
nitrogen dioxide, carbon adsorption filters vs.,
      208
nitrogen mustard, 63, 64, 65, 66, 67, 68, 81,
      469, 490, **490**, 491, **491**, 492, **492**
   decontaminant/neutralizer for, **316**
   dose values for, 82
   open air aerosol toxicity of, 100
nitrogen oxide, 65
nivalenol, 467
NN-dialkyl phosphoramidic dihalides, 518
NN-dialkylaminoethane ols, 518
NN-dialkylaminoethyl 2 chlorides, 518
NN-diemthylaminoethanol, 518
NN-diethylaminoethanol, 518
NNdialkylaminoethyl 2 chlorides, 469
Nocardia asteroides (nocardiosis), 34, 392, 423
   disease and dose curves for, 440, **448**
   disease progression in, 94
nonionic surfactants, in decontamination, 313
nonlethal chemical weapons agents and, 62
notexin, 453, **461**
noxiustoxin, 467
nuclear power plants, 115, 139–140, **139**
numerical integration, in ventilation modeling,
      158

O mustard, 65, 469, 493, **493**
   decontaminant/neutralizer for, **316**
   dose values for, 82
occupancy levels
   commercial buildings and, 123
   simulation of building attack scenarios and,
      231

occupancy vs. risk, buildings and attack
      scenarios and, 116–118
Occupational Safety and Health Administration
      (OSHA), 61, 298
odors as indicator of attack, 295
office buildings (*see* commercial buildings)
Omega 7 terrorist group, 18
open-air toxicity of aerosolized toxins, 99, **100,
      101**
optimization of defense for, 212, **212**, 212
ornithosis (*see* Chlamydia psittaci)
otitis media (*see Streptococcus pneumoniae*)
outdoor air intake access security, 289, 291,
      292
outdoor dispersion systems, 97–99
outdoor releases, 140–141, **142**
overload attack simulation, 242–244, **245**
oxidizing agents in decontamination, 314
oxytocin, 453, **461**
ozone decontamination, 166, 311, 315–322, **318,
      319, 320, 322, 328**, 337–340, 377, 380
ozone generators, 315, **318**
ozone sensors, 318–321, **319**

palytoxin, 453, **462**
   lethal doses for, 91
   open air aerosol toxicity of, 101
pandemics, 33
Paracoccidiodes spp., 26, 34, 392, 424, **424**
   disease progression in, 94
   disease and dose curves for, 440, **448**, 440
   dose values for, 81
paracoccidioidal granuloma (*see
      Paracoccidioides*)
paralytic shellfish poisoning (PSP), 46
parrot fever (*see* Chlamydia psittaci)
particle detectors, 263–267, **265, 267**, 273–274
particle diffusion coefficient, filtration and, 174
passive dissemination, 99, 100, **103**, 107–108,
      109, 110
   aerosol agents and, 107
   indoor, 146
passive solar exposure, in decontamination,
      328, 330, **331**
pathogen input data for simulation of building
      attack scenarios and, 215, **216**
pathogenic microorganisms for, **28–29**
Peclet number, filter efficiency, 174
Pentagon, 123–124, 127
peptide hormone, 48
perfluoroisobutylene, 65, 66, 68, 469, 494, **494**
   dose values for, 82
   open air aerosol toxicity of, 100
perfluoroisobutylenephenyldichloroarsine, 518
performance criteria selection, 349–350, **350, 351**
performance curves for filters, 169–171, **171**
personnel training in, 309–310
pertussigen, 467
Petri dish for culturing biological agents in,
      258–259, **260**
phoneutriatoxin, 48
phosgene, 13, 63, 65, 66, 68, 469, 495, **495**
   carbon adsorption filters vs., **208**
   decontaminant/neutralizer for, **316**
   dose response curve for, 83, **83**
   dose values for, 82
   inhalation dose for, **88**
   open air aerosol toxicity of, 100
   simulation of building attack scenarios and,
      214

phosgene oxime, 65, 66, 68, 469, 496, **496**
    decontaminant/neutralizer for, **316**
    dose values for, 82
    open air aerosol toxicity of, 100
phosgenephosphorus oxychloride, 469, 497,
    497
phosphorus oxychloride, 65, **317**
phosphorus pentachloride, **317**, 469, 498, **498**
phosphorus trichloride, 65, **317**, 469, 499, **499**
phosphorus trichloridepinacolyl alcohol, 469,
    518
photoacoustic gas analyzer in, **255**
photocatalyic oxidation (PCO) decontamination,
    166, **328**, 330, 336–337, **337**, 373
photoionization detectors in, 252
physical means of decontamination, 312–313
physical security measures, 288–293
piezoelectric biosensors, 261, 262
pitch, 5
plague (*see Yersinia pestis*), 437
plasma fields as air purifier, 166
pneumococcus (*see Streptococcus pneumoniae*)
pneumonia (*see Francisella tularensis*;
    Influenza A virus; *Mycoplasma
    pneumoniae*; *Nocardia asteroides*;
    Streptococcus pneumoniae)
pneumonic plague (*see Yersinia pestis*)
pneumonitis (*see Chlamydia psittaci*)
poison arrow alkaloid, 46
poison dart frog (*Phylobates* spp.), 43
poison gases, 5–6, 12–15, **14, 62**
    filtration and, 185–188, **187**
poisoned arrows, 3, 5, 46
poisons, 5, 6
poisons, naturally occurring, 3
Poland, 15
police stations as, 125
polytetrafluoroethylene (ePTFE) fibers,
    179, **180**
Pond, Peter, 10–11
Pontiac fever (*see Legionella pneumophila*)
portable gas detectors in, **252**, 252–253, **253,
    254**
Post Office, 125–126, 146
    (*see also* mailrooms and CBW agents)
potassium hydroxide, 314
potassium permanganate, in GAC filters, 204
Powassan encephalitis, 39
powders, aerosolization and, 104–106, **106**
power usage, cost and optimization, 365–366
PPLO (*see Mycoplasma pneumoniae*)
pressurized release devices, 294
prisons, 126
profile (importance, visibility) of building vs.
    risk, 116–118
protective clothing, 298–299, 372, **373**
protozoa as disease organisms, 23
prussic acid, 68
    (*see also* hydrogen cyanide)
Pseudomonas fluorescens, pulsed electric field
    (PEF) vs., 346
Pseudomonas mallei, UVGI and filtration vs.,
    202
Pseudomonas pseudomallei, UVGI and filtration
    vs., **202**
psittacosis (*see Chlamydia psittaci*)
pulsed electric fields (PEF) air purifier, 166,
    328, 346, 377
pulsed filtered light, **328**, 343–345
pulsed light disinfection, 166, 341–345, **342,
    343**, 377

pulsed microwaves, 340–341
pulsed UV (PUV), 341–343, **342, 343**
pulsed white light (PWL), 328, 341–343, **342,
    343**
purging by outside air, 246–247, **247**
purging systems in, 277–278, **278**
Putnam, Israel, 9

Q fever/Query fever (*see Coxiella burnetii*)
QL, 65, 469, 515
quaternary ammonium, 323
quicklime, 5
quinuclidine-3-OL, 469, 518

radiation view factor, UVGI systems and,
    188–189, 190–195, **190, 191**
radiological agents, 140
radiological releases, 3
Rajneeshee cult, 20, 111, 128
Raman spectroscopy/Raman scattering,
    253–254, 267
recirculation of air through filters, 182–183,
    186, 200, 201
red tide, 43
reflective materials in UVGI systems, 193–195,
    194, 195, 357–358, **357**
relative risk assessment for buildings, 118–119,
    119, 150
release rates, simulation of building attack
    scenarios and, 220, 222
religious places of worship, 137–138, **138**
remediation (*see* decontamination and
    remediation)
removal rate of contaminants in ventilation
    systems, 158–160
    simulation of building attack scenarios and,
    236, **237**, 238
renin, 467
residences and homes as, 131–132
residential UVGI system, **198**
respiratory infections, indoor air quality and,
    387–388
response time, 293–297
response times of detectors, 256–257
restaurants as, 128, **128**
Reston, Virginia, and Ebola decontamination,
    311
retrofitting
    filtration and, 177
    ventilation systems and, costs associated
    with, 154
Revolutionary War, 9–12
Rhizopus, 52
Rhodesia/Zimbabwe, anthrax outbreak in, 17,
    18, 97
Rhodesian Central Intelligence Organization
    (CIO), 17
ricin, 20, 46, 87, 138, 453, **462**
    biosensors and assay times for, 262, **264**
    lethal doses for, 91
    mean lethal concentration (LC50) in, **88**
    open air aerosol toxicity of, 101
Rickettsia candensis, 39–40
Rickettsia prowazeki (typhus), 6, 15–16, 32, 35,
    36, 39–40, 392, 425, **425**
    disease and dose curves for, 440, **449**
    disease progression in, 94
    dose values for, 81
    open air aerosol toxicity of, **102**
    size of microbe, vs. filtration, **182**
    UVGI and filtration vs., **202**

Rickettsia rickettsi (Rocky Mountain Spotted
    Fever), 39–40, 392, 426, **426**
    dose values for, 81
    disease and dose curves for, 440, **449**
    disease progression in, 94
    UVGI and filtration vs., **202**
Rift Valley Fever virus, 37, 39, 34, 392, 427,
    427
    disease and dose curves for, 440, **449**
    disease progression in, 94
    size of microbe, vs. filtration, **182**
risk or vulnerability assessment, 113–119
    commercial buildings and, 119–120
    cost of prevention vs. relative risk in, 211
risk factors for buildings, 116–119
Rogers, Robert, 10–11
Roman use of chemical warfare, 5–6
roof access and security, 289, **290**
roridin, 467
Rudolph, Eric, 129–130
Russian spring-summer encephalitis, 39

Salmonella spp., 17, 20, 24, 26, 111, 136, 262
Salmonella typhi (typhoid fever), 15–16, 41,
    128, 392, 428, **428**
    biosensors and assay times for, **264**
    disease and dose curves for, 440, **449**
    disease progression in, 94
    dose values for, 81
    piezoelectric biosensors and, 261
sand, 312
Sao Paulo fever (*see Rickettsia rickettsii*)
sarafotoxin, 453, **462**
sarin, 14, 63, 65–67, 131–132, 141, 469, 500,
    500
    decontaminant/neutralizer for, **317**
    dose response curves for, 83, **84**
    dose values for, 82
    open air aerosol toxicity of, 100
    simulation of building attack scenarios and,
        214
    SNL foam vs., 323
    supertropical bleach (STB) to neutralize, 314
saxitoxin, 46, 453, **462**
    lethal doses for, 91
    open air aerosol toxicity of, 101
schools, 132–134
sea wasp toxin, 47, **47**, 453, **463**
Sears Tower, Chicago, 122
secondary system operation and, 278–279, **279**,
    280
security and emergency procedures, 287–310,
    288, 385
    building automation systems/BACnet and,
        292
    casualty distribution throughout building
        during attack, 295–296
    commercial buildings and, 122
    disabling devices in, 301–303, **302**
    drawings, blueprints and, 292–293
    emergency response for, 297–301, **299**
    evacuation procedures and, 294, 299–300,
        303–304, 309–310
    government buildings and, 124
    illnesses and absences as indication of
        biological attack, 295, 296
    incident recognition and response in,
        293–297
    indicators of agent present (mist, oily films,
        dead insects, etc.), 294–295

security and emergency procedures (*cont.*):
    inspections and, 308–309
    malfunctions or changes in equipment
        operation, 295
    medical emergency response in, 306–307
    metal detectors in, 292, **293**
    nuclear power plant, 139–140
    odors as indicator of attack, 295
    personnel training in, 309–310
    physical security measures in, 288–293
    security protocol in, 307–309
    sheltering in place in, 304–306, **305**
    stadiums, 137
    video cameras in, 291
    water pipes and, 291–292
security bars, 293
security protocol in, 307–309
self-contained breathing apparatus (SCUBA),
    298
Selous Scouts, 17
September 11 terrorist attack, 20, 367–368, 386
sesqui mustard, 65, 469, 501, **501**
    decontaminant/neutralizer for, **317**
    dose values for, 82
shells or cylinders for dispersion, 98
sheltering zone isolation and, 279–281, **281**,
    304–306, **305**
shiga toxin, 91, 453, **463**
Shigella dysenteriae (dysentery), 41, 111, 128,
    392, 429, **429**
    biosensors and assay times for, **264**
    disease and dose curves for, 78, 440, **450**
    disease progression in, 94
    dose values for, 81
shigellosis (*see Shigella dysenteriae*)
showers, 301, 307
shutdown systems, 274–275, 298–299
sick building syndrome, 387, 430
    (*see also Stachybotrys chartarum*)
Silver Dome, Pontiac, Michigan, **137**
simulants of chemical weapons agents and, 70,
    70
simulation of building attack scenarios,
    211–247
    air cleaning and disinfection systems in, 214,
        220, **221**, **223**, **224**, 230, 231, 233
    air intake release in, 236–238, **238**, **239**
    airflow rate and, 231
    anthrax, smallpox, tuberculosis as agents in,
        214, 215, 224–226, **225**
    apartment building, 231
    auditorium, 231
    baseline conditions for, 211–214
    building input data/parameters for, 215, **216**
    carbon adsorption and, 214
    chemical agents in, 214
    chlorine, sarin, phosgene as agents in, 214
    commercial building, 231
    computed data for, **217**
    CONTAMW program for, 214–215, 226–229,
        227, **228**, **231**, **232**, **233**, 238–242, **240**,
        241, **242**
    continuous release attack in, 214
    cost of defense system in, 212–213
    dilution ventilation and, 215
    diminishing returns principle in, 234, 236,
        236
    dosage calculation in, 219–220
    elevator shafts and stairwells in, 227
    fatality rate computation in, 222, 233

simulation of building attack scenarios (*cont.*):
    filtration and, 215–216, **216**, 222–226, **225,**
        226, 230, 231, 233
    gas phase filtration and, 214
    government building, 231
    high efficiency particulate air (HEPA) filter
        and, 225
    large-sized buildings and, 231–232, **235**
    lethality vs. concentration of agents in, 226,
        226, 233
    medium-sized buildings and, 231, **234**
    minute-by-minute concentrations/calculations
        in, 216–220, **217, 219**
    modeling of, 214–226
    multizone buildings (using CONTAMW),
        238–242, **240, 241, 242**
    occupancy levels and, 231
    overload attack, 242–244, **245**
    pathogen input data for, 215, **216**
    purging by outside air in, 246–247, **247**
    release rates in, 220, 222
    removal rates, 236, **237,** 238
    risk vs. cost of attack in, 211
    schematic of multistory office building used
        in, 213, **213**
    several model buildings in, 230–238
    sizing and optimization of defense for, 212,
        212
    steady state concentration in, 220
    sudden release attack, 244–246, **245, 246**
    ultraviolet germicidal irradiation (UVGI)
        and, 215–216, **216,** 222–226, **225, 226,**
        230, 231, 233
    ventilation system and, 214
sinusitis (*see Streptococcus pneumoniae*)
size distribution, biological weapons agents
    and, 50–52, **51**
size or floorspace of buildings in, 113–114
sizing and optimization of defense for, 212, **212**
smallpox (Variola), 6–12, **7, 11,** 12, 16, 19, 35,
    36, 372, 385, 392, 432, **432**
    disease and dose curves for, 440, **450**
    disease progression in, 94
    dose value for, 80, 81
    removal rates, 236, **237**
    simulation of building attack scenarios and,
        214, 215
    size of microbe, 181, **182**
    sudden release attack simulation of,
        244–246, **245, 246**
    UVGI and filtration vs., 200, **202**
smoke detection/control, 282–283, **282, 283,** 1,
    282
smoke generator for aerosols, **105**
snake venom toxins, 43
SNL foam as disinfectant, 311, 323, **324**
Socrates, 5
sodium hydroxide, 314
sodium hypochlorite, 314
solid-state detectors, 252
soman, 14, 63, 65–67, 469, 502, **502**
    decontaminant/neutralizer for, **317**
    dose values for, 82
    open air aerosol toxicity of, 100
    SNL foam vs., 323
    supertropical bleach (STB) to neutralize, 314
somatostatin, 453, **463**
sonic generators, 334–336
sorbents, 312, 313
South Africa, bioweapon research in, 17–18, 111

South American blastomycosis (*see*
    Paracoccidioides)
Soviet Union, 14, 16, 19
special facilities, 138–140
spectroscopy detection systems and, 251,
    253–254
spider toxins, 47, **48**
spills of hazardous materials, 303, 312
spinning top atomizers, 104
spores of fungi/bacteria
    anthrax, 27, 30, 27
    filtration vs., 168–169
    forced germination of, for decontamination,
        313
sprayers, 97, 104, 294, 308
sprinkler systems, 144, 301
St. Louis encephalitis, 39
stachybotriotoxicosis (*see Stachybotrys*
    chartarum)
Stachybotrys chartarum (bleeding lung
    disease), 34, 52, 392, 430, **430**
    disease and dose curves for, 440, **450**
    disease progression in, 94
    UVGI and filtration vs., **202**
stadiums as, 137, **137,** 145
stairwells, 161, **162,** 227
Staphylococcus aureus, 47
    ozone disinfection and, 338–340, **338**
    pulsed light treatment vs. UVGI in, 342–343,
        344, 345
    UVGI vs., 57–59, **57, 58**
Staphylococcus enterotoxin B (SEB), 47, 453,
    463, 464
    biosensors and assay times for, **264**
    lethal doses for, 91
    open air aerosol toxicity of, 101
    steady-state concentration, 220
steam treatments, in decontamination, 313,
    329
Streptococcus pneumoniae, 34–35, 392, 431, **431**
    disease and dose curves for, 440, **450**
    disease progression in, 94
    UVGI and filtration vs., **202**
Streptococcus pyogenes, UVGI vs., 56, **57**
Streptomyces, as microbial air scrubber, 332
substance P, 48, 453, **464**
    lethal doses for, 91
    open air aerosol toxicity of, 101
    sudden release attack simulation, 244–246,
        245, 246
sulfur, 5
sulfur dioxide, 65, 469, 515
    carbon adsorption filters vs., **208**
    decontaminant/neutralizer for, **317**
sulfur monochloride, **317,** 470, 516
sulfur mustard, 65–67, 470, 503, **503**
    decontaminant/neutralizer for, **317**
    dose values for, 82
sulfur trioxide, 65, **317,** 470, 516
Sun Tzu, 3
Sung Dynasty, China, chemical warfare in, 3
sunlight, in decontamination, 313, **328,** 330,
    331
    (*see also* passive solar exposure)
supermarkets as, 135–136
supersonic nozzles, 334–336
supertropical bleach (STB), 314
surface acoustic wave (SAW) detectors, 254
surface plasmon resonance (SPR) biosensors, 262
surfactants, in decontamination, 313

surgical face masks, 372, **373**
sylvatic plague (*see Yersinia pestis*)
symptoms of chemical agents, 294–295

T2 mycotoxin, 47, 453, **464**
    lethal doses for, 91
    open air aerosol toxicity of, 101
tabun, 14, 63, 65–67, 470, 504, **504**
    decontaminant/neutralizer for, 317
    dose values for, 82
    open air aerosol toxicity of, 100
    SNL foam vs., 323
    supertropical bleach (STB) to neutralize, 314
Taipin snake (*Oxyuranus* spp.), 43
taipoxin, 43, 91, 453, **464**
tampering, 295
Tartar invasions, 6
TB (*see Mycobacterium tuberculosis*)
tear gas, 62, 63, 66
terrorist attacks, 2, 19, 385–386
    first recorded, 5–6
tetanus toxin, 91, 453, **465**
tetrodoxin (TTX), 91, 453, **465**
textilotoxin, 91, 453, **465**
theaters as, 136
thermal disinfection, in decontamination, 328, 329
    (*see also* heat treatment)
thiodiglycol, 65, **317**, 470, 505, **505**
thionyl chloride, **317**, 470, 518
Third-World countries and bioterrorism, 19–21
thyroliberin (TRF), 453, **465**
tick-borne typhus (*see Rickettsia rickettsii*)
tick-borne encephalitis, 40
ticks, 25, 26, 37–40
timers, 303
titanium dioxide, 336
titanium oxide, 330
titanium tetrachloride, 65, **317**, 470, 517
Togavirus, 35, 37
Tortona, Battle of, 6
totally encapsulating chemical protective suits (TECPS), 298
toxicity
    biological weapons agents, 28–29, 49
    chemical weapons agents and, 63, **64–65**, **66**
    lethality vs. concentration of agents in, 226, **226**, **233**
toxins, 14, 23, 41–47, **42**, **44–45**, 52, 111, 127, 453–466
    aerosolization delivery for, 101–109
    biosensors for, 262
    dispersion and delivery in, 99
    dose curves for, 453–466, **455–466**
    mean lethal dose (LD50) in, 90, **91**
    open air toxicity of aerosolized agents, 99, **100**, **101**
    UVGI and filtration vs., **202**, 202
training (*see* personnel training)
transmission routes for diseases, 25–26, **25**
trichothecene toxins, 47, 453, **466**
triethanolamine/triethylamine, 65, **317**, 470, 506, **506**
triethyl phosphite, 65, **317**, 470, 507, **507**
triethylamine, **317**, 470
trikabuto poison, 467
trimethyl phosphite, 65, **317**, 470, 508, **508**
tuberculosis (*see Mycobacterium tuberculosis*)
tularemia (*see Francisella tularensis*)
Turks, 6

two-dimensional angular optical scattering (TAOS), 266
Type A viral hepatitis (*see* Hepatitis A virus)
typhoid fever (*see Salmonella typhi*)
typhus (*see Rickettsia prowazeki*)
typhus abdominalis (*see Salmonella typhi*)

U tube UVGI lamps, 192, **193**
Ukraine, 6
ultra low penetration air (ULPA) filtration, 168, 176
ultrasonic atomizers, 104
ultrasonication, in decontamination, **328**, 334–336
ultraviolet germicidal irradiation (UVGI), 50, 52–59, **54**, **55**, **57**, **59**, 165, **166**, 169, 188–200, 278–279, 327, 340, 383–384, 387
    annualized cost (AFC) of, 359
    applications for, 196–200
    biaxial lamps in, 192, **193**
    biosensors in, 273–274
    calculation of intensity field, 196, **197**
    carbon adsorbers and, 364
    decontamination and remediation in, 312
    "detect to alarm" principles in, 271, **271**, **272**, **273**
    "detect to isolate" principle in, 271, **271**, 273–275, **274**, **275**
    "detect to treat" principle in, 271, **271**, 275–276
    dimensions of typical system, 355, **356**
    diminishing returns principle in, 234, 236, **236**
    dose (duration of exposure) vs. efficiency of, 196, 199–200
    ductwork costs, 358
    dust spot efficiency (DSP), 199
    economics and optimization of, 354–363
    emergency systems for, 278–279
    fan costs, 358–359
    field average intensity for, source code for, 525–529
    filtration and, 179, 200–203, **200**, **201**, **202**, 359
    food processing and, 127
    gamma irradiation and, 346–347, **347**
    ghost lamp calculation of view factor in, 191–192, **191**
    hospital use of, 129
    human exposure limits to UVGI and, 369, **371**
    in-duct systems using, 188, **189**
    intensity of field in, 192–195, 199–200
    intensity of field used in, 188
    inverse square law (ISL) to compute intensity of, 188
    kill rate of, combined with filtration, 201–202, **203**, 215, **216**, 222–226, **225**, **226**
    large-sized buildings and, 231–232, **235**
    mailrooms and, 369–372, **371**, 374–377, **375**, **376**
    maintenance costs of, 361–362
    mathematical modeling of, 189–196
    medium-sized buildings and, 231, **234**
    minimum efficiency reporting value (MERV) rating and, 199
    mounting of lamps in, 200
    museum use of, 134
    operating costs of, 355–358, **356**, **357**

ultraviolet germicidal irradiation (*cont.*):
  overload attack and, 244, **245**
  particle detectors in, 273–274
  performance criteria in design of, 196–197, 349–350, **350**, **351**
  power cost of, 355–356, **356**, 360, 365–366
  radiation view factor in, 188–189, 190–195, **190**, **191**
  rating value (URV) for, 197–198
  recirculation vs. efficiency in, **200**, 201
  reflective materials to boost efficiency, 193–195, **194**, **195**, 357–358, **357**
  residential system, **198**
  response of microorganisms to, 188
  simulation of building attack scenarios and, 215–216, **216**, 222–226, **225**, **226**, 230, 231, 233, 234
  size of system vs. kill rates for, 519–524
  sudden release attack simulation, 244–246, **245**, **246**
  U tube lamps in, 192, **193**
  velocity of airflow through vs. efficiency of, 196, 199–200, 360–361
  ventilation systems and, 154, 161–162
  water-based systems using, 188
ultraviolet laser induced fluorescence (UV LIF) in, 267
ultraviolet light, 112, 330
  (*see also* passive solar exposure; sunlight)
Unabomber, 132
undulant fever (*see Brucella*)
UV resonance Raman spectroscopy detection systems and, 267
UVGI rating value (URV), 197–198

vaccines/vaccination (*see* immunization and vaccination)
Vaccinia virus, 35
valley fever (*see Coccidioides immitis*)
van der Waal's forces, 203
vapor clouds, water spray to disperse, 301
vapor protective suits, 298
variable air volume (VAV) ventilation, 120, 149, 152, 154, 276
Variola virus (*see* smallpox)
variolation, smallpox, 9
vasoactive intestinal polypeptide (VIP), 467
vasopressin, 49, 453, **466**
vector-borne diseases, 24–27, 34–40
vegetation air cleaning, in decontamination, **328**, 331–332
velocity of airflow vs. efficiency
  filtration and, 177, 181
  UVGI systems and, 196, 199–200, 360–361
Venezuelan equine encephalitis (VEE), 25, 35, 37, 40, 399, 433, **433**
  disease and dose curves for, 440, **450**
  disease progression in, 94
  dose values for, 81
  open air aerosol toxicity of, **102**
  size of microbe, vs. filtration, **182**
venoms, 453–466, **455–466**
ventilation systems, 113–114, 151–164
  air change rate (ACH) in, 155
  air intake releases in, 141–143, **142**, **143**
  all-outside-air, 152, 154, 165
  auditorium and theater, 136–137
  buildings and attack scenarios and, 148–149
  calculus modeling methods in, 156–157
  commercial buildings and, 120–122

ventilation systems (*cont.*):
  computational fluid dynamics (CFD) in design of, 154, 163
  computational modeling methods in, 157–162
  concentration of contaminants in, 156–162, **156**, **158**, **160**, **161**
  constant volume airflow, 120, 152
  CONTAMW computer program for, 161, 162–163
  dedicated outside air systems (DOAS), 152
  department stores and malls, 134–135, **135**
  dilution, 165, **166**, 167, 327
  dilution rates in, 154–155
  economics of, 351–354, **352**, **354**
  effectiveness of (air mixing), 155–156, 159
  elevator shafts and stairwells in, 161, **162**
  emergency response and, 298–299
  filtration and, 154, 161–162
  food processing facilities and, 127
  forced air, 153
  government buildings and, 124
  healthcare facilities, 152
  hospital, 152
  hotel, 132
  indoor aerosolization release in, 146–147, **148**
  indoor passive releases in, 146
  internal releases into ducts in, 147–150, **149**
  life cycle costs of, 351–354, **352**, **354**
  mailrooms and, 368–369
  modeling of, 153–163
  museum, 134
  natural, 151, 153
  nightclubs, 129
  outdoor air used in, 153–154
  post office, 126
  removal rate of contaminants in, 158–160
  retrofitting of, costs associated with, 154
  shutdown systems for, 298–299
  simulation of building attack scenarios and, 214
  types of, 151–152, **152**
  ultraviolet germicidal irradiation (UVGI) in, 154, 161–162
  variable air volume (VAV), 120, 149, 152, 154, 276
  zones of, 161
  (*see also* air cleaning and disinfection systems)
venturi style aerosol nozzles, 104, **108**
verotoxin, 467
verrucologen, 467
vesicants (*see* blister agents)
VHA (*see* Hepatitis A virus)
Vibrio cholerae (cholera), 16–18, 26, 41, 111, 392, 434, **434**
  disease and dose curves for, 440, **451**
  disease progression in, 94
  dose values for, 81
video cameras, 291
viruses as disease organisms, 25, 26, 32, 34, 388
  carbon adsorption vs., 204
  disease progression curves for, 93, **95**
  filtration and, 177
  ozone disinfection and, 338–340, **338**
  ultraviolet germicidal irradiation (UVGI) vs., 52–59, **54**, **55**, **57**, **59**, 201
volatile organic compounds (VOCs), 203, 336–337, 387
vomiting agents, 62, 63, 66

VX, 65, 66, 86, 470, 509, **509**
  chloramine B to decontaminate, 314
  decontaminant/neutralizer for, **317**
  Decontamination Solution 2 (DS2) to
    neutralize, 314
  dose values for, 82
  open air aerosol toxicity of, 100
  supertropical bleach (STB) to neutralize, 314

walking pneumonia (*see Mycoplasma
  pneumoniae*)
warfarin, 17, 453, **466**
Warsaw Ghetto Uprising, 14–15, **16**
washing, to decontaminate, 312–313
Washington, George, 9–10, **10**
water-borne agents/diseases, 24, 25, 40–41, 88,
  110–112
water pipe security, 291–292
water/well poisoning, 6, 15–19, 111, 291–292
weathering, in decontamination, 313
West Nile virus, 39
Western equine encephalitis (WEE) virus, 35,
  37, 392, 399, 435, **435**
  disease and dose curves for, 440, **451**
  disease progression in, 94
  dose values for, 81
  open air aerosol toxicity of, **102**
  size of microbe, vs. filtration, **182**
White Aryan Resistance, 386
White House, 123–124
whooping cough (*see Bordetella pertussis*)

wild olive, 3
woolsorters disease (*see Bacillus anthracis*)
World Trade Center attack, 119, 386
World War I, 13, **14**, 97, 98, 314
World War II, 14–16, **62**, 98

X rays, in decontamination, 347

Yellow Fever virus, 39, 392, 436, **436**
  disease and dose curves for, 440, **451**
  disease progression in, 94
  size of microbe, vs. filtration, **182**
Yersinia pestis (bubonic plague; plague) 16, 20,
  35, 138, 392, 437, **437**
  biosensors and assay times for, 262, **264**
  disease and dose curves for, 440, **451**
  disease progression in, 94
  dose values for, 81
  open air aerosol toxicity of, **102**
Ypres, chlorine gas attack at, 13
Yugoslavia, 19

Zimbabwe, anthrax outbreak in, 17, **18**, 97
zinc chloride, 314
zoned buildings
  multizone building attack simulation,
    238–242, **240, 241, 242**
  sheltering zone isolation and, 279–281, **281**,
    304–306, **305**
zoned ventilation systems, 161
Zyklon B, 15

ABOUT THE AUTHOR

Wladyslaw Jan Kowalski, P.E., Ph.D., is a Research
Associate at the Pennsylvania State University
Department of Architectural Engineering, where he is
involved in the research and development of air-cleaning
technologies and immune building systems. He has published numerous studies on the subjects of microbial filtration, UVGI system design, and ozone disinfection, and
lectures and consults on the subject of bioterrorism
defense.